Signal and Image Processing Sourcebook

Signal and Image Processing Sourcebook

Robert L. Libbey

Library of Congress Catalog Card Number 93-81166
ISBN 0-442-30861-2

Printed in the United States of America

Page composition by Nancy Sugihara
Graphics by Shawn Wallace

Van Nostrand Reinhold
115 Fifth Avenue
New York, New York 10003

International Thomson Publishing
Berkshire House
168-173 High Holborn
London WC1V 7AA, England

Thomas Nelson Australia
102 Dodds Street
South Melbourne 3205
Victoria, Australia

Nelson Canada
1120 Birchmount Road
Scarborough, Ontario MIK 5G4, Canada

16 15 14 13 12 11 10 9 8 7 6 5 4 3 2 1

To Helen and Carol

To Helen, my dear wife, I thank you for your patience, understanding, and ever-present encouragement during the writing of this book. I assure you that you are number one in my life although it may seem that you are, at times, competing with science and technology.

To Carol, my daughter, what a pleasure it has been to have you contribute to this book. Your metamorphosis since the troubles with that first college course in BASIC to becoming one of the innovators in your field has been very rewarding to me. Congratulations and thanks for your help.

Contents

Introduction

"I usually choose a book based upon the contents of its introduction."
One of the reviewers of this book

This handbook is about the principles and techniques of signal processing. It traces the rudiments and methods of signal processing (SP) from early signal fires used to send simple messages, to Morse's coding method of intelligently changing the current in copper wires to send information, through signal processing in audio, radio, telephony, TV, radar, computers, space, medicine, and the use of SP techniques modeled after the processing of the human brain. One of the major contributions of this handbook is to provide the reader with examples, in many diverse applications and in many different technical disciplines, that relate back to the same overriding principles of signal and image processing.

The contents of a handbook can be codified and presented in several different ways. This book's table of contents uses a chapter list of the major technologies, such as audio, radio, television, etc., as its fundamental indexing method. The content of each chapter shows examples of the principles, techniques, and types of SP growth that are germane to that particular technology. A second indexing method, with equal validity, can provide a catalog of the above mentioned *overriding principles*, combined with their techniques and applications, of signal proc-essing. The following sections in this introduction provide such a listing. The reader can observe the major topics of signal and image processing and see how their principles and techniques grow and diverge as they are applied to the technologies of audio, radio, TV, etc., including those SP techniques modeled after the processing of the human brain. This second listing, or catalog, of the contents of the book is introduced with the archetypical graphic model of a *transfer function* and, in turn, the reader is also aided by a framework graphic termed "The Tree of Signal Processing." This presents the key words that aid in weaving the handbook's varied tapestry of the composite principles and technology of signal processing.

Overriding Principles—The Elementary Models

The fundamental synonym for signal processing is *change*. Throughout the following sections, the SP "black boxes"contain equations, programs, or circuits that alter or sometimes reconstitute the incoming signal. Signal processing, with its ability to perform these programmed changes, can be modeled in two rudimentary ways —with or without feedback. Figure I.1 shows two fundamental graphical paradigms. In both (a) and (b) in Figure I.1, the changing or transfer function may be a somewhat esoteric equation (often the ratio of two polynomials), a specific process, or a function that may be carried out by a computer, or be represented by the hardware in discrete or integrated circuits.

Overriding Principles—The Transfer-Function Model

The preliminary graphic models, introduced in Figure I.2, are echoed and expanded in Chapter 1 as a prototype analog-to-digital-to-analog processing system and in several noise-reducing systems. Chapter 1 also introduces and describes the much-used idea of the *nonlinear* transfer function. Chapter 2 uses this same "in-out" model to show the processing for single sidebands, the superheterode receiver, and FM stereo. Chapter 3 uses variations of this transfer function concept to present the system principles of black and white and color television, as well as to describe several high-definition systems. Chapter 4 uses it consistently to show the system

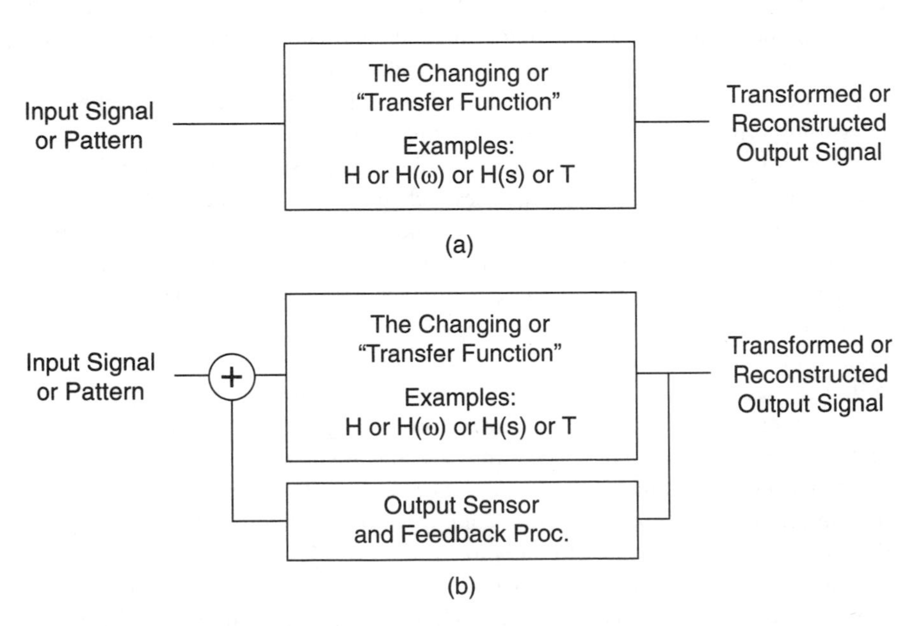

Figure I.1 Two graphical models for signal processing. Note that in both (a) and (b) the paramount idea is changing: (a) is the fundamental transfer function paradigm while (b), with its feedback, can be considered a subset.

operation of both devices and complete networks. Chapter 6 uses a much more complex model for image processing that includes both a transmission and a processing transfer function. Later generations of the transfer-function model are used throughout the remainder of the book. It is especially useful in describing feedback in human and artificial learning systems in Chapter 10.

Overriding Principles—Modulation

Certainly the fountainhead of most, if not all, signal processing is the diverse family of SP schemes termed *modulation*. With the family of modulation techniques, described in Chapter 2, it is possible to systematically and predictably change a first signal's (the carrier's) amplitude, frequency, phase, duration, position, or coding with a second or input signal. This modulation family requires an understanding of the signal's bandwidth, spectrum, and the relationships in modulation that are governed by the Nyquist Limit. Chapter 3 develops the principles and the relationships of frequency, phase, and transient response that are crucial for the successful use of any modulation technique. It also shows the fundamental links between bandwidth, rise time, and transient response. Chapter 6 uses these overriding relationships to describe the use of RF signals to modulate the spin of the body's hydrogen nuclei to help in the process of magnetic resonance imaging.

Chapters 2 and 3 are heavily dependent upon the various techniques of modulation. The use of both AM and FM in radio, with their superheterodyne configurations, as well as stereo broadcasting is detailed. Further, the basic AM is expanded with the use of the balanced modulator and vestigial sideband transmission for TV. The use of QAM or quadrature amplitude modulation is described for several high-definition TV systems. Sharing a communication channel by two or more signals with multiplexing is also introduced in Chapter 3.

Chapters 4 and 8 not only expand upon the aforementioned analog modulation techniques but introduce several digital modulation schemes. These chapters first detail the use of several types of analog modulation and demodulation and then show the transition from analog to digital telephone modulation and transmissions. The primary modulation technique of pulse code modulation (PCM) is described as it is used for voice telephone networks, computer and data networks (including fiber optics), cellular telephones, and microwave and satellite communications. The somewhat less used, but still important, quasi digital technique of delta modulation is also described.

Overriding Principles—Signal Filters

The second most important SP tool, after modulation, is filtering. The early radio and telephone designers soon learned the need for better defined and executed electrical filters.

Filters for high-, low-, and mid-range tone controls and for noise suppression are introduced in Chapter 1. Chapters 2, 3, and 4 show many filter techniques and

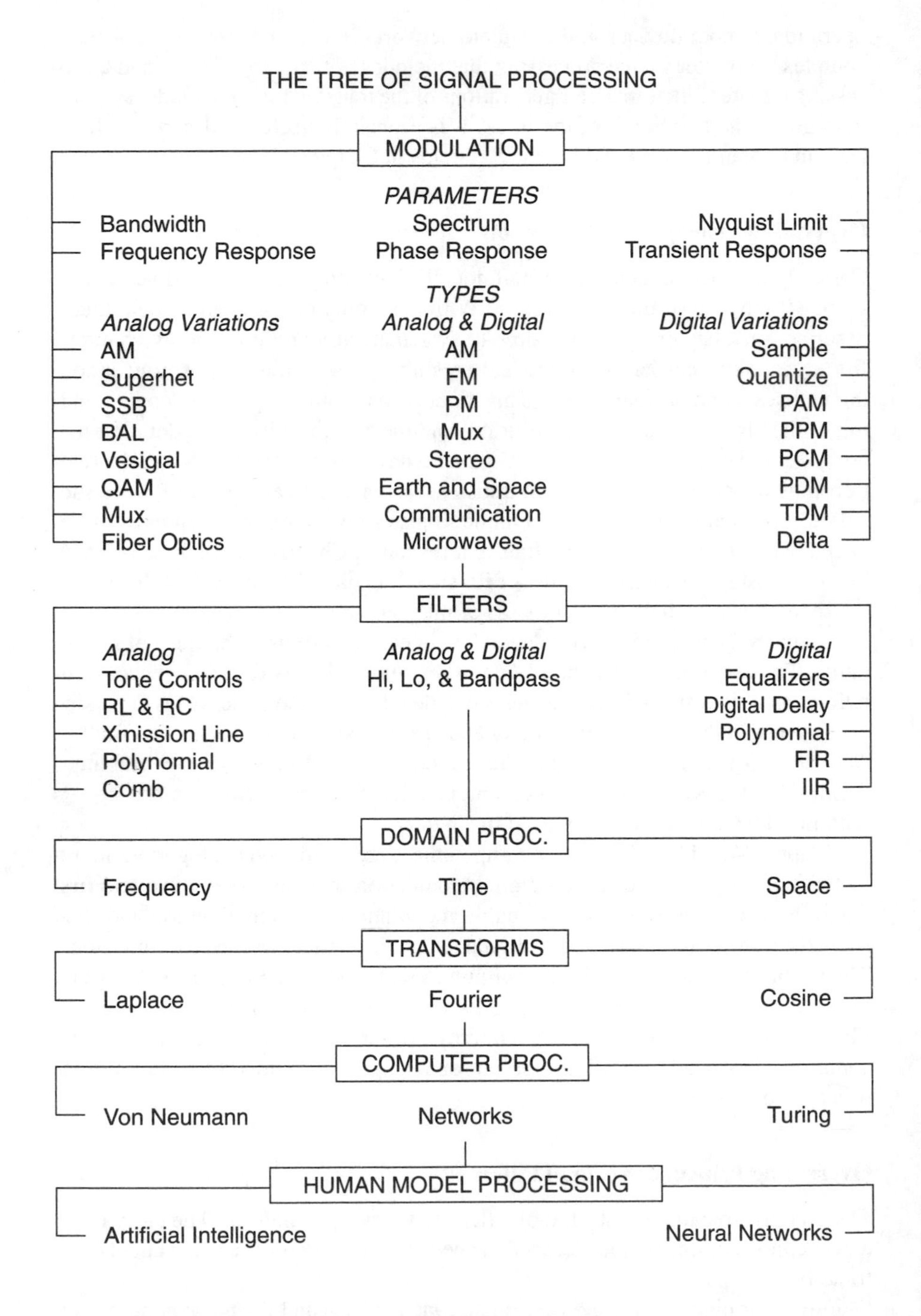

Figure I.2 The tree of overriding principles used in this book.

a wide variety of applications. Chapter 3 introduces the important techniques of analog comb filters, which also provide a background for the subsequent introduction of digital filters. Chapter 5 provides more detail and applications for both analog and digital comb filters and introduces the idea of a matched filter. In fact, this is the primary chapter for filter principles and techniques. The important use of filters in integrated processors is described in Chapter 9.

Overriding Principles—Domain Processing

This book provides examples of signal processing in the frequency domain, the time domain, and the space domain. Audio and radio signals were primarily visualized and described in the frequency domain. However, TV, especially color TV, required the inclusion of transient and phase response in its designs. While the audio and radio chapters primarily present the early principles in the frequency domain as "sin ω" or "cos ω," later descriptions, of the more complex forms of modulation require the addition of phase (angle) information as "sin ω ± ϕ." The television chapter introduces techniques for analyzing the time domain rise time and transient response of a pulse and relates them back to an analogous response in the frequency domain. Chapter 6 shows the frequency response as three-dimensional rather than the more familiar two-dimensional plot. It also introduces the ideas and techniques of using the two- and three-dimensional space domains. Chapter 6 also describes the techniques of analyzing complex images with the two- and three-dimensional (analog) Fourier transform, the (digital) discrete Fourier transform (DFT), and the more common fast Fourier transform (FFT).

Overriding Principles—Domain Transforms

Chapter 3 introduces the Laplace transform as a convenient method for solving the differential equations of physical systems. It then illustrates the techniques of analyzing periodic waveforms with a Fourier series and nonperiodic waveforms, such as single pulses, with the Fourier integral. Chapters 3 and 6 describe the ideas of the Fourier series, the Fourier integral, and the Fourier transform. They also explain both the continuous and the discrete forms of the FFT. Chapter 3 also introduces the Cosine transform as a method of changing coded picture samples into coefficients that require fewer bits to store or transmit.

Overriding Principles—Computer Processing

Chapters 7 and 8 present many detailed processing techniques that are used in both individual and networked computers. Examples of 4-, 8-, 16-, and 32- bit microcomputers show the hardware developments as well as the progression of computing ideas and the increased influence of faster, larger, and much less expensive memories; of software and graphics; and of speed of operation on microprocessor design. Also, for historical perspective and as an aid to understanding some of the concepts in artificial intelligence, Chapter 10 details the background of John Von Neumann's arithmetic computer and Alan Turing's symbolic computer.

Overriding Principles—Human Model Processing

Chapter 10 details the ideas, background, and growth of computer processing based upon some of the models of the reasoning and learning processes in the human brain. The development of artificial intelligence is presented with two very broad approaches—the use of artificial thought processing based upon highly developed control strategies and processing based upon a wealth of background knowledge and the methods and rules for the use of that knowledge. The use of such knowledge, and its rules, is then illustrated in an example using an "Expert System."

Neural Networks are also presented as methods of processing based upon models of the human brain. Whereas artificial intelligence processing is primarily based on control or knowledge and has a very limited ability to learn, neural networks are predominantly used because of their outstanding ability to learn. Chapter 10 presents both a textual and a mathematical background for several representative neural networks and examples of how they are applied. Chapter 10 gives detailed examples of how to use both expert systems and neural networks.

Acknowledgments

This book was influenced by my students and by my activities as a practicing engineer. To these bright and energetic students and to my many engineering colleagues, I thank you for the joy and intellectual stimulation you have given me throughout the years. This book is also founded upon the research, developments, products, and literature of many fine organizations. These contributions are also gratefully acknowledged.

I especially thank Mike Carrol, Michael Isnardi, Terry Smith, and Glenn Wrightmier of the David Sarnoff Research Center; Stella Luna of the NASA Lyndon B. Johnson Space Center; Brian Nicklas of the Smithsonian Institution; former colleagues Andrew Inglis and Dan White; and Karl Rookstool of Bell Atlantic. Bill Bauer, Alex Walker, and Carol Libbey furnished information for Chapter 8. Drs. Harry Cooperman, Steven Falkowski, Mark Frawley, Lawrence Liebman, Mary Moore, and Warren Werbitt gave valuable counsel for Chapter 6. The staff of the Cherry Hill, New Jersey library, especially Helen Brombaugh and Ann Moore, and Natalie Mamchur of the Moorestown, New Jersey Martin Marietta library were always there to help on the literary research—a heartfelt thanks to you all.

The following organizations provided information and, in many instances, granted permission to use parts of it in the text. Apple Computer, Bell Atlantic, Dolby Laboratories, The FCC, GE Medical Systems and their James Hildenberger, Hewlet-Packard, IBM, IEEE, NeuralWare, Picker International, Rutgers University and their Prof. Casimir A. Kulikowski, Scientific American, Siemens Medical Systems, Toshiba America Medical Systems, Intel, Motorola, and Texas Intruments.

Finally, this book truly would not exist without the efforts of Marjorie Spencer and the many, many long and understanding telephone conversations (and the wonderful dinners) of Dr. Alan Rose of Intertext Publications.

1

Audio Signal Processing

1.1 FEEDBACK AMPLIFIERS

Audio Signal Processing—An Introduction

Most of us either perform audio signal processing or experience its result every day of our lives. We use the volume or the tone controls on the radio or stereo to change (process) the signal for our listening requirements. Likewise, before the signal has even reached the receiver, it has been processed at the recording or radio studio, possibly by a land or satellite link, and certainly it was inevitably slightly changed, again processed, by the transmitter. Even the microphone may process the audio signal by altering its bandwidth or its exact frequency makeup—its *spectrum*.

This chapter creates a background for more complex signal processing by illustrating some of the classic analog audio processing as well as some contemporary digital processing techniques. Another important part of this chapter is the introduction of electrical analogs for mechanical and acoustical "circuit" elements. The reader can learn to mentally switch between these disciplines and, in doing so, broaden the understanding of each of them.

Amplifier Distortion

The ideal amplifier will faithfully increase the amplitude of a given input signal without disturbing (distorting) the *relative* amplitudes or *relative* phase or time relationships of the overall signal and its frequency components in any way. The magnitude of the signal will have been increased, but the spectrum will stay exactly the same.

Unfortunately, any practical real-world amplifier may distort the input signal. This signal distortion is usually specified in terms of the change in the amplitude of the input signal's harmonics, a change in their relative phase, the signal's rise time, etc. Thus, amplitude distortion can be defined as the process of not amplifying all of the signal components by the same amount. Besides amplitude or phase distortion, there are other more exotic determinations, such as intermodulation distortion (the signal components modulate each other), that can also be measured or specified.

The classical way to reduce the signal distortion introduced by the amplifier is to feed back some of the out-of-phase output signal and, after adding it (comparing it) to the original in-phase input signal, to use the combination of the two signals as the new amplifier input signal. (See the Black reference.) Unfortunately, there is a price to pay for this distortion reduction—the reduction in overall amplifier gain. Likewise, the feedback amplifier must be very carefully designed if it is to be stable for all reasonable input signals and still provide the desired reduction in distortion.

For an amplifier with very simple resistive feedback and an open-loop (nonfeedback) gain of A:

Closed-loop gain:

$$A_{fb} = \frac{V_{out}}{V_{in}} = H \ or \ H_{(s)} \tag{E 1.1}$$

where H(s) is the complex transfer function (some books use the symbol K for the gain).

If the feedback is a simple two-resistor network and the feedback ratio is $R_1/R_2 = \beta$:

$$A_{fb} = \frac{A}{1 + AR_1/R_2} = \frac{A}{1 + A\beta} \tag{E 1.2}$$

If the acronym %Dist is used to symbolize the harmonic distortion of an amplifier *without* feedback and $\%Dist_{inv}$ and $\%Dist_{non}$ are used to symbolize the resulting (reduction in) distortion for both inverting and noninverting *feedback* amplifiers, the following equations give the reduction in harmonic distortion using the processing technique termed voltage feedback—see Figure 1.1.

Reducing distortion with feedback:

$$\%Dist_{inv} = \frac{\%Dist}{1 + A} \tag{E 1.3}$$

$$\%Dist_{non} = \frac{\%Dist}{1 + A\beta} \tag{E 1.4}$$

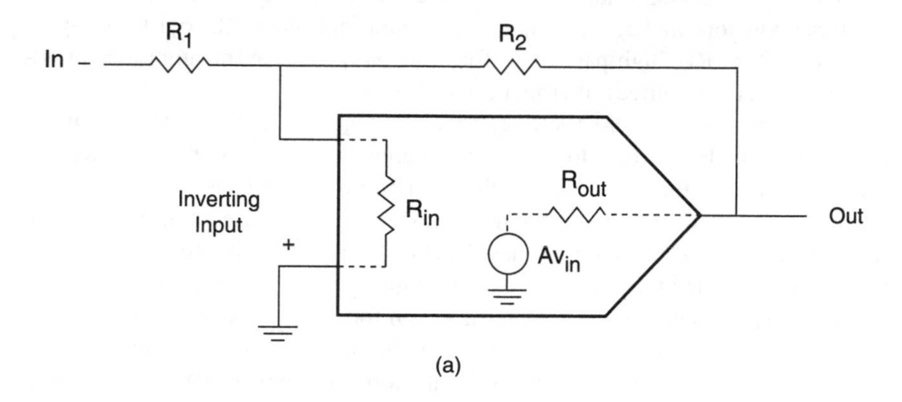

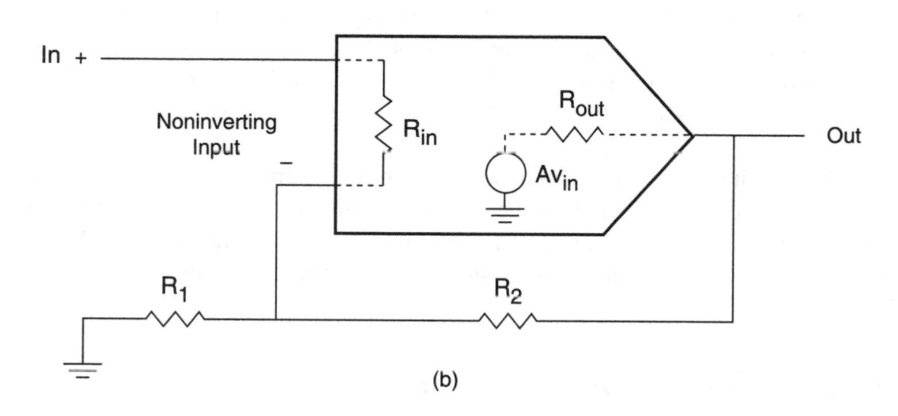

Figure 1.1 The equivalent circuits of amplifiers using resistive feedback in the (a) inverting mode (sometimes termed state) and (b) in the noninverting mode. (See the Libbey reference.)

1.2 FILTERS AND TONE CONTROLS

Simple Analog Filters

A filter is a device that controls the amount of material or information that will pass through it. A paper coffee filter will block the coffee grounds but will allow the water to pass. Electrical filters, such as tone controls, will reject some of a signal's frequency components and pass or even enhance others. By using such a filter, the users can select the band of frequencies that they find most useful or pleasing.

The simplest analog filters, such as tone controls, use resistors and usually capacitors. Inductors are larger and more costly than capacitors. RC (or LR) low-pass filters or CR (or RL) high-pass filters form the basic passive (no amplifiers or additional power is required) filter models.

These low-pass or high-pass filters act as voltage dividers in a manner similar to two resistors. However, when one of the elements is a capacitor or inductor, the "AC resistance"—the reactance—of the capacitor or the inductor varies with frequency and thus the dividing ratio varies with frequency. The paramount design factor is termed the cut-off frequency. This cut-off point is the frequency where the reactance value of the capacitor (C) or inductor (L) equals the value of the resistor (R). The equation for the voltage division for a two-resistor circuit is termed simple—it requires only relatively simple mathematics to solve—while the equation for a filter circuit involving a resistor and a reactance (C or L) is termed complex. Complex numbers and equations require solutions based upon techniques derived from the mathematical principles of the Pythagorean theorem.

The design of tone controls for analog audio equipment has produced many innovative circuits. One outstanding example is the combined low-pass and high-pass circuit known as the Baxandall circuit which was first described in a 1952 issue of the British technical magazine *Wireless World* (see the Tremaine reference, p. 290).

Although it uses very few components, this circuit configuration will either cut or boost both the low and high frequencies in the audio spectrum. There have been many variations of this circuit in audio instruments since its introduction. A very rudimentary configuration of this circuit is shown in the following sections.

For a simple voltage divider using two resistors:

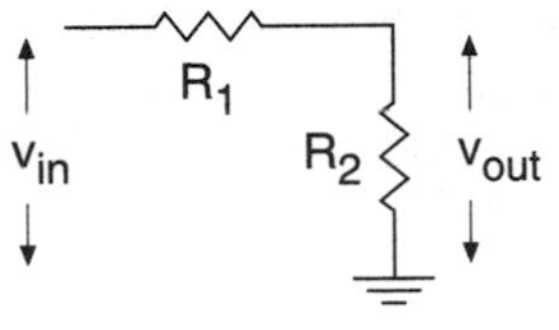

$$V_{out} = V_{in} \frac{R_2}{R_1 + R_2} \qquad \text{(E 1.5)}$$

For an RC low-pass filter configured as a voltage divider:

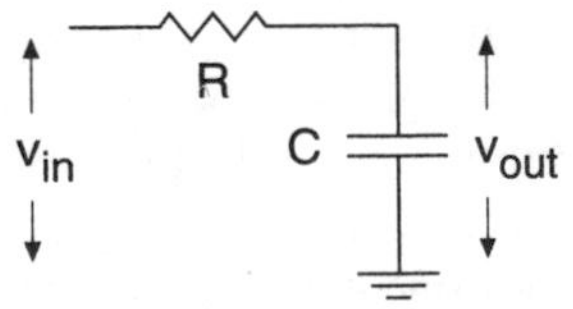

$$V_{out} = V_{in} \frac{X_C}{\sqrt{R^2 + X^2}} \qquad \text{(E 1.6)}$$

For a CR high-pass filter configured as a voltage divider:

$$V_{out} = V_{in} \frac{R}{\sqrt{R^2 + X^2}} \qquad \text{(E 1.7)}$$

The Baxandall Tone-Control Circuit

The Baxandall circuit is a clever combination of the low-pass and high-pass filters as presented before. For an example of how the elementary circuit models are configured to give both a low- and high-frequency cut and boost, the following derivation is presented in two steps. First, two circuits are presented that show rudimentary bass cut and boost circuits and then a third circuit is given to show how these can be combined to give a combination cut and boost. A second set of circuits also presents the rudimentary treble (high-frequency) tone control. Finally, these introductory circuits can be coupled to produce a combined bass and treble cut and boost.

In Figure 1.2, circuit (a), it is assumed that the magnitude of e_{in} is constant (with frequency) and that the resistance of the control VR is five to ten times greater than the series resistor R. The *bass-boosting action* can be analyzed by assuming three conditions. [1] If the signal input is a very low frequency, the reactance of

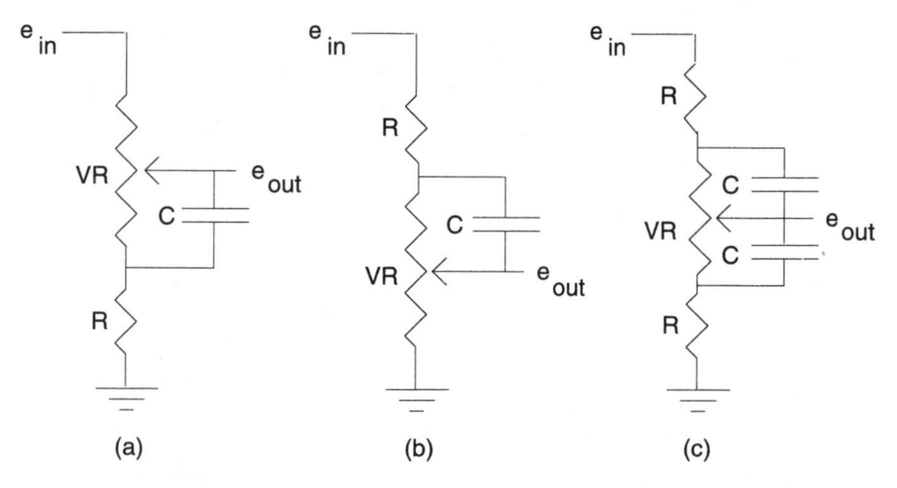

Figure 1.2 Baxandall bass-control circuits: simplified bass boost and cut circuits. Circuit (a) shows a prototype for a bass boost control, (b) shows the prototype for a bass-cut control,. and (c) is a boost-cut circuit derived by combining (a) and (b).

capacitor C will be high and thus its shunting of VR will be minimal. The voltage output will be approximately the same as it would be if the C were not present. [2] As the signal frequency is increased, the reactance of C will become less and therefore its shunting of part of VR will start to reduce the output. [3] As the frequency is increased further, a point will be reached where the reactance of C will equal the resistance of R. This is the cut-off or cross-over frequency. Above this frequency, the reactance of C is so small that it is almost an AC short circuit and the output is determined by the voltage across the combination of R and any of VR as the control arm is moved up. Significant bass boosting only occurs whn the arm is set to the upper portion of VR.

Figure 1.2 (b) shows a *bass-cut circuit.* Again, we assume a similar situation. [1] If the signal frequency is very low, the reactance of the shunt capacitor C will be high and it will not shunt VR. The output will be about the value determined by the value of R (and the setting of the arm in VR). [2] As the signal frequency is increased, the reactance of C will decrease and the shunting effect of C will increase the output voltage. [3] When the reactance of C equals the value of R—the cut-off point—the output will almost be maximum. Any further increase in signal frequency will produce about the same output. If the components and ratios (and the analysis) are inverted, the circuits can be made to boost and cut frequencies in the treble region.

In all of the circuits, the designer must tailor the component values as well as their ratios to produce the required response. Note that the circuit configurations in both Figures 1.1 and 1.2 are the same. The difference in the circuits is the ratio of the Rs to the Cs. If more detail about the analysis and design of these circuits is desired, it is recommended that the reader use a circuit analysis program such as PSpice©.

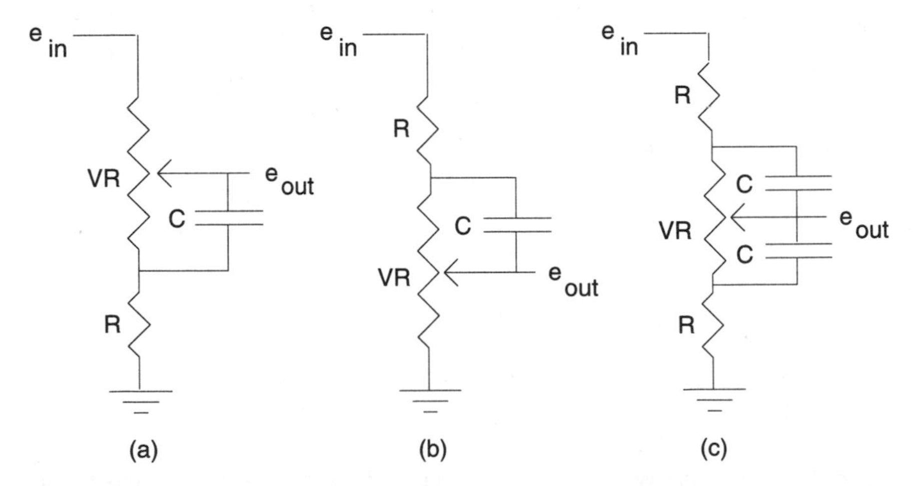

Figure 1.3 Baxandall treble-control circuits: simplified treble cut and boost circuits. Circuit (a) is a prototype treble-cut control, (b) a treble boost circuit and, (c) is a combination treblecut and boost circuit using only one control.

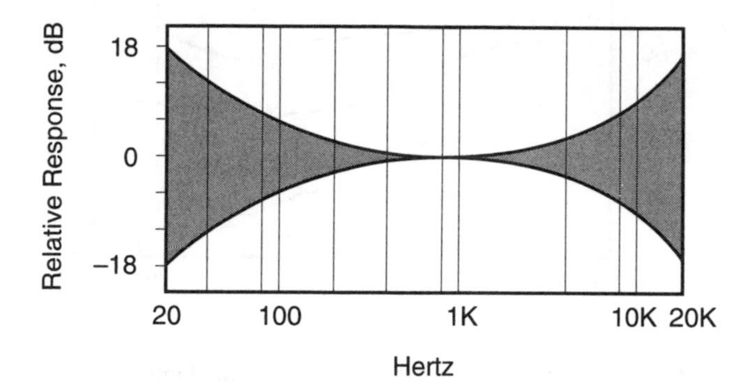

Figure 1.4 The response of combined bass and treble tone controls: the approximate range of a tone control system that can be obtained by using a combination of the circuits shown in 1.2 (c) and 1.3 (c). When the variable controls are both set at their center positions, there is no bass or treble cut or boost—the output is termed "flat."

Filters (Equalizers) That Reduce the Effect of Noise

Simple filters, such as the RC low-pass and high-pass configurations in E 1.6 and E 1.7 are used to improve the noise performance of analog phonograph records and FM broadcast radio. Their use is based on the principle that there is only a small amount of energy in both the signal and the noise at the higher audio frequencies (see the appendix on noise). Originally, when the high-frequency noise was excessive in shellac records, it was discovered that the effect of this noise could be significantly reduced by a technique known as *preemphasis.* This technique used a CR filter to increase (emphasize) the amount of high-frequency energy being recorded. The filter, termed a preemphasis network, was configured to increase steadily the amplitude of the incoming "flat" signal above about 1,000 Hz. This preemphasis or *preprocessing* in recording required a similar deemphasis or *postprocessing* in the record's playback amplifier. For a time, different manufacturers would use different preprocessing techniques. Finally, in the 1950s, the members of the record industry agreed on an industry standard. This general technique of audio preprocessing and postprocessing is used in recordings, motion pictures, audio and video tape, and FM radio.

1.3 HEARING, LOUDNESS, AND LOUDNESS CONTROLS

One of the useful forms of audio signal processing is the establishment of low-frequency enhancement to compensate for the reduced sensitivity of the human ear. Much of the pioneering work on the characteristics of human hearing was done by Flectcher and Munson of the Bell Telephone Laboratories. (See the Beranek and

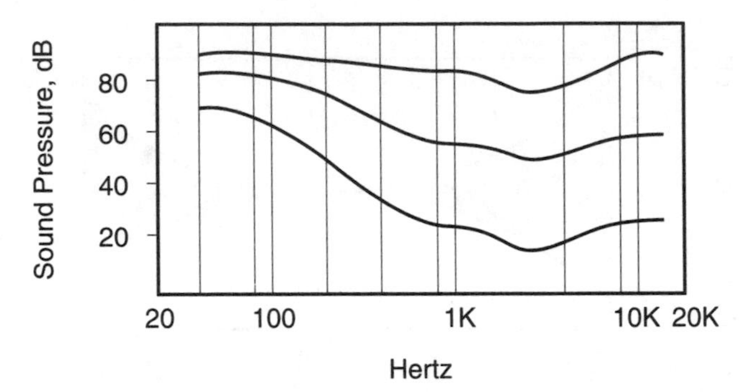

Figure 1.5 Loudness contours: a simplified sketch of the sound intensity required to give the perception of equal loudness. Note that these curves (hearing contours) are in sound pressure that is the *reciprocal* of sensitivity.

Tremaine references.) These researchers and others found that the *relative* response of the ear—its sensitivity—decreases at both the lower- frequency and the higher-frequency ends of the audio spectrum as a function of the overall loudness of the signal. Thus, since the sensitivity of the average ear drops off in the lower- and higher-frequency regions, it is often necessary to increase the loudness or sound pressure in these regions.

If a recorded sound is reproduced at a level that is lower than it was originally played, it will be perceived as lacking intensity (loudness) in both the lower frequency (bass) and the higher-frequency (treble) regions of the spectrum. This lack of loudness effect is especially noticeable and annoying at low frequencies because the effect is very nonlinear. It becomes increasingly noticeable as the listening level is decreased.

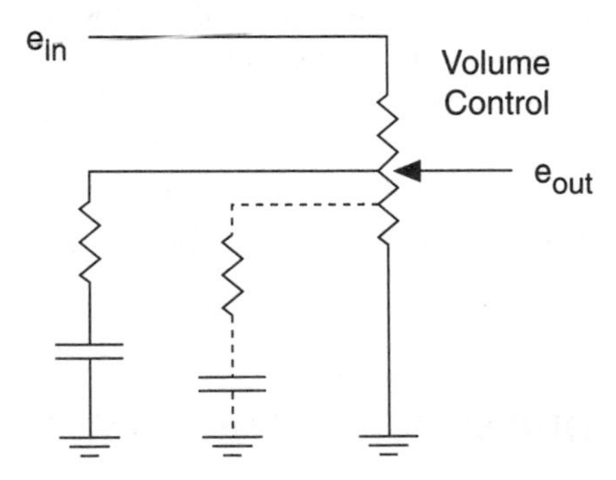

Figure 1.6 A volume-control circuit with two bass-boost circuits added for loudness control. Most designs only use one RC boost circuit, as shown in the solid-line circuit. The addition of a second boost circuit, shown as a dashed line, will give more precise loudness (versus frequency) at lower control settings.

In sound-reproducing systems, the lack of loudness effect can be solved by using appropriate boosting circuits. Since the high-frequency (treble) loudness sensitively drops off in a nearly linear manner (see the loudness contour lines in the above references), circuits are usually only included to boost (enhance) the bass spectrum. In general, an average listening level is established, through listener tests, and an appropriate bass-boost loudness circuit is incorporated. In most cases, the loudness components are part of the volume-control circuit (see Figure 1.6). If the control is set above the predetermined average level, the loudness boost is decreased. If it is set below the average level, the boost will be more dominant. For a more precise loudness versus control, two boost circuits are sometimes used.

1.4 MICROPHONES

Microphones—An Introduction

The actor/comedian George Burns often recalls this anecdote about early radio broadcasting. When the art and technology of radio were very young, if a program's rating was sagging, one of the first changes that was tried was to use a different type of microphone. Possibly without knowing all the technical comparisons, both the artists and the technical people were aware that most of the early microphones colored the sound in different ways.

Microphones and loudspeakers are fundamentally more mechanical than they are electrical. The fact that a microphone will generate an electric output voltage and, likewise, a loudspeaker needs an electric input voltage is only part of a much larger explanation of how they operate. Both a simple microphone and a simple loudspeaker can be analyzed using electrical analogs for the mechanical components and functions that form the true make-up of these devices. The ability of one scientific discipline to model the devices and functions used in different but allied disciplines is one of the most powerful tools for the understanding and development of technical systems.

Simple Microphones—Mechanical and Acoustical Circuits That Process

A microphone is a device for converting the changes in air motion caused by speech, music, thunder, etc., into an electrical signal that is (hopefully) a faithful representation of those changes in the air (or other fluid) motion. Most microphones are designed to sense the changes in the sound (air) pressure, although some respond to changes in velocity and some respond to the changes in both pressure and velocity. One model of a microphone is a simple electric generator. The air motion produced by the original sound source will move the diaphragm. In a moving-coil microphone, this diaphragm is usually directly connected to a coil in a magnetic field. When the diaphragm is moved, the moving coil will generate a voltage (EMF). The design can make this output EMF proportional to changing air pressure or velocity. In a carbon microphone, the moving diaphragm changes the resistance of the carbon unit that in turn will change the small current proportionately to the change in air pressure. In a condenser (now electret-type) microphone, the motion of the diaphragm changes the capacitance of the unit that, in turn, changes the flow of current.

The three most important mechanical parameters affecting the operation of a moving-coil microphone are the *mass* of the diaphragm and the coil, the *compliance* (stiffness) of the suspension, and the *friction* in the suspension and the surrounding air that retards free movement. The mechanical mass is the analog of electrical inductance—they both store energy and, in theory, do not dissipate energy. Likewise, the mechanical stiffness of the suspension is the analog of electrical capacitance. Any friction in the suspension or air is the analog of electrical resistance. Both friction and resistance dissipate energy. The three elements of mass, stiffness, and friction can combine to form mechanical filters just as the electrical elements of inductance, capacitance, and resistance can. A combination of mass and stiffness can form a resonant circuit, just as inductance and capacitance can. Any friction in the mechanical circuit will damp the resonance just as resistance will do in an electrical circuit. Also, in mechanical circuits, force or pressure is the analog of electrical voltage (electromotive force). There are also acoustical component analogs. When air is forced through small holes or slits, there is a frictional loss caused by acoustical resistance. When a quantity of air is moved, there is a reluctance to the starting and stopping of this motion that is analogous to the reluctance to the starting and stopping of the flow of current in an inductor—thus, there is acoustical inductance. Likewise, both a spring and an enclosed volume of air resist being compressed, which is analogous to the reluctance of a capacitor to being charged—thus, there is acoustical stiffness or capacitance.

For a simple dynamic microphone, the movement (velocity) of the voice coil resulting from a sound-pressure input is:

$$vol = \frac{f_M}{r_{M1} + j\omega m_1 + \dfrac{1}{j\omega C_{M1}}} \tag{E 1.8}$$

where:

vol = the velocity of the diaphragm, in centimeters per second
f_M = the sound driving force, in dynes
r_{M1} = the mechanical resistance of the suspension system, mechanical ohms
m_1 = mass of the diaphragm and voice coil, in grams
C_{M1} = compliance (1/stiffness) of the suspension system, in centimeters per dyne

ELECTRICAL	MECHANICAL	ACOUSTICAL
R_E Resistance	R_M Friction	R_A Air Friction
L Inductance	m Mass	M Air Mass
C_E Capacitance	C_M Spring	C_A Air Stiffness

Table 1.1 The equivalent element symbols for three different systems.

The voltage output (v) of the microphone (generator) is:

$$v = Bl(vol) \quad \text{(E 1.9)}$$

where:

B = the flux density in the air gap cutting the coil, in gauss
l = the length of the voice coil, in centimeters
vol = the velocity of the diaphragm and the coil, in centimeters per second

For clarity, Figure 1.7 does not include the acoustical/mechanical elements or the microphone enclosure. The circuit models of a microphone and its enclosure, as well as a loudspeaker and its enclosure, can get very complex since the complete circuits will contain acoustical elements and mechanical elements as well as the electrical elements.

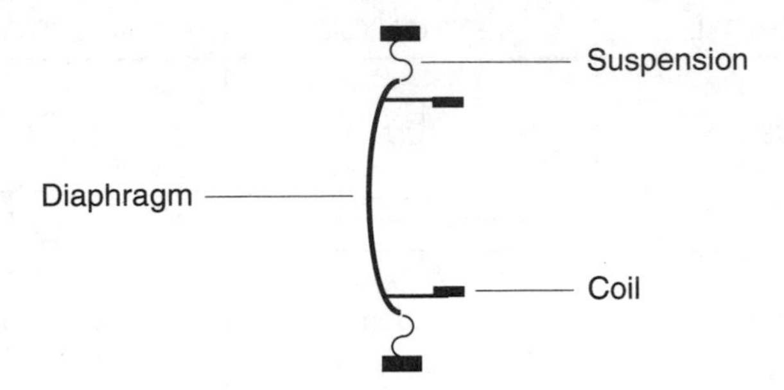

Figure 1.7 The elementary mechanical schematic for a dynamic (moving- coil) microphone without including any enclosure circuit elements.(See the Olson references.).

The observer, experienced in electronics, will see that Figure 1.8 resembles an RLC series resonant circuit. In fact, the three mechanical elements, shown as an elementary microphone, behave in a manner exactly analogous to an electrical circuit. In electrical series resonance, the input electrical force—a voltage—will produce a peak in the series current, as a function of frequency, because of the reaction between the inductance and the capacitance. The magnitude of the peak will be determined by the value of the resistor. In the mechanical (microphone) circuit, the input sound pressure (force) will cause a displacement with a magnitude that is a function of the interaction between the mechanical mass and the mechanical stiffness. The friction (mechanical resistance) will control the magnitude of the motional peak at resonance.

When the magnet and the case are added, the model of a pressure microphone becomes more complicated. These elements will constrict the air flow and thus change the mechanical-acoustical relationships. While the conceptual model shown in Figure 1.9 is still not as complex as an actual working model would be, it does show how designers add elements to compensate or improve the response.

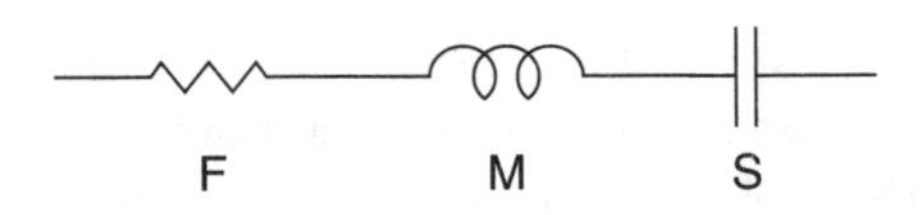

Figure 1.8 An elementary electromechanical symbolic circuit for a dynamic microphone. F is the mechanical friction of the suspension, M is the mass of the diaphragm and the coil, and S is the stiffness of the suspension.

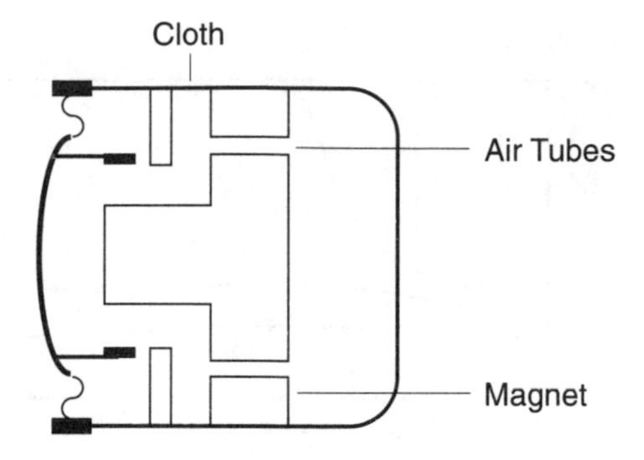

Figure 1.9 An elementary prototype dynamic microphone showing the addition of a microphone housing and the permanent magnet (see also Figure 1.7). The pieces of cloth are used primarily for acoustical resistance damping, although they do possess mass that must also be considered. Further acoustical damping may be obtained by using air vents or tubes. (See the Olson reference.)

NOTE: One of the early pioneers in the theory and practice of using analogs or "analogies" was Dr. Harry F. Olson of the (then) RCA Victor Laboratories and later at the David Sarnoff Research Center. Dr. Olson's *Dynamical Analogies* and his many other books helped developed the ideas and the technology that provided a foundation for understanding the relationships between several allied disciplines in science. They are still very worthwhile references.

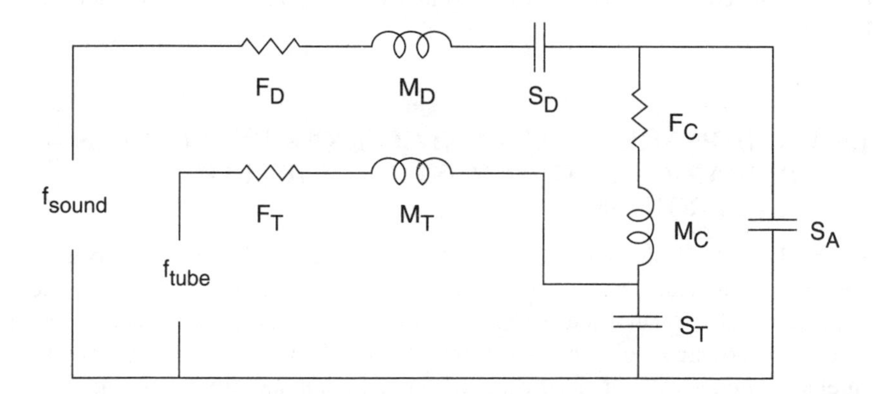

Figure 1.10 The electromechanical circuit of a dynamic microphone: the symbolic circuit for a slightly more realistic dynamic microphone is shown in Figure 1.9. The f_{sound} and the f_{tube} are the input sound forces at the diaphragm and the openings of the tubes. The mechanical and acoustical elements have subscript symbols that include C for cloth, D for diaphragm, T for tube, and A for the air directly behind the diaphragm.

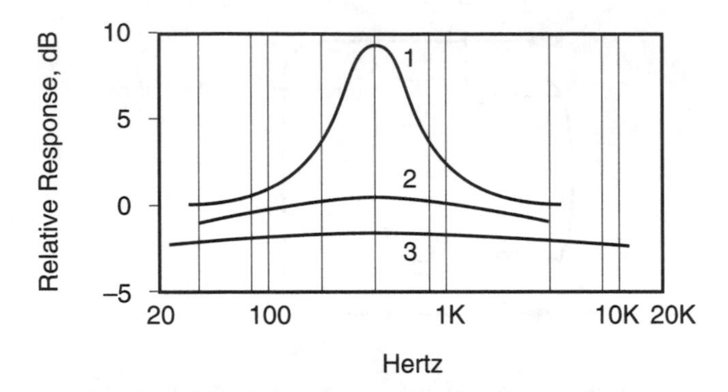

Figure 1.11 Microphone resonance curves: the seminormalized responses of the equivalent circuits for simple pressure microphones. Curve 1 refers to Figure 1.8 and assumes very little mechanical or acoustical damping. Curve 2 could be the same equivalent circuit with the value of the resistor increased to increase the damping. Curve 3 represents a broader response that could be obtained by using a more realistic acoustical design, as illustrated in Figures 1.9 and 1.10.

Using the electromechanical circuits of Figures 1.8 or 1.10, it is possible to develop and plot resonance curves that are a function of the system damping, i.e., the friction in the electrical, mechanical, and acoustical equivalent circuits. The circuits in the two figures are still more simplified than most actual device designs. Once an elementary equivalent circuit is developed, it is very convenient to analyze it using computer-aided design and analysis programs such as PSpice©. See Figure 1.11.

1.5 LOUDSPEAKERS AND LOUDSPEAKER ENCLOSURES—MECHANICAL AND ACOUSTICAL CIRCUITS THAT PROCESS

The study of the design and operation of loudspeakers and speaker enclosures is not only fun, it is an excellent model for other electromechanical devices in electronics as well as general science and medicine. A loudspeaker can be considered as a piston moving air or as a pump pumping air. Likewise, it can be thought of as an electric motor with an electrical audio signal input and driven air output. Unfortunately, in speakers and other electromechanical devices, these simple, unified and linear models do not very well describe the operation in the real world. A somewhat more accurate model of the loudspeaker can be constructed if the spectrum of operation, and thus the model, is separated into three parts: [1] the lower-range or low-frequency operation, [2] mid-range or mid-frequency operation, and [3] high-range or high-frequency operation.

Low-Frequency Operation

In general, in the low range most speakers operate quite closely to the model of a piston. Its equivalent circuit resembles that of the microphone shown in Figure 1.6. The dominant model parts are the mass of the cone, the stiffness of the suspension, and the friction of the suspension. For example, an 8-inch round loudspeaker will have a low-frequency resonance of 90 to 100 Hz. In the region of about 50 to 100 Hz, the cone motion can be predicted by assuming a simple series resonant equivalent and a simple motor model. The cone motion is likewise reasonably predictable in the frequency range of 3 to 400 Hz.

Mid-Range Operation

As the operating frequency range is increased, the concepts of a piston air pump or of a simple electric motor do not accurately describe the motion of the speaker cone. In the mid-frequency range, there is usually a problem with the operation of the rim or end suspension of the cone. In the piston model, it is assumed that the usually corrugated rim "hinge" simply allows the cone to move back and forth and has no direct connection with the acoustical output of the speaker. Unfortunately, this is not always the case. As the input signal frequency increases and the signal wave length decreases, there is a point—sometimes several points—where the cone acts as a transmission line and the rim suspension acts as its termination. If the rim suspension is not designed correctly, this imperfect termination will cause a standing wave in the cone that translates to some peaks and usually one severe valley (dip) in the sound output. These peaks and valleys can be reduced by special suspension shaping and by using termination materials such as leather, rubber, or a semi-liquid plastic or foam. Thus, the audio signal may have some unexpected and unwelcome processing and degradation if the rim suspension is not carefully designed and constructed of the proper material(s).

High-Frequency Operation

In the high-frequency portion of the speaker's spectrum, the cone can produce more unwanted effects caused by *cone breakup*. The cone no longer vibrates as a unit but degrades into multiple output sources. Some of these sources are directly related to the signal current in the voice coil and some are unrelated parts that vibrate at their own natural resonances. These sources were first triggered by the input electrical signal but now resonate in independent modes. Sometimes these independent sections produce four symmetrical areas on the speaker cone and sometimes they are configured in concentric bands radiating out from the area of the voice coil. (See the Corrington and Kidd reference). At the very higher audio frequencies, most of the acoustical energy is radiated from the area of the voice coil and the major portion of the cone does not significantly move. In fact, many loudspeakers add a second cone, usually called a whizzer, in the area of the voice coil.

This small whizzer-cone is self-supporting and does not need a rim suspension. To a first approximation, its operation conforms to the model of a miniature piston.

The Equations of Loudspeaker Motion

For a cone-type loudspeaker, when considered as a simple piston, the velocity of the cone is:

$$vol_c = \frac{e_{amp}\, Bl}{(R_{amp} + (R_E + jX_{LE})) + (R_M + j(X_M)} \tag{E 1.10}$$

where:

$$R_M = \frac{B^2 l^2}{R_{amp} + R_E} + R_{MS} + 2R_{MR} \tag{E 1.11}$$

and

$$X_M = \omega M_{MD} + 2X_{MR} - \frac{1}{\omega C_{MS}} \tag{E 1.12}$$

For reference, the loudspeaker power output (watts):

$$W = \frac{2e_{amp}^2 B^2 l^2 R_{MR}}{(R_{amp} + R_E)^2 (R_M^2 + XM^2)^2} \tag{E 1.13}$$

The symbols are the same as those listed for E 1.8 plus:

e_{amp}	=	the voltage output of the audio amplifier
R_{amp}	=	the equivalent internal output resistance of the amplifier
X_{LE}	=	the reactance of the electrical inductance
X_M	=	the reactance of the mechanical reactance (the mass)
R_{MS}	=	the resistance (friction) of the suspension
M_{MD}	=	the mass of the diaphragm

Using equations E 1.10 through E 1.13, plus Figure 1.12, it is possible to start to appreciate the relationships between the electrical, mechanical, and acoustical factors that affect the operation of a loudspeaker. This same type of analysis is used

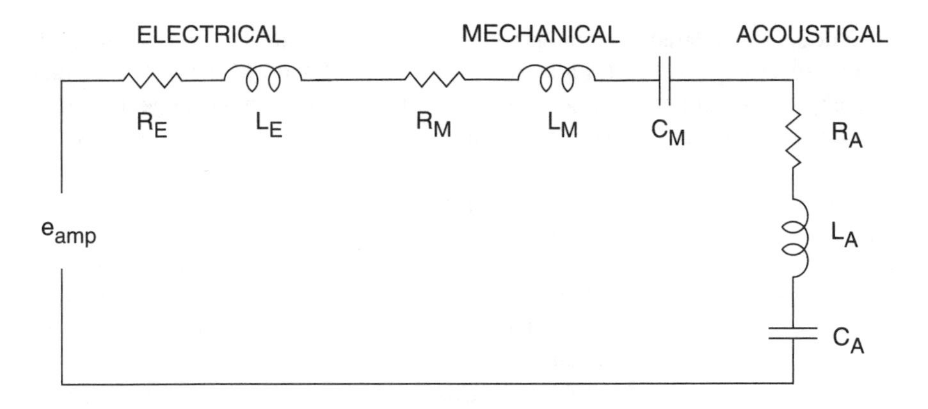

Figure 1.12 The Three-Domain Circuit of a Loudspeaker: one of many equivalent circuits that can be used to model, and thus analyze, the operation of a cone-type loudspeaker.

in designing the so-called "voice coil" type of mechanism of the head and arm for a computer floppy-disk player and recorder or the pickup-head mechanism for a CD player.

1.6 AUDIO PROCESSING FOR INTELLIGIBILITY, SOUND REALISM, AND PERSONAL TASTE

An Introduction

The use of preemphasis and its corollary, deemphasis, made a major contribution to increase the high-frequency signal relative to the noise—termed improving the signal-to-noise ratio—in phonograph and other recordings. However, particularly before the development of the vinyl (LP) record, there was still considerable objectionable noise, especially after the records were played many times. Three other analog signal-processing techniques were developed that produced improvements in the signal-to-noise ratio and thus made the records much more listenable. These techniques included compression and expansion, dynamic bandwidth limiting, and truly dynamic noise reduction.

Analog Compression and Expansion—Processing with Nonlinear Elements

One of the characteristics that adds to the realism of both live and rcorded music is a wide volume range—the soft passages should sound soft and the loud passages should sound loud. When music or speech is recorded or transmitted, there is usually a noticeable degradation in this range from soft to loud. This was significant in older recordings and is still a problem for both AM and FM radio. In general, the compression is more dominant in the upper part of the dynamic range. Because this compression is usually nonlinear, there is also an increase in harmonic distortion.

Some of the unwanted effects of compression can be overcome if the receiver or playback amplifier is designed to expand the dynamic range and thus incorporates a gain characteristic (its in-out *transfer function)* that is the converse of the overall transmission or recording compressed-gain characteristic. Many of the first attempts at expanders were somewhat less than satisfactory because they did not produce the correct transfer function.

In their prototype form, both a compressor and an expander can be modeled as a linear amplifier with a nonlinear diode or even a light bulb in its output or feedback circuit. For a compressor, the diode or light bulb circuit is configured to produce a nonlinear roll off, or slump, as the input signal amplitude increases—see Figures 1.14 and 1.15. For the expander, the diode circuit will produce a gain, peaking as the amplitude of the input increases. It should again be emphasized that, at least in theory, the expander gain characteristic should increase with signal amplitude the same amount that the compressor characteristic decreases with signal amplitude.

The Equations for Compression and Expansion

For reference, the equation of a *linear* amplifier, where A is the gain:

$$V_{out} = A\, V_{in} \quad or \quad A = \frac{V_{out}}{V_{in}} \qquad \text{(the transfer function)} \qquad \text{(E 1.14)}$$

For compression, the equation for the *upper* portion of the in-out gain curve (transfer function) will be:

$$V_{out} = A^k\, V_{in} \qquad \text{(E 1.15)}$$

where k represents nonlinear in-out gain for compression, $k < 1$.

For expansion, the equation for the *upper* portion of the in-out gain curve will be:

$$V_{out} = A^k\, V_{in} \qquad \text{(E 1.16)}$$

For expansion, $k > 1$.

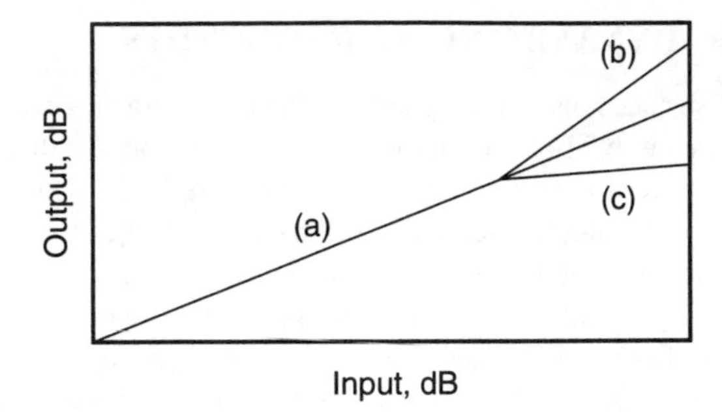

Figure 1.13 The amplifier response curves for expansion and compression: the normalized gain of amplifiers with (a) linear response, (b) with expansion, and (c) with compression.

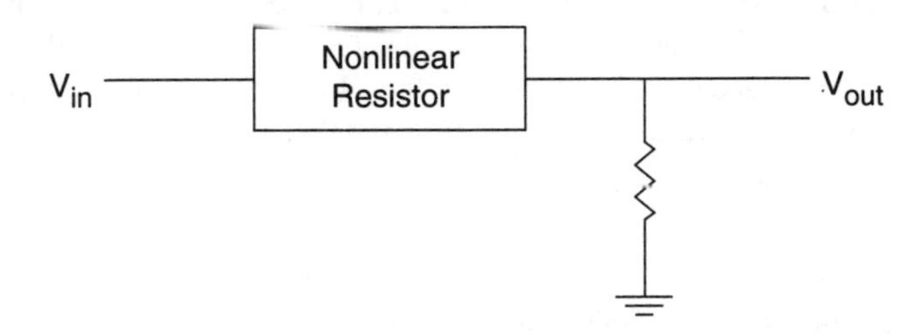

Figure 1.14 The prototype compression model: a simplified example circuit used to illustrate signal compression. It is assumed that the value of the nonlinear resistor, which could be a lamp element, a diode, or some other semiconductor element, increases in magnitude as the amplitude of the input signal voltage increases. In practical circuits, the nonlinear element is usually in an amplifier feedback circuit.

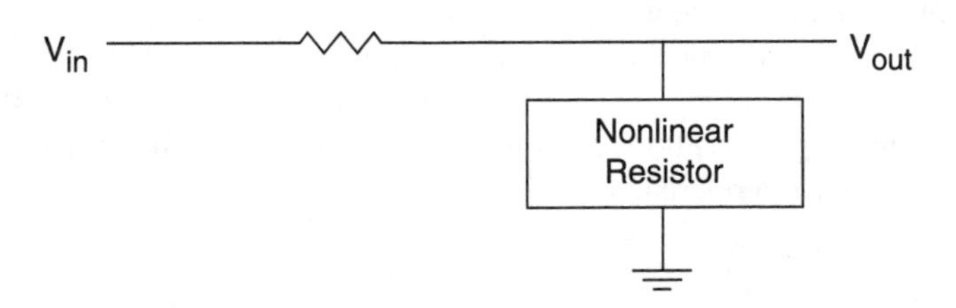

Figure 1.15 The prototype expansion model: a simplified example circuit used to illustrate signal expansion. Again, it is assumed that the value of the resistance element changes nonlinearly with the amplitude to the input signal.

1.7 ANALOG DYNAMIC NOISE REDUCTION

The problem of surface noise has plagued the recording industry ever since Edison's "Mary had a little lamb." In general, most of this objectionable disturbance occurs in the upper portion of the audio spectrum. One simplistic solution to this high-frequency noise problem is to attenuate (cut) these high frequencies. Unfortunately, although a simple high-cut, low-pass filter will reduce the noise, it will also reduce any high-frequency signals. A more clever approach is to cut the high-frequency noise only when there is no high-frequency signal. Thus, a technique is required that will look at the signal and measure its high-frequency content. Although there were a number of reasonably successful developments that solved this problem, only two will be used for illustration.

The underlying principle of most of the early dynamic noise suppressers was to (1) determine, on a continuing basis, whether the signal contained high frequencies and (2) use that information to control a changeable (dynamic) high-cut filter. The information about the presence of high frequencies was usually obtained with a separate high-pass or band-pass filter and a signal rectifier. Any high-frequency signals would produce an AC output from the filter that was then rectified (peak detected) to produce a resulting DC control signal. This DC voltage was used to control the dynamic high-cut filter.

The H. H. Scott Dynamic Noise Suppressor

Scott's circuits were based upon the principles stipulated in the preceding paragraph. As a refinement, his system included both high-frequency and low-frequency noise suppression. The filters included a DC controllable *reactance tube* in those early days that acted as a variable capacitor. If there were incoming low frequencies or high frequencies, the DC control voltages would increase the passband of the appropriate filters. If the frequencies were missing, the resulting control voltage would reduce the bandwidth of the appropriate filters and thus reduce the amount of noise in the output.

In their simplified form, the Scott circuits were controlled LC tuned filters. The details of the reactance tube are not germane to this work—see any text on electronic circuit design. One form of a tuned circuit, with its in/out transfer function, is included here for a reference.

The transfer function of a tuned circuit:

$$H = \frac{V_{out}}{V_{in}} = \frac{L/C}{\sqrt{(L/C)^2 + R^2(X_L - X_C)^2}} \qquad \text{(E 1.17)}$$

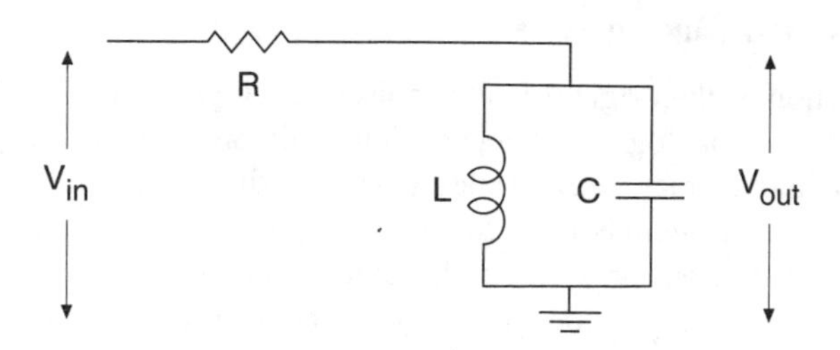

Figure 1.16 The basic principle of the scott dynamic noise suppressor: a simplified LC resonant circuit represented by E 1.13 for its in/out transfer function H. When this, or any similar resonant circuit, is used in a Scott noise suppressor, part of the element C would be a reactance tube (or transistor). The value of its equivalent capacitive reactance will be controlled by the amount of input signal present.

The Harry Olson Dynamic Noise Suppressor

Olson's circuits were based upon the premise that very small noise amplitudes would not be heard. However, if their amplitude increased, these noise signals could be passed through a nonlinear circuit element that could be configured to prevent their amplification—it would clip them. The Olson system needed special bandpass filters both before and after the nonlinear elements, as shown in Figure 1.17.

The equation for the transfer function of the Olson amplifiers cannot be put in a simplified form. It is derived from the simple, linear input/output relationship shown in E 1.14 and the linear part of Figure 1.13. However, this linear relationship is purposely changed for the noise—see the transfer function curve in Figure 1.18.

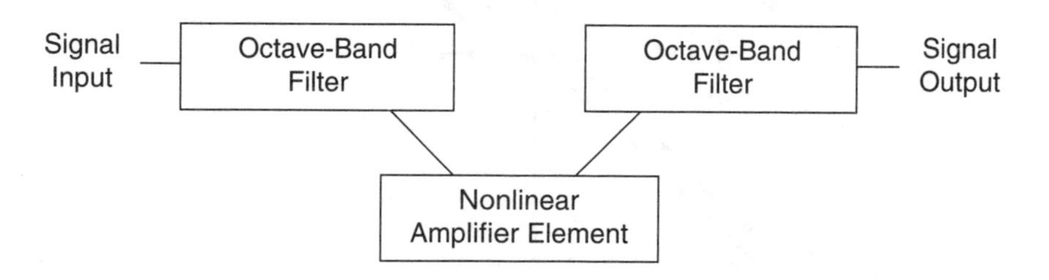

Figure 1.17 The components used in Olson's analog noise suppressor. The usual implementation of this system would contain at least two sets of these components tuned for different noise frequencies. The second or output octave-band filter is used to remove signal distortion caused by the nonlinear component. Often one channel would be an octave (two to one in frequency) somewhat below 3,000 Hz and the other channel would be an octave above 3,000 Hz. (See the Langford-Smith and the Frayne and Wolfe references.)

The Dolby Noise-Reducing System

The introduction of the original Dolby noise-reducing system was the first of a series of innovative analog and (later) digital audio products for the home, commercial sound, motion pictures, and television. This first product represented a new way of thinking about audio noise reduction—preprocessing. Not only did this new system include variations on some of the older techniques such as nonlinear and multiband processing, it provided a method to reduce greatly the problems of the previous incompatibilities between the characteristics of the recording or transmission medium and the playback medium.

The Dolby system is based on what Dr. Dolby called the "differential network." This network (shown in Figure 1.19), which includes bandpass filters as well as nonlinear amplifiers, is used in both pre- and postprocessing. However, its exact use in recording and playback is not the same. In the pre-processor for recording, low-level signal components that have been passed through the filters and the nonlinear amplifiers (the differential network), are added back to the signal itself—the input signal is pre-distorted. This predistortion amplifies these low-level signals in a nonlinear, disproportionate manner. At playback, this same differential network subtracts both these same low-level signals and the noise. Thus, the low-level signals are restored to their original correct amplitudes and the noise is reduced (See Figure 1.20).

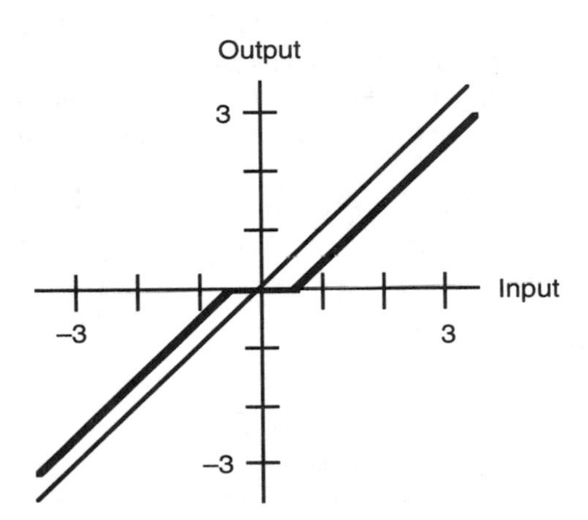

Figure 1.18 The graphical representation of the nonlinear amplifier's transfer function in Figure 1.17. Note that, for reference, the thin diagonal line shows a *linear* transfer function—a 0.1, 1.0, or 2.0 volt, etc., input signal will produce a 0.1, 1.0, or 2.0 volt output signal. However, the bold line shows that if the input is in the range of about ± 0.6 volts, the output is zero. Thus, any low-level noise will not be amplified. This technique is sometimes termed "coring." (See any text on circuit theory or the Dorf reference.)

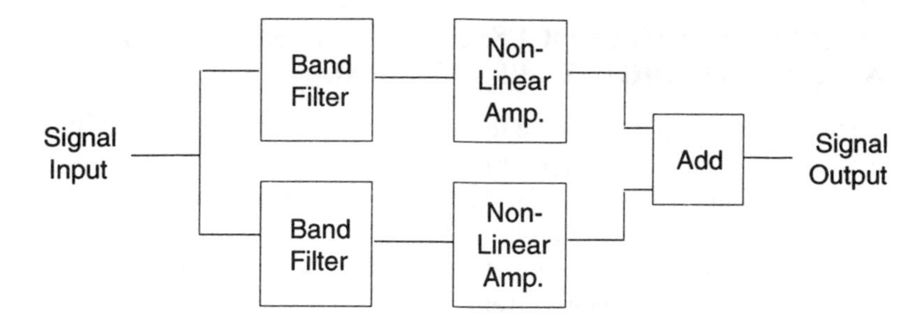

Figure 1.19 A simplified diagram of the Dolby differential network and noise-reducing system. In a typical implementation, there were usually four, rather than two, filters and noninear amplifiers. (See the Dolby reference.)

For recording, if x is the signal input, y is the signal output, and G_1 is the in/out transfer function of the differential network:

$$y = [1 + G_1(x)]\, x \qquad \text{(E 1.18)}$$

For playback, if this same y is now the input to the playback system, G_2 is identical to G_1, and z is the final output:

$$z = y - z\, G_2(z) = \{1/[1 + G_2(z)]\}\, y \qquad \text{(E 1.19)}$$

or,

$$z = \{[1 + G_1(x)]/[1 + G_2(z)]\, x \qquad \text{(E 1.20)}$$

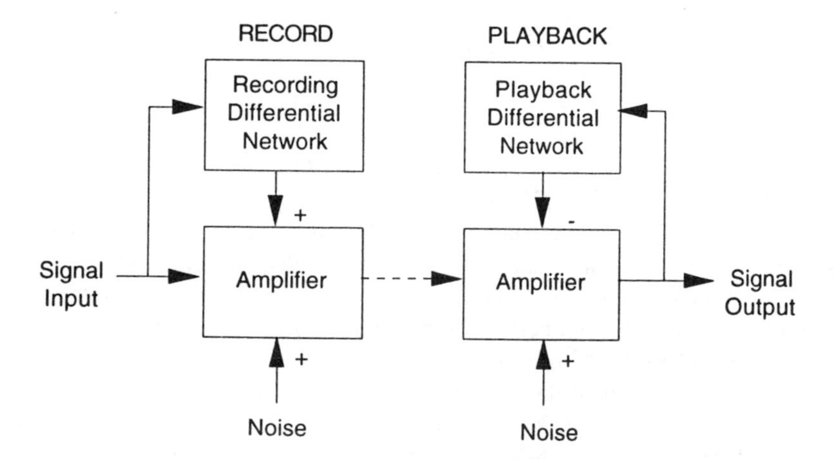

Figure 1.20 The elementary Dolby analog noise-reducing system.

1.8 DIGITAL AUDIO PROCESSING—A BRIEF REVIEW OF ANALOG-TO-DIGITAL PROCESSING

Digital signal processing became the technique of choice in the 1980s. When the microprocessor was invented in the early 1970s, it was primarily adapted for control circuits and some calculations. However, it was the decade of the eighties that brought forth widespread digital processing.

A fast and accurate analog-to-digital (ADC) converter is the key to digital signal processing. (Digital-to-analog converters are much easier to build and use.) Most real-world signals originate in the analog domain. These analog signals must be rapidly and accurately transferred to the digital domain by the ADC. As corresponding digital signals, they may be manipulated arithmetically, filtered with a digital filter, converted with a fast Fourier transform, etc. After these transformations, the new signal forms can be converted back to the analog world (domain) by using a digital-to-analog converter (DAC).

1. When analog signals are converted to the digital domain, thye become a series of *digital numbers*—usually, but not necessarily, *binary numbers*. The understanding of digital processing requires the understanding of digital numbers.

2. All conversions and processing in the digital domain are accomplished by using sampling. The most important rule in digital sampling involves *Nyquist's theorem*. The sampling or quantizing frequency *must* be at least twice the frequency of the highest input signal frequency. It is accepted practice to use an *anti-aliasing* analog filter before the input to the ADC to insure that there are no input frequencies above one-half of the quantizing clock frequency.

Analog-to-Digital Converter—ADCs

Although there are several types of ADCs, many audio and video converters use a very high-speed design termed a flash converter. Flash ADCs consist of four major component groups including [1] a very stable DC analog reference voltage that is used by the comparators, [2] a series of precision resistors used to divide accurately the input DC analog reference voltage, [3] voltage comparators used to compare the level of the input signal with that of the divided reference voltage, and [4] a logic network—an encoder—that converts the outputs of the comparators to the digital (usually binary) numbers. If there is no (zero) signal input, the digital output of the encoder is 0000 for a 4-bit ADC and 0000 0000 for an 8-bit ADC. If the input is about one-half of the DC reference, the corresponding binary numbers would be 1000 and 1000 0000. If the analog input signal voltage is about equal to the DC reference, the 4-and 8-bit binary numbers would be 1111 and 1111 1111.

There are several other factors that must be considered when choosing an appropriate ADC. Is the device linear? Is the digital output corresponding to an input of 0.5 volts exactly twice that of an input of 0.25 volts, etc? Also, does the ADC contribute more noise than dictated by its n-bit (4-bit, 8-bit, etc.) value? Likewise, does the ADC contribute unwanted processing that corresponds to harmonic distortion in the analog domain?

The process of converting the analog signal to digital numbers is called quantizing. It is assumed that a 4-bit ADC has 16 quantization levels (including zero) and an 8-bit ADC has 256 levels (255 plus 0). A 16-bit ADC will have 65,536 quantization levels. In real-world converters, these levels are not all equal. A good ADC should not have any levels that deviate more than ±0.25 of one quantization level.

One of the major characteristics of, and ADC use for, audio is its ability to faithfully convert signals with a wide dynamic range. For illustration, assume an 8-bit ADC configured to accept a maximum signal of 1.0 volt peak-to-peak (vpp). Since this 1-volt signal is utilizing all of the ADC range, it will use all 256 levels or the complete eight bits. Now assume that the input analog signal is reduced to 0.5 vpp. This 0.5 volt signal is only using one-half of the ADC range, which is 128 levels or the equivalent of seven bits. A 0.25 vpp signal would utilize six bits and a 0.125 vpp signal would utilize five bits. From the preceding, it can be argued that an 8-bit ADC will not accurately digitize a wide dynamic range audio signal. For these and other reasons, including noise, audio ADCs are usually at least 12 bits with 14- or 16-bit units quite common.

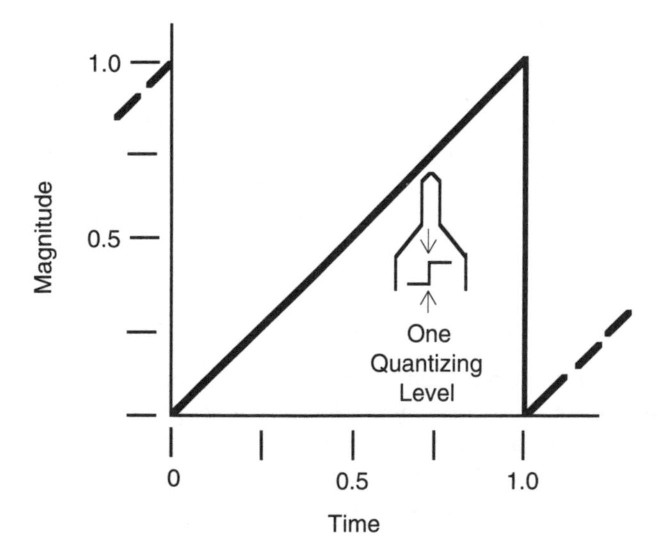

Figure 1.21 Analog-to-digital quantization: a saw-tooth waveform used as an input test signal for an ADC.

Quantizing the Analog Signal

The two functions of an ADC are [1] to divide the analog signal into equal levels or quanta and [2] to convert these quanta into digital codes. One common method of assessing the linearity of an ADC is to use a very linear sawtooth as a precise input test function. If the digital output of this ADC under test is fed directly into a nearly perfect DAC (a valid assumption), the resulting analog output can be used to display and measure the quantization levels and their linearity with a high-gain oscilloscope. For an 8-bit, 255-level ADC, configured for a 1-vpp input, each quantizing level is about 0.00392 volts (about 4 millivolts).

The Elementary Equations for ADCs

For an n-bit flash ADC, where c equals the number of comparators:

$$c = n^2 - 1 \qquad \text{(E 1.21)}$$

For an n-bit flash converter, the number of quantization levels (q) is:

$$q = 2^n \text{ (including zero)} \qquad \text{(E 1.22)}$$

The theoretical best sine wave signal-to-noise ratio for an n-bit ADC:

$$SN = (6n + 1.8) \text{ dB} \qquad \text{(E 1.23)}$$

Figure 1.22 illustrates the prototype for a flash ADC. There are, of course, several variations including some with built-in reference voltages. Also, some allow for a *minus* reference instead of the ground at the bottom of the reference resistors. Likewise, specific designs may change the way the reference resistors are connected to the comparators. For an n-bit converter, there are fundamentally $2^n - 1$ dividing resistors and comparators.

Digital-to Analog Converters

DACs are, in general, simpler and easier to produce than ADCs. An elementary DAC is composed of a resistor network and an analog operational (output) amplifier. Each of the digital inputs uses a resistor, termed a weighting resistor, to scale its input digital value for addition to the other inputs to form the final composite analog output signal. The accuracy of a DAC such as the one shown in Figure 1.24 is, for the most part, dependent upon the accuracy of this resistor network.

The use of these weighting resistors and the operational amplifier will be illustrated with a simple 4-bit DAC. It is assumed that the digital input to this DAC

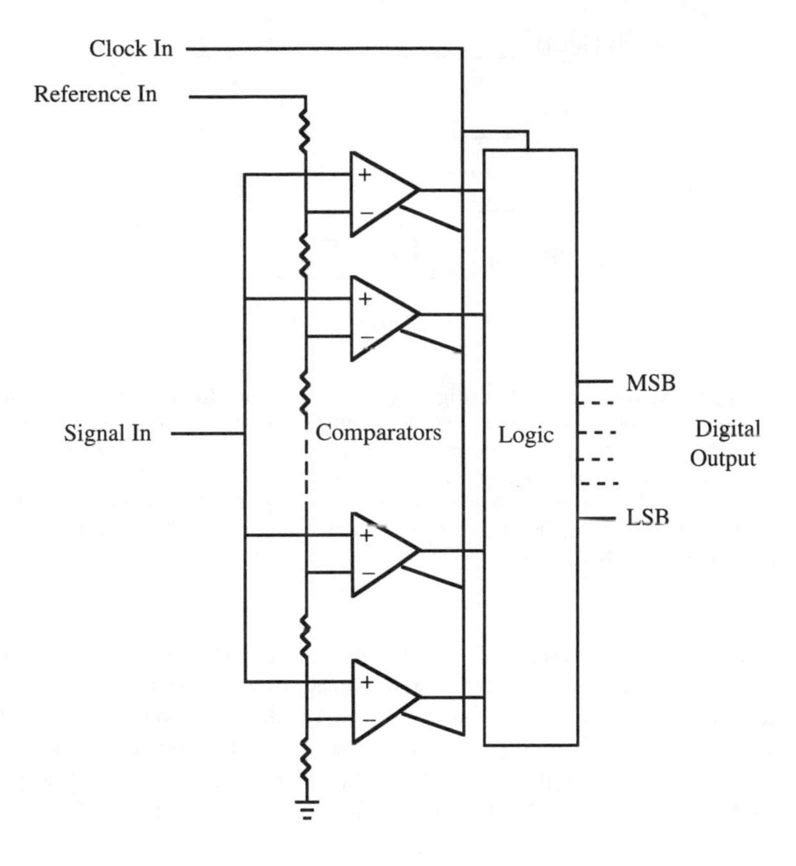

Figure 1.22 The elemental components of an *n*-bit flash ADC. The clock input is often termed enable or convert. Many flash converters include the choice of using either a built-in reference or one that is supplied externally.

comes directly from a "perfect" 4-bit ADC similar to the one in Figure 1.22 and that the original analog input waveform was a ramp, as shown in Figure 1.21. The output of the ADC (and thus the input to the DAC) is a series of 4-bit numbers that can be represented with four waveforms (See Figure 1.23). These waveforms can be developed by reasoning, using the amplitudes of the original input sawtooth. At the start of the sawtooth, in the lower left-hand corner of Figure 1.21, the magnitude is zero, which is represented by the digital (output) number 0000. At the end of the sawtooth, in the upper right-hand corner, the output is unity and can be represented by the sum of 16 quantizing levels (including zero) and the digital number 1111. The halfway point on the waveform, which is one-half of the full output or eight quantizing levels (again including zero), is 0111. The one-quarter value (four levels including zero) is 0011 and the three-quarter value (twelve levels

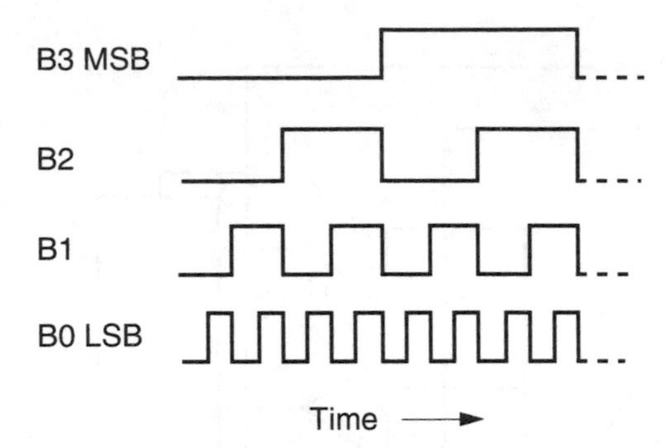

Figure 1.23 The waveforms of the digitally encoded outputs from a 4-bit ADC resulting from a saw-tooth analog input. Note that, at any time point, the waveform values (high = 1, low = 0) add to give the value of the encoded digital output. For instance, at the beginning of the waveforms they are all low or zero and the digital number is 0000. At the end (of the input saw-tooth) all the output waveforms are all high and the digital number is 1111. At the one-half time point, the three less significant bits are high and the MSB is low, giving the digital number of 0111.

including zero) is 1011. At any given time period, the four ADC output coded numbers (the DAC input) can be weighted and then added to give the approximate value of the original analog input. It can be seen that if the MSB (most significant bit) value is normalized to a value of unity or one, the next lower bit will be normalized to a value of one-half, the next lower bit to one-fourth, the next lower bit to one-eighth, etc. Figure 1.25 shows a design for a more accurate ADC.

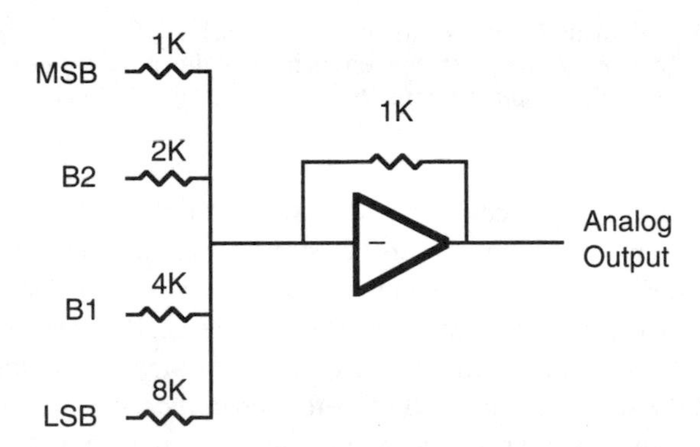

Figure 1.24 An elementary 4-bit DAC consisting of a resistor weighting network and an operational output amplifier. The resistor values select the gain, and thus the weithting, for the MSB input equal to one. The weighting for B2 is one-half, B1 is one-fourth, and the LSB weighting is one-eighth. There are many more sophisticated circuits that give the same or similar outputs.

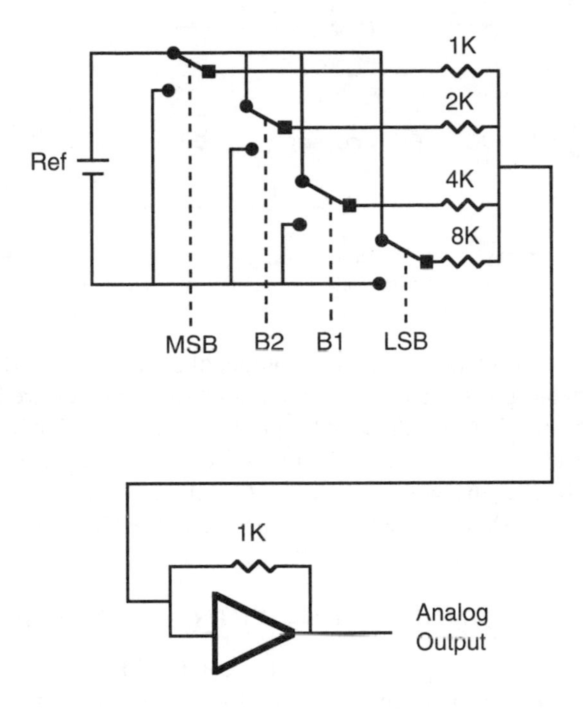

Figure 1.25 The diagram of an improved 4-bit DAC. In this configuration the input digital signal is used as switching controls. The stable reference is divided by the weighting network, rather than the digital inputs, as shown in Figure 1.24. The switches are shown in positions to produce a 1111 code. Different manufacturers' resistor weighting network schemes will often vary from these prototypes. (See the Gordon reference as well as the data and application notes from various ADC and DAC manufacturers.)

ADC—DAC Systems

Real-world systems will generally use a signal processor in addition to the ADC and DAC. It is usually necessary to also incorporate some additional circuit functions not yet shown. First, when using any ADC, it is necessary to use an input low-pass filter to assure that the signal entering the ADC does not contain frequencies above one-half of the clock frequency (the Nyquist frequency). Many times this filter, termed an anti-aliasing filter, can be a simple RC. Nevertheless, its presence is mandatory. Another circuit component that is usually required, especially at high data rates, is a latch at the input of the signal processor and another latch at the input of the DAC. A second mandatory circuit function is a low-pass analog

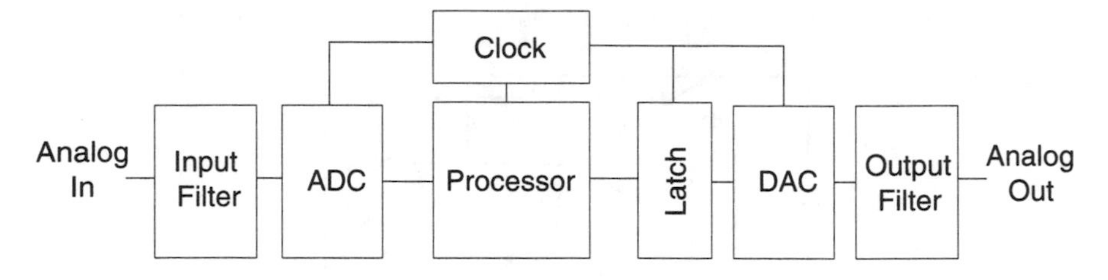

Figure 1.26 A complete prototype analog-to-digital-to-analog data processing system. Even though it is not shown, it is assumed that the processor also includes an input latch.

filter, termed a reconstruction filter, at the output of the DAC. The addition of the weighted digital bits gives a semisegmented output signal the required *smoothing*. This reconstruction filter, like the ADC input filter, can be a simple RC with its cut-off just slightly above the highest frequency of the original analog input signal. It can also be seen that any practical DAC will require a switching clock input. The clock to the DAC is usually the same one as used by the ADC.

1.9 DIGITAL AUDIO SYSTEMS—THE COMPACT DISC

The development and mass production of the compact disc (CD) has given the world an almost perfect source of recorded music. Gone are the ticks, pops, and hisses of surface noise. There is no inherent groove wear and both the signal-to-noise ratio and the dynamic range can approach or exceed 90 dB.

NOTE: The following is not a complete, detailed explanation of the operation of a CD player—it is not meant to be. Rather, it serves as an excellent model for a complete, contemporary digital-processing system. Likewise, the CD record and playback system is extremely well documented if more detailed information is desired.

The elementary CD player, shown in Figure 1.27, represents a truly complete digital signal-processing system. For instance, even at the start of any modern recording, there is both electronic and artistic processing. These may include digital reverberation, multiple mixes and dubs, and equalization of both the frequency and phase (transient) responses. When the final two-channel (stereo) digital signals are placed on the disk, several more processing steps are incorporated. The recorded signal must contain many codes, including clocking and sync information and an equalizing code. In addition, there must be control information for the motor and laser pickup, as well as the two channels of 16-bit audio. The digital information placed on the disk also includes beginning and ending information, selection indexing information, and there is the potential for selection title information. Since this digital system (like most digital systems) has the potential for many errors, there must be exotic error correction processing in the player.

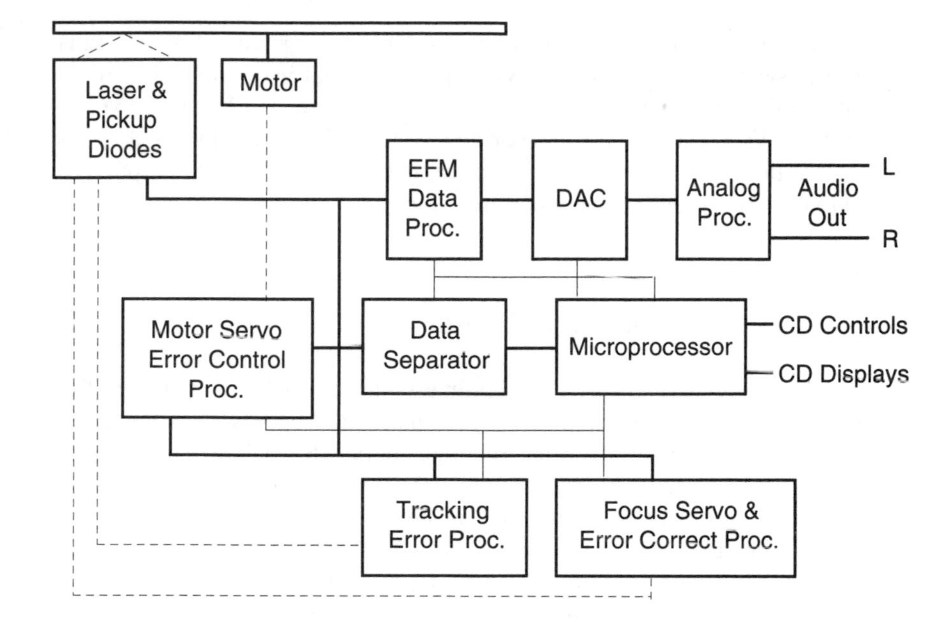

Figure 1.27 A simplified block diagram of the CD player control system. There are six major processing subsystems: [1] the EFM music data, [2] the motor, [3] the pickup tracking, [4] the pickup focus, [5] the CD controls and display, and [6] the master controller—the microprocessor. Each of these subsystems is responsible for retrieving its pertinent data from the CD, reprocessing it into a usable form, and doing very sophisticated error correction. The CD data path is shown bold, the microcontroller path is shown lighter, and the control circuits back to the motor and the pickup are shown dashed.

Audio Information Processing

Before the original 16-bit data words are placed on the CD, they are first processed by dividing each of them into two 8-bit bytes. Likewise, because the timing and data rates required for playing the CD dictate that the transitions from a 0 to a 1 or from a 1 to a 0 are extremely important, the form of the data in these bytes must also be changed. This necessitates changing these new 8-bit bytes to an EFM (eight to fourteen modulation) format. (See parity checking in Chapters 3, 4, and 8.)

Clocking the EFM Format

The 14-bit EFM code on the CD includes the reprocessed digital audio bytes plus some clocking information. A review of Figure 1.26 will reinforce the concept of hard wiring the system clock to the ADC, the processor, and the DAC. In systems such as FM radio, satellite communications, and the CD, there is no direct connection between the transmitter (recorder) and receiver (player). Therefore, a stable, reliable clock signal must be reconstituted. The EFM also includes timing bits that are used to synchronize an internal, 8.6436 MHz crystal in a voltage-controlled oscillator (VCO). Using a divide-by-two phase-locked loop (PLL), the EFM synchronizes the VCO to the phase of the original 4.3218 MHz recording clock. Thus, the internal playback clock and the bit stream coming off the disk are properly locked to each other both in phase and frequency.

CD Bit Packing

There is one further modification needed to the original audio bit stream. When an 8-bit audio byte is converted to EFM, it is possible to produce codes that would be invalid, and thus unusable, based on certain established coding rules for the CD. To prevent the violation of these rules, three more packing bits are added to the EFM bit stream. The playback signal processor continually looks at EFM data and its packing bits to see if an error has occurred. Thus, the complete EFM data word contains a combination of fourteen signal and clock synchronization bits plus three packing bits for a total of seventeen bits. It must be remembered that each EFM is derived from only one-half of the 8-bit audio data, so to encode each complete 16-bit audio word, there must be a second corresponding EFM downstream. Another one of the major jobs of the playback signal processor is to reconstruct each audio data word in its proper 16-bit orientation and time sequence.

The bytes of audio data for both channels do not follow each other in the data format. Rather, they are interleaved—placed downstream—to make it easier to use an error-correcting technique called *interpolation*. This is used in addition to the more common error-correcting technique of parity checking.

CD Data Synchronization

Each pair of EFM codes, along with some other codes to be described later, is contained in a bit stream called a *frame*. These frames repeat and repeat as the disk plays. To ensure that the processor always knows where it is, each frame starts with a unique frame synchronizing bit pattern. The sync pattern is composed of 14 "1s," followed by 14 "0s," plus two more "1s," followed by three packing bits—for a total of 27T bits. (T is one bit time that is equal to one clock cycle.)

The Subcode Byte

There is one subcode byte in each CD data frame. The subcode is used for different control purposes depending on where they are found in the data stream. They may be used alone or combined (again, depending on where they are) for indexing, a catalog of the selections on the disk, start address information, play time information, information on whether preemphasis was used, etc. It is the job of the CD microprocessor to continually interrogate the data stream and then sort the bits or bytes for their proper music or control functions.

Disk Motor Control

The disk rotation must be essentially constant to avoid excess errors. The motor controller (servo) includes a *timebase corrector* (see Chapter 3) that electronically stabilizes the disk rotation to give it a constant velocity.

Laser Focus Control

The data code on the disk is obtained by reflecting the beam of a laser onto four pickup diodes. The disk has a mirror finish except where there are code "bumps" (sometimes called pits), which are decoded as 1s or 0s. However, due to the extremely small dimensions and tight tolerances, this laser-optical system only operates properly if the laser beam is in precise focus. To assure this focus, a laser focusing lens can be moved to compensate for mechanical errors. When the laser beam is received, a special circuit connected to the four pickup diodes checks for this focus. If the beam is not in focus, a signal is sent to the focus servo to move the lens to again establish focus. Incidentally, the coil and magnet configuration that moves the lens strongly resembles a miniature loudspeaker. In fact, the loudspeaker principles discussed earlier in the chapter apply equally well to the focus coil design.

The Pickup-Tracking Control

One of the most interesting servos in the CD player is the head tracking control system. The "groove" on the CD disk is not a physical groove at all but a series of very small bumps or dots on the disk surface that contain the data. The pickup does not physically contact the record. At the start of play, the pickup sensor head must find the data dot pattern and then follow it while keeping the correctly focused. If the user signals for another selection, the head must advance or fall back to a new approximate data position and, by the very rapid and precise servo action, again capture the path of the data pattern and follow it. The goal of the tracking pickup and servo is to keep the laser beam positioned in the middle of the data bumps. The servo signal for tracking will move the laser from side to side while the servo signal for focus will move the laser up and down.

Digital Audio Systems—Dolby Surround Sound with Pro Logic

The idea of *surround sound* is derived from the various early attempts to provide a more realistic sound environment to movie audiences. Starting in the 1950s, several systems included multichannel (multitrack) audio reproduction. Although many of these systems produced spectacular aural effects, when it came to the practical problems of combining normal dialog with surround music, the final result was often unrealistic or confusing.

As one of the solutions to these problems, Dolby Laboratories developed the Motion Picture (MP) sound encoder—see Figure 1.28. This approach has three

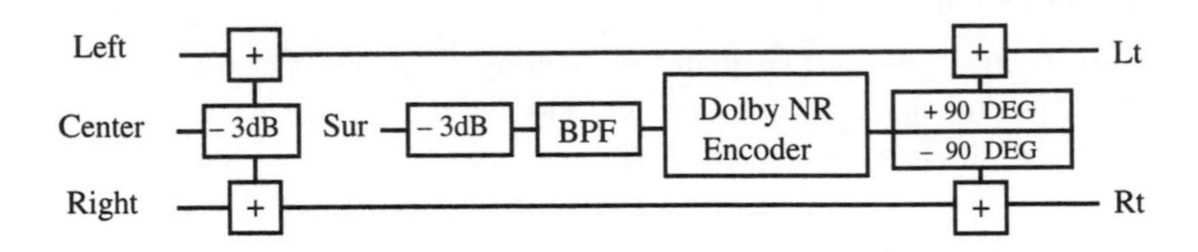

Figure 1.28 A simplified schematic of the Dolby motion picture sound encoder. The center channel is composed of the attenuated left and right input channels. The surround (Sur) channel is band limited to 7 kHz, noise reduced, and matrixed into the resulting Lt and Rt channels.

distinct advantages. First, by using a center channel as well as the normal left and right, the "hole in the middle," especially when dialog is included, is eliminated. Second, a back surround channel is included which provides cinematic realism. Third, the system is so configured that small theaters, with only single-channel reproduction, can use this system for normal mono sound without any additional decoders.

With the advent of large-screen TV receivers and VCRs with two-channel audio came the need to develop a system, patterned after the motion picture's MP system, that would provide surround sound for the home viewer. The primary new component in this home system is Dolby's Pro Logic (receiver) adaptive matrix. Before giving the details of Pro Logic, some background on the requirements and goals for a home system for television viewing—and listening—are in order.

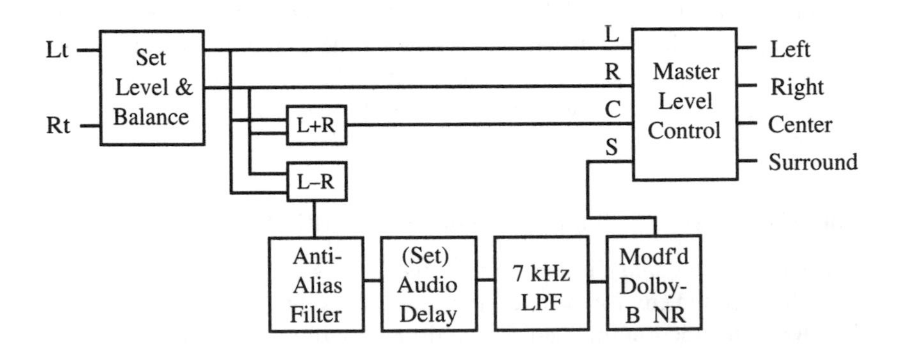

Figure 1.29 An early prototype of the Dolby surround sound decoder. The channel separation in this type of decoder is termed "passive," since there are no special logic circuits to determine the dominant sound source. (Modf'd Dolby-B NR is an abbreviation for Modified Dolby-B Noise-Reducing system.)

The early 1970s produced a four-channel analog audio system that is remembered as quadraphonic (Quad) sound. It came in several flavors. For the purist, there were discrete four-channel tape machines that usually had channel separations of 25 dB or more. The conventional method of recording and listening was to have two microphones or speakers in the front of an auditorium (or listening room) and two in the rear. In addition to the four-channel tape machines, there were several four-to-two channel matrixing schemes that could be used for phonograph records.

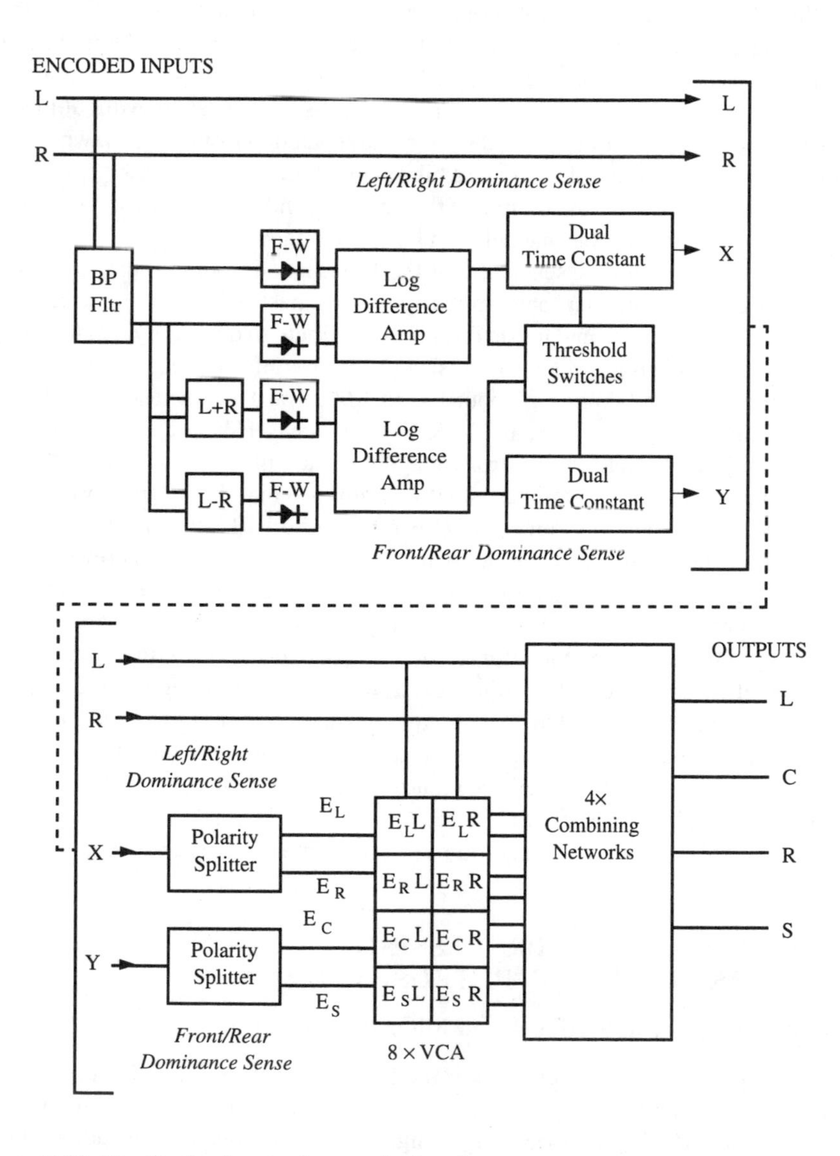

Figure 1.30 The Pro Logic adaptive matrix decoder.

The major problem with the four-to-two analog matrixing methods was termed "pumping"—the jumping of the perceived sound source. Likewise, this was a system conceived for music, and the addition of dialog would have produced more pumping problems.

The Pro Logic decoder is based upon the premise of a dominant sound source. In most instances the dominant sound source in TV and movie program material is the actor(s) giving the dialog. If the actors move around the set, the sound should follow them but the movement should be graceful, without pumping. When there is no dialog and the important sound is the background music, this should be presented in full-dimension stereo. Any general special sounds such as rain, traffic, etc., can be relegated to the side or rear surround-sound speakers without making them the dominant source. Using this idea, the separation from channel, while stilla factor, is less important. If you have a highly intelligent and very fast "leprechaun" with a joystick, which can ensure that the proper sound source is always dominant, then the separation of the channels can be less demanding.

Figure 1.30 shows the schematic of the Pro Logic decoder. In the top section, the control signals are first band limited by the bandpass (BP) filter. Since the lower or bass region of voice and music does not contribute to directivity and the higher parts of the treble region can give misleading phase information, the signals used for control are band limited to about 100 to 7,000 Hz. Then the four band-limited signals, including L, R, L+R, and L–R, are full wave (FW) peak detected to produce four instantaneous DC control voltages. The use of the logarithmic difference amplifiers sets the range for these control signals so that they more closely duplicate the way loudness is perceived. (Recall that the "dB" dimension is logarithmic.) The signals are then continually inspected for the duration to prevent overly rapid directivity changes and to assure that any changes are graceful. The two X and Y signals are converted to dual polarity to increase their directivity information. For instance, when the left/right voltage deflects upward, the dominance is on the left; when it goes downward, the dominance is to the right. Likewise, two control signals are created to show dominance to the center or to surround.

The four control signals, E_L, E_R, E_C, and E_S are used to control an eight-unit VCA (voltage controlled amplifier) matrix. The four dual-polarity signals (eight controlled audio signals) are then algebraically mixed with the original encoded L and R signals to create the final output Left, Center, Right, and Surround audio signals.

1.10 AUDIO SIGNAL PROCESSING—THE CHAPTER IN RETROSPECT

Audio Signals with Feedback and Filter Processing

The subject of audio was chosen for the first chapter because of its universal familiarity and the fact that it was among the very first electronic technologies to incorporate meaningful signal processing. The chapter opens with audio amplifiers and a description of the dominant early audio processing developments in controlled feedback. The work of Black, as well as Bode, Nyquist, and others, shows the introduction of the pioneered techniques required to determine and stabilize a desired amplifier gain with a predictable bandwidth. Researchers also learned more about the process of hearing and the idea of "loudness" and developed tone controls, filters, and equalizers that could (nearly) optimize the way audio signals were transmitted and, likewise, adjust the "tone" to suit the tastes of the individual listener.

Audio Signals with Electromechanical and Acoustical Processing

The concurrent development of the microphone and the loudspeaker, along with their electromechanical and acoustical environments, is introduced next . These advances increased our knowledge of both the theoretical and applied relationships between the domains of electrical, mechanical, and acoustical components and their interactions. The work of Harry Olsen and many others in the field of what Olsen called "Dynamical Analogies" not only provided for a better understanding of microphones and loudspeakers and their enclosures, but it also created a firm theoretical basis for, as one example, the design of modern computer disk drives and their head mechanisms.

Audio Recording and Playback and the Need for Noise Processing

The subject of noise is first introduced by illustrating that analog phonograph records, from Edison's cylinders to the vinyl LP, all produced some unwanted surface noise. The first crude attempts to reduce this noise incorporated simple low-pass electrical or acoustical filters. Later, slightly more sophisticated methods involved band pass and band elimination filters coupled with nonlinear processing. A major advance in surface noise reduction is illustrated with Ray Dolby's introduction of preprocessing the signal as it was being recorded. Since the postprocessor in the playback system "knew" how the signal was preprocessed, it had a priori knowledge of some of the original signal's noise characteristic and could do a much more accurate job of reducing the system's (the record or tape) noise. It is further shown that the Dolby Laboratories have upgraded their systems with digital audio and digital control in their Pro Logic system. This represents an excellent example of the use of digital sampling, control, and processing to produce a modern, very high-quality stereo and surround-sound system for both motion picture and home use.

Digital Audio Signal Processing—The Modern Compact Disc

Presentations of the recording and playback systems for the modern CD represent almost an introductory textbook on how to design and use digital signal processing and control techniques. As examples, it is shown that the stereo audio is very cleverly interleaved into two 16-bit channels that produce extremely wide frequency range, with very low distortion, and a dynamic range of 90 dB. Likewise, the digital audio stream is composed of data packets, with start and stop and play-time information, data synchronizing and data correction information, address information, and motor and head control information. In fact, it is shown that the CD data stream and the entire data system closely resemble the data streams of complex computer information transfer systems or the data packets used in cellular telecommunications. Anyone who is new to contemporary digital data and control techniques should upgrade his or her skills by rereading this chapter and consulting some of the references on CD technology.

Archival and Cardinal References

Beranek, Leo L. *Acoustical Measurements*. New York: John Wiley & Sons, 1949.

Black, H. S. "Stabilized Feedback Amplifiers." *Bell System Technical Journal*, January 1934.

Brewer, Bryan, and Key, Ed. *The Compact Disc Book*. New York: Harcort Brace Jovanovich, 1987.

Corrington, M. S. "Transient Testing of Loudspeakers," *Audio Engineering*, August 1950.

Corrington, M. S., and Kidd, M. C. "Amplitude and Phase Measurements in Loudspeaker Cones." *Proc. IRE*, September 1951.

Dolby, Ray M. "An Audio Noise Reduction System." *Journal of the Audio Engineering Society*, October 1967.

Frayne, John G., and Wolf, Halley. *Elements of Sound Recording*. New York: John Wiley & Sons, 1949.

Langford-Smith, F. (ed.). *Radiotron Designer's Handbook*. Sydney, Australia: Wireless Press, 1952.

Olson, Harry F. *Acoustical Engineering*. New York: D. Van Nostrand Co., 1947.

———. *Dynamical Analogies*. Princeton, NJ: D. Van Nostrand Co., 1958.

Tremaine, Howard M. *Audio Cyclopedia*. Indianapolis, IN: Howard W. Sams & Co., 1969.

Contemporary References

Benson, K. Blair (ed.) *Audio Engineering Handbook.* New York: McGraw-Hil,l 1988.,

Benson, K. Blair, and Whitaker, J. (eds.), *Television and Audio Handbook for Technicians and Engineers.* New York: McGraw-Hill, 1990.

Davidson, Homer L., *Troubleshooting and Repairing Compact Disc Players.* Blue Ridge Summit, PA: Tab Books, 1989.

Dorf, Richard C. (ed.). *The Electrical Engineering Handbook.* Boca Raton: CRC Press, 1993.

Gordon,, Bernard M. "Linear Electronic Analog/Digital Conversion Architectures, Their Origins, Parameters, Limitations, and Applications." *IEEE Transactions on Circuits and Systems*, no. 7, July 1987.

Immink, K. A. S. "Coding methods for high-density optical recording." *Phillips J. Res.*, vol. 41, 1986.

Libbey, R. L. *A Handbook of Circuit Mathematics for Technical Engineers.* Boca Raton FL: CRC Press, 1991.

Marchant, A. B. *Optical Recording. Reading,MA*: Addison-Wesley, 1990.

Mead, William. "Mulit-dimensional Audio for Stereo Television." *Broadcast News*, July 1987.

Tohlmann, Ken. Principles of Digital Audio. Indianapolis, IN: Howard W. Sams & Co., 1989. (Distributed by Prentice Hall)

General References

IEEE Transactions of Acoustics, Speech, and Signal Processing. The IEEE, 345 East 47 St., New York, NY 10017-4929; (212) 705 7900.

Journal of the Acoustical Society of America. 500 Sunnyside Blvd., Woodbury, NY 11797; (516) 349-7800.

Journal of the Audio Engineering Society.. Audio Engineering Society, 60 East 42 St., Room 2520, New York, NY10065.

Journal of the Society of Motion Picture and Television Engineers, 595 West Hartsdale Ave., White Plains, NY 10607; (914) 761-1100.

2

Radio Signal Processing

2.1 AMPLITUDE MODULATION AND DEMODULATION

Radio Signal Processing—An Introduction

To modulate is to change. In music, it can mean to change gracefully and tunefully from one key signature to another. In radio, modulation usually means to use information such as speech, music, or digital codes, to change the form of a basic transmitted wave shape—the carrier. To demodulate is to change back—to change back or return to the original forms of the carrier and the information.

Modulation is used as a convenient technique to transport information such as speech, music, or digital data over a long distance. Guglielmo Marconi first modulated an electromagnetic (radio) carrier wave circa 1900 by systematically turning it on and off using Morse code. The slightly later inventions of Alexanderson, De Forest, and Fleming made it possible to modulate with, and thus transmit, speech and music. By using the reverse process—demodulation—one can retrieve the original music or data signal from the modulated carrier.

There are several forms of modulation that are available to the designer of communication equipment. Each of these techniques has its particular advantages and disadvantages for its required bandwidth, required power, its performance in the presence of noise, the complexity and cost of the modulation and demodulation circuitry, etc. Likewise, each of the major types of modulation, such as AM (amplitude modulation) or FM (frequency modulation, has a similar or secondary technique with its modified set of advantages and disadvantages.

Amplitude Modulation and Demodulation

Amplitude modulation is the simplest both in concept and technique. Both the modulaton and demodulation circuits can be simple and inexpensive. The original Marconi Morse code on and off procedure is the simplest form of AM. It utilized the full carrier *on* for a dot or the longer dash. The time between the dots or dashes or between words was a carrier *off* period. The off periods were, of course, very vulnerable to noise (static).

Signals such as speech and music can continually change the amplitude of the carrier as a function of their amplitude. Loud signals change the amplitude more than soft signals. The frequency components of the input (modulating) signal determine the rate at which the carrier is modulated. Thus, the modulated carrier possesses information corresponding to both the amplitude and frequency of the original modulating signal.

Amplitude modulation—sometimes termed mixing—is the process of multiplying the carrier by the input signal. This multiplication produces an output that, in its simplest form, contains three signals, including: [1] the frequency of the original carrier, [2] an upper frequency sideband that is equal to the sum of carrier frequency plus the input signal frequency, and [3] a lower sideband frequency that is equal to the difference of the carrier frequency minus the signal frequency. For example, if the unmodulated carrier is a sine wave with a frequency of 1 MHz and the signal is a sine wave with a frequency of 1 kHz, the modulated signal will contain [1] a signal component of 1,000,000 Hz, [2] an upper sideband of 1,001,000 Hz, and [3] a lower sideband of 999,000 Hz. These three new signals, termed the AM spectrum, are combined in such a way that they cannot now be separated by a filter. Most real-world input signals are not single-frequency sine waves and thus the upper and lower sidebands will be the composite of the carrier multiplied by many input signal components.

Amplitude demodulation is accomplished by a number of methods. The simplest one uses a diode rectifier (detector), a load resistor, and a capacitor, acting with the resistor, to form an RC low-pass filter. This nonlinear diode envelope-detector circuit works as a multiplier something like the multiplier used in the original modulation. This simple diode detector produces two simultaneous actions. It acts [1] as a half-wave rectifier and thus passes only one-half of the modulated wave shape, and, in doing this, [2] the carrier and the input signal are unmixed or separated. This latter action makes it possible to filter out the carrier frequency, leaving a replica of the input modulating signal.

Let it be assumed that v_C is the carrier signal and v_S is the input modulating signal:

$$v_c(t) = A_c \sin \omega_c t \qquad \text{(E 2.1)}$$

where A_C is the carrier peak value

$$v_s(t) = A_s \sin \omega_s t \qquad \text{(E 2.2)}$$

where A_s is the signal peak value

For AM, the amplitude of the carrier envelope is increased or decreased as a function. It varies or changes because of the amplitude of the input signal. Thus, the amplitude of the composite modulated envelope is:

$$e_{env} = A_c + v_s(t) = A_c + A_s \sin \omega_s t \qquad \text{(E 2.3)}$$

NOTE: E 2.3 can be misleading. Although this equation shows simple addition, it is the equation for the amplitude of the modulated envelope, not the equation for the modulation process. The following E 2.4 and E 2.5 equations are for the modulation process and they show multiplication.

The total output signal voltage for the AM wave is:

$$v_{out}(t) = e_{env} \sin \omega_c t$$

$$= (A_c + A_s \sin \omega_s t) \sin \omega_c t$$

$$= A_c \sin \omega_c t + A_s (\sin \omega_s t)(\sin \omega_c t) \qquad \text{(E 2.4)}$$

Equation 2.4 can be rearranged, using a trigonometric identity, to give the following terms:

$$v_{out}(t) = A_c \sin \omega_c t + \frac{1}{2} A_s \cos(\omega_c - \omega_s) - \frac{1}{2} A_s \cos(\omega_c + \omega_s)$$

The upper sideband

The lower sideband

The carrier frequency

or, if the letter "m" is used to indicate the *modulation index* or (again, with some manipulation) the per cent modulation,

$$v_{out}(t) = A_c \sin \omega_c t + \frac{m}{2} A_s \cos(\omega_c - \omega_s)$$

$$- \frac{m}{2} A_s \cos(\omega_c + \omega_s)\ , \qquad \text{(E 2.5)}$$

where

$$m = \frac{A_s}{A_c} \qquad \text{(E 2.6)}$$

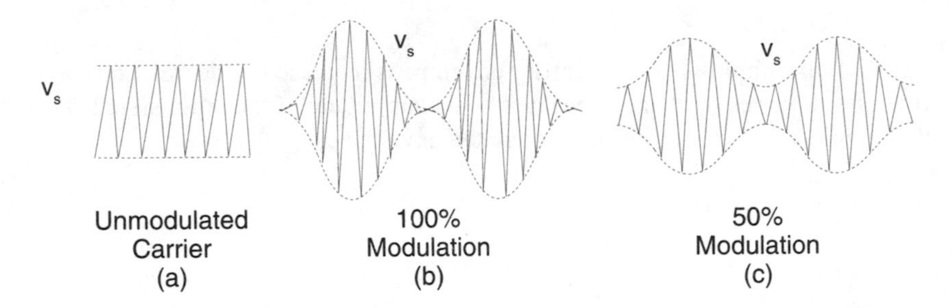

Figure 2.1 Amplitude modulation: (a) shows an unmodulated carrier, (b) the overall envelope when the carrier is modulated 100 percent and (c) the envelope for a carrier modulated 50percent. The dotted lines are used to highlight the contour of the modulation envelope—they do not actually exist nor do they show on an oscilloscope display.

Figure 2.1 indicates amplitude modulation at 50 and 100 percent. Modulation over 100 percent is theoretically possible in some circuits but such overmodulation produces severe harmonic distortion. Figure 2.2 shows the prototype amplitude modulator and Figure 2.3 shows a simple amplitude demodulator (detector).

Non-Linear Circuits—A Brief Review

Chapter 1 introduced the concept of the nonlinear resistors and described their use in compressors and expanders. The use (and misuse) of nonlinear components, including resistors, amplifiers, and other circuit elements, is crucial to the understanding of general circuit theory and to the signal processing known as modulation and demodulation.

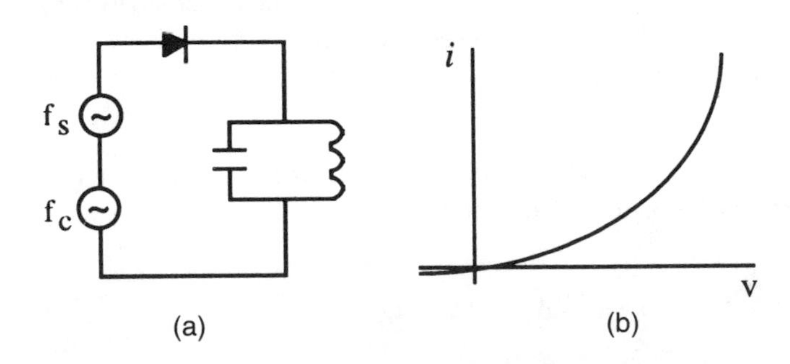

Figure 2.2 A very simple amplitude modulation circuit. In (a), the two input signals are mixed (multiplied) using the nonlinear characteristics of the diode. This circuit is rarely used but it shows the principle involved. The LC circuit is tuned to the carrier frequency, but its bandwidth must be broad enough to also pass the upper and lower sidebands; (b) shows the general in-out, *nonlinear* characteristics of a diode. Practical circuits use a tube or transistor biased into their nonlinear region.

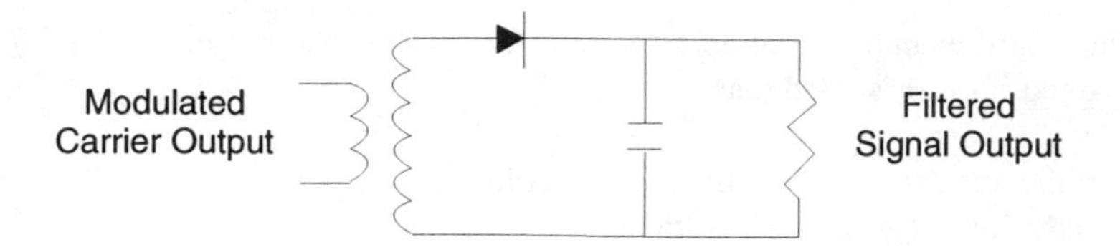

Figure 2.3 A common envelope detector. The modulated carrier is half-wave rectified (detected) and the RC filter removes the residual carrier, leaving the original signal. The process uses the *nonlinear* diode characteristic.

As a first example, assume a normal, linear resistor. Using Ohm's law, the current through the resistor is proportional to the voltage imposed across it. To be more exact, the current is equal to that imposed voltage multiplied by a proportionality factor we recognize as the value of the reciprocal of the resistor. Thus, the current-voltage—the i-v—relationship in a linear resistor is:

$$i = v\,k = v\frac{1}{r}$$

where $k = 1/r$ (E 2.7)

As a second example, if the resistance changes with voltage, the resistance is termed nonlinear and the current is now an exponential (power) function of the input voltage:

$$i = v^{\kappa} k = v^{\kappa}\left(\frac{1}{r}\right) \qquad \text{(E 2.8)}$$

where the κ is an exponent that makes the term a "power" or nonlinear function.

If there are two input voltages, v_1 and v_2, they can be defined as:

$$v_1 = V_1 \sin \omega_1 t \text{ and } v_2 = V_2 \sin \omega_2 t: \qquad \text{(E 2.9)}$$

$$i = (v_1 + v_2)^{\kappa} k \qquad \text{(E 2.10)}$$

And, using the appropriate trigonometric identities, E 2.9 and E 2.10 combine in this form:

$$i = (V_1^{\kappa} \sin^{\kappa} \omega_1 t + V_2^{\kappa} \sin^{\kappa} \omega_2 t + 2V_1V_2 \sin \omega_1 t \sin \omega_2 t)\,k \qquad \text{(E 2.11)}$$

The κ exponents are usually whole numbers such as 2 or 3 when the nonlinear resistor is used for expanding. The κ exponent will be less than unity when the nonlinear resistance element is used in compression circuits — see Chapter 1.

For a third example, assume a vacuum-tube or transistor amplifier. E 2.7 can be adapted if it is assumed that:

i = the output current of the plate or collector
v = the input grid or base voltage
k = the transconductance of the tube or transistor

For a linear (output) resistance in the tube or transistor amplifier:

$$i = a + vk = a + bv \tag{E 2.12}$$

where a is the nonsignal or quiescent (DC) plate or collector current and the term "k" has been changed to the more traditional term "b" where $b \equiv k$.

If the tube or transistor is operating biased so that its voltage input versus its current output is not completely linear (a rather common case), the circuit acts in a nonlinear manner. The input-output relationship will not be a straight line, which means the input-output curve (transfer characteristic) can be represented with one or more exponential terms. The current will often be proportional to a voltage squared term and may be a voltage cubed term. A typical current output would be in the form:

$$i = a + bv + cv^2 + dv^3 + \ldots \tag{E 2.13}$$

where a is the DC term, b is the first-order or linear term, c is the portion of the nonlinear transconductance causing the second-order or v^2 term, and d is the portion of the transconductance causing the third-order or v^3 term, etc. See Figure 2.4.

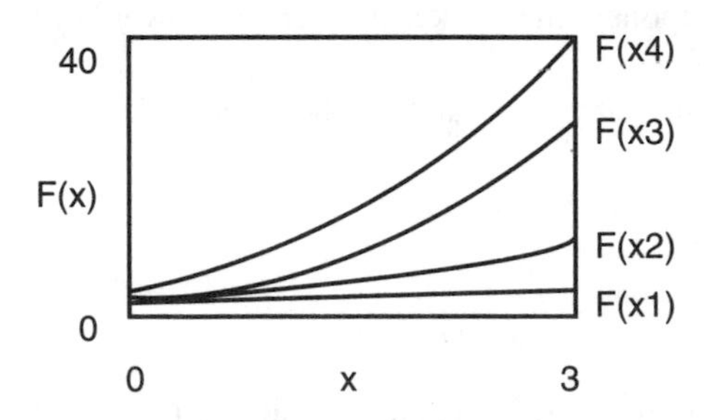

Figure 2.4 A graph of common functions found in amplifiers and modulators. The "a" or DC term has been set to unity and the "x^n" or "n" can be interpreted as different functions of gain.
F(x1) = 1 + x (for reference, a linear function)
F(x2) = 1 + x^2 (a second-order function)
F(x3) = 1 + x^3 (a third-order function)
F(x4) = 1 + x + x^2 + x^3 (a polynomial function)

The equation of a nonlinear amplifier producing only first- and second-order terms is:

$$i = a + bv + cv^2 \tag{E 2.14}$$

Like E 2.10, it can be expanded to show interesting signal combinations:

$$\begin{aligned} i = a &+ b\,(V_1 \sin \omega_1 t + V_2 \sin \omega_2 t) \\ &+ c\,(V_1^2 \sin^2 \omega_1 t + V_2^2 \sin^2 \omega_2 t + 2V_1V_2 \sin\omega_1 t \sin\omega_2 t) \end{aligned} \tag{E 2.15}$$

and, using more trigonometric identities and further expansion:

$$i = [a + \frac{1}{2} cV_1^2 + \frac{1}{2} cV_2^2] \tag{E 2.16a}$$

$$+ [bV_1 \sin\omega_1 t] \tag{E 2.16b}$$

$$+ [bV_2 \sin\omega_2 t] \tag{E 2.16c}$$

$$- [\frac{1}{2} cV_1^2 \cos 2\omega_1 t + \frac{1}{2} cV_2^2 \cos 2\omega_2 t] \tag{E 2.16d}$$

$$+ [cV_1V_2 \cos (\omega_1 - \omega_2)] \tag{E 2.16e}$$

$$- [cV_1V_2 \cos (\omega_1 + \omega_2)] \tag{E 2.16f}$$

where:
(a) represents the DC component of the output current,
(b) represents the sin ω_1 or carrier term,
(c) represents the sin ω_2 or input modulating,
(d) represents the harmonics of the carrier and modulating signal
(e) represents the lower sideband, and
(f) represents the upper sideband.

This entire example, including the text and E 2.13 through all of E 2.16, shows that a nonlinear amplifier is really a modulator and produces quite complicated output signal combinations that can be described as useful signal processing or unwanted distortion products—depending upon the application.

2.2 SINGLE SIDEBAND MODULATION AND DEMODULATION

An Introduction

Single sideband (SSB) techniques are special cases of amplitude modulation. By using very specialized modulation and demodulation processing, the same information can be transmitted as with conventional AM but the total transmitting power can be significantly reduced.

For conventional AM with a sine wave input modulating signal, the required transmitter power may be 150 percent of the power required to transmit the carrier alone. If the sidebands can be deleted before transmission, the transmitted power can be reduced to only one-third of the transmitt power required compared to 100 percent AM. Further, if only one sideband is transmitted, the power is halved so that for a suppressed-carrier, single-sideband transmission, only one-sixth of the 100 percent AM power is required.

For SSB, both the modulation and demodulation circuits are more complex and costly than for conventional AM. There are three model suppressed-carrier, single-sideband modulating systems. Likewise, there are a number of model SSB demodulating systems. The choice of a demodulating system is sometimes determined by how it detects and processes the phase information. A very specialized form of SSB modulation and demodulation, termed *vestigial sideband* is covered in Chapter 3.

Single Sideband Modulation and Demodulation

The previous review section on nonlinear functions developed the ideas of the outputs resulting from signal multiplication. Usually, but not necessarily always, this is accomplished with some nonlinear power-function circuits that produce a multitude of output signal combinations. Once these new combinations are produced they can be selected and processed, depending upon the final required output(s).

The first requirement in SSB modulation is to produce modulated sidebands void of the carrier. SSB modulation usually starts with a balanced modulator. A balanced modulator uses the standard audio signal and the carrier as inputs, but produces an output with only the upper and lower sidebands—it has suppressed the carrier. Early copper-oxide balanced modulators were used in the first carrier telephone circuits (see Figure 4.9). Modern applications use integrated circuits.

By using a balanced modulator, the problem of producing a true SSB is now reduced to removing one of the output sidebands. There are three model systems that are used to remove the unwanted sideband: [1] a filter, often analog, [2] a phase shifter circuit, and [3] the advanced system, using four balanced modulators. This advanced system is termed the Weaver or "third method" in many texts.

Unwanted Sideband Removal with a Filter

If it is assumed that the SSB modulation was created with some form of a balanced modulator, the output signal will include the two sidebands. Often, for operational and receiver reasons, it is prudent to provide the user with a choice of using either the upper or lower sideband. In theory, this can be accomplished by providing a selectable bandpass filter for both the upper and lower sidebands. However, for frequencies above 100 kHz, it is difficult to provide analog filters that have the required frequency cut-offs and are stable with time and temperature. Special

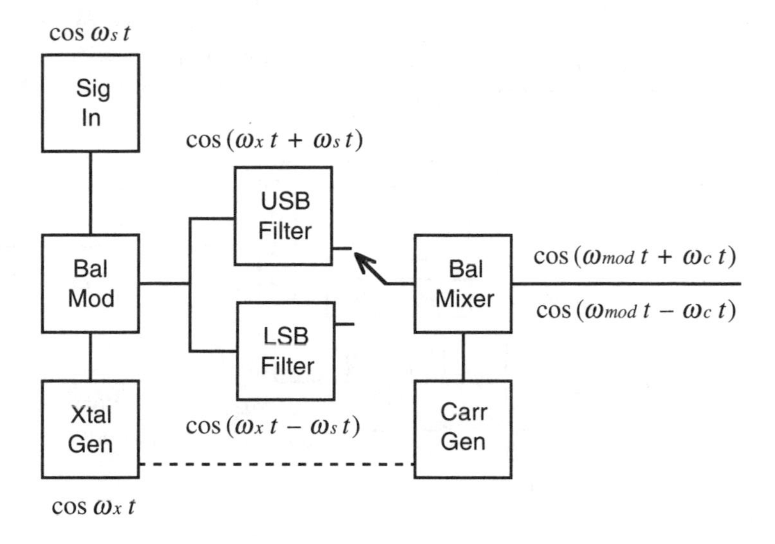

Figure 2.5 SSB generation using sideband filters: the filter method. The input signal and the signal from a very stable crystal oscillator are combined in a balanced modulator to form sum and difference outputs. The operator can select the upper or lower sideband filter. The output of the selected bandpass filter is, in turn, fed into a balanced mixer to form the final upper and lower SSB signal. Although it is not shown, there will be a final bandpass filter (a tuned circuit) at the output of the mixer. The dashed line shows that the first crystal generator and the final carrier are usually phase locked.

crystal, ceramic, mechanical, or surface acoustic-wave filters have helped to solve some of these problems but even so, as the frequency range increases, these filters also become difficult or expensive. Figure 2.5 presents a block diagram of single sideband generation using balanced modulators and analog filter circuits.

Unwanted Sideband Removal with Phase-Shift Circuits

While discrete-circuit analog filters or the special filters mentioned above operate quite well at lower frequencies, successful high-frequency, sideband elimination filters usually require more sophisticated circuit techniques. By combining two additional balanced modulators with precise 90° phase shifters, a much more stable band-elimination system can be obtained.

In most SSB balanced modulator systems, the carrier is generated using a very stable crystal oscillator. In addition, the phase-shift method uses two balance modulators and two precision 90° phase-shift circuits. The combination of the balanced modulators and the phase shifters perform the equivalent trigonometric calculations that cancel either the upper or lower sideband. Although these phase-shift circuits are somewhat more stable than the analog bandpass filters, there are still some problems. The audio shifter must produce exactly 90° for the entire audio spectrum, which is difficult to maintain with time and temperature. See Figure 2.6.

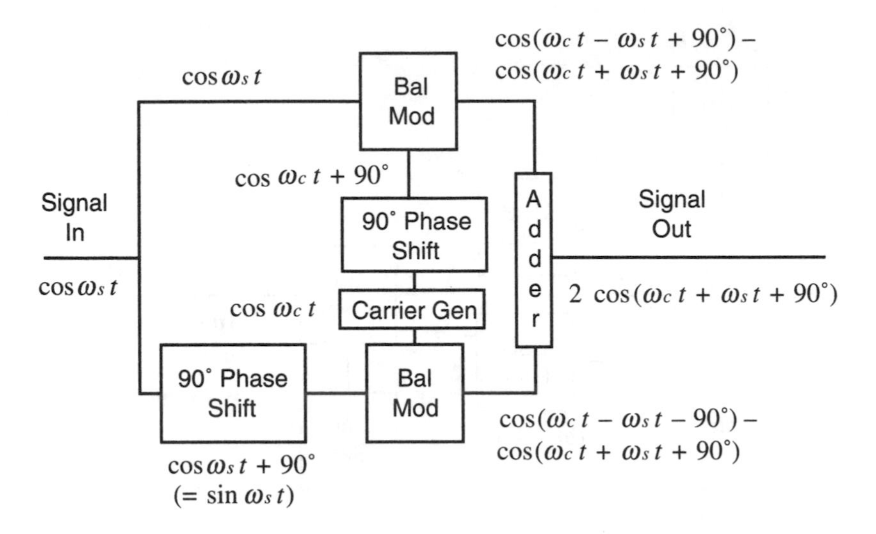

Figure 2.6 SSB generation with phase-shift circuits: the phase-shift method, the technique of suppressing the unwanted side-band in an SSB system using phase shift. The phase shifters are configured positive and thus give the $\omega_C + \omega_S$ sideband. If they had been set to produce a negative 90° shift, the added output would be the $\omega_C - \omega_S$ sideband.

Unwanted Sideband Removal by the "Third Method"

This is a further sophistication of the previous systems. Although it does contain two phase shifters and two low-pass filters, the phase shifters are used only at the carrier frequency and the low-pass filters are noncritical circuits. This third method is successfully used in communication systems up to and including 10 MHz and above. See Figure 2.7.

2.3 SIGNAL PROCESSING IN THE SUPERHETERODYNE RECEIVER

The RF Stage(s)

The function of the RF (sometimes termed tuned radio frequency or TRF) stages is to amplify the small signal received by the antenna enough to be detected. The total equivalent gain of these RF stages may be in the order of 1 million times. For example, if the signal at the antenna terminals is 1 micro-volt (μv) and the required RF input to the detector is 1 volt, the receiver needs an RF gain of 1 million or 120 dB. If this very high gain is produced by simply cascading three or four RF stages, severe problems may arise. First, the circuits may be very unstable. Any external mechanical or acoustical shock or vibration may cause the circuits to oscillate.

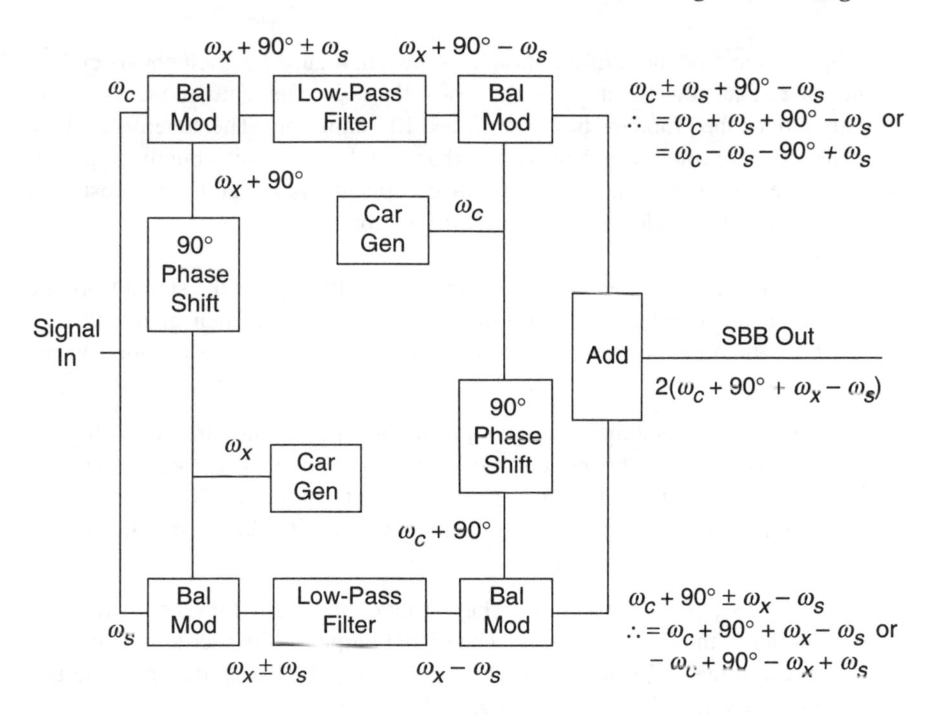

Figure 2.7 SSB generation with advanced or additional phase-shift circuits: The advanced phase-shift or "third" method. This is one of several implementations of this method given in the literature. The lower-frequency Crystal Generator (Xtal Gen) and the higher-frequency Carrier Generator (Car Gen) are often phase locked. Note that this is an excellent example of an analog electric circuit that can be very rigorously presented as a device that solves mathematical equations to produce an output. (See texts on communication circuits such as the Kennedy and the Roddy and Coolen references.)

Second, when selecting (tuning) a new input signal frequency, all the amplifiers should change their tuning together—they should track each other. However, because of electrical and mechanical tolerances, this is almost impossible. This lack of proper tracking will cause the overall gain and bandwidth of the cascaded amplifiers to vary with frequency (tuning). Third, these cascaded RF stages may not reject unwanted frequencies or frequency products very well. This results in the amplification and detection of frequencies either not wanted or at the incorrect spot on the dial.

The Superheterodyne Circuit Configuration

The superheterodyne (sometimes termed "superhet") is a group of circuits that are used in various combinations of AM, SSB, FM, TV, and PCM receivers. It is almost universally used for the RF amplifier portion of devices ranging from home radio and TV sets to all types of communication and radar receivers.

The superheterodyne configuration, as shown in Figure 2.8, includes special intermediate frequency (IF) amplifier stages—all tuned to the same frequency— that greatly reduce the problems of the cascaded RF amplifiers. The superheterodyne circuits are more complicated and costly than simple RF circuits, but the improvements in performance certainly justify the complications and additional cost. The superheterodyne include these four circuit functions:

1. An input tuned RF amplifier. This input amplifier takes the signal from the antenna and amplifies it with one or two RF stages. Their gain, although very important, is small enough that most of the limitations outlined before do not exist.
2. A mixer that combines the signals from the input RF amplifier and a tracing local oscillator. (The input portion of the mixer may also produce some RF gain.) The output of the mixer always contains the IF. The frequency of this IF is usually equal to the frequency of the tracking oscillator minus frequency of the input signal.
3. A tunable, tracking oscillator. The tuning of this oscillator is coupled to the tuning of the RF amplifier(s). The signal output of the oscillator is usually equal to the frequency being tuned by the input RF amplifier plus the frequency value of the IF amplifier.
4. The IF section is a cascaded RF-type amplifier with all stages tuned to the same frequency. The IF amplifiers take the multiplied output of the mixer and amplify that one IF frequency. All the amplifiers in the IF stages are used to amplify this same frequency. The output of the IF section goes to the receiver's detector.

The input tuned RF amplifier is used to improve the sensitivity (gain) of the receiver and to improve its image-frequency rejection. An image frequency is the other input frequency that the input amplifier might pass. For instance, if a common broadcast receiver is tuned to receive a signal (f_s) of 900 kHz, the tracking oscillator (f_{os}) will be 1,355 kHz. The output of the mixer will include the IF (f_{if}) and will equal $f_{os} - f_s = f_{if}$ = 455 kHz. If a very strong input frequency of 1,810 kHz, the other frequency, overrides the tuned circuits of the input RF amplifier(s) and reaches the mixer, it will also produce the IF by $f_s - f_{os} = f_{if}$ or 1,810 – 1,355 = 455. The higher the gain of the input tuned RF stage (and the better the RF shielding), the more the other or image frequency will be rejected.

The Superheterodyne Circuit Equations

Using the following equations, the fundamental frequencies involved in a superheterodyne receiver are:

f_s = the input signal that is received (collected) by the antenna (the input to the RF amplifier)

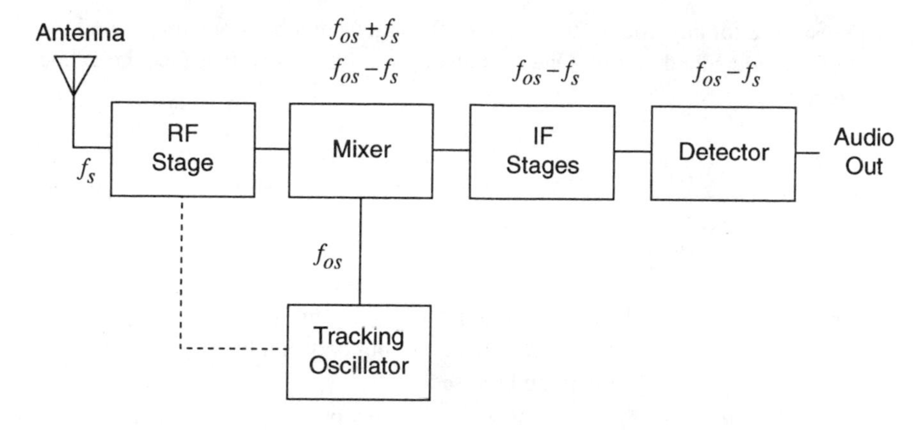

Figure 2.8 The superheterodyne receiver with the RF and IF signals. When selecting (tuning) stations, the tuned frequency of the local oscillator will track the tuned frequency of the RF amplifier(s).

f_{os} = the frequency of the local tracking oscillator

f_{if} = the frequency of all the double-tuned amplifiers in the intermediate-frequency amplifier system

f_{im} = the other or image frequency, besides the normal input frequency, that can override the tuning in the RF amplifier and also produce a mixer output at the frequency of the tuned IF amplifiers

$$f_{if} = f_{os} \pm f_s \qquad \text{(usually } f_{if} = f_{os} - f_s \text{)} \tag{E 2.17}$$

$$f_{im} = f_s \pm 2f_{if} \tag{E 2.18}$$

The image rejection ratio of a single-tuned RF stage, i.e., the ratio of the gain (response) of the input signal frequency to the gain (response) of the image frequency, is:

$$A_r = \frac{1}{\sqrt{1 + k^2 Q^2}} \tag{E 2.19}$$

where Q is the "figure of merit" of the RF LC resonant-tuned circuit

$$k = \frac{f_{im}}{f_s} - \frac{f_s}{f_{im}} \tag{E2.20}$$

The response for any double-tuned RF or IF stage is much more complicated than that of a single-tuned circuit. One of many approximate formulas (see the Libbey reference) is:

$$A_{res} = g_m k \frac{\omega_o \sqrt{L_s L_p}}{k^2 + \frac{1}{Q_p Q_s}} \qquad \text{(E 2.21)}$$

where:

g_m = the transconductance of the amplifier
k = the coefficient of coupling between the primary and the secondary
ω_o = the resonant center frequency
Q_p = the Q of the primary LC circuit
Q_s = the Q of the secondary LC circuit
L_p = the inductance of the primary coil
L_s = the inductance of the secondary coil

The approximate image rejection for a double-tuned (dt) circuit can be calculated by substituting a new "Q" in E 2.19. For a double- tuned circuit, the approximate equivalent Q_{dt} is:

$$Q_{dt} = \sqrt{Q_p Q_s} \qquad \text{(E 2.22)}$$

2.4 FM AND PM AS ANGLE MODULATION

An Introduction to FM

Frequency modulation, and the similar technique of phase modulation, can be categorized as being subparts of the general class of angle modulation. Angle modulation may be visualized by thinking of the familiar clock face and its rotating hands or vectors, as shown in Figure 2.9. Using the common convention, a single rotating hand or vector is used to indicate the single or combined parameters of angular motion, amplitude, and phase. Again, by accepted convention, the starting or 0 degree point is usually at three o'clock. Assuming a counter clockwise rotation, twelve o'clock will indicate a positive rotation of 90-degrees, nine o'clock would indicate a 180-degree rotation, six o'clock a 270-degree rotation and back at three o'clock the vector has rotated a full 360-degrees or one cycle. For reference, the length of the vector will indicate the amplitude of the wave or the instantaneous per cent of amplitude modulation, and the rate of rotation will indicate its angular velocity or frequency ($\omega t = 2\pi ft$).

For frequency modulation, it is normally assumed that the amplitude of the rotating vector is constant—there should be no amplitude modulation in FM. The instantaneous,rate of rotation, and thus the instantaneous angular velocity, will change in FM.

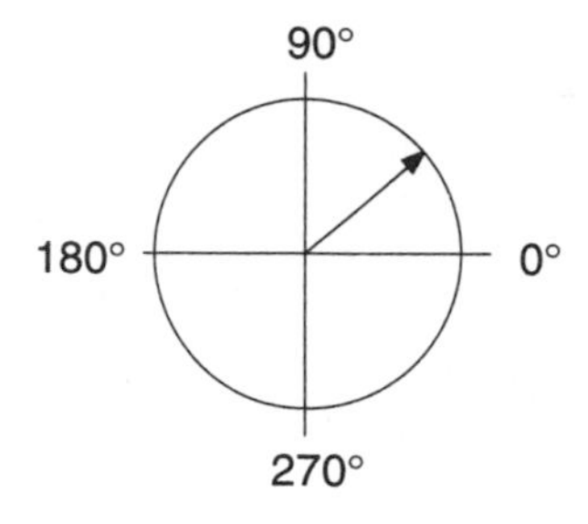

Figure 2.9 The rotating vector or clock face presentation illustrating the angular motion, i.e., the rotational amplitude, and the phase of sine waves. The usual convention is that the three o'clock position is considered the 0° or reference position, and a positive rotation of the vector is counterclockwise.

For phase modulation (PM), both the concept and the clock face diagram are more subtle. It was stated that by convention, the starting or 0 degree point for a waveform is assumed to be at three o'clock. This may or may not describe the real-life, working wave shape. A wave can start at any point we choose. A different starting point on the vector clock face can be the three o'clock point plus or minus n degrees or radians. The carrier frequency $\omega_C t$ is now modified to $\omega_C t \pm \phi$, where the ϕ is n degrees or radians, the angle that is advanced or retarded from any arbitrary starting reference point such as three o'clock. For FM, the instantaneous change is in the "$\omega_C t$" part of E 2.23. The angular frequency is changing as a function of the input modulating signal f_S. For PM, the instantaneous change is in the "ϕ" part of the equation. Phase modulation keeps shifting the wave shape to a new clock face reference point.

Frequency Modulation

When a carrier is frequency modulated, its amplitude remains constant. However, the instantaneous frequency is changed as a function of input signal amplitude. Likewise, the rate at which the carrier frequency is changing is a function of the input signal frequency. The resulting wave shape is reminiscent of an accordion—see Figure 2.10. The incoming signal will repeatedly squeeze the carrier together (make its frequency higher) and then stretch it out (make its frequency lower). The change up and down from the center carrier frequency is termed deviation.

The spectrum resulting from this frequency modulation may be much more complex than the spectrum of AM—it depends upon how much the carrier frequency is squeezed and stretched, i.e., how much the FM carrier is deviated. If the instantaneous change up and down from the carrier center frequency is relatively small, the spectrum of the FM is very similar to that of AM. If the input signal produces large up and down swings, the resulting spectrumis quite complicated and requires some advanced mathematics (such as Bessel functions) to define all of the

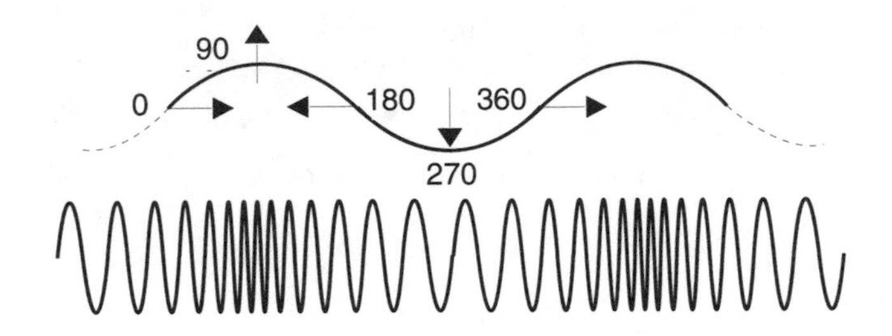

Figure 2.10 The elementary model of a carrier as it is frequency modulated by a sine wave input signal. The top input modulating signal is shown with vectors indicating the phase positions in degrees. The lower waveform is the carrier as it is progressively compressed and expanded—the frequency is increased and then decreased—by the input sine wave.

spectral components. Small up and down swings from the center frequency are termed low deviation and larger swings are called high deviation.

For an* unmodulated *carrier wave:

$$e_c = \sin(\omega_c t + \phi) \text{ or } = \sin \omega_c t \quad if \quad \phi = 0 \qquad \text{(E 2.23)}$$

where:

ω_c = the carrier angular frequency = $2\pi f_c$
ϕ = the constant (for an unmodulated signal) angle ± the established reference point

For a frequency* modulated *wave, the instantaneous frequency is:

$$f_i = f_c + ke_m = f_c + kE_{max} \sin \omega_m t \qquad \text{(E 2.24)}$$

where:

f_i = the instantaneous carrier frequency
f_c = the unmodulated carrier frequency
k = the deviation constant
e_m = the modulating (sine wave) signal

For FM, when the constant $\phi = 0$ and the input signal is sinusoidal:

$$e = \sin\left(\omega_c t - \frac{\Delta f}{f_m} \cos \omega_m t\right)$$

$$= \sin(\omega_c t - m_f \cos \omega t) \qquad \text{(E 2.25)}$$

where: $m_f = \dfrac{\Delta f}{f_m}$ is the modulation index or the deviation ratio

For FM, the frequency swing (deviation) is:

$$\Delta f = m_f f_m \quad \text{(E 2.26)}$$

For phase modulation, using E 2.25 for a reference,* phase *deviation is:

$$\Delta\phi = m_\phi f_m \quad \text{(E 2.27)}$$

where: m_ϕ is the phase-deviation constant

For phase modulation, using E 2.23 for reference, with a sinusoidal signal:

$$e_\phi = \sin(\omega_c t + \Delta\phi \sin \omega_m t)$$

$$e_\phi = \sin(\omega_c t + m_\phi \sin \omega_m t) \quad \text{(E 2.28)}$$

where: m_ϕ is similar to m_f

FM can be generated in principle by accurately changing the value of one of the components in an LC-controlled oscillator. For instance, if it were possible to change the capacitance of a capacitor in an LC tuned circuit as a function of some input audio signal, the result should be FM. In a rather crude and somewhat impractical way, this idea can be implemented by using a condenser microphone for part of the C in a tuned oscillator's LC circuit. A soft sound would only change the capacitor, and thus the frequency a small amount (low deviation), while a loud sound would increase the FM (higher deviation). High-pitched sounds would produce a high rate of change in the frequency modulation and low-pitched sounds would produce low rates of change of the FM. Practical FM is usually produced by circuits called reactance-tube or reactance-transistor modulators. Likewise, FM can be produced by using a *variable* (varactor reactor) diode as part of an LC tuned circuit. In a varactor diode, its equivalent capacitance is changed by changing the voltage across it. See any text or handbook on circuit design including the Dorf and Roddy and Coolen references.

The form of the spectrum of a frequency modulated signal carrier, as well as the content of the spectrum, will change depending upon the degree of modulation —the deviation. For instance, for a low deviation, the form of the spectrum resembles the spectrum of AM with a single upper and lower sideband. For higher deviation, several other sideband frequencies will be generated. In fact, in some cases, a spectral sideband will be generated that will have the same frequency as the carrier. It should be remembered that this generated frequency is a modulation product at the frequency of, and in addition to, the carrier. Likewise, there are sometimes sideband frequencies at harmonics of the input modulating frequencies even though there were no harmonics in that original input modulating frequency. A higher mathematical technique called Bessel functions is used to describe the spectrum of an FM signal as a function of the deviation or modulation index. The modulation index is defined as the change of the carrier frequency (the modulation) divided by the carrier frequency. For FM, this modulation index may be greater than one, whereas for AM, the modulation index must never be greater than one.

Bessel functions, shown in Figure 2.11, are used in FM technology to plot curves showing the value of the sideband frequencies versus the modulation index. Bessel functions are used in many other branches of science, including optics and astronomy. The curves will show the amplitudes of the carrier frequency or J_0 component, the first sideband or J_1 component, the second sideband or J_2 component, etc. Most textbooks on communication also present tables showing the sidebands (some books call them side frequencies) versus a range of modulation indexes m_f. With the excellent mathematical computer programs such as *Mathematica®*, *MathCAD®*, or *Theorist®*, it is usually more informative to plot these functions for any specific problem or application.

Frequency Modulation Transmitters

The implementation of the principles of FM into working, reliable transmitter systems represents some very creative design. The two major problems are to create an original, very stable, non-modulated FM carrier and to create a modulated sig-

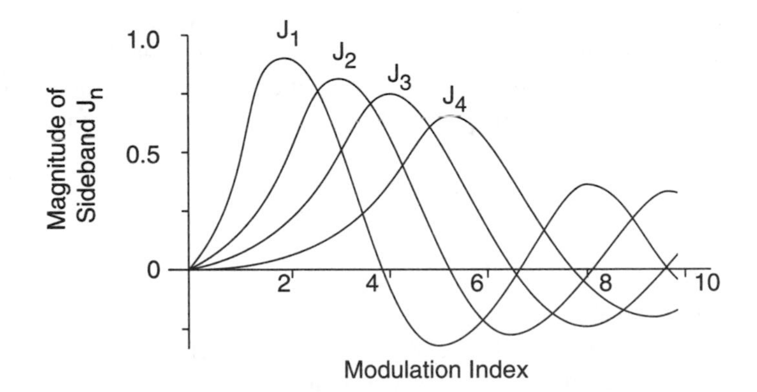

Figure 2.11 Bessel function plots: a plot of J_0 and three sidebands, J_1–J_3, for a sinusoidally modulated FM carrier. It must be remembered that the J_0 component is not the carrier but a component of modulation at the carrier frequency. The information in the graph is often presented in table form or it can be conveniently plotted using such programs as Mathematica©, MathCad©, or Theorist©.

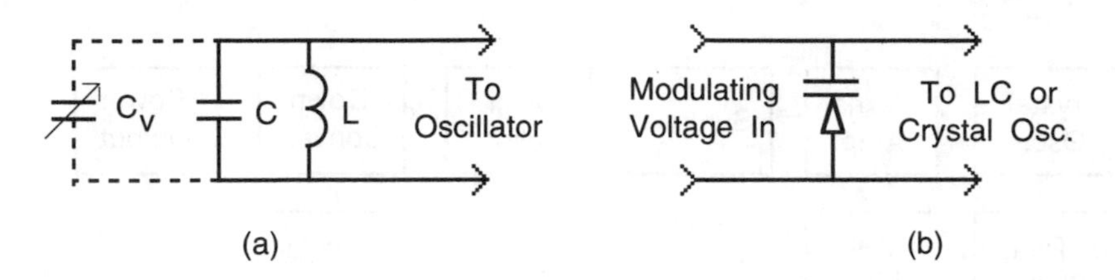

Figure 2.12 Frequency modulation techniques: elementary methods of modulating an FM transmitter. (a) shows an additional variable capacitor C_v in the LC section of an oscillator. In theory, this might be a condenser microphone. A more practical method is to use a reactance tube or transistor circuit or a *varactor diode.* (b) shows one way a varactor diode can be used to modulate an LC or a crystal oscillator circuit.

nal with high output power and high deviation but not to product unwanted distortion products. Although a high-deviation FM signal is quite complex and contains many extra sideband products not found in AM, it is still necessary to create only these inherent frequencies and not to introduce extra distortion products which the discerning listener can hear.

Direct FM Generation

There are two general models for simple FM generation and transmission. The simplest is to design a transistor or tube LC oscillator and to frequency modulate it with a reactance element—see Figure 2.12. This approach does work, but it has the two disadvantages mentioned above. However, the effective deviation can be increased by a frequency-multiplier circuit. The circuit will multiply both the carrier and the FM sideband's deviation ratio *x* times. There may be a problem if the designer or user does not want to multiply the carrier frequency. Also, this approach will not produce a stable unmodulated carrier frequency. Some designs have solved this carrier stability problem by using a crystal-stabilized automatic frequency control (AFC) loop, as shown in Figure 2.13.

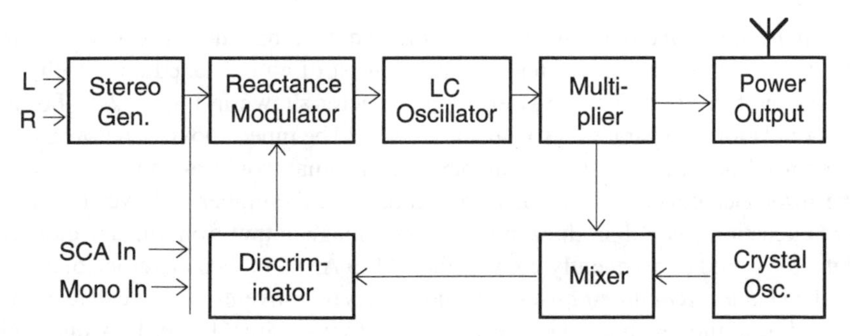

Figure 2.13 A directly modulated stereo transmitter with a crystal oscillator feeding an AFC loop to assure frequency stability.

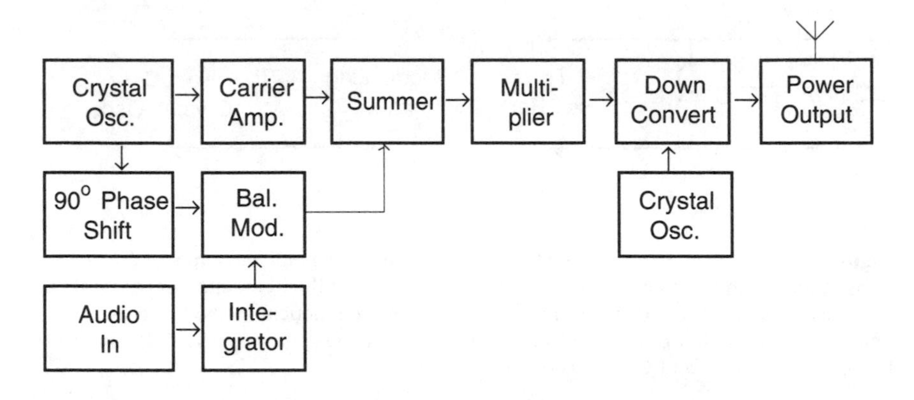

Figure 2.14 The Armstrong FM transmitter really uses phase modulation (PM) in its first 90° phase-shift loop. However, the result is converted back to FM by using the audio integrator. The final crystal oscillator and down converter produce a lower carrier frequency with lower deviation. The Armstrong-type transmitters could also accommodate the use of a stereo encoder box.

Indirect FM Generation

Major Edwin Armstrong, the inventor of FM, devised a transmitter system that would solve some of the problems that have been noted. His system, usually termed the indirect or Armstrong method and shown in Figure 2.14, uses a phase modulator to modulate a crystal-controlled signal. The system also uses a frequency multiplier to increase the effective deviation and a crystal-controlled reverse multiplier or down converter to reduce the carrier frequency for transmission. (See any text on radio communications or, for an excellent archival reference, try to obtain a copy of Nilson and Hornung.)

2.5 FM DEMODULATION

The *demodulation* circuits for FM are more complex than those for AM. The simplest conceptual FM demodulator would consist of an LC tuned circuit plus the diode and filter part of the AM envelope detector shown in Figure 2.3. For FM demodulation, the parallel resonant circuit would be tuned above or below the frequency of the carrier so that the incoming FM signal would ride up and down on the resonance curve. As indicated in Figure 2.15, the higher or lower frequency FM excursions would produce more or less voltage output from the detector. The detector would consequently convert the FM to AM and then envelope detect it.

The *balanced double-tuned* or Round-Travis (the inventors) slope detector is a more linear implementation of the slope-detector principle. It uses two tuned circuits—one tuned slightly above the carrier frequency and one tuned slightly below the carrier frequency. The slopes of the two tuned circuits are combined to

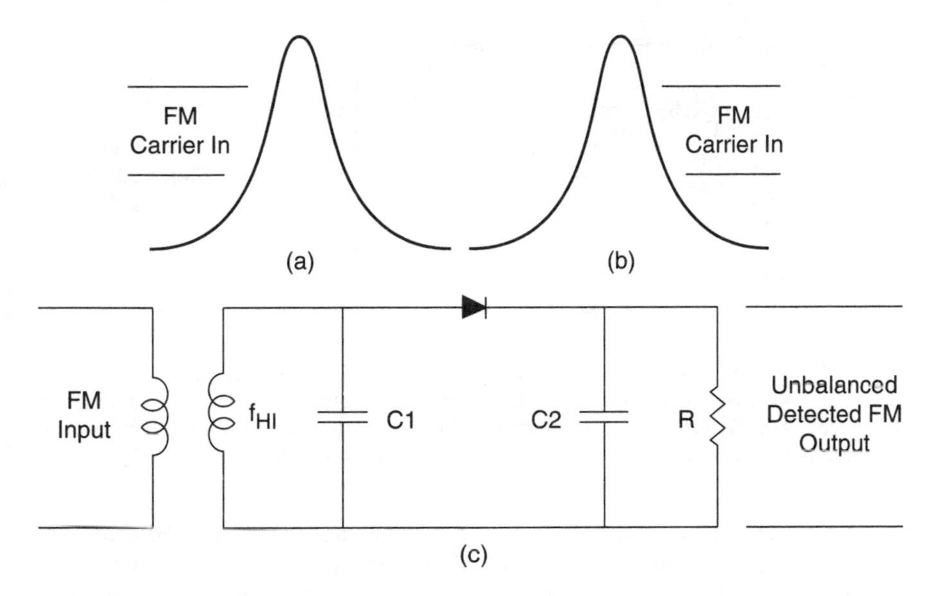

Figure 2.15 The response settings for an elementary FM demodulator using only one tuned circuit: (a) shows the resonance peak set to a frequency slightly higher than the carrier so that the FM will ride up and down on the left-hand side of the resonance curve; (b) shows the resonance peak set slightly lower than the carrier center frequency and therefore the FM carrier will ride up and down the right-hand side of the resonance curve—note that in both cases the slope of the curve is nonlinear; (c) shows a simple (but impractical) tuned-circuit FM detector based upon the principles of (a) or (b).

produce a more linear and more stable FM demodulation. The FM-to-AM-to- envelope detection process is similar to the single-tuned configuration. The symmetrical detection areas above and below the carrier frequency have produced the common reference to the "S" curve—see Figures 2.16 and 2.17.

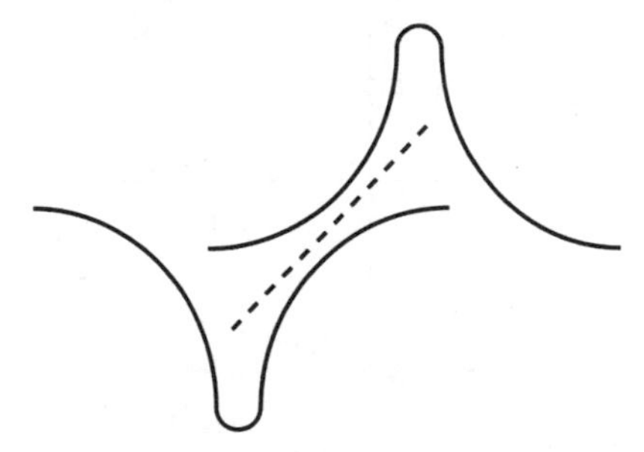

Figure 2.16 Two tuned-circuit FM deodulators: an exaggerated sketch to show the combined response of two tuned circuits. The dashed line illustrates the approximate combined response. If the dashed line were extended, it would roll off at the top and bottom extremes to show a reversed "S" curvature.

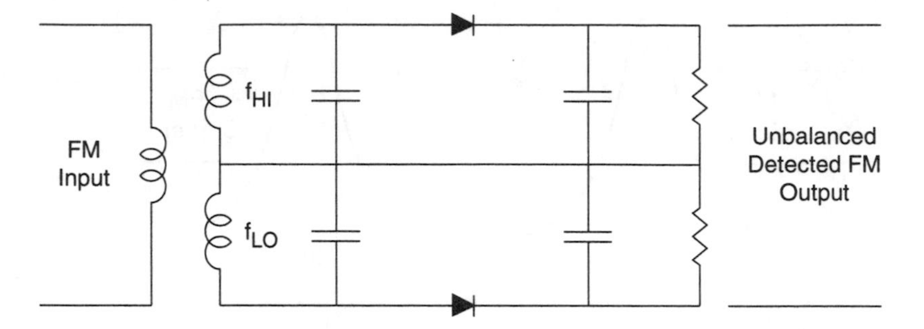

Figure 2.17 A doubled-tuned FM demodulator. This circuit requires that one of the LC tuned circuits be adjusted above the carrier frequency and one be adjusted below the carrier frequency. Although the circuit configuration does give a more linear demodulated output, its setup procedure is too complex to be used in practical receivers.

The Foster-Seeley *phase discriminator* is a more advanced and workable form of the slope detector. By adding two additional parts to the double balanced slope detector and slightly changing the circuit, an FM demodulator was produced that is more linear and, most importantly, only needs to have the primary and secondary LC circuits tuned to the one carrier frequency.

The *ratio* detector is another variation on the balanced slope detector. The ratio-detector circuit has the advantage of not requiring prelimiting. In all the previous FM detectors described, it was tacitly assumed that the input FM carrier was always at a constant amplitude. By definition, frequency modulation changes the frequency of the carrier but does not affect its amplitude. However, when the FM carrier is transmitted and processed in the receiver, some changes in the carrier amplitude do occur. Receivers that use the previously mentioned slope-type detec-

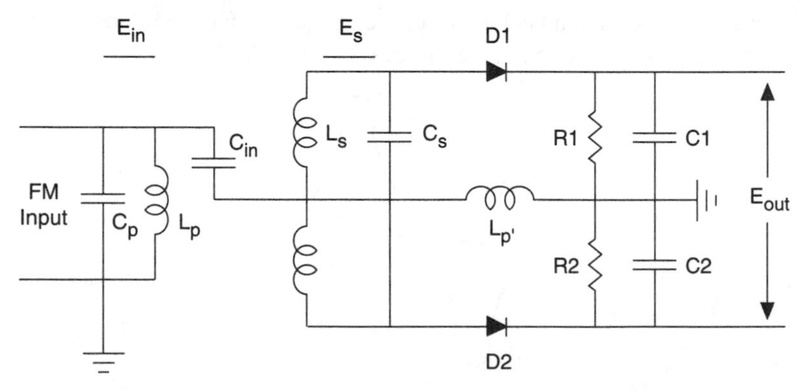

Figure 2.18 Double-tuned discriminator FM demodulator: the circuit of the center-tuned Foster-Seeley FM discriminator. Since the reactance of C_{in} is very small, the entire voltage that appears across the primary tuned circuit (E_{in}) also will be across $L_{p'}$. By fundamental transformer theory, at resonance, the (E_s) voltage across the secondary windings L_s will be 90° out of phase from the primary winding L_p and therefore also $L_{p'}$.

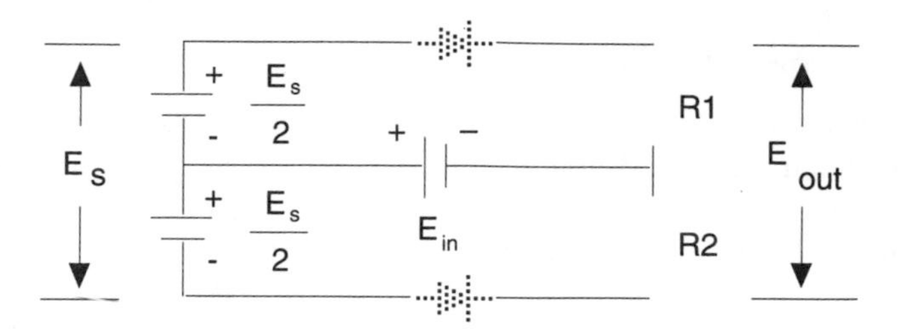

Figure 2.19 An elementary DCmodel to demonstrate the principals of the FM discriminator.

tor must include automatic RF and/or IF gain control circuits and circuits that limit (set) all the input RF or IF signals to a constant amplitude. By a clever reversal of one of the diodes plus the addition of a capacitor and a reconfiguration of the output terminals, the ratio detector makes its demodulation processing almost completely immune to changes in the carrier level. Most contemporary FM receivers use integrated circuit (IC) demodulators. One of the more standard circuit configurations is the quadrature detector. The fundamental configuration requires only one tuned circuit, and ICs such as the Signetics CA3089 give the user exceptional performance with a minimum of discrete parts.

As stated earlier, the Foster-Seeley discriminator circuit is a distinct improvement over the ratio detector but its operation is difficult to understand. Figure 2.18 shows the detector circuit and Figure 2.19 can be used as a simplified model. In this model, the secondary voltage E_s of Figure 2.18 is represented by two batteries in series and the primary input voltage E_{in} is represented by a single battery connected to their center tap. Note that if the shaded diodes represent short circuits, the voltage across R1 is $E_{in} + E_s / 2$ and the voltage across R2 is $E_{in} - E_s / 2$. E_{out} represents the algebraic addition of the voltages across R1 and R2. In Figure 2.18, the voltages across R1 and R2 are the vector sum of the two secondary voltages and the voltage across $L_{p'}$.

2.6 FM STEREO

An Introduction

Like the introduction of color TV, when FM stereo was first considered, it also had the cardinal rule that its signal must be compatible with existing single channel (monophonic) FM receivers. Of the several systems that were considered, the result was to transmit two channels consisting of left (L) + right (R) and L – R. If the receiver did not have a stereo decoder—a mono receiver—it would detect and use the combined L + R signal. If the receiver was built for stereo, the decoder (dermatrixing circuit) could algebraically add and subtract the two transmitted stereo signals to produce a 2L left signal and a 2R right signal.

The Stereo Transmitter Encoder

When stereo FM broadcasting began, it did not require a new FM transmitter. All that was required was to add an input encoding box (see Figure 2.13) that would accept the left and right stereo signals and create the two new encoded L + R and L – R signals. The encoding technique that was adopted is vaguely reminiscent of the subcarrier synchronizing signal—the burst—used in color TV. For FM stereo, a 19-kHz pilot tone is used to synchronize the coding and receiver decoding of the stereo signals.

The Stereo Receiver Decoder

Unlike the two or three basic detectors used for monophonic FM, there are almost as many stereo-decoding circuits as there are (or were) stereo-receiver or tuner manufacturers. Stereo broadcasting rapidly caught the fancy of the public and there

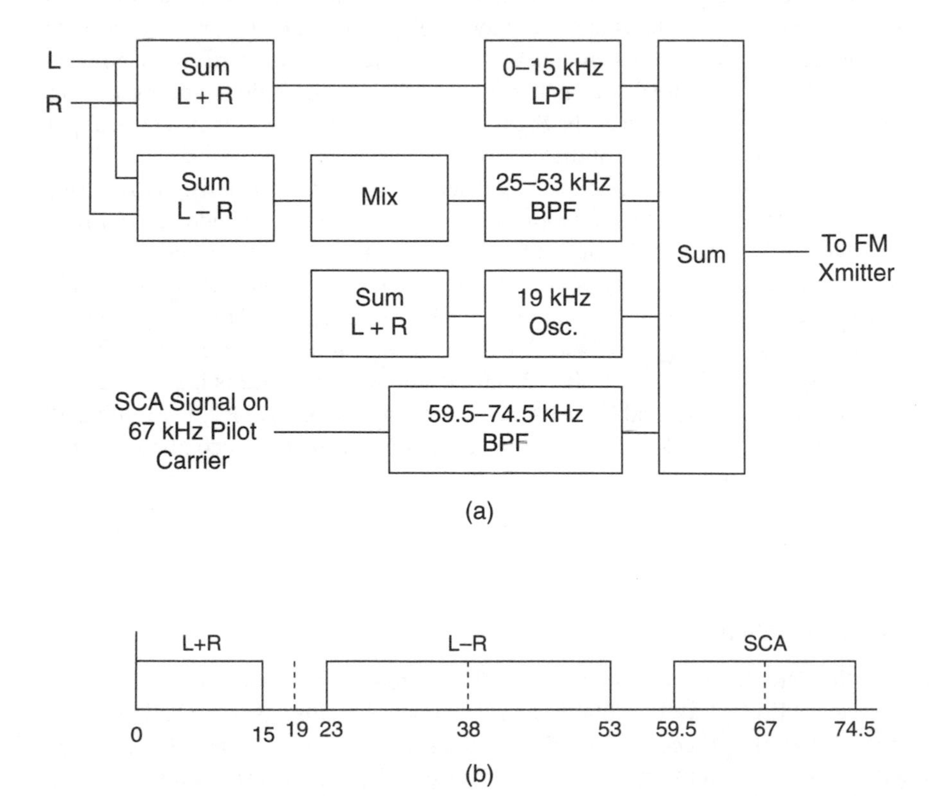

Figure 2.20 A prototype stereo signal encoding box: the transmission scheme for FM stereo. Note that, including SCA, four signals are combined on the encoded carrier. Part (a) of the figure shows the components of the transmitter and part (b) shows the signal's bandwidth allotment.

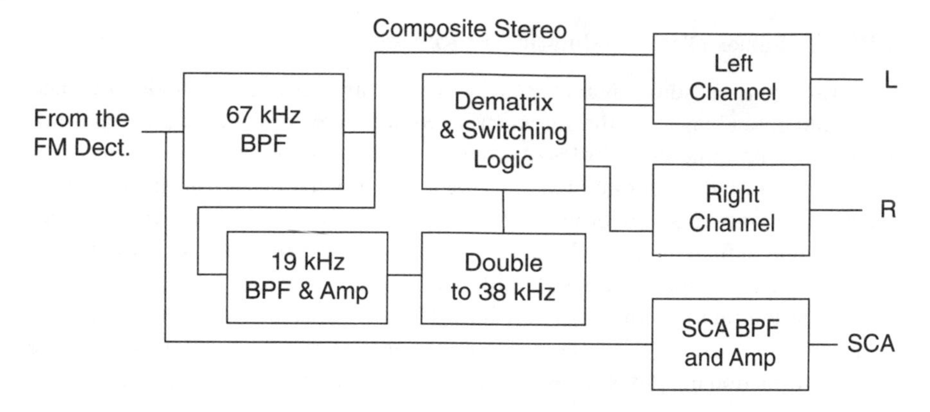

Figure 2.21 A prototype stereo receiver decoder: the fundamental scheme for decoding FM stereo. The circuit details vary quite widely from manufacturer to manufacturer. Most, if not all, of contemporary decoders are included in integrated circuits.

were literally dozens of sets, each with their own decoding scheme, to choose from. As shown in Figure 2.21, the fundamental decoding principle is to use the transmitted 19-kHz pilot tone to create a synchronous 38-kHz switching signal that will repeatedly toggle back and forth to switch on the left and right channels. The various decoding schemes differed in the way they routed the switching signals and the type and number of filters used to keep the stereo distortion low and to ensure good channel separation. Almost incidentally, the subsidiary communication authorization (SCA) or background music signal was added after the introduction of FM stereo. Note that the bandwidth of the stereo signals and the SCA signal, as shown in Figure 2.20 (b), is still within the 75-kHz deviation limit.

2.7 RADIO SIGNAL PROCESSING— THE CHAPTER IN RETROSPECT

The Radio Carrier and Modulation Processing

The chapter on audio processing is followed by a similar major area of signal processing—radio. It is shown that the discovery that man-made electromagnetic waves could be transmitted and received, without wires, over great distances provided a giant leap in mankind's ability to communicate. At first the modulation was simply the on and off pulses of Morse code. However, the subsequent development of continuous amplitude modulation, and later frequency modulation, provided for the transmission and reception of both speech and music and not only opened the age of radio but set the stage for television, radar, commercial wireless communication, etc.

Different Forms of the Modulation Processing

As illustrations of radio signal processing, the techniques of radio modulation and demodulation, along with the filters they required, are shown to expand and improve. Two prominent examples of inventions that provided for the expansion of the art of radio and telecommunications include the superheterodyne and the single sideband circuits. The "superhet," with its superb rejection of interfering frequencies, is still the fundamental RF configuration in most terrestrial and space receivers. Likewise, the example single-sideband circuits, besides creating an entire branch of the radio communication industry, will later be shown to have been adapted to provide the fundamental circuits for the creation and detection of the color information in television as well as for the coded pulse information in radar.

Advances in the Theory of Radio Processing

Some cardinal principles are introduced with the development of radio circuit analysis and the introduction and growth of the theory of radio signal processing. The ideas, the mathematical definitions, and the analysis of frequency response, bandwidth, noise, nonlinear circuits, Bessel functions, etc., were introduced or expanded in these fledgling days of radio. One particularly important outcome that is illustrated was the concept that the processing in a circuit, such as a single-sideband generator, actually solved equations as the signal progressed through the circuit. Such developments as FM stereo or advanced radio communications are a direct outcome of the ideas and techniques that came from the early research and developments in radio theory and analysis.

Archival and Cardinal References

Feldman, Leonard. *FM Multiplexing for Stereo.* Indianapolis, IN: Howard W. Sams & Co., 1972.

Landee, Robert W., Davis, Donovan C., and Albright, Albert P. *Electronic Designers Handbook.* New York: McGraw-Hill, 1957.

MIT staff. *Applied Electronics.* New York: John Wiley & Sons, 1949.

Nilson, Arthur R., and Hornung, J. L. *Practical Radio Communication.* New York: McGraw-Hill, 1943.

Terman, Fredrick Emmonds. *Radio Engineering.* New York: McGraw-Hill, 1947.

Contemporary References

Dorf, Richard C. (ed.). *The Electrical Engineering Handbook.* Boca Raton, FL: CRC Press, 1993.

Kennedy, George. *Electronic Communication Systems*. New York: McGraw-Hill, 1985.

Libbey, R.L. *A Handbook of Circuit Mathematics for Technical Engineers*. Boca Raton FL: CRC Press, 1991.

Roddy, Dennis, and Coolen, John. *Electronic Communications.* Reston, VA: Reston Publishing Company, 1984.

3

Television Signal Processing

"Television doesn't really have to work,
it just has to look as if it works. . ."

Robert A. Dichert—Television Pioneer
and Visionary, 1935–1992

3.1 CREATING RUDIMENTARY TELEVISION PICTURES

Historical Illuminating and Scanning Techniques— Putting a Frame around the Picture

From the perspective of the viewer, television pictures are the result of the pickup and transmission of direct and reflected light. Just as the eye or a photographic camera "sees" the reflected light from an object or a scene, so does the TV camera observe it for the viewer. If the resulting picture is pleasing and gives the viewers the illusion of reality, then they are happy and the set is working. The TV system designers will also have a second perspective. They are interested in creating hardware and software that will produce the best possible picture within the limits of the broadcasting standards, the available technology, and the economic constraints imposed on any given product. Likewise, from a more esoteric viewpoint, a third group, the research scientists, are continually grappling with the mathematical and physical limitations of bandwidth, noise, viewer perception, and colorimetry.

The idea of transmitting pictures dates back more than a century. In 1884, the German scientist Paul Nipkow proposed a method to electrically analyze or scan a picture. Nipkow's pickup system involved [1] light shining on an object, [2] a special scanning wheel (the Nipkow disk) with several holes placed in a spiral configuration, and [3] a photo tube pickup. The display system, shown in Figure 3.1, also involved a neon flash tube, and a second disk that was used as a synchronized scanner for the viewer. The two scanning disks were driven by electric motors that were synchronized by the 60 Hz power line.

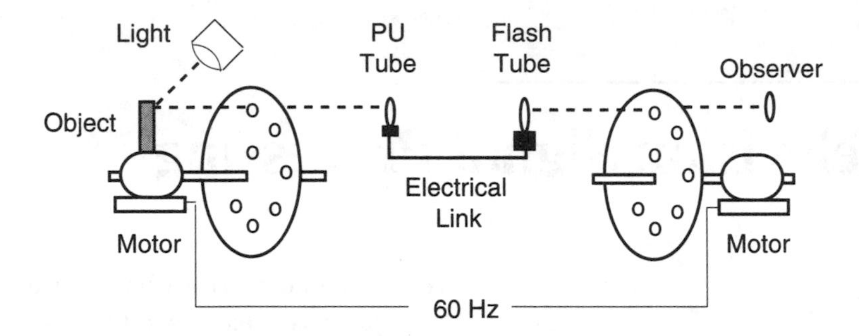

Figure 3.1 A simple television system as proposed by Paul Nipkow. This very early idea anticipates several important concepts that will be used universally 100 years later in all-electronic TV. These precepts include scanning the scene and then transmitting each small area in a serial fashion and synchronizing the pickup scanning with the display scanning.

Other Proposed Scanning Techniques

In the first third of the twentieth century there were a number of scanning proposals. Some of them (such as spiral scanning in some radar displays) were adapted to later electronic technology and used. Many of them ended up as only laboratory curiosities. One of the factors, seemingly unrelated to the choice of a scanning system, was the shape of the pickup or display tube. Since most early cathode ray tubes (CRTs) were round, some method of circular or spiral scanning seemed to have merit, since it would utilize more of the display area than a rectangular scan. This type of scan might make more sense in oscilloscopes rather than TV kinescopes or picture tubes. Other scan systems that have been proposed include sine-

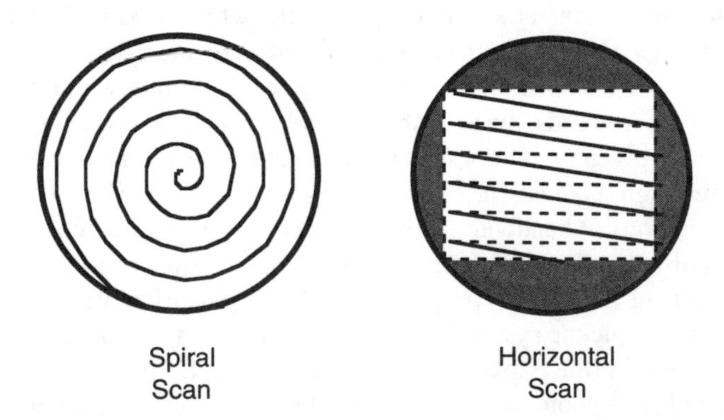

Figure 3.2 A simplified comparison of the maximum picture area available from a circular or spiral scan and a rectangular scan using a round picture tube. The dashed lines in the horizontal scan are really invisible retrace lines.

cosine and vertical orientations. It is interesting to note that few researchers can find any technical reason not to use vertical (top to bottom) instead of horizontal (left to right) scanning except that "it just doesn't seem the natural thing to do."

Scanning, Retrace, and Blanking

Horizontal scanning is the process of [1] repeatedly moving the beam across the focused scene on the face of a pickup tube, or [2] repeatedly moving the beam across the face of a display (picture) tube. Once the beam is swept across the tube face to produce one line on either a pickup or a display tube, it must very rapidly be returned to its approximate starting point to begin the next line sweep. Rapidly returning the beam to produce the next line involves several processes. First, as has been stated, this return or retrace time must be much faster than the normal horizontal scan time. Second, during this beam retrace time, the beam must be turned off or *blanked.* This ensures that the retrace will not interfere with the pickup proccess and will not be seen by the viewer—see Figure 3.31. Also, during this blanked retrace time, all circuits must be resynchronized to assure that each line is timed exactly the same as each other line.

As the horizontal scanning is taking place, there also must be a concurrent vertical scanning process. The vertical scanning assures that the next horizontal line is moved slightly and precisely down the face of the pickup or display tube. Once the bottom horizontal line is completed (using some version of an analog or digital counter), the vertical scan processor must rapidly return the retrace to the top of the tube (raster) ready to control the progress of another set of horizontal lines. During this vertical retrace time, the pickup or display beam must be vertically blanked.

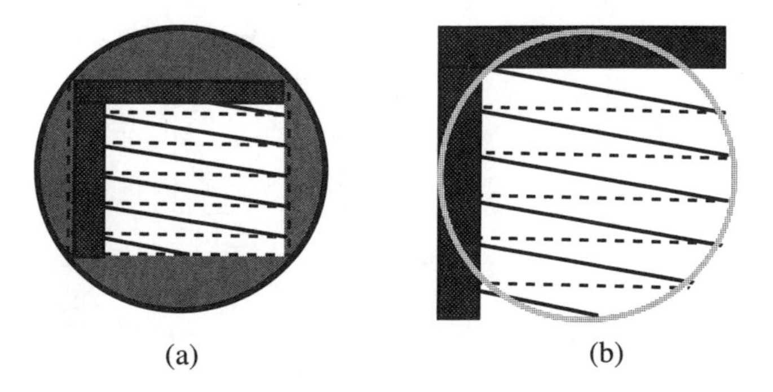

Figure 3.3 A graphic showing how the use of blanking will take away part of the picture area. (a) uses the horizontal scan from Figure 3.2 to show the loss of active picture area to blanking. (b) shows that some of the active area can be regained by a technique termed "overscanning."

Electrostatic and Electromagnetic Scanning

Scanning circuits are used to move the beam across the face of a pickup or display TV tube to assure that the original scene is sampled and displayed as uniformly and as accurately as a given TV system will allow. Historically, there are two fundamental methods of scanning, i.e., moving the beam of electrons. These include electrostatic and electromagnetic deflection circuits (see Figure 3.4). For 5- to 7-inch CRTs, electrostatic scanning is usually preferred. These tubes have their "deflectors" (deflection plates) built in and the external circuitry required to generate the scan voltage is relatively simple and inexpensive. For display tubes that are larger than 8 inches (or even for small tubes in portable TVs), it is more realistic to use magnetic deflection. This is especially true for three-beam color tubes or camera pickup tubes.

The negatively charged electron beam of display tubes is focused (made small so it will not diverge) by the electron gun assembly and is attracted to the center of the positively charged faceplate of the tube. The voltage (charge) on the deflection plates attracts or repels the electrons in the beam and thus deflects them. The current in the deflection coils forms instantaneous magnetic poles and moves the electron beam just as the current in the coils of a m otor will move the wires, containing a flow of electrons, in the armature. For a given tube size and geometry and a given deflection angle D, the relationship between the voltage on the deflector plates V_d and the anode or accelerating voltage V_a is for electrostatic deflection:

$$D = K_1 \frac{V_d}{V_a} \tag{E 3.1}$$

where K_1 includes electron constants and the dimensions of the display tube

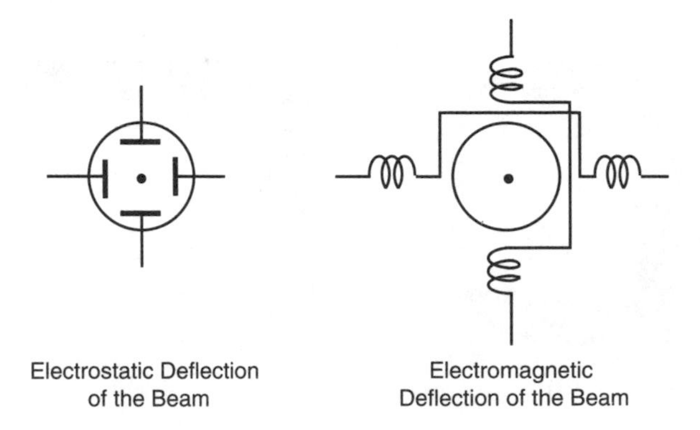

Figure 3.4 The elements of electrostatic and electromagnetic deflection. The beam-deflection plates are built inside tubes using electrostatic deflection. Tubes requiring electromagnetic scanning require external deflection coils. In actual systems, especially for color display tubes, the coil structure (the yoke) is much, much more complex than is indicated by this schematic drawing. The center dot represents the beam of electrons emerging from the electron gun structure at the back of the tube.

For magnetic deflection with a current I in the deflection coils:

$$D = K_2 \frac{I}{\sqrt{V_a}} \tag{E 3.2}$$

where K_2 is similar and also includes magnetic constants.

It can be seen that magnetic deflection is proportional to the square root of the coil current whereas electrostatic deflection is directly proportional to the deflector voltage. This means that with the higher anode voltages of the modern large-diameter kinescopes, not only would the resulting power be much more for electrostatic deflection but the components required to generate the higher voltages would be specialized and costly.

Elementary Pickup and Display Tube Models—Interacting with the Negatively Charged Electron Beam

In their very elementary forms, a TV pickup tube and a display tube employ the general configuration of the old elementary triode or multigrid vacuum tube, which included a filament and cathode (the electron gun), one or more control grids, and a plate or anode. There are two common vacuum-tube type pickup devices, the Image Orthicon ("image orth") and the Vidicon as well as semiconductor charge coupled devices (CCs).

The Vidicon shown in Figure 3.5 makes an excellent representative model for TV pickup devices. The electron gun of the Vidicon emits an electron beam that is manipulated by three different electromagnetic systems, including a beam alignment coil, a focusing coil, and the deflection coils. There are four control grids and a photoconductive anode or target. The original acceleration on the beam is quite low. As the electrons near the target, the potential on the fourth grid retards the beam so when it does strike the photoconductive surface, its velocity is so low that although it does charge the target, there is no secondary emission. (Secondary emission occurs when a single impinging electron dislodges other unwanted electrons from a material.) Thus, the target is uniformly charged by the very low velocity scanned beam. The reflected light from the program scene is focused on the end of the tube and strikes the other side of the photoconductive surface.

NOTE: Pickup and display tubes are excellent examples of how TV as we know and enjoy it today resulted from the creative and dedicated work of individuals and companies around the world—TV was not a single country development. Many countries, including the United States, England, France, Germany, Holland, and later Japan, all contributed significant pickup and display tube developments.

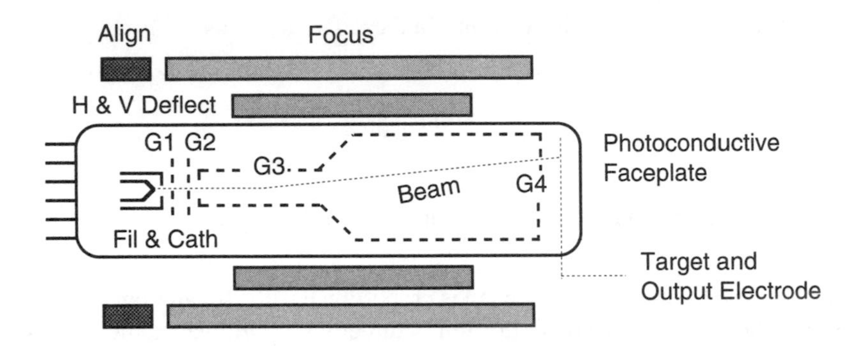

Figure 3.5 A sketch of the elementary Vidicon pickup tube. The four grids and the three sets of coils are shown to illustrate the complexities that are involved in modern TV pickup tubes. The parts are not necessarily in their exact relative position and this schematic drawing is not to scale. See various manufacturers' literature for exact details.

The Prototype Electronic Scanning and Picture Transmission System

From the foregoing, it can be gleaned that one of the most important aspects of a successful video system is to have the display raster precisely synchronized with the raster scan of the pickup. The three signals that must be accurately cloned in time are the horizontal (H), the vertical scan (V)—sometimes termed the sweep—and the accompanying H and V blanking (b). One simple system that directly connects the pickup and display H, V, and b signals is shown in Figure 3.6. For a normal transmitted picture, the synchronizing (scanning) signals must be serialized for transmission, just as the video must be scanned and transmitted in a serial format. Packing the correct signals in their correct time slot for transmission and then unpacking them and reconstituting the timing and display components is the imaginative "trick" that makes television either work or, at least, certainly appear to work.

An Introduction to the Concepts of the Fundamental Television Signal, Including Its Spectrum and Bandwidth

Any communication system is dependent upon its spectrum details (its frequency make-up) to convey specific information. It is usually limited, i.e., it must somehow compromise the amount or speed of information transmitted because of its limited bandwidth. The development of any coherent theory is usually based upon x-y coordinates in the raster space. The following series of equations presents one method of approaching the problem of how to determine the brightness waveform

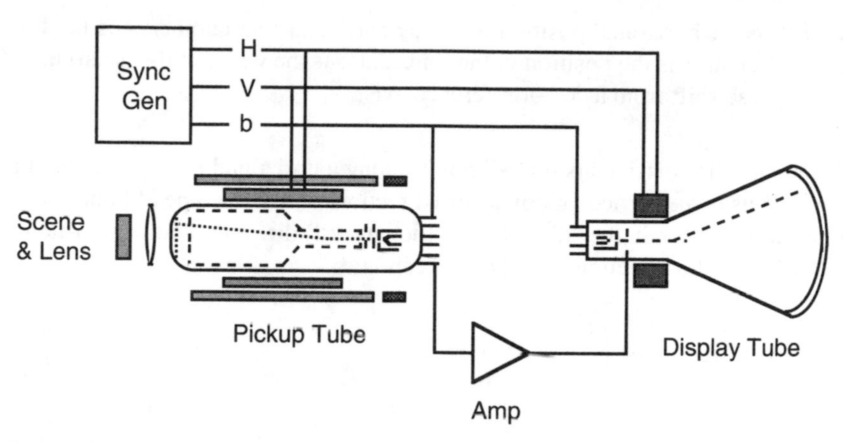

Figure 3.6 A simple prototype closed circuit electronic scanning system. The details of standard scanning systems will be described later in this chapter as will the exact techniques of recreating the synchronizing pulses in the receiver.

and the frequency spectrum of a raster with "n" lines, "w" width, and "h" height. This development also assumes that the electron beam spot size (the aperture) is considered infinitely small and there is no finite time required for H and V retrace. Although this method of analysis is somewhat inaccurate, it does provide one of the methods to attack the very involved problem of finding the spectrum and the required bandwidth of a scanning system.

It can be postulated that any x-y trace, and thus any part of the picture, can be reduced (either literally or in principle) to a Fourier series. The classical Fourier series is written:

$$f(t) = \frac{1}{2} a_0 + \sum_{n=1}^{\infty} a_n \cos n\omega_0 t + \sum_{n=1}^{\infty} b_m \sin n\omega_0 t \qquad \text{(E 3.3)}$$

If there is no DC offset, there will be no a_0 term, and if the wave shape is made symmetrical about a "y" axis, the sin term will drop out to give:

$$f(t) = \sum_{n=1}^{\infty} a_n \cos n\omega_0 t \qquad \text{(E 3.4)}$$

For a single raster horizontal scan line (see Figure 3.2), the Fourier series for the brightness of any single scan line can be given as:

$$B(x,y_1) = \sum_{k=0}^{\infty} a_k \, cos \left(\frac{2\pi kx}{w} + \theta_k \right) \qquad \text{(E 3.5)}$$

where x is the horizontal position, y_1 is any particular line number, k is the Fourier coefficient, x is the position in the line, and w is the width of the (scan) line. θ_k is any phase shift from a "0" or reference axis.

E 3.5 gives the coefficients for "k" points along a horizontal line. What about the coefficients in the vertical or downward direction? The same type of Fourier equation can be written in the "h" or height (downward) direction that was written for the "w" or width direction. For the h direction:

$$B(x,y) = \sum_{l=0}^{\infty} a_l \cos\left(\frac{2\pi\, ly}{h} + \theta_l\right) \tag{E 3.6}$$

E 3.5 and E 3.6 can be multiplied trigonometrically to produce an equation that describes the Fourier coefficients for the entire two-dimensional surface as:

$$B(x,y) = \sum_{k=0}^{\infty} \sum_{l=-\infty}^{\infty} \frac{b_{kl}}{2} \left[2\pi\left(\frac{kx}{w} + \frac{ly}{h}\right) + \theta_{kl}\right] \tag{E 3.7}$$

It is especially interesting to note that the horizontal and vertical terms $\left(\frac{kx}{w} + \frac{ly}{h}\right)$ can be modified (see the Anner reference) to produce the spectrum of the brightness wave given in E 3.7. Probably the most important result of all these equations is finding that the spectrum of a raster system is built upon multiples of the (horizontal) line frequency resembling the model in Figure 3.7. This is true even with the assumptions of an infinitely small aperture and no H and V retrace time.

The Checkerboard Pattern Model—Calculating the TV Picture Bandwidth

The analysis of a video raster, resembling a checkerboard, presents another, more intuitive, method of determining the system spectrum and bandwidth.

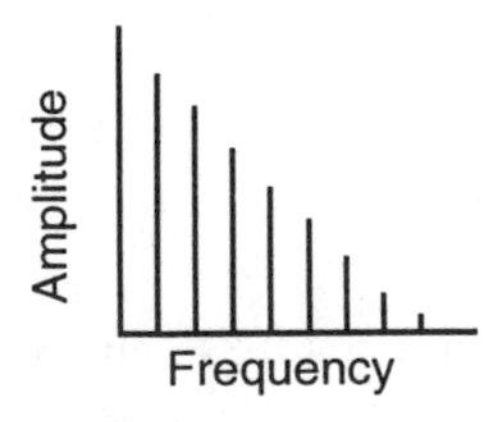

Figure 3.7 The elementary spectrum of a TV video signal. The spectrum or Fourier series is built from multiples of ω_0, which, for the monochrome part of the signal, is the horizontal line frequency.

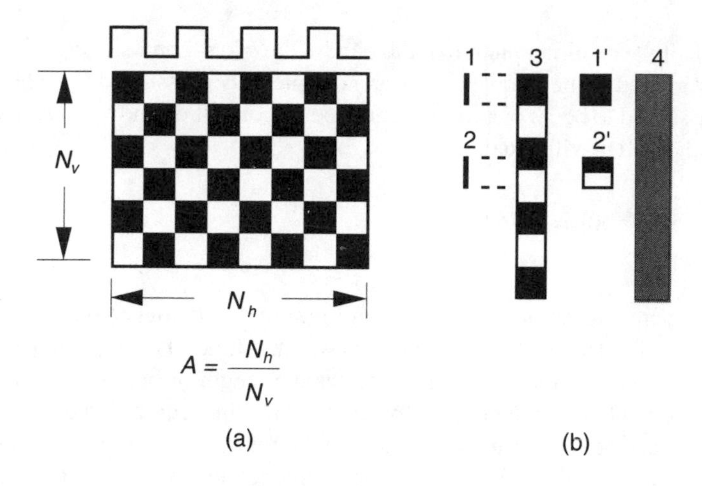

Figure 3.8 The checkerboard bandwidth-approximating model. (a) A checkerboard pattern that can be used to approximate system bandwidth. (b) If the aperture alignment is incorrect, the picture detail will be impaired.

If an assumed aperture that is infinitely narrow and exactly one square high scans the checkerboard of Figure 3.8 (a), it will produce square waves of brightness. Since the checkerboard has N_v vertical and N_h horizontal squares and it takes two squares to produce one high-low square-wave cycle, the number of square-wave cycles per picture f_p will be:

$$f_p = \frac{1}{2} N_v N_h \tag{E 3.8}$$

If the number of complete pictures (checkerboards) per second is now defined as p_s and is multiplied by the above square waves per picture f_p, the result will be a fundamental frequency f_f that can be defined in cycles per second or Hertz.

$$f_f = \frac{1}{2} N_v N_h p_s \tag{E 3.9}$$

Likewise, since the size of $N_v = N_h$, the subscripts can be discarded and:

$$f_f = \frac{1}{2} N^2 p_s \tag{E 3.10}$$

These are very important, fundamental formulas. They state that if it is assumed that a picture is made up of some arbitrary size of black and white square dots, the bandwidth of the resulting picture or the bandwidth of a second's worth of pictures is directly related to the frequency f_f. For example, if N is assumed to be 500 lines and f_p is assumed to be 30 pictures (frames) per second, the bandwidth f_f in cycles per second or MHz will equal:

$$f_f = \frac{1}{2}\left(500^2\right)(30) = 3.75\,\text{MHz} \qquad \text{(E 3.11)}$$

It is of interest to retrace some of the foregoing derivations. Note that both N_v and N_h can be divided into "k" square waves where k_h times the length of one square wave would give the line length and k_v times the length of one square wave will give the picture height or the number of lines. Thus, this length of the black-white square wave can be interpreted as a fundamental unit of time or, using its reciprocal, a fundamental unit of frequency. The frequency unit very closely resembles (but certainly does not equal) the Fourier term ω_0 in E 3.3.

The Checkerboard Pattern Model—An Analysis of Its Limitations

Several simplifications that were made using the checkerboard model must be recognized and dealt with if a more accurate model of the system spectrum and the scanning process is to be developed. First, it was assumed that the horizontal and vertical retrace times were zero. This erroneous assumption can be easily rectified by subtracting these times from both the active horizontal and vertical scan times. Likewise, in the simple model, the conventional aspect ratio "A," that is, the ratio of the horizontal picture size to the vertical was either assumed to be 1:1 or not specified. In most calculations (not including recent wide-screen pictures) an aspect ratio of $A = N_h/N_v = 4/3$ is used.

There are two other simplifications that arc even more important that must be dealt with. The first of these is the use of the narrow, one-square-high aperture. Although the electronic beam in both monochrome and color kinescopes is quite small and well focused, it still somewhat limits the detail in the picture. Part (b) of Figure 3.8 illustrates how a large or misaligned aperture can fuzz a picture. Note that if a perfectly aligned aperture, depicted as 1, scans the black area 3, it will produce a black spot, shown as 1'. However, if the aperture is misaligned as in 2, it will produce a spot that is half black and half white, as in 2'. If an aperture were as badly aligned as shown in 2, and it scanned the column under 3, it would produce a fuzzy or blurred effect, as shown in 4. Another way of describing column 4 is to say that the blurred picture is caused by a limiting of the bandwidth. This band-limiting property of a finite (larger) aperture was studied by several early researchers and its effect was determined to reduce the bandwidth to about 70 percent of its original calculated value. This 70 percent or 0.7 is usually known as the Kell factor and was named after the RCA research scientist R. D. Kell. Bandwidth equations such as E 3.11 usually include the 0.7 Kell factor.

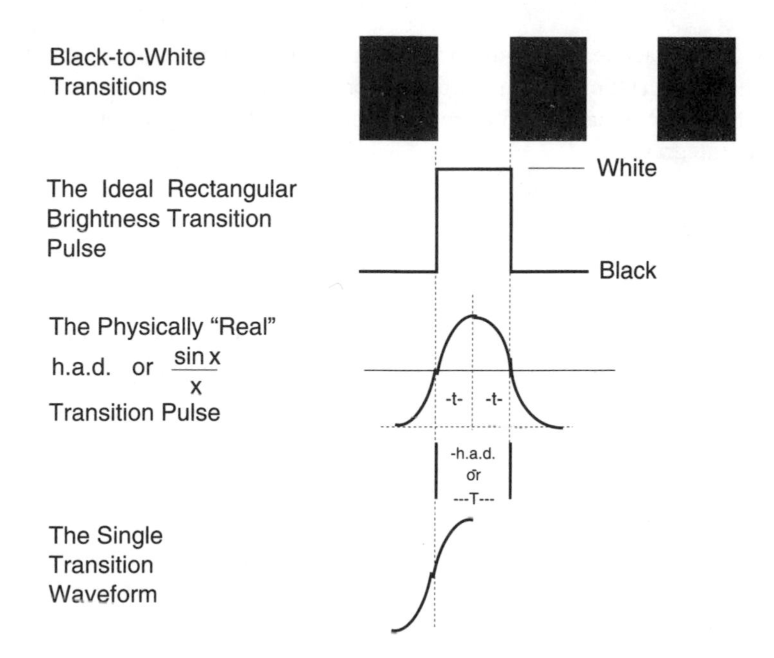

Figure 3.9 A graphic showing the ideal pulse black to white brightness transition and the less than ideal (but real-world) $\frac{\sin x}{x}$ brightness transitions.

Probably the most unsound assumption using the simple checkerboard model is the idea that scanning across the picture, the brightness could instantaneously change from black to white or from white to black. With very few exceptions, mechanical or electrical systems or circuits cannot change instantaneously. These desired sharp transitions are most often slowed down by the inertia (the physical reluctance to change) in mechanical or electrical components. In electrical circuits, the possible rate of change is dependent upon the system bandwidth. Although the rapid change or transient problems can be quite involved and require the application of higher mathematics for a complete solution, a first approximation relationship between sharp transitions can be derived from the concepts shown in Figure 3.9.

An Analysis of Abrupt Brightness Changes Using a Single Pulse and the Fourier Integral

Ideally, when a scene changes from black to white or back to black the camera, the transmitter, the receiver, and the kinescope should all duplicate these rapid changes and produce voltage or current waveforms that closely resemble the rectangular pulse in Figure 3.9. However, the bandwidth limitations and other more esoteric physical constraints limit the rise or fall time of these transitions. The analysis of single black to white pulse transitions using the Fourier integral (f), rather than the aforementioned simple Fourier series, is much more representative and accurate for TV signals. For reference, the more familiar trigonometric form of the Fourier integral is:

$$\boldsymbol{f}\,\{f(t)\} = \int_{-\infty}^{\infty} A(\omega) \cos \omega t \, dt + \int_{-\infty}^{\infty} B(\omega) \sin \omega t \, dt \qquad \text{(E 3.12)}$$

E 3.12 can be converted, using Euler's rule (see the Libbey reference) to its more convenient exponential form that yields:

$$\boldsymbol{f}\,\{f(t)\} = \sum_{n=-\infty}^{\infty} C_n e^{j n \omega_0 t} \qquad \text{(E 3.13)}$$

where the $C_n e^{jn\omega_0 t}$ part is the spectrum of the sin x/x (often termed sin x/x, sinc, $\sin^2$, or, in television, "T") pulse. Also, the T pulse is sometimes described as ± t as in Figure 3.9.

Since a form of the pulse's spectrum is known, with some more mathematical manipulations (again, see any reference on the Fourier integral), it can be shown that if the half-amplitude duration (also known as h.a.d. or "T") time is known, it is possible to obtain the system bandwidth by defining the cut off frequency (f_c) as:

$$f_c = \frac{1}{h.a.d.} \quad or \quad h.a.d. = \frac{1}{f_c} \qquad \text{(E 3.14)}$$

As an example, if the h.a.d is 0.1 microseconds (0.1×10^{-6} sec), the cut off frequency is:

$$f_c = \frac{1}{h.a.d.} = \frac{1}{0.1 \times 10^{-6}} = 10 \text{ MHz} \qquad \text{(E 3.15)}$$

Scanning and Brightness—Some General Conclusions

The previous sections have presented some of the introductory techniques with their generalizations and problems. One of the major principles that was presented is the limitation of sine wave thinking. The early experimental approaches to telephony and television were usually based upon the intuitive use of sine waves and bandwidth. However, many times the pictures had defects that could not be explained using these concepts. Later, especially as color television started to be developed, it became necessary to upgrade both the experimental and analytical approaches to include the ideas of phase and phase delay and rise time. These and other subtleties became the part of circuit theory know as transient analysis. THE landmark paper that congealed the previous thinking and produced a firm foundation was the written by Corrington and Murakami (see reference).

3.2 BLACK AND WHITE (MONOCHROME) TELEVISION

B&W TV—The Beginnings of Commercial Broadcasting

Serious attention was given to standards for all-electronic Black and White (B&W) as early as 1929. In the United States, the first all-industry standards were recommended by the National Television Committee in the early 1940s. In July 1941, the Federal Communication Committee issued its "Standards of Good Engineering Practice Concerning Television Broadcast Stations" and commercial TV was started. Of course, in December 1941, war was declared and commercial operations ceased. In 1945, the Federal Communications Commission (FCC) issued a new and expanded set of rules and commercial TV truly began.

B&W TV—Synchronizing Lines, Fields, Interlace, and Flicker

After several iterations before 1940 and during the war, the scan standards that were adopted in 1945 included 525 interlaced lines, 30 pictures or frames per second (f_p), 60 fields (two frames) per second (f_f), and an aspect ratio of $A = N_h/N_v = 4/3$. The B&W standards were based on many inputs, including the experiences of the motion picture industry. It will be remembered that old-time motion pictures were characterized by an annoying 16 frame-per-second (fps) flicker, which could be remedied by a unique system that showed each frame twice. Although the standard frame rate for modern motion pictures is 24 fps, numerous tests comparing the brightness of television pictures to those of 16- and 35-millimeter motion pictures resulted in the f_p of 30.

The 525 horizontal line frame is presented as two 262.5 interlaced fields. The principle of interlaced scanning is similar to the moving picture technique of presenting each frame twice. Each television frame is presented twice although its lines are slightly displaced from one to another. There is a first field and a second field. In the receiver picture tube, the first field starts in the upper left-hand corner with line number one. The next first-field line is number two, then three, etc., until the bottom line number 263. This bottom line only travels halfway across the bottom face of the tube. After the rapid vertical retrace, the second field again starts at the top of the tube. This is the second half of line number 263 and it starts in the middle of the tube. These second field lines continue down to the bottom of the tube until the last line, number 525, ends at the extreme lower right-hand corner.

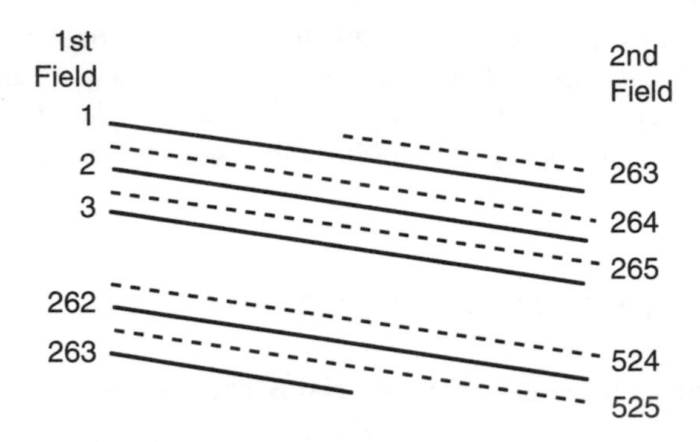

Figure 3.10 The principle of interlaced fields. Using this technique, there are two displays, i.e., two fields for every television frame. Although the field lines are displaced, the characteristics of the eye integrate them into one flicker-free picture. (No retrace lines are shown.)

The television 4/3 aspect ratio very closely duplicates those of the two motion picture standards. (It is interesting to note that several wide-screen motion picture standards were used and some wide-screen TV systems seem imminent.) The number of lines was first determined to be 500 but an interlaced system requires an odd number. Also, the number 525 can be derived by the factors $3 \times 5 \times 5 \times 7$, which is more convenient with simple digital counting systems.

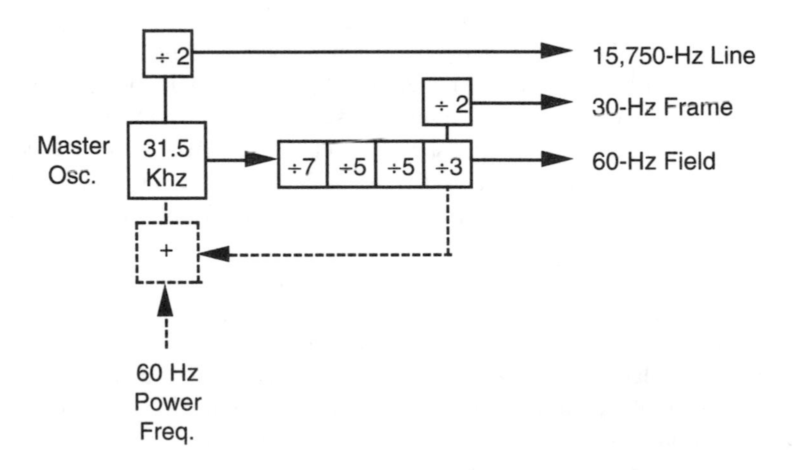

Figure 3.11 A rudimentary B&W synchronizing generator. Note the dotted circuit that would permit locking the master oscillator to the 60 Hz power line. This circuit, which was devised to reduce power-line interference, actually caused more problems than it solved and was soon abandoned.

The actual sync generator produces several more pulses than are shown in Figure 3.10. There are, of course, horizontal and vertical blanking pulses. Likewise, there were a number of equalizing pulses that were created to ensure that the vertical system would start the odd and even fields of the interlaced receiver picture correctly. As the receiver circuits became more sophisticated, some of these pulses were not needed but, since they are in the standards, they are still transmitted.

B&W TV—The Transmitter and Transmitting Standards

The bandwidth allotment for a TV channel in the United States is 6 MHz. For B&W, this 6 MHz must include the TV picture carrier, the video sidebands, and the channel sound. (Later, the color information was also added to this bandwidth.) If the video carrier was modulated with a 4.0-MHz video signal, a normal AM signal would require a bandwidth of about 8 MHz—not including any bandwidth for sound. One possible way to save bandwidth would be to use single-sideband AM for the video. Chapter 2 gives a discussion of three methods of creating SSB, including using filters. However, it would be a difficult task to design an SSB that would operate at the required high power levels in TV transmitters. Likewise, it would be almost impossible to make such a filter linear phase, a must requirement for color TV.

After many experiments, as well as a detailed mathematical analysis, a technique termed vestigial sideband (VSB) was adopted. With VSB, one of the sidebands (for TV it is the lower one) is partially removed. Thus, there is a small amount or vestige remaining. The VSB filter is easier to apply than an SSB filter and it can be made linear phase. There are several interesting signal-processing subtleties that have to be incorporated in the VSB system. Since some of the lower sideband remains in the television signal, it could affect the output of the video AM detector in the receiver. This could put an unwanted (step) signal in the detected response corresponding to the first or vestige part of the bandpass. There are two workable methods to remove this step. First, an additional roll-off filter, placed in the transmitter, could compensate for the presence of the partial lower video sideband. Second, an appropriate roll-off filter could be placed in the receiver. Both methods have definite advantages and disadvantages. The latter scheme was adopted (as shown in Figure 3.12). The receiver compensation is usually done by a unique tailoring of the response of the receiver's IF amplifiers. This entails a special IF broadband response with the carrier frequency offset from its normal mid-band position.

The sound in TV is FM with its carrier 4.5 MHz above the frequency of the picture carrier. As with the video, there were several proposals about how to handle the TV sound transmission. Although FM is usually more noise resistant than AM, many early system designers were concerned about the extra bandwidth that FM usually required compared to AM. For a compromise, the TV sound FM signal deviation is limited to ± 25 kHz rather than the ± 75 kHz used in normal FM broadcasting.

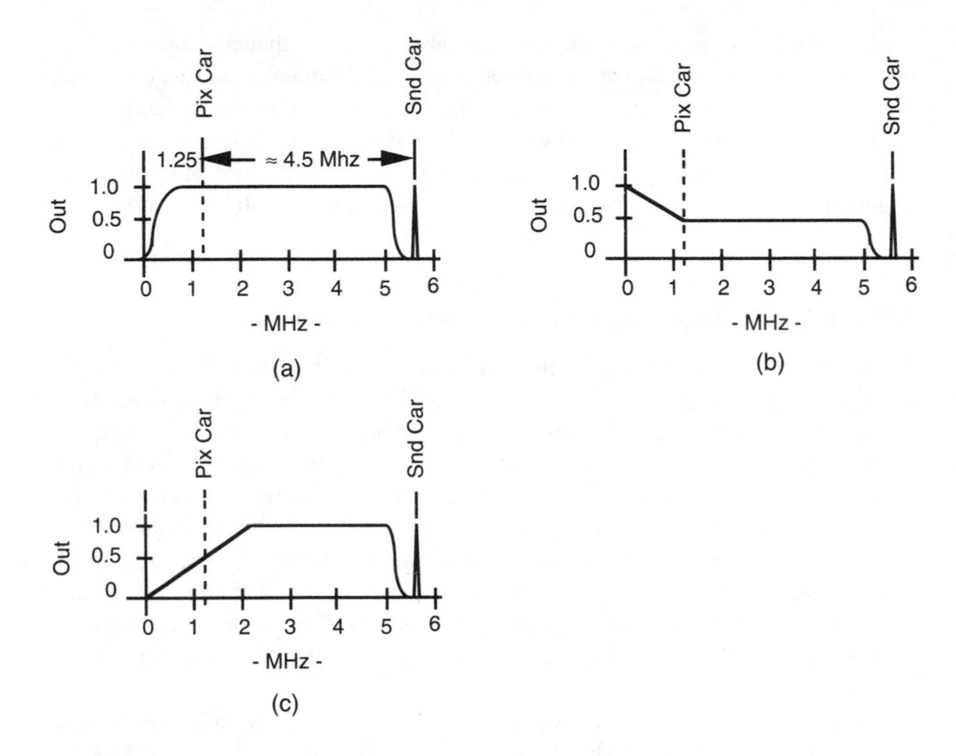

Figure 3.12 (a) The components and the frequency allotment for a 6-MHz vestigial sideband U.S. television channel. (b) The nonlinear output of the video detector, as a function of frequency, if the receiver IF is not properly designed and aligned to compensate for the effect of the vestige of the lower video sideband. (c) The receiver response when the IF is properly designed and adjusted to compensate for the effect of the lower vestige sideband. This produces a combined transmitter-receiver (transfer) response that is about flat and has linear phase.

There are several methods of configuring a TV transmitter. In general, the major difference is where the video modulation is inserted, i.e., does the transmitter use a low-level or high-level modulation. The choice depends on how the designers choose to deal with the problems of nonlinear amplification that lead to signal distortion. A prototype B&W transmitter is shown in Figure 3.13.

B&W TV—The Polarity of Modulation

The envelope of the amplitude modulated TV transmitter is somewhat different from the envelope (see Figure 2.1) of a broadcast (audio) transmitter. For instance, theTV signals, unlike the semi random audio signals, have continually repeating synchronizing pulses that give them a periodicity. Likewise, the presence of noise af-

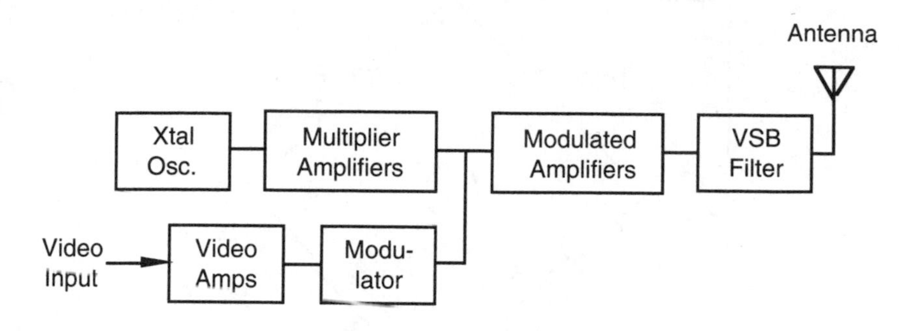

Figure 3.13 The block diagram of an amplitude-modulated vestigial sideband B&W TV transmitter.

fects the eye rather than the ear. In the United States, the modulating wave is upside down, i.e., negative, with the sync pulses above what is termed the black pedestal. Any noise will add to the existing modulation (envelope) and this increased modulation will go toward the black level, making the noise pulses appear black rather than white—see Figure 3.14. Likewise, with negative sync, there is an average saving in the usually costly power used by the transmitter. The average duty cycle of negative sync transmissions is about 20 percent lower than the average duty cycle of positive sync as used in some European systems.

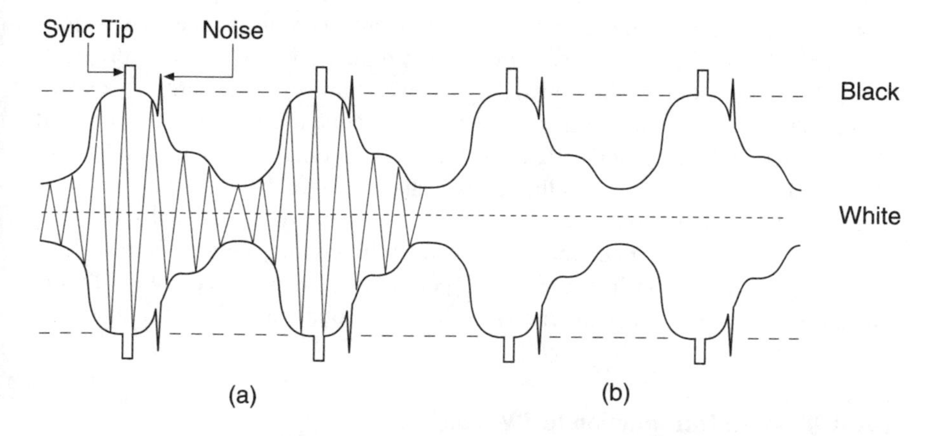

Figure 3.14 (a) Two lines of a symmetrically amplitude modulated TV waveform showing the sync tips and a noise pulse. (b) The envelope only of the waveform. Note that the noise pulse is moving in the direction of maximum modulation that is the black level—a black display. Very low or zero modulation would be displayed as a white raster on the receiver picture tube.

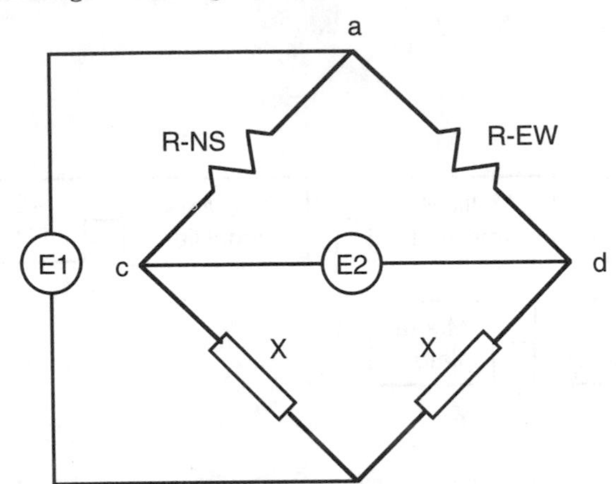

Figure 3.15 The Wheatstone-type bridge that is used for the diplexer. The voltage E1 causes currents in R-NS and R-EW (and the two reactors labeled X) but, because of the bridge balance at points a and b, there is no interacting current in E2. Likewise, E2 causes currents in the resistors and the reactors but, because of the balance at a and b, there is no interaction with E1. E1 can be the sound transmitter, E2 the TV transmitter, and the two Rs make up the equivalent resistance of the antenna.

Besides the vestigial sideband filter, there is another signal processing device that must go between the transmitters and the radiating antenna. The "diplexer" permits both the picture transmitter and the sound transmitter to feed the same antenna without interfering with one another. The diplexer is another form of the four-element Wheatstone bridge circuit, such as a diode rectifier bridge, an FM demodulator as described in Chapter 2, or a telephone phantom circuit as described in Chapter 4, etc. The elementary circuit for a diplexer is shown in Figure 3.15. The R-NS (north-south) and the R-EW (east-west) make up the carefully crafted elements that are specially built for the allotted frequencies for a TV channel. It should be remembered that the elements are not components as we normally visualize them. The conductors and the reactors may be large, hand-formed elements that look like plumbing or maybe abstract art. However, these unusual shapes become well-behaved circuit elements at several hundred megahertz. There is also a similar device termed a triplexer that will permit an additional, separate FM radio transmitter to use the single antenna as well as the sound and picture transmitters of the TV station.

B&W TV—An Introduction to TV Receivers

The B&W TV receivers follow the basic configuration of the superheterodyne (see Chapter 2) with the proviso that there are two signals in the input—the sound and the picture—that must be processed, i.e., amplified, demodulated, and filtered. Some early designs had completely separate IF amplifiers and demodulators for both the sound and the picture while most later designs combine the video and sound in one

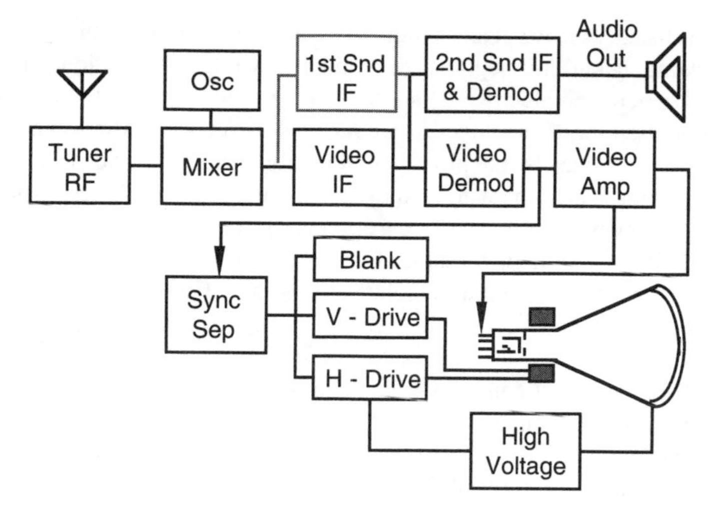

Figure 3.16 The block diagram of an elementary B&W TV receiver. In earlier designs, the signal for a first IF amplifier was taken directly from the output of the mixer (see the dotted lines).

IF circuit. The video detector must include circuits that also extract the incoming sync signal and process it for the deflection circuits and for the picture tube high-voltage (flyback) driver—see Figure 3.16.

It is noteworthy that the original high-voltage supplies were conventional 60 Hz power line designs. Even with voltages in the order of only 5,000 volts, these supplies put an inordinate strain on the rectifiers, the insulation of the transformers, and other components. Also, these power line supplies were extremely dangerous and soon became known as man killers. The horizontal-frequency driven high-voltage supplies were inherently current limited and thus were much safer. Likewise, they adapted well to higher voltages required by the larger B&W and color picture tubes.

3.3 COLOR TELEVISION

Color TV—An Introduction to Its Increased Demands

In its simplest conceptual forms, color TV seems to be at least twice, and possibly three to four times, as difficult to transmit and receive as B&W. A first, rudimentary model would dictate that color transmission would require a second band in addition to that for the original mono transmission. Further speculation might also conclude that, in addition to mono, color transmission would require three additional bands for the three red, green, and blue primary colors Although there are fallacies in both the two- and the four-signal arguments, the general ideas will serve well as starting points.

Color is usually analyzed and synthesized by the use of three building-block or primary components. Although there are many possible combinations, the primaries of red, green, and blue (RGB) are most often used. (The photographic industry has long used these three dyes to produce color pictures.) Almost any color or hue can be synthesized by the proper combination of RGB. Besides its characteristic hue, a color is also characterized by its saturation and brightness. The saturation denotes how penetrating or vivid a color is. Highly saturated colors have little or no white light—a unique mixture of the RGB primaries. Dull colors contain a large amount of white. Brightness relates to the intensity of the color. On a TV screen, brighter color, like a brighter BW, requires additional driving voltage or current.

One of the two most common early proposals for color included one camera tube for pickup and a single kinescope for display. The color signal was created and displayed with a rotating Nipkow-like wheel with RGB color filters. This approach, including its later embellishments, became known as the CBS field-sequential system. (Sometimes we forget that this motor-driven system was officially adopted by the FCC in October 1950 and was the rule of the land for about three years.) The CBS system was very interesting and, in demonstrations, provided excellent color picture quality. Likewise, around 1950, it appeared that future circuit and display tube technology might permit an all-electronic (no rotating wheel) receiver based on the CBS standards.

The original field-sequential system required a total bandwidth of about 10 MHz. Later, the FCC mandated that any color transmission must be within the B&W standard of 6 MHz. To conform to this standard, the horizontal resolution was reduced from 525 to 405 lines. This, coupled with a field rate of 72 fields per second (and several other dissimilarities), made the CBS system completely incompatible with the standards for B&W transmission. In general, it was this incompatibility rather than any obvious lack of picture quality that doomed field-sequential TV. Many members of the technical and manufacturing community, and later the FCC, felt strongly that it would be unwise and unfair to have a second TV system that would literally make obsolete millions of B&W receivers.

Color TV—The NTSC Color Transmission System

The color TV system that is in common use today in theUnited States was created by a very knowledgeable and dedicated group from the industry. The work of this National Television Standards Committee (which involved well over 100 people) is an outstanding model for what intelligent interindustry cooperation can accomplish.

The work of the committee was based on information gathered or derived from many sources. There were numerous subcommittees that thoroughly researched many pertinent subjects such as modulation, filter characteristics, colorimetry, etc. Members of the committee also witnessed demonstrations of proposed all-electronic

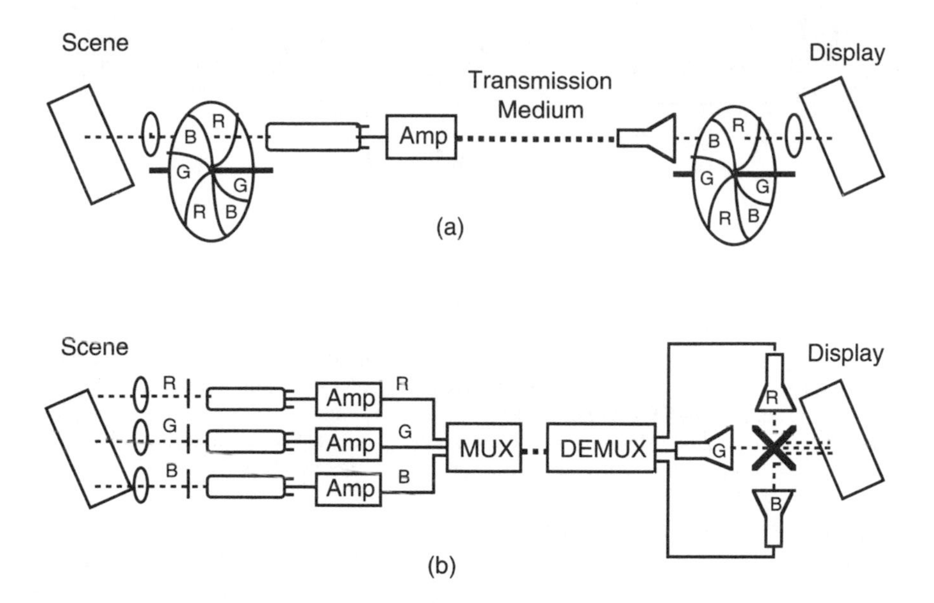

Figure 3.17 (a) The elementary pickup and display components of a color TV system that uses motor-driven scanning disks with color filters to separate a composite color scene into its RGB primaries. This was the basic scheme of the CBS field-sequential system. (b) The elementary three-tube pickup and display components in an early all-electronic TV system similar to one of the original proposals of RCA. Each pickup tube had a color filter used to obtain (filter out) one of the primary colors. The display tubes had built-in color filters so that they emitted only one color of light. The large X represents a special "dichroic" mirror that combines the three primaries into a composite color picture that was projected onto a screen. In a later variation, a similar dichroic mirror, in a reversed orientation, was substituted for the color filters to derive the RGB primaries for the pickup tubes.

systems by the Radio Corporation of America, Hazeltine Corporation, Philco Corporation, Color Television Incorporated, Alan B. DuMont Laboratories, and General Electric.

In the introduction to this section, two very elementary models were derived that indicated that the addition of color might double or quadruple the required TV bandwidth. Likewise, it was indicated that in the CBS field-sequential system, the number of horizontal lines was decreased from 525 to 405 in order to meet the 6-MHz bandwidth limit of the older B&W standard. The question that now arises is how did the NTSC create a workable and aceptable color system that could be accommodated in a 6-MHz bandwidth when others had apparently failed? The NTSC color system relies on two major principles: [1] an acceptable color or chroma signal can be generated in a bandwidth that is less then than the B&W or mono signal, and, [2] there is additional band space between the cracks in the B&W spec-

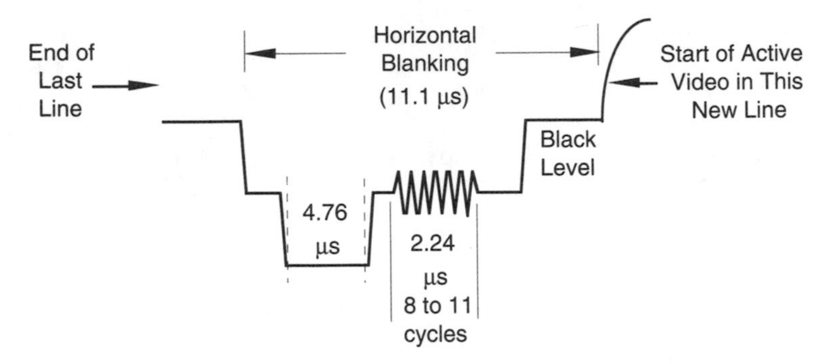

Figure 3.18 The color horizontal-blanking interval including the 3.58 MHz reference burst. The total time for each horizontal line is approximately 63 μs. Of this, approximately 11 μs are blanked out and used for horizontal and subcarrier synchronization. The blanking pulse (not shown) is a separate signal from the horizontal picture synchronizing pulse. It is used to shut of the beam in the picture tube during the horizontal resynchronization.

trum. Figure 3.7 shows a discrete line spectrum based on some generalized frequency ω_0. For a standard mono signal this ω_0 is the horizontal line frequency of 15,750 kHz. Any mono signal will be composed of some combination of the series 15,750, 31,500, 47,250 kHz, etc. There are many possible additional frequency series that would produce another line spectrum and could be interleaved or fitted between the cracks. The additional interleaved signal that carries the color information is know as the color subcarrier.

The color subcarrier, together with the band-limited color signal (later it will be shown that there is more than one color signal), forms the cornerstone of the NTSC system. This color subcarrier is uniquely related to the horizontal—it is an odd multiple of one-half the horizontal frequency. The subcarrier is related in frequency, but not in phase. To assure synchronization between the subcarrier at the transmitter and that of the receiver, a pilot tone or reference "burst" is transmitted—see Figure 3.18.

Color TV—Generating Coded Color Signals in the Studio

Once it is postulated that the additional band space for a color (sub)carrier signal can be provided by interleaving, the questions of how to code and then modulate the color carrier must be answered. The answers to these questions are best found by again referring to the idea of color primaries and outlining the use of vectors for the mathematics of modulation.

It was mentioned earlier that there can be several sets of color primaries. For instance, TV, which uses reflected and projected light, has adopted the additive RGB primaries. If it is desired to create different colors with paint or in printing, the subtractive primaries are magenta, cyan, and yellow. These and other possible sets of primaries are physical primaries—one can create and see them. There are also different sets of mathematical primaries. These mathematical primaries are usually derived by mathematical or vector transformations from the physical primaries. It must be emphasized that the mathematical primaries are certainly more than just academic abstractions. Color TV relies on the fact that these mathematical primaries can first be derived from RGB. These new primaries can then be added, subtracted, modulated, etc., for serial transmission, and finally converted back to the original RGB primaries to compose the displayed picture.

The Y or Luminance Signal

A prototype three-tube pickup system, as shown in Figure 3.17 (B), will capture the complete original picture, including both the monochrome (luminance) and the color or chrominance information. After the pickup and amplification, the NTSC processing will derive a single luminance or "Y" signal from the three RGB signals. This creates a true black and white signal that, for one thing, is completely compatible with the older B&W sets. Exhaustive experimentation and viewing tests established that a signal composed of 30 percent red, 59 percent green, and 11 percent blue will produce white. Thus, to produce true B&W only, the outputs of the camera amplifiers must be appropriately scaled (divided). (Signal networks that add, subtract, divide, and sometimes invert, are usually termed "matrix" or matrixing networks.)

The R-Y, B-Y, and G-Y Primaries

In addition to the combination of signals that create Y, three other signals can be created, termed "color difference." Again referring to Figure 3.17 (B), it is possible to obtain color-only signals by subtracting out Y. Thus, three R-Y, G-Y, and B-Y signals can be created. Since the wide bandwidth Y signal is eliminated, the signals have significantly reduced bandwidth. The derivation of R-Y will illustrate the processing required to create a color-difference primary. (Remember that a polarity inversion may be part of any matrixing process.) To derive R-Y:

		R	G	B	
Red	=	1.00	0.00	0.00	
Minus Y	=	– (0.30	0.59	0.11)	
R – Y	=	0.70	– 0.59	– 0.11	so, using this method:

$$R - Y = 0.70R - 0.59G - 0.11B$$
$$B - Y = -0.30R - 0.59G + 0.89B$$
$$G - Y = -0.30R + 0.41G - 0.11B$$

Since there are three equations and three unknowns, it is possible to derive any one from the other two. Thus, the G-Y primary is rarely directly computed.

It is most common to place the two color-difference primaries 90° from each other, both mathematically and physically. After doing so, a resultant vector can be constructed that represents the combined amplitude of both the R-Y and the B-Y components. Figure 3.19 (a) shows the R-Y and B-Y components drawn in quadrature with a resulting vector "a." If either one of these primaries is reduced in amplitude, both the hue and the color saturation will be changed. Such a resultant is shown in "b." If the R-Y and B-Y signals are used to modulate a color carrier in quadrature, the hue information will be contained in the modulated signal phase and the saturation will be contained in the modulated signal amplitude.

The I and Q Primaries

It is more common to use a still different set of mathematically derived primaries for the transmission and detection of the color information. These are termed I (in phase) and Q (quadrature). The I and Q primaries are so oriented that there is no modulation for a B&W picture—thus, there is no color interference. The I and Q primaries can be derived from the original RGB primaries or matrixed from R-Y and B-Y.

The I and Q primaries also contain RGB as:

$$I = -0.60R - 0.28G - 0.32B = 0.74(R-Y) - 0.27(B-Y)$$
$$Q = 0.21R - 0.52G + 0.31B = 0.48(R-Y) + 0.41(B-Y)$$

Both the I and Q signals are band limited (bandpass filtered) before they are used to modulate the subcarrier. The bandwidth of the filtered Q signal is (including both sidebands) ± 1.3 MHz, and the bandwidth of the I signal is ± 0.5 MHz.

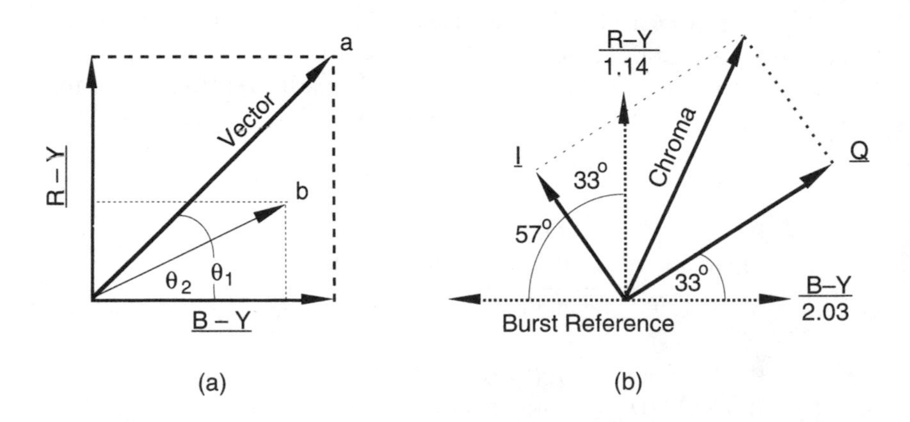

Figure 3.19 (a) The R–Y and B–Y primaries in quadrature. These two color-difference signals can be added vectorially to produce vectors representing different hues and saturations. (b) The I–Q primaries. Another more common set of color primaries that can be used to modulate the color subcarrier for the transmission of hue and saturation information. Note that the I and Q primaries, although shifted from the usual x-y coordinates, are also in quadrature.

When the subcarrier is transmitted, the vestigial sideband filter reduces the power of the upper sideband of both the I and Q signals about one-half.

Color TV—Color Modulation

In theory, either the color-difference or the I and Q primaries can be used to modulate the color TV transmitter. However, the I and Q signals allow a wider chrominance bandwidth with more color information. The modulation of the subcarrier is usually accomplished with a double-balanced modulator (see Chapters 2 and 4). This circuit will produce the sidebands but suppress the 3.58-MHz carrier. Figure 3.19 (a) shows that if the magnitude of one of the R-Y or B-Y coordinates is changed, the value of the angle between the resultant vector and the reference (x-axis) is also changed, i.e., θ_1 changes to θ_2. The same thing happens when the I and Q primaries (also in quadrature) are used to modulate the subcarrier. In the modulation process, this translates into phase modulation. This phase modulation is in addition to the amplitude modulation that results when the magnitude of the components, and thus the vector, is changed. The hue information is carried in the phase modulation and the saturation information is carried in the amplitude modulation. Figure 3.20 shows the use of balanced modulators to produce the I and Q chroma signals.

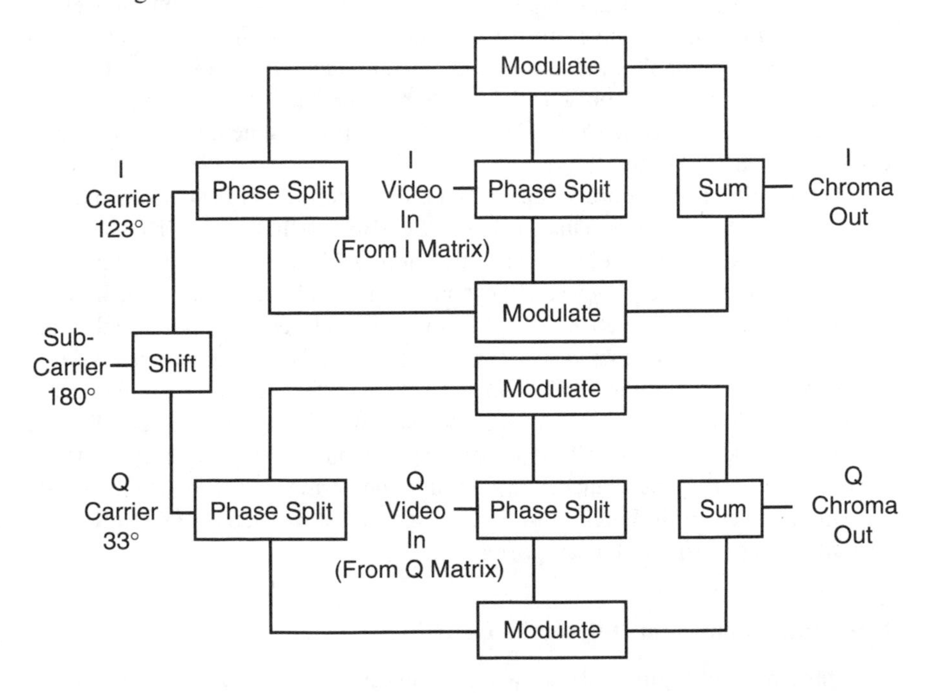

Figure 3.20 A block diagram of the I-Q modulation process using double-balanced modulators. The incoming subcarrier is shifted to provide a properly phased carrier for both the I and Q (derived) primaries. The outputs contain hue and saturation information.

Color TV—The Transmitter and the Studio and Camera Links

A color transmitter resembles, but certainly does not duplicate, the elements pictured in Figure 3.13. Color would, at least, require the addition of a subcarrier generator and a slight rearrangement of the function blocks. However, any new block diagram would not give a useful insight into the subtle changes required for color. As has been indicated before, color TV requires processing circuits that provide excellent transient response and its corollary, linear phase response. Whether it is the vestigial sideband filter, the amplitude modulators, or the linear amplifiers, color transmitters require much more attention to the details of circuit design than those for B&W.

Very quickly after the introduction of color, the technical community, including the broadcasters, the equipment manufacturers, and test instrument developers, produced a series of standard test signals that could be used to test each element in the studio link, as well as the transmitter. Although there are many such signals described in the literature, a few reflecting the elements of the NTSC system will be described.

The T-Pulse Test Signals

The sin x/x or T pulse, as presented in Figure 3.9, is the fundamental building block for a series of tests that have proven invaluable for the design, installation, and maintenance of color TV studio, transmission, and receiving equipment. Using E 3.15, the standard (one) T pulse is defined to have an h.a.d. of 125 ns that equates to a cut-off frequency of about 8 MHz. The T pulse is convenient for checking studio signal-distribution amplifiers.

The literature is not entirely consistent on the definition of bandwidth and cut-off frequency for a T pulse. This work uses the IEEE definitions. A T pulse is defined to have an h.a.d. of 125 ns. The spectrum amplitude response at 4 MHz is about 0.5 of its zero-frequency (DC) response and at 8 MHz, the response has continued to roll off down to about zero. See the IEEE reference. For a 2T pulse, the spectrum amplitude is down to 0.5 at 2 MHz and down to about zero at 4 MHz.

There is also a common 12.5T, with an h.a.d. of 1.56 μs, that is usually modulated with both luminance and chrominance signals. This modulated 12.5T pulse is used to check not the overall frequency response but the amplitude and phase of the chroma. There is also a modified step function, termed a T step (see the single transition waveform in Figure 3.9) that is used to test the transient conditions of rise time and overshoot of an amplifier.

Differential Phase and Differential Gain

The three parts of Figure 3.21 show a progression of more complex waveforms that are used to test TV transmission and receiving equipment. The (a) ramp can

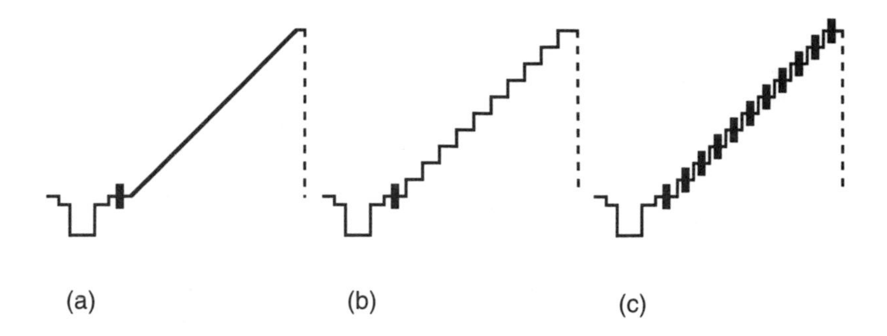

Figure 3.21 (a) A one-line ramp used to test the linearity of a TV amplifier. This is really a monochrome signal—the blank rectangle only indicates the (color waveform) position of burst. (b) A stepped waveform to test for some amplifier monochrome nonlinearalities in a color signal. This test would check the low-frequency differential step response of an amplifier and possibly detect serious overshoots (glitches) or ringing. (c) The waveform, somewhat derived from the previous two, that can be used to test for an amplifier's differential gain and differential phase distortions of subcarrier.

be used to test the general linearity of device gain or Δout/Δin. This is a low-frequency test signal and it indicates very little about transient response. The (b) stair step not only gives information about the gain or Δout/Δin as a function of input-signal magnitude, but it also indicates some minimal information about rise time and overshoot. The (c) signal combines the attributes of the other tests and, by adding bursts of subcarrier, also performs special tests at higher frequencies around the 3.58 MHz frequency. The use of the (c) test signal is unique to color television. It is used to perform two tests relating to the subcarrier called differential phase and differential gain.

The determination of differential phase (Δθ) and differential gain (ΔG) are two very imaginative and often-used tests developed for broadcast color TV equipment. These tests are based on the principle that an amplifier's gain and phase response is not always consistent as the signal level is increased. Thus, when a precise stair step is used to test the linearity of an amplifier, the relative gain (or phase) may not be the same in the incremental step from (for example) 80 percent to 90 percent as it is in the step from 20 percent to 30 percent. By using an input stair step, augmented with a burst of subcarrier on each step, both the incremental (differential) gain and differential phase of the burst can be compared and precisely measured. Typical numbers for one studio amplifier (there may be ten to fifteen or more used to originate a program) are in the range of 0.1 percent ΔG and 0.1° Δθ. The overall transmitted color signal will average about 3 percent ΔG and 3° Δθ.

3.4 THE COLOR RECEIVER

The Color Receiver—The Basic Upgrades from B&W

The color TV receiver departed materially from the configuration of the B&W receiver, as presented in Figure 3.16. The obvious changes included the additional color demodulators, dematrixing circuits, the three video amplifiers, and the additional circuitry that is required to process the burst and subcarrier signals. The sync circuits are much more complex and robust because they must accommodate the larger three-gun picture tube. Likewise, all the signals are derived from the subcarrier frequency rather than some harmonic of 60 Hz, as indicated in Figure 3.11.

The Color Receiver—An Introduction to Design Choices

Receiver designers and manufacturers have a *very* wide latitude in the strategies they may use to create their final color set. The basic requirements are that the set is compatible with any B&W transmission and that it provides a good or pleasing picture, a requirement dictated by the marketplace rather than the FCC. For instance, not all receivers use the precise decoding matrix that the NTSC equations tacitly suggest—brand X may produce an average color balance that, by the designer's choice, tends to be slightly more red or blue than brand Y. This is just one example of the many ways receivers are designed to produce pictures that are assessed as more pleasing by the viewer. It must also be remembered that, until the proliferation of cable, many sets were in fringe or partial snow areas. Different manufacturers took various methods to improve their sets' performance in weak-signal areas. Many, if not all, of the performance compromises that set manufacturers incorporated involved the added cost of the parts and labor required compared to the value that most customers would get in improved picture quality. It is interesting that integrated circuits have, in many cases, made these compromises unnecessary and have thus significantly improved the quality of most color receivers. Likewise, digital circuitry in the receivers and in the remote controllers has made the contemporary sets more convenient for us all.

The Color Receiver—The Intermediate Frequency (IF) Amplifiers

Referring back to Figure 3.16, the B&W receiver somewhat resembled a superheterodyne radio receiver with the addition of the video IF and the video sync and deflection circuits. A color receiver is likewise a superheterodyne. However, in addition to the B&W circuitry, there is the color demodulation and decoding circuitry, three video amplifiers, and a method of deriving the color synchronization from the burst. The shape (frequency response) of the color IF is extremely important if [1] the best color fidelity is produced and [2] there is not sound in the picture interference caused by the sound carrier. Throughout the years, different manufacturers have varied their design and alignment procedures in an effort to provide the best IF response consistent with excellent sound-carrier rejection. Many have peaked their IF response in the region of the chroma to provide the proper (demodulated) chroma bandwidth. Not only were there the usual double-tuned IF transformer bandpass filters, but most sets also included a tuned sound trap to eliminate any remnant of the sound carrier before the video detection.

In recent years, the surface acoustic wave (SAW) filter has been used to replace many of the parts in the more conventional IF amplifier chain. The SAW filter is somewhat akin to the 3.58-MHz piezoelectric crystal except it is constructed in such a way that the input electric current is converted to mechanical (acoustical) vibrations. The way the surface of the piezoelectric material is fabricated determines the gain and phase response (and the bandpass) of the filter. It is a very small, relatively inexpensive component with no required adjustments. The SAW is configured as a very sharp cut-off, bandpass filter for the 45.75-MHz IF acoustic waves. At the output end of the SAW, the vibrations are reconverted to an electric current. Since the SAW filter is manufactured to extremely tight tolerances, it is pretuned to the TV IF and no further alignment is necessary or usually provided.

Color Receiver Processing—Wide-and Narrow-Band Color

In contemporary color receivers, the demodulation and dematrixing of the color subcarrier is most often accomplished deep down in the recesses of an integrated circuit. Before the proliferation of integrated circuits for TV receiver functions, the design of most circuit functions, including chroma demodulation, was significantly influenced by the omnipresent cost of production. The following paragraphs present two of many model circuits that were used for color demodulation.

Wide-Band Color Demodulation

In the transmitted chroma signals, not counting the effect of the vestigial sideband filter, the I signal has a bandwidth of about ± 1.3 MHz and the Q signal has a bandwidth of ± 0.5 MHz. As indicated before, if the IF is carefully designed and properly aligned, the use of I and Q signals will give the widest range of color hue and brightness provided by NTSC. Using conventional analog techniques as shown in Figure 3.22, the best color display can be obtained by [1] bandpass filtering the composite chroma signal around 3.58 MHz to eliminate most or all of the luminance information and [2] demodulating and properly filtering the I and Q signals.

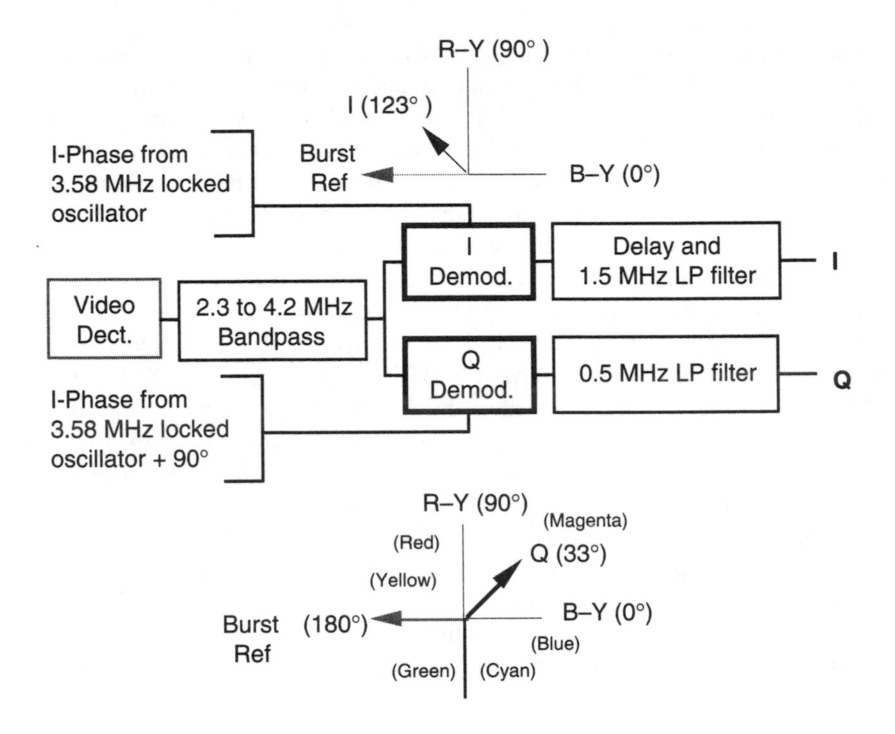

Figure 3.22 A demodulation scheme using conventional analog components to obtain wide-band chroma signals. This technique is similar to the balanced-modular technique shown in Figure 3.20. Note that the input reference signals are derived from the 3.58-MHz oscillator locked to the transmitted burst. They are 90° apart to assure that the I and Q signals are in exact quadrature. Also, the I low-pass filter contains the proper delay to maintain the exact 90° I-Q relationship. In most receivers, the color-difference signals will next be derived to the drive elements (usually the cathodes) of the color picture tube. For additional perspective, the bottom diagram shows the relative positions of some representative colors. The B-Y vector is at 0°, magenta at 61°, R-Y at 90°, red at 104°, yellow at 167°, reference burst at 180°, green at 241°, cyan at 284°, and blue at 347°.

Narrow-Band Color Demodulation

In the earlier section on color matrix arithmetic, it was pointed out that the I and Q signals contained the R-Y and B-Y color-difference signals and vice versa. Using this information, circuit designers reasoned that it should be possible to retrieve these color-difference signals directly from the output of the video detector without first demodulating I and Q. In principle this certainly is feasible. However, because of signal interferences, the bandwidth of the color-difference signals must be limited to about 0.5 MHz with some resulting lack of color definition. In some earlier

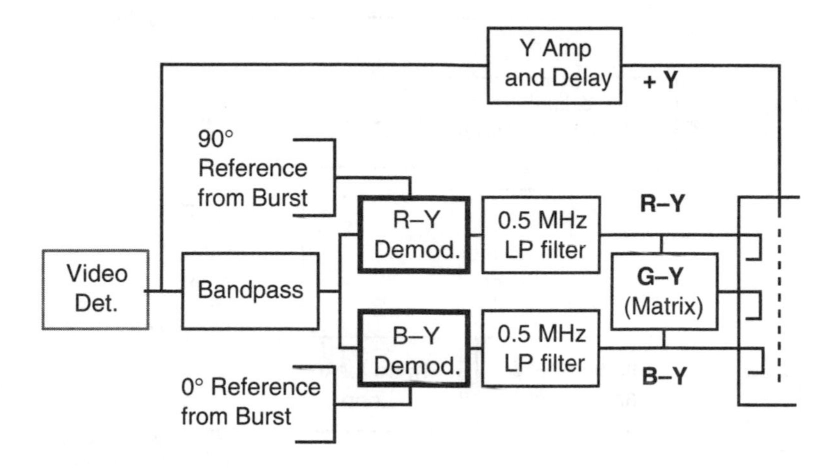

Figure 3.23 Narrow-band chroma demodulation used to derive the color-difference color primaries. The three different primaries plus the Y (luminance) can be used to directly feed the picture tube. Note that, like wide-band demodulation, the reference phases from a burst-locked oscillator are necessary.

receivers, especially in the bottom of the line price leaders, the small loss of color fidelity was considered acceptable. Figure 3.23 indicates one method of directly demodulating R-Y and B-Y.

3.5 IMPROVEMENTS IN COLOR— THE MATURING INDUSTRY

Improved Color—Early Hardware Limitations

The previous explanation of wide- and narrow-band color demodulation is indicative of a number of compromises that at first seemed inherent in NTSC. However, many of the problems—now viewed from the perspective of hindsight—were not inherently systemic, but resulted from the very limited early hardware available to implement the system. The two technical improvements that permitted NTSC and other television systems to be used to their potential were the development of integrated amplifying and processing circuits along with the developments of high-density semiconductor memories. A valid argument can be made that the later advances in studio videotaping, as well as the much improved color cameras, finally demonstrated the latent potential of the NTSC system. Notwithstanding, it was the receiver's improved processing circuitry and picture tubes that ultimately gave the consumer the look of improved performance.

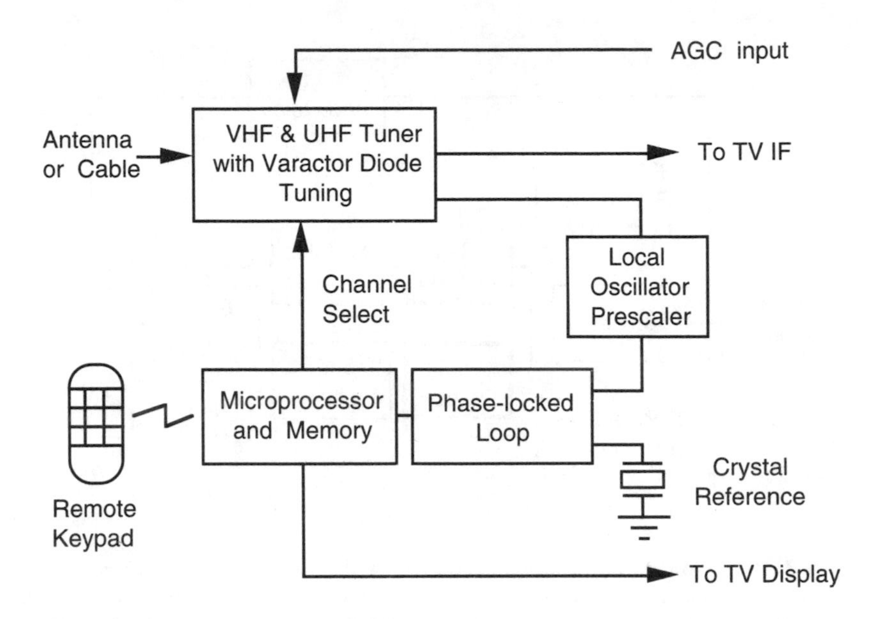

Figure 3.24 A simplified schematic of a microprocessor remote-controlled tuning system using varactor diodes and a phase-locked loop.

Improved Color—Receiver Signal Processing and Control in the Tuner and RF Sections

The early years of color were difficult for tuner designers, especially UHF tuner designers. Vacuum tubes—and then early RF semiconductors—did not excel at 400 to 800 MHz. Not only were the amplifiers noisy, the local oscillators were microphonic and drifted excessively. Likewise, the channel switching was originally accomplished with wafer switches with contacts that often became noisy and intermittent. It was actually fifteen to twenty years after the advent of color and the growth of UHF that receiver tuners were primarily designed with technology rather than art.

The perfection of reliable varactor diodes (a high-frequency p-n diode whose nonlinear characteristic can be used in VHF and UHF oscillators or as a voltage-controlled capacitor for tuning) was a breakthrough for tuner design. It simplified the switching and permitted automatic fine-tuning (AFT). AFT not only eliminated a large, difficult to adjust properly, coaxial tuning dial, but channel selection became much more precise and convenient. Later, the simplicity of the voltage-controlled varactor diode was combined with the technologies of the phase-locked loop frequency synthesizer, the microprocessor controller, and the digital remote controller to give the user the ability to rapidly select most VHF and UHF channels whether they were received using antennas or cable—see Figure 3.24.

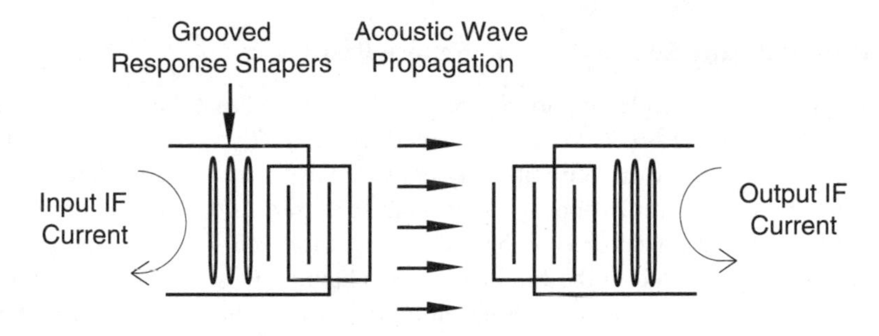

Figure 3.25 A conceptual drawing of an SAW bandpass filter as used in color receiver IF amplifiers. The precise response is a function of the input and output electrode configuration as well as the grooves that reflect the acoustic waves and thus shape the filter response.

The IF Amplifier

The SAW filter, as shown in Figure 3.25, provides the correct IF bandwidth along with assuring sharp but phase compensated cut-off characteristics.

It should be remembered that a prototype SAW filter is created with interdigital electrodes or fingers etched on a slab of piezoelectric material. By carefully controlling the shape, length, and spacing of these fingers, as well as arrays of input and output grooved reflectors, the bandpass amplitude and phase response can be accurately controlled. As with other processing improvements discussed in this section, after the SAW filter was perfected, new processing ICs were developed to optimize its usage.

Chroma Processing—Dynamic Fleshtone Correction

Even with all the circuit and technique improvements in the studio-transmitter chain, there are still times when the relationship between the reference burst and the color subcarrier is not transmitted correctly. This, of course, either produces improper colors or, most often, the exact color balance or tint is not displayed correctly. When it is remembered that some program material may go through literally dozens of distribution and processing amplifiers, several generations of tape dubbing, a satellite link, and one or two frame synchronizers, one can understand why it is sometimes difficult to maintain the accuracy of the amplitude and phase of the original reference burst. When the integrity of burst is compromised, it is most noticeable in the fleshtones (especially the face) of the performers.

Dynamic (auto flesh) fleshtone correction is usually included in contemporary receivers. Its circuitry is usually based on the principle that the fleshtone color vectors fall on or very near the (123°) I-vector axis (see Figure 3.22). The auto-flesh circuits usually compare the phase of (I) chroma with the reference subcarrier. If there is a phase difference, the reference phase is momentarily moved (phase modulated) to make the I-vector move to its correct 123° position. (See the Harwood reference.)

Improved Picture Sharpness—Luminance Peaking

In an earlier section, referring to Figure 3.9, it was noted that the maximum transient rise time of the NTSC system is limited to about 125 ns. This is the theoretical maximum—real-world circuits may restrict it even more. Some dark-to-light or light-to-dark transitions appear fuzzy because of transmitted picture or receiver limitations. These transitions can be artificially improved if upper-band peaking is dynamically inserted in the luminance (Y) channel. It is desirable to improve the rise time but not to get a resulting blooming of the white areas on bright pictures. Early circuits used viewer-adjustable RLC peakers. A later solution used a tapped delay line (a so-called transversal) filter combined with limiting diodes to limit kinescope blooming.

Improved Chroma and Luminance Separation—The Comb Filter

With the proliferation of both analog and digital ICs, more and more of the earlier performance limitations were overcome. One of the major practical limitations to the NTSC system in receivers has been the complete separation of the interleaved luminance and chrominance signals. In theory, both of these have a discrete line spectrum (see Figure 3.26) and they are, again in theory, separated from one another by one-half the line frequency of 15,750 Hz or 7,875 Hz. Unfortunately, these two spectra tend to cross the bounds into one another and cause false colors or distortions, called dot crawl around color edges. Up to a few years ago, most receivers limited the luminance bandwidth to about 3 MHz rather than the 4.2 MHz allowed by the NTSC system. This limits the horizontal resolution to about 250 lines—about three-quarters of the resolution of the broadcast signal.

For many years, broadcast studio equipment had used a technique termed comb filtering to separate the luminance and chrominance signals. A comb filter relies on the fact that the luminance signal is completely contained in a line spectrum derived from the line frequency (15,750 Hz) and, likewise, the chroma also has a line-frequency spectrum, centered around 3.58 MHz, but is offset 7,875Hz and is

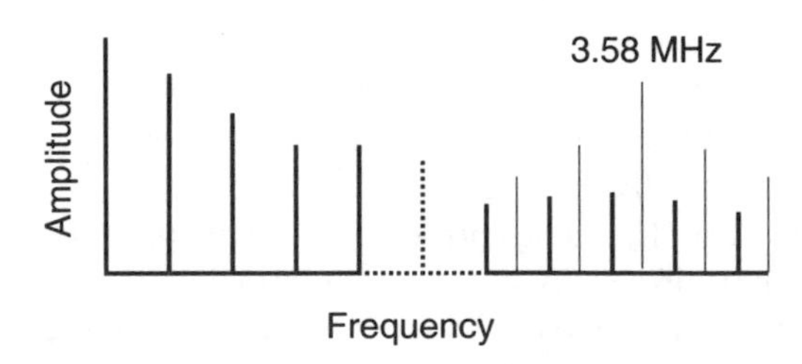

Figure 3.26 The line spectra of the NTSC luminance (darker lines) and chrominance (lighter lines). Each one has a 15,750 Hz base-line spectrum but the chroma is offset by a one-half line frequency—7,875 Hz.

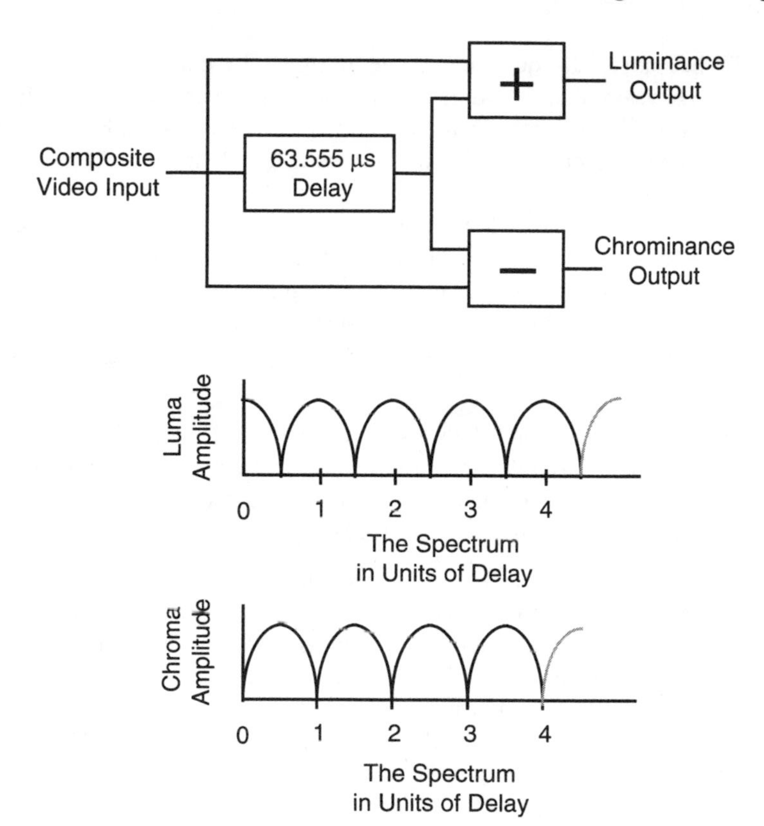

Figure 3.27 The scheme and the output spectra of a one-line-delay video comb filter. Although this is usually considered an analog technique, it works well as an introductory model for digital filters.

in the cracks of the luminance line spectrum. Broadcast line-rate comb filters used a precise and costly delay line of exactly 63.555 μs in circuits as shown in Figure 3.27. Although receiver designers were well aware of the combing technique, only recently have similar delay lines (usually semiconductors) been economical for receiver production. The comb filters produce near-perfect chrominance and luminance separation and thus permit a wider luminance bandwidth with the resulting restoration of horizontal resolution.

3.6 DIGITAL TELEVISION AND AN INTRODUCTION TO HIGH-DEFINITION TELEVISION

What Is Digital TV?

It is difficult to precisely define the phrase digital television. If the definition includes broadcast and receiving equipment with some digital circuits, then partial digital TV has existed for quite some time. If digital TV connotes a transmission and receiving system that uses digital techniques and components in the RF and IF sections, as well as for video processing, then it is in the future. Based on these two conflicting statements, working definitions will include:

1. *Digital TV* will include the majority of the circuits used after the receiver's video detector to process and display standard NTSC interlaced pictures. It should be understood that the driver circuits to the picture tube and the output deflection circuits will still be analog. Future flat-panel displays may use digital-driver circuits. Figure 3.28 shows a model for digital video processing.
2. *Progressive Scan Digital TV* (to be covered in the next section) will include, in addition to the above, digital synthetic processing circuits in the receiver only. This technique effectively doubles the number of picture lines that are received from a standard NTSC transmission.
3. *High Definition TV (HDTV)* will include techniques and circuits in both the transmitter and the receiver that will increase the picture definition in both the horizontal and vertical directions. HDTV will also include multichannel CD-quality sound. It will be assumed that quasi-digital techniques may be used in the RF portion of the signal. HDTV will include a different aspect ratio (a wider screen) than NTSC for added realism. HDTV will be discussed in the last section of this chapter.

Digital TV, like most digital circuits, requires a realignment in perspective. As an example, a carbon-composition potentiometer is simple and effective as an analog volume control. If it is replaced with discrete digital components, the digital circuits may seem comparatively cumbersome and uneconomical. However, when a digital control is used with other highly integrated circuits in a digital system, not only is it cost effective but the versatility of digital control and processing usually wins by a wide margin.

The Elements of Digital TV Processing—The Digital Clock

For a synchronous NTSC digital processing system, the obvious choices for the master digital clock are 2X, 3X, or 4X subcarrier. Twice subcarrier would just equal the Nyquist limit but would not adequately sample the upper chroma sidebands or the upper part of the luminance band. 3X works well and has been used in some systems. However, it is often awkward to divide down from a 3X number. Four times subcarrier or 14.32 (14.31818) MHz is the clocking frequency that is used in most NTSC digital processing. Also, with a 4X clock precisely phase locked to the transmitted reference burst, the demodulation of the 90° R-Y and the 0° B-Y signals (see the vector diagrams in Figure 3.22) becomes a very simple task. Some systems cheat a wee bit and use some quasi-analog analog processing to obtain the burst to phase lock the 14.32-MHz master digital clock.

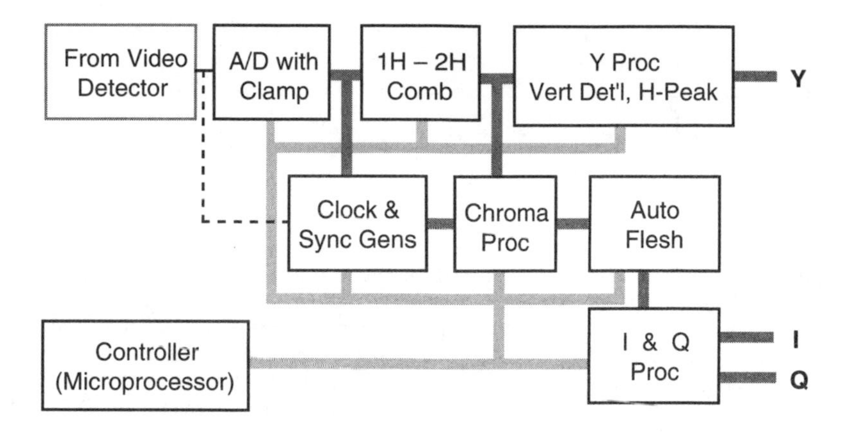

Figure 3.28 An elementary digital video processor. The control and signal busses usually consist of eight bits. However, the internal block processing may require wider data streams. Not shown are the D/A converters, their required low-pass smoothing filters, and the matrixing to derive the required kine signals.

The Elements of Digital TV Processing—The Analog-to-Digital Converter

For video speeds, a flash converter (see Chapter 1) is almost mandatory. The choice of the number of bits is usually eight but 7-bit converters have been used. Seven bits is often considered adequate for picture quality but may limit the dynamic range. With the limited headroom of seven bits, an improperly adjusted set or an improperly modulated transmitter can cause a degraded display, with such problems as picture blooming.

A second approach to using a 7-bit A/D converter is termed dither. Dithering is a technique that can subjectively trick a system into operating as if it had at least one-half more bits and, sometimes, even one more complete bit. Thus, a dithered 7-bit converter can appear under certain conditions to be operating as a 7.5 or 8-bit A/D. It will be remembered (see Figure 1.22) that a flash ADC uses a resistor chain to divide the reference voltage into 2^n voltages for the input comparators. Thus, an elementary 8-bit ADC will have 256 voltage divisions (some designs vary this by one or two units) and an elementary 7-bit ADC will have 128 resistor divided steps. Using a reference voltage of 1.0 volts, the steps in an 8-bit ADC are about 4 mv (1 volt/256) and the steps in a 7-bit unit are about 8 mv. If the reference voltage for a 7-bit unit is alternated up and down 4 mv (one-half bit), the dither will tend to fill in the cracks and make the 7-bit unit seem to operate as an 8-bit ADC (see Figure 3.29). The rate to dither is somewhat controversial because, it must be remembered, this dither is like amplitude modulation and produces unwanted sidebands in the output spectrum. Line-rate dither is probably most common.

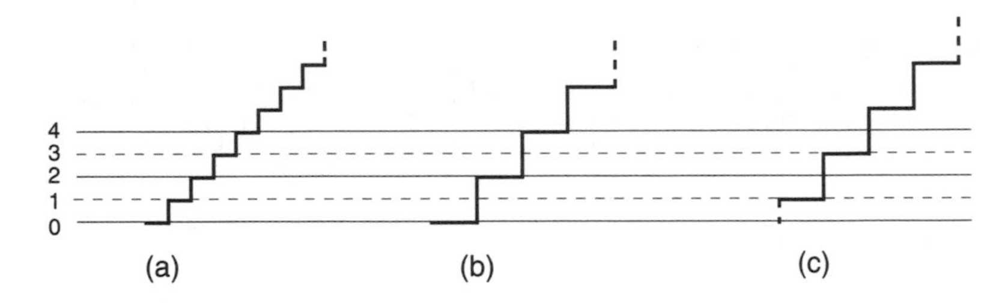

Figure 3.29 An illustration of the use of dithering to give a subjective increase in the bit count in a video analog-to-digital converter. (a) shows the 4-mv quantizing steps in a typical 8-bit ADC. (b) shows the 8-mv steps in a comparable 7-bit ADC operating in a similar manner. (c) is this same 7-bit unit with the comparator reference voltages offset one-half bit or 4 mv. Note that the (c) voltages seem to fill in the cracks left out by (b) in lines 1, 3, etc. There are other methods of accomplishing dither including alternating the DC level of the input analog video.

The Elements of Digital TV Processing—Digital Comb Filters

An analog comb filter is made up of some material, often glass with the appropriate driver and pickup, that will delay the TV signal a precise amount—typically one line or 63.555 μs. This can be accomplished in the digital domain with any *n*-stage "bucket brigade" such as D-type flip-flops, charge-coupled devices (CCDs), or any other integrated device that can successively clock the *n*-bit digital words down the line. As pointed out earlier, the comb filter is the prototype for digital filters. A digital comb filter needs *n* stages of Z^{-1} clocked delay elements. If the clock is running a 4X subcarrier or about 14.43 MHz (about a 70 ns clock period), 910 successive Z^{-1} delays will produce an exact line delay of 63.5+ μs. Many receiver digital-processing schemes include a two-line comb that gives additional separation between the luminance and chrominance signals.

The Elements of Digital TV Processing—Y-Channel Processing

Digital processing of the Y can, in its simplest form, duplicate most of the more advanced techniques of analog luminance processing. Digital circuitry for Y processing is certainly more versatile but it also presents some housekeeping details that, if not carefully attended to, can make the whole process untenable. For instance, a one- or two-line comb filter will remove some low-frequency parts of the luminance and insert them in the chroma. If the maximum separation and bandwidth is desired, this residual luminance must be removed (filtered) from the chrominance and added back to the Y channel. When user controls are added, often in the form of digital multipliers, care must be taken not to truncate many of the overflow bits or the desired effect of the multiplier control will be negated. In nonlinear circuits used for peaking, the designer usually has the choice of using a digital filter or storing the desired transfer functions in RAM for a look-up table. Figure 3.30 indicated one scheme for digital Y processing.

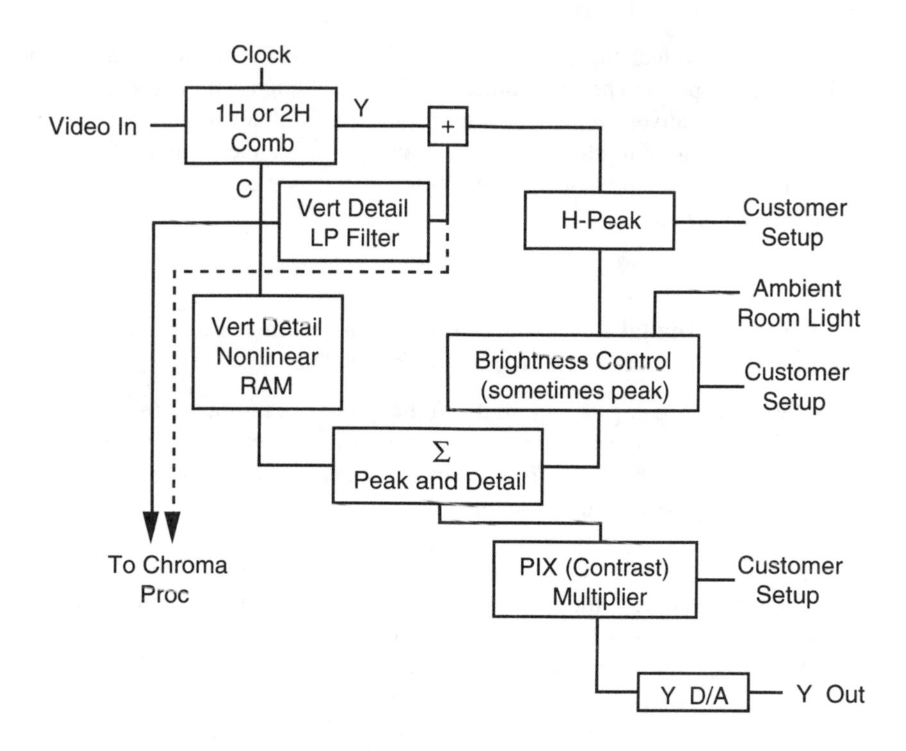

Figure 3.30 One of many schemes for digital Y processing. Note that digital techniques, in very highly integrated circuits, can significantly expand the designer's processing choices. For instance, the output of the vertical detail low-pass filter, used for luma channel enhancement, may, at times, also be used in the chroma processor.

Digital processing can significantly improve the way the system acts on noisy or improperly modulated signals. Improper modulation usually comes from the aforementioned multiple paths of studio, network, and satellite paths. It is very difficult to keep a constant black level with old movies. Digital circuits can average and store references for the timing, shape, and DC levels of sync. These can be used to keep the picture much more stable when some of the transmitted pulses are corrupted by noise. The relationship between the picture saturation (video gain), picture brightness (kine bias), and the picture-tube beam currents can be constantly monitored and adjusted to reduce the problems of picture blooming and to give a much more constant average brightness when going from channel to channel.

The Elements of Digital TV Processing— Chroma Processing and Hue Correction (Auto Flesh)

Like luminance processing, the designers have a larger bag of digital tools to use for chroma processing. The use of ROM look-up tables permits the implementation of precise nonlinear functions that would be difficult in the analog domain.

The noise-reduction technique of coring (see Figure 1.18) or the use of abrupt or hard limiting (clipping) to protect from excessive modulation is more feasible, with more precise alternatives, than in the analog domain. For hue correction, one very attractive technique is to store in ROM a perfect reference-replica of I-phase subcarrier. As has been indicated, digital filters with sharp cut-off and with linear phase can maximize the available I and Q bandwidths.

The Elements of Digital TV Processing—The Insertion of a Second Channel's Picture in the Main Display or Pix-in-Pix

One of the most intriguing uses of digital signal processing for modern TV receivers is pix-in-pix. From the start, many designers stumbled on this idea because its implementation is much more difficult than it first appears. Some of the basic limitations that must be overcome include:

1. The second picture is coming from an entirely different channel than the one the main set is tuned to and is displaying. This necessitates a second tuner for the inserted picture. The pix-in-pix set manufacturers have to either completely duplicate the video and sound RF, IF, and detectors, or they must assume the customer will supply the second channel's video and sound from a separate source such as a video tape recorder (VCR).
2. The second picture will not be transmitted in synchronism (including the color reference) with the main picture. This means that the pix-in-pix processing must include quite an elaborate system of counting, reading, and storing—then recounting and writing the original stored picture information so that the inserted picture is displayed in sync with the main picture. In most implementations, the inserted picture is much smaller than the main picture. This also means that some of the pixels in the incoming video must be eliminated, i.e., digitally decimated.
3. The customer will also wish to use the set's picture as the insert and the second signal as the large main display.
4. Just as the pictures must be switched, so must their sound. A desirable feature is to enable the viewer to listen to the sound from either source regardless of which source is used for the main, large display.

A Model for Pix-in-Pix Development—The Frame Synchronizer

Since about the middle 1970s, the broadcast industry has had a device available that permits almost any remotely originated NTSC program material, be it live or on tape, to be synchronized with local originations using the master or plant sync system. Although there were certainly some TV-signal synchronizers in use before the above-mentioned era, the availability of 4K by 8, and soon 16K by 8, dynamic RAMS made possible relatively small, reliable frame synchronizers.

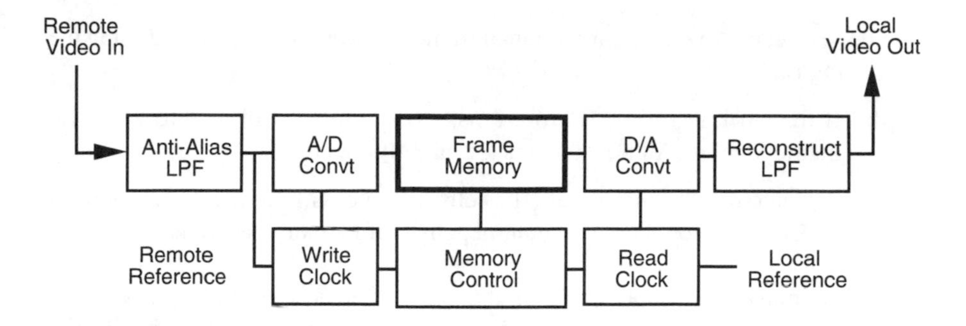

Figure 3.31 The prototype frame-store synchronizer. Incoming data are written into the memory at a time and with a clock derived from its subcarrier frequency and phase. Data are read out of the memory based on the timing of the local or plant sync. The two major demands are that the incoming video does not write precisely when and where the outgoing video is to be read and that the chroma of the outgoing video emerges with the correct phase.

The trick in designing a synchronizer or pix-in-pix system is to provide a memory control that can sense the frame position of the incoming signal in time and accurately compare it with a similar position of the plant signal. If the timing information on these two signals is known, then the synchronizer can routinely write the nonsynchronous input signal into memory using its own timing. The controller can then ensure a readout of this input with timing that is in step with the video that is in use in the local main picture operation—see Figure 3.31.

Although the tolerances on the subcarrier frequency and phase are very tight, station A or camera chain A will not have a subcarrier frequency and phase exactly the same as station B, camera chain B, or tape machine B. Likewise, a remote sync generator (say, at a cold football stadium) may tend to drift from one extreme of the tolerance limit to the other more often than the master generator at the plant. Statistically, it can be shown that the signals derived from sync generators A and B will become exactly synchronous every few hours. When this happens, it may cause a frame synchronizer to crash. That is, the read clock will attempt to read memory cells at the exact time and location that the write clock is attempting to write a new frame. The solution to this crash problem is twofold. First, if there is a write-read contention, the memory will freeze. This will inhibit memory writing and allow a continuous reading of the now stationary, frozen picture. Second, a temporary delay must be inserted in the video of either the incoming or synchronized video to artificially eliminate the overlap or crash situation. In a few seconds, as the signals drift apart in time, the memory controller may eliminate the additional delay.

NOTE: In most cases the memory is not exactly one field in length. The designer must, of course, use an integral number of chips, which usually dictates a larger memory than is needed. One design alternative is to add extra memory for temporary delay. An elementary minimal frame memory can be designed using the following rules:

1. If the clock is 14.32 MHz or 70 ns, a 63.5 μs line will require 910 8-bit memory cells. (All the numbers are slightly rounded.)
2. A frame of 525 lines times 910 cells per line will require a minimum of 477,770 cells or about one-half megabyte of memory per frame.

A last, much more subtle problem that affects the design of picture freeze in synchronizers and pix-in-pix displays is the positional phase of subcarrier. To understand completely the problem of chroma positional phase, it is necessary to go through some minor arithmetic gymnastics. Each picture line is 63.5555 μs long and contains 455/2 or 227.5 cycles of subcarrier—an odd number of subcarrier-cycle inversions. Each 262.5-line field contains 59,718.75 cycles of subcarrier, another fractional number, and each complete frame contains 119,437.5 cycles of subcarrier—still an odd number of subcarrier inversions. Continuing to play this game through, it is not until a second frame is completed that the cycle count (2 × 119,437.5 or 23,8875) becomes a complete whole, even number of subcarrier inversions. Thus, a synchronizer or pix-in-pix cannot arbitrarily read a given frozen-picture line, field, or frame from memory and be assured that it will synchronously merge with the subcarrier signal coming form the plant. To ensure the proper subcarrier phase in the proper field, it is often necessary to change the phase of chroma by 180°. This is termed chroma inversion. One technique to produce two new, properly phased frames (four fields) from the one stored frame includes the following:

NEW FRAME 1	Field 1	Read the stored (frozen) field 1 with no alternation.
	Field 2	Read the same stored (frozen) field 1 with chroma inversion.
NEW FRAME 2	Field 3	Read the same stored (frozen) field 1 with chroma inversion.
	Field 4	Read the same stored (frozen) field 1 with no chroma inversion. The process is continually repeated as long as a frozen picture readout is needed.

The chroma inversion process can be done with either analog or digital components. An elementary chroma inverter can be constructed by using a delay line and a subtracter circuit, as shown in the Figure 3.27 comb filter, plus an inverter. More complex chroma inverters usually use two analog or digital delay lines and continually average the chroma in three-line sequences.

The Elements of Digital TV Processing—Improving the Received Picture Definition with the Technique of Progressive Scan

One elementary definition of improved TV picture definition is simply more lines. If, by magic, the camera could produce a picture with more horizontal lines, the transmitter could send it, and the receiver could process it, supposedly the viewer would have more realistic and enjoyable TV reception. At a first approximation, this is true. However, the final section of this chapter will detail some of the major problems that must be overcome before the camera, the transmitter, and the receiver can accommodate more lines and several other technical advances that are necessary to make them truly produce HDTV. Before proceeding to the description of HDTV and some of the radical departures it will take from the NTSC system, it seems prudent to investigate a very interesting interim development that a number of manufacturers have incorporated in their sets—progressive scan.

Referring back to Figure 3.10 in the NTSC system, a complete TV frame is made up of two interlaced fields. As the lines of each field are scanned down, there are blank or skipped spaces that are subsequently filled with the lines of the next field. The most intriguing high-definition scheme is to find a way to fill in these blank lines and make each field 525 lines rather than 262.5 lines. Such a system could be built into the receiver without requiring any change in the studio and transmitter standards or equipment.

The sections on comb filtering and the addition of a second, nonsynchronous auxiliary picture should alert the reader that there are many subtle problems associated with such a design. The general approach that has been taken is to average the surrounding picture elements (and, of course, to chroma invert when needed) to produce a new fill-in element.

The major problem with synthesizing additional picture elements and thus new picture lines is picture motion. Either scheme (a) or (b), as shown in Figure 3.32, will most often give a pleasing, additional detail for a stationary color picture. However, if there is motion in the picture, and thus changes in the position of the

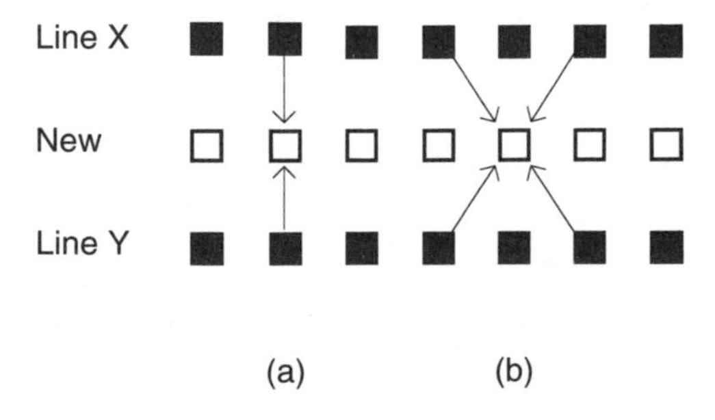

Figure 3.32 An illustration of two techniques for synthesizing an additional fill-in line from two adjacent lines. In (a), each new picture element is created by finding the average or mean value of the immediate upper and lower picture elements. In (b), the new elements are created by processing four of the surrounding normal picture elements.

colored objects and their edges, these general averaging techniques can, at times, produce smearing that detracts from, rather than adds to, the detail of a picture. Such picture distortions can be minimized, and thus the progressive scan effect enhanced, if the concurrent lines in the previous stored field are processed along with the current lines.

3.7 HIGH-DEFINITION TELEVISION—HDTV

An Introduction to HDTV—Its Goals and Limitations

Figure 3.33, which is really a variation on part of Figure 3.17, shows a direct color transmission system that has become to be known as RGB. This is the type of color-signal transfer (although not usually camera originated) that is most common in computer systems. Also, like computer systems, the final signal fidelity is dependent upon the definition produced by the pickup or other source, the bandwidth and transient response of the transmission medium, and, especially, the quality of the display device.

It is particularly important to realize that a perfect transmission medium can be seriously compromised if the signal definition of the pickup or display devices does not equal the precision of the transmitter and receiver electronics.

As a definition touchstone or ultimate goal, the model for HDTV is often given as the quality of 35-mm motion-picture film. This definition is again somewhat obtuse because photographic lines are not defined in the same way as are TV lines. (See the Fink reference.) Assuming an approximate rationalization between the lines of 35-mm film and a TV picture display, the film will display the equivalent of somewhere between 900 and 1,200 "real" TV lines. The term "real" is used in reference to the perceived resolution of NTSC. Although there are a total of 525 lines per frame, 41 (20.5 lines per field) lines are used by the nondisplayed vertical interval. This gives 484 active (displayed) lines per frame or 242 active lines per field. However, because of the NTSC band limits, the actual real horizontal resolution is not 484 lines per frame but in the order of 320 to 350. This is usually stated as the Kell factor, as described in an earlier section on brightness. (See also the excellent chart on page 3.6 of the Benson and Fink reference.)

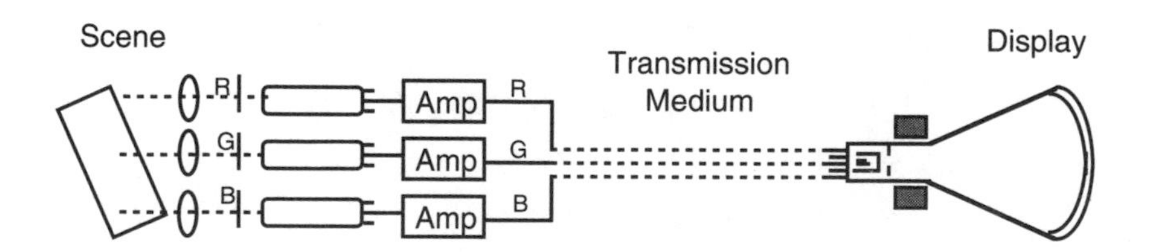

Figure 3.33 The rudimentary RGB transmission system. The quality of the final display is dependent upon the resolution of the originating and displaying devices as well the bandwidth, phase, and transient characteristics of the complete transmission medium.

NOTE: A completely precise and universally accepted definition of resolution has eluded engineers and scientists since the inception of TV. Because resolution is a perceived human response, any definition must be somewhat subjective. Figures 3.8 and 3.9 indicate some elementary background. Many test patterns (remember the Indian?) have used complex line and circle structures, as well as rectangular and triangular wedges, to relate picture scan lines to resolution and bandwidth. Even these were usually based upon some majority subjective agreement among assembled observers. Certainly the addition of color complicated the problem of defining video resolution. The following sections will indicate that motion also confuses the definition of resolution. Another factor affecting the choice of bandwidths for fundamental or derived color primaries is the acuity and sensitivity of the human eye. The Otto Schade, Sr., work provides one of the best early references since, with text and photographs, it compares the resolution of still photographs, movies, and TV. See also the various comparisons in the Fink reference.

With these constraints and limitations noted, it is now possible to provide an introductory list for an HDTV model. Such a rough specification can be used as a guide only when investigating some of the proposals that have been presented to the FCC for a U.S. high-definition standard.

1. It would be desirable to have an HDTV standard that is compatible with the existing, lower-definition, NTSC color receivers. This compatibility rule served the United Sates well when the switch to color was made. In Europe, when the noncompatible 625-line PAL and SECAM color systems were introduced, the existing 405-line and 819-line B&W transmissions continued for many years. However, many of the proposals provide for separate or simulcast channels.

2. An HDTV system should give the best possible resolution consistent with the above requirement and also the existing channel system in the United Sates Counting VHF and UHF, there are 68 channels available for terrestrial (over the air) broadcasting. In any given heavy usage (metropolitan) region, many of these are denoted as taboo channels, which means that at present they cannot be used because of interference with other assigned channels. Because of the extremely crowded spectrum, it is not likely that additional space will be allotted for HDTV.

3. An HDTV system should be compatible with existing home VCRs. Here compatible is defined as the ability of an existing VCR to record and play back the HDTV signal, although at a lower (NTSC-type) definition. The VCR manufacturers would certainly prefer a high-definition system that would permit them to convert to high-definition recording and playback by just changing their electronics, while keeping their existing tape transport and head designs.

4. An HDTV system should be compatible with older, current, and future motion pictures. It should be recognized that movies form a significant portion of American TV program material. Certainly it appears that this will continue in the future. One of the items that is almost certain to be included in an HDTV standard is wide screen or a new aspect ratio. NTSC (and pre-1953 movies) has an aspect ratio (the ratio of the picture width to height) of 4:3 or 1.333. However, the movie industry has, over the years, tended to produce wider-screen movies. Many movies are now produced with an aspect ratio of 1.85:1 and CinemaScope pictures have a ratio of 2.35:1. It is interesting to note that most wide-screen movies use film with an equal vertical and horizontal resolution but, in widening the picture, the special (anamorphic) lenses reduce the inherent horizontal resolution of the film. At this writing, it is assumed that the HDTV aspect ratio will be 16:9.
5. An HDTV system will certainly impact the cable TV industry. Some proposals are targeted to the wider bandwidths of cable (including fiber) or satellite channels. Gross HDTV bandwidth, including enhanced chrominance and luminance could, simplistically, require as much as six to eight times the NTSC bandwidth or between 24 and 32 MHz. (See the Benson and Fink reference.) Although most theoreticians and designers consider this extremely pessimistic, especially with the available compression techniques, HDTV is certainly going to require significant additional bandwidth. To some pundits, the additional bandwidth of cable, fiber, or satellite transmissions seems very appealing. Likewise, any new high-definition standard should be able to accommodate new techniques and technologies that will be forthcoming in the twenty-first century.
6. An HDTV system will often require a much larger, and thus more costly, display. The eye's ability to resolve detail is limited by distance. If the viewer is back more than about three times the picture height from the viewing screen, any increased detail in HDTV will be lost!

The Categories of the Proposed HDTV Systems

There were about twenty original proposals for a U.S. HDTV system. Most of them were derived from either the concepts or the details of NTSC and they suggested some mode of compatibility with that system. However, there were a few that completely deviated from NTSC. Their proponents felt that their particular system would have such a significant technical or economic advantage that compatibility would be secondary.

The HDTV proposals can be grouped into the following five categories.

1. A single-channel, NTSC-compatible system with the high definition signal existing within the present FCC-mandated 6-MHz channel bandwidth.

2. A single-channel noncompatible simulcast system. These proposals assume separate HDTV channels in addition to the existing (at least for the present) NTSC transmissions.

3. A two-channel system with one channel transmitting a compatible 6-MHz signal and the second channel transmitting the additional information required to produce a high-definition picture. In a special receiver, this second or augmented 3-MHz channel would be combined with the first channel to produce the composite HDTV picture. Some of the proposals presuppose that the augmented channel, with a 6-MHz bandwidth, could be shared with two HDTV stations.

4. A two-channel system, similar to the one above, with the first channel sending an NTSC-compatible signal and using the complete 6-MHz bandwidth of the second channel. As above, the signals from the two channels would be combined in the receiver to produce the high-definition picture.

5. A satellite-only system that could be NTSC compatible but would not have any FCC channel restrictions on its overall HDTV bandwidth.

Models for HDTV—The NHK and the SMPTE 1125-Line HDTV Systems

Before representative models of the above five types of HDTV systems are discussed, it is prudent to provide a foundation by presenting two important pioneering techniques. The NHK (Japan Television) system dates back at least to 1978. In fact, up to about 1985, in most references, it was assumed that the initials "HDTV" referred to the NHK, or later, the MUSE or MUSE-E system. (MUSE is an acronym for Multiple Sub-Nyquist Encoding and there have been several subsequent versions following this original version of their analog HDTV system.) The original MUSE or HDTV used a satellite transmission medium, FM, and required a composite (luminance and chrominance) bandwidth of about 30 MHz. It used 1,125 lines and 60 (not 59.94) fields per second with an aspect ratio of 16:9. For about the past ten years, this has been far more than a laboratory curiosity. It is truly a complete system with working high-definition cameras, tape recorders, and receivers. Likewise, the designers of the original MUSE system and its later generations of systems, have documented their developments well in the technical literature. It has been used in dozens of world wide demonstrations, including coverage of the 1988 Olympics. Daily broadcasts are conducted in Japan. Since MUSE is FM, it is not even remotely NTSC compatible. Likewise, because of its system FM modulation, it is incompatible with the de facto (Scientific Atlanta) standard of U.S. satellite transmission.

Because of the extended bandwidths of the various MUSE systems (especially the first development), new techniques of bandwidth compression were required. The general techniques of compression for transmission and the subsequent required expansion in the receiver work well to produce the desired high definition for stills, but not for moving images. With moving images, the horizontal resolution of the (nominal) 1,125-line definition is significantly reduced (about 60 percent), sometimes almost back to that of the 525-line (again, not including the Kell factors) NTSC. Thus, technologies had to be developed to control and artificially enhance the band expanded horizontal resolution for received picture motion.

In addition to the MUSE 1,125-line system, the Society of Motion Pictures and Television Engineers (SMPTE) developed its SMPTE 240M-1988, 1,125-line, 60-field studio production standard. Since the motion picture producers are also so involved in the creation of TV programs, it is natural for them to have a universally accepted standard for HDTV program production. Like the NHK, the SMPTE system entails a very wide baseband system. Because it is a system designed to be used by a variety of motion picture and TV broadcasting program originators, the SMPTE standard is documented in detail, including wave shapes and their timing, clock frequencies, the derivation of color primaries, etc. Ever since 1988, when they were first demonstrated at the convention of the National Association of Broadcasters (NAB), there has been an adequate variety of cameras, tape machines, etc., that operate using this high-definition standard. Also, there are down-converters that can transform from the SMPTE standard down to the lower-resolution NTSC or PAL standards.

Representative Models for an HDTV Standard—The NHK/MUSE System Family of NHK MUSE and MUSE-6

All the variations of the MUSE system use the same studio equipment and techniques and are directly traceable back to the original 1,125-line HDTV system. In MUSE, as well as all the proposed HDTV systems, the two major challenges are bandwidth compression and motion (resolution) control.

The MUSE systems use what NHK calls Multiple Sub-Nyquist Encoding to perform the compression. This is a very clever tcchnique of using aliasing rather than trying to prevent it or get rid of it, as is done in most signal processing. The input baseband video signal is over 20-MHz (about 24-MHz) and uses an A/D sampling clock of 48.6-MHz. At the output of the A/D, the signal is low-pass (digitally) filtered with a 24.3-MHz cut-off filter and then each field is again sampled, but now with a 24.3-MHz clock and then filtered with a 12.15-MHz LPF. By using this unorthodox trick, the aliased high frequencies are folded back into the low-frequency portion of the spectrum. The resulting signal's frames are again sampled, this time with a 16.2-MHz clock and filtered with an 8.1-MHz LPF. The repeated subsampling compresses the nominal incoming video bandwidth of 24-MHz down to about 8-MHz which results in only presenting every third picture element. For stationary pictures, this results in a final horizontal resolution, in the received and reconstituted 1,125-line picture, of 600 to 750 real Kell lines. To detect and correct for the problems of any moving portions of the input video, the motion is sampled once every frame. The horizontal motion can be detected with an accur-

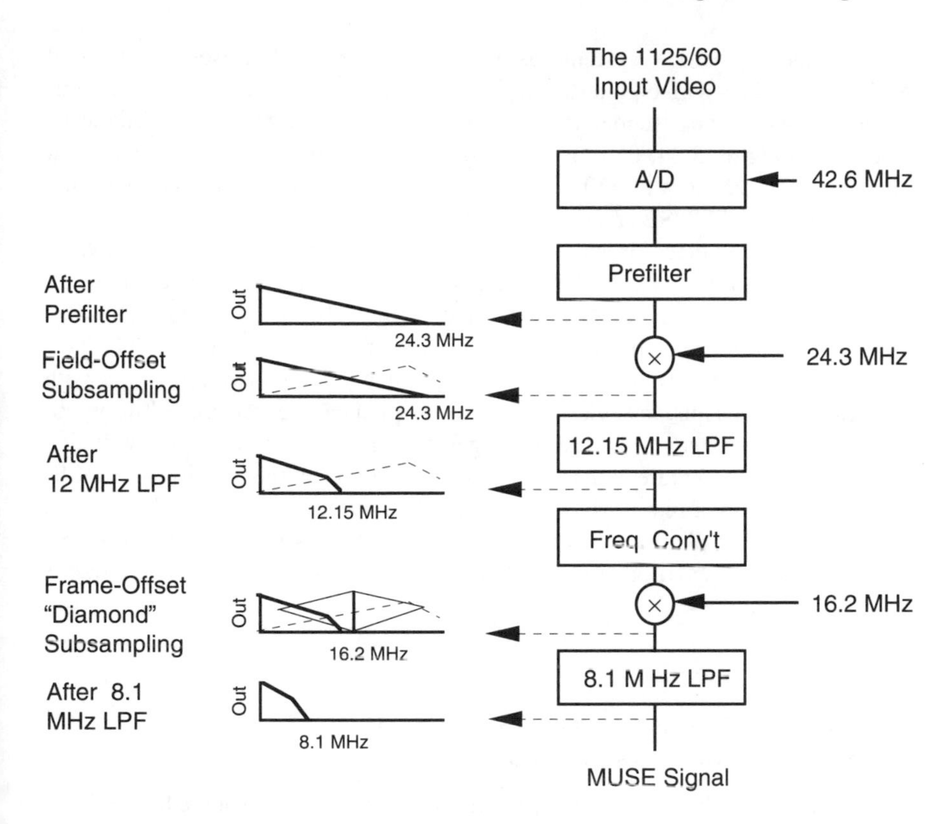

Figure 3.34 The derivation and compression of the MUSE signal. By using a system of progressively subsampling and filtering to alias the high frequencies into the lower portions of the spectrum, the nominal 24-MHz video bandwidth is compressed into a band of about 8 MHz. Note that the frame bandwidth will change depending on which particular frame is being sampled—thus the term "diamond" subsampling.

acy of about ± 7 clocks/frame or 0.19 picture widths/sec and the vertical motion ± 3 lines/ field or 0.23 picture heights/sec. The overall horizontal resolution for the motion is about one-half of stationary pictures. However, because of the eyes' lack of ability to accurately resolve moving scenes, this reduced motion resolution is not considered a significant limitation.

NOTE: The information about this HDTV system, and the descriptions that follow, are derived from the public domain proposals prepared by the companies involved and presented to the FCC. The information is, of course, abstracted for presentation here but every effort was made to be accurate. For detailed information, consult the complete proposals or see the Prentiss and Benson and Fink references.

The generic MUSE-6 system has a 6-MHz bandwidth but uses the standard MUSE 1,125 lines with 60 interlaced fields. If it is to be used for NTSC, the originating video is first transcoded (their wording) to a 59.94-Hz field rate. Likewise, referring to Figure 3.34, there is an another step inserted in the processing that changes the 1,125 lines to 750 lines. This reduction of lines, combined with the other conventional MUSE processing, produces a signal that fits into a 6-MHz bandwidth. The motion detector is used to control the conversion of the 60- to 59.94-Hz field rate and the 1,125 to 750 line rate. After this processing is complete, a final processing section is added, not shown in Figure 3.34, that converts the MUSE-encoded luminance and chrominance to an NTSC-type 525 line, Y, I, and Q. This final processing will produce a transmitted signal that is completely compatible with NTSC receivers. Thus, in review, a MUSE-6 signal can be transmitted in two 6-MHz versions. One can be received using a standard NTSC receiver and will give a compatible color picture with a 4:3 aspect ratio. Also, another version of MUSE-6 can be transmitted in a 6-MHz bandwidth that, with a special MUSE receiver, will give a 750-line picture with a 16:9 aspect ratio. There is also a version of MUSE that requires one 6-MHz and one 3-MHz channel and another version that requires two 6-MHz channels.

Representative Models for an HDTV Standard—The Sarnoff ACTV Series and ADTV

The David Sarnoff Research Center, which was the research arm of the former RCA Corporation, has proposed a family of high-definition systems to the FCC. The original Sarnoff proposals included the ACTV-I and ACTV-II. These two systems, which included mostly analog technology, would enable the broadcast industry to convert to HDTV in a series of planned steps. A later, all-digital proposal, called ADTV (Advanced Digital TV) was made by a combined group that includes, in addition to the David Sarnoff Research Center, North American Philips, Thompson Consumer Electronics, Compression Laboratories, and NBC. These systems can be simply defined as follows:

ACTV-I — The signal is completely compatible with present NTSC receivers and will also produce a 525-line progressive-scanned picture with a 16:9 aspect ratio in ACTV-I receivers.

ACTV-II — By using a second augmented channel in addition to the ACTV-I service, a high-definition, 1,050-line, 16:9 aspect ratio picture can be received on an ACTV-II receiver. Since this service also presupposes ACTV-I transmissions, it is also compatible with NTSC and ACTV-I receivers.

ADTV Advanced Digital Television is a fully digital system that delivers high-resolution TV in a single 6-MHz channel. It is designed to be simulcast on a second channel in addition to the existing NTSC signal. Although the proposal was submitted as a 1,050 line, 59.94 field, 16:9 aspect ratio system, because of the way the system is coded and implemented, its scanning parameters are not restricted and it could perform equally well using scans stipulated by other systems such as 1125-MUSE or the SMPTE standard.

ACTV-I

Both ACTV-I and ACTV-II transmissions assume a wide-screen input video with an aspect ratio of 16:9, 525, 1,050 (or 1,125) progressively scanned or interlaced lines at 59.94 fields/frames per second. If the input is progressively scanned, it can be transcoded for NTSC compatibility. The essential ACTV-I system shown in Figure 3.35 is composed of four primary processing components. Component 1 is the standard NTSC kernel signal and can be visualized as the center portion of a wide-screen input signal. The left and right sections or side panels of the wide picture are processed separately.

Signalwise, these side panels are each 1 μs wide and carry their low-frequency information. When the complete ACTV-I signal is processed in a standard NTSC receiver, the normal receiver overscan will hide these side panels. Component 2 of the transmitted signal carries the high-frequency part of the side panels. Some of the high-frequency information extends to up to 5 MHz but, because it is modulated on an alternate interleaved subcarrier, it does not interfere with either the chroma or sound carriers. Besides adding side-panel highs, the processing in a wide-screen ACTV-I receiver uses the information in both components 1 and 2 to feather the boundary between the main center picture and the side panels.

The component 3 information included additional luminance detail in the 5-to 6-MHz region, which is transposed (and filtered to eliminate the upper sideband) down to the 0- to 1-MHz region of the 6-MHz bandwidth. Some of the horizontal detail is time compressed and transmitted during the active line time in component 1. Component 4 is termed a helper signal and is used to transmit additional vertical and temporal (the motion as presented by a sequence of frames) information. By using a five-tap vertical-temporal filter that might be made up of a five-unit comb filter with field or frame delays rather than line delays, the vertical detail can be averaged and restored. This is the form of motion-detail restoration detail that is used in the ACTV system. The helper signal is 750 kHz low-pass filtered and quadrature amplitude modulated (QAM) on the main RF carrier. Without belaboring the details, the circuitry of the ACTV-I receiver is the inverse of the transmission processing. See the proposal documents to the FCC or the Benson and Fink reference.

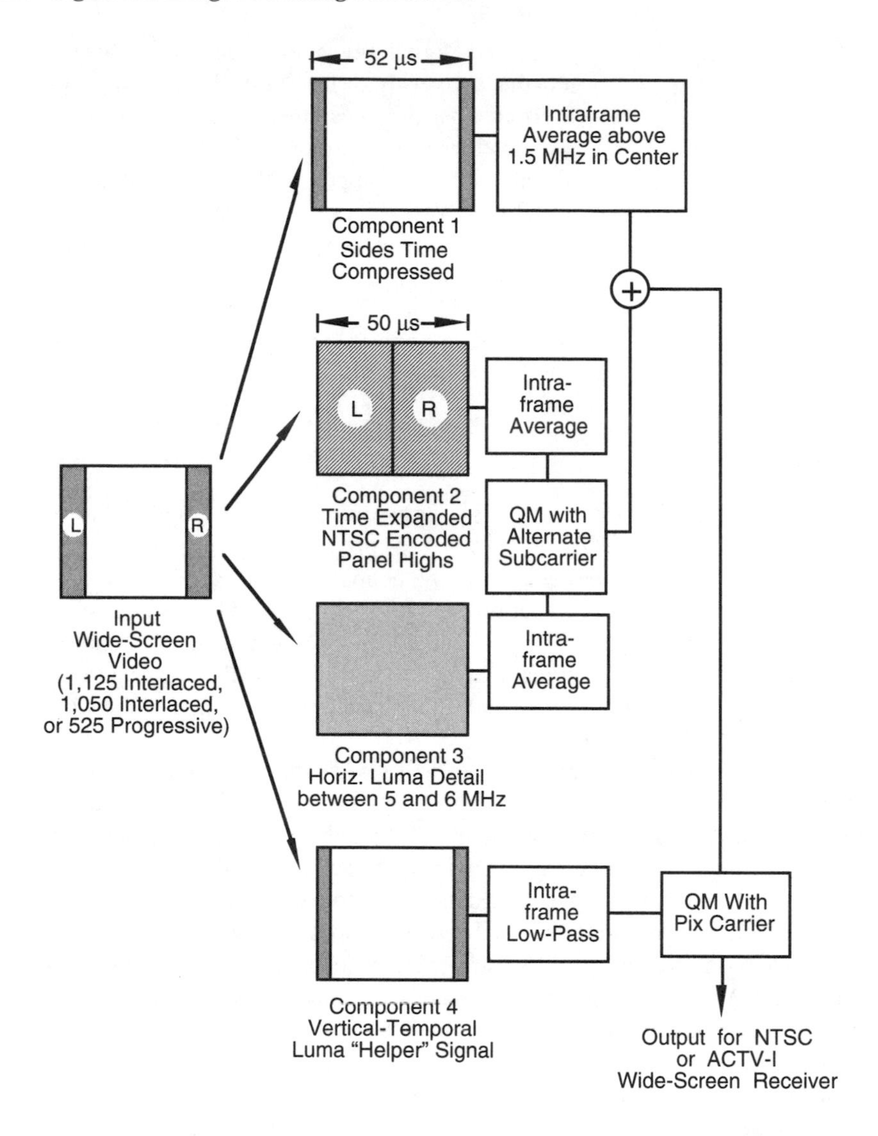

Figure 3.35 The use of components in the processing of the transmitted signal for the NTSC-compatible Sarnoff ACTV-I wide-screen system.

ACTV-II

This is termed an augmented channel that provides the information to upgrade ACTV-I into a full-blown HDTV system. It provides a wide-screen 16:9, 1,025-line picture. The second or auxiliary channel supplies the additional coded detail

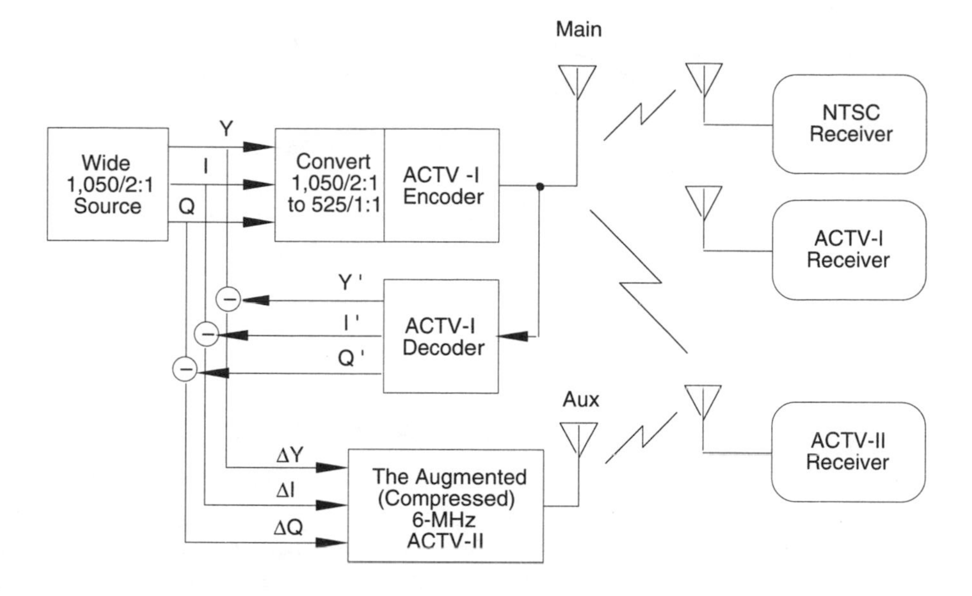

Figure 3.36 The scheme of the processing, transmission, and reception of the signals for the Sarnoff ACTV-I and ACTV-II signals.

to upgrade the presentation from the ACTV-I's 525-line progressive scan picture to a 1,050-line interlaced picture. The actual luminance resolution is 600 to 800 lines and the actual chrominance resolution is about 100 lines. The motion (dynamic) resolution is about 250 lines—see Figure 3.36.

The auxiliary transmitted sends the additional resolution information in the form of difference (ΔY, ΔI, ΔQ) information. Like all HDTV systems, ACTV-II must use compression to reduce the wide-band (18 MHz) signals to the normal U.S. TV channel bandwidth of 6-MHz. The compressed auxiliary signal is divided into odd and even lines, each of which is transmitted in a 3-MHz band using quadrature modulation of the carrier placed in the center of the augmented channel.

New Generation Thinking—An Introduction to the All-Digital Proposals

There are two predominant techniques that characterize the all-digital proposals —the use of mathematical and graphical modeling and layered architecture. Although one or the other of these techniques may have been used in the design of some analog systems, they dominate the approach and implementation of all the later digital proposals.

The use of modeling went far beyond the conventional use of circuit or system modeling. For instance, an early modeling program that was used at the Sarnoff laboratories was dubbed light-to-light. This program was developed to study the performance of an overall TV system from the (usually nonlinear) response of the

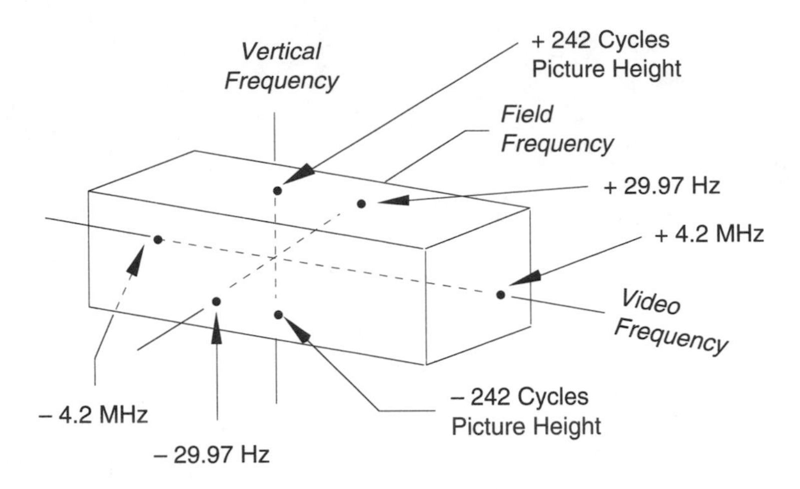

Figure 3.37 One example of the many multidimensional, conceptual figures that are created to study the static and dynamic spectrum of TV signals. This three-dimensional graphic shows vertical or picture line frequency on the normal Y axes, video frequency on the X axes, and field frequency on the Z axes. This volume illustrates the Nyquist plus and minus (mathematical) frequency limits. Both the frequency and the spectrum of the picture, sound, and subcarriers can also be shown in such a three-dimensional form. Interestingly, an interlaced scan graphic would be twisted about the video-frequency axis due to the offset of the interlaced lines. (See the Isnardi reference.)

pickup device to the similar light output response of the display device. The computer simulator included the dynamic transfer and modulation or demodulation characteristics of all the system elements of the studio and transmission system as well as those of the receiver. More contemporary simulators were used to find the spectrum of both static and dynamic signals. This involves analyzing signals in two-dimensional x-y space plus their analysis of the three dimensions of space and time (or frequency). More complex Nyquist volumes show the amplitude and positions of any subcarriers that are used in the signal. Graphics such as these, coupled with the mathematical model programs that create them, are used to find any holes in the spectrum that can be used for more information transfer. See Figure 3.37, the Benson and Fink reference as well as the fine paper by Isnardi.

The second all-digital technique, layered architecture, is reminiscent of the multilayered Open System Interconnection (OSI) and the Integrated Services Digital Network (ISDN) models for telecommunications detailed in Chapter 8. As in these models, hierarchical divisions provide a multielement system shell that not only facilitates system and interface planning, but makes it possible to alter or augment one of the layers without interfering with the structure or operation of the rest of the layers of the overall system. Even in the late fall of 1992, four of the HDTV contenders submitted documents to the FCC that detailed changes, improvements, or additions. It is to their credit that the original layered architecture permits these and future upgrades without affecting the integrity of the complete system.

Representative Models for an HDTV Standard—The Sarnoff/Advanced Television Research Consortium's ADTV

Advanced Digital Television is a fully digital system that delivers high-definition in a 6-MHz bandwidth. It is designed to be simulcast with existing NTSC service. It is designed with a four-layered architecture, as shown in Figure 3.38. As with the OSI/ISDN structures, the four modules in the ADTV system can be updated without affecting the functions of the other three. Notwithstanding, the four layers are interrelated and some of them are directly affected by the operation of each other. An example of this is the rate controller that resides in the Data Prioritization layer but also interacts with the Compression and Transport modules. Figure 3.39 illustrates how the four layers of ADTV are implemented as a warning system.

The ADTV Compression Layer

The usual input to the video processor is in the form of 1,050/59.94/2:1 and is then changed to the Y, U, V format.

(The U color primary is the same as B-Y and V is the same as R-Y, as shown in Figure 3.22. See also the Sandbank reference.) The luminance or Y signal is sampled to produce picture frames of 1,440 *active* pixels by 960 *active* lines. The U and V color information is sampled at one-half the Y sampling rate, producing color difference frames of 720 active pixels by 960 active lines.

Further processing is carried out to vertically decimate (break down into shorter sequences—see the Lynn and Fuerst reference) the U and V components into 480 active lines per frame. To recover the necessary U and V definition, the necessary vertical interpolation—to "un-decimate" or digitally fill in—will be carried out in the receiver's video postprocessor.

The keystone to this or any of the HDTV systems is the video compression and expansion processes. Each submitted system relies heavily on the particular compression/decompression algorithm that they determine to be the best to produce the concept of an ideal HDTV system. ADTV uses MPEG (actually their upgraded MPEG++), which is a compression technique developed by the Moving Pictures Expert Group, a committee within the International Standards Organization (ISO). By using MPEG++, data prioritization will allow the most important video data to be transmitted with the greatest reliability to prevent disruptive transmission errors. This data prioritization, which is accomplished in the Data Prioritization layer, also allows the smooth transmission of other data such as (four-channel) audio, computer, and auxiliary video that can be compatible with ISDN. (See Chapter 8.)

The video compression operations include motion estimation, motion-compensated predictive coding, and variable-length coding (VLC) based upon the Discrete Cosine transform. (See most texts on digital signal processing or, for video, the Sandbank reference.) The picture frames are coded as independent (I) interframes, (P) predicted frames, and (B) bidirectional frames. The P frames are coded by a motion-compensated predictive coder using previous I or P frames. The B frames

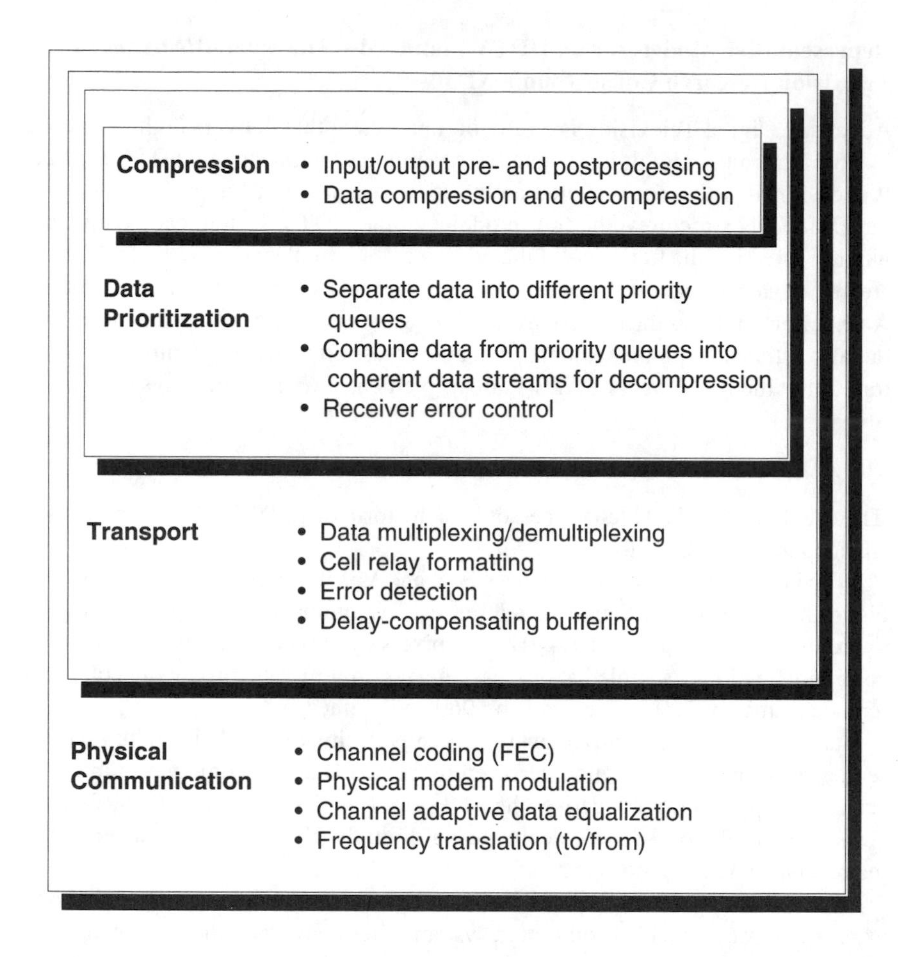

Figure 3.38 The four layers of the Sarnoff/Advanced Television Research Consortium's ADTV system. It should be noted that these layers encompass both transmission and reception. Thus, some items, such as receiver control, may refer to only one part of the system.

are coded by a bidirectional, motion-compensated predictive cder using previous I or P frames.

The ADTV Data Prioritization Layer

The main function of the Priority Processor is to identify the channel protection requirements of each piece of data. The Priority Processor knows the type of each data element. In the Priority Processor, Prioritized Data Transport (PDT) is accomplished by using what is termed a fast packet cell transport format. In a manner similar to data exchange in cellular telephones (see Chapters 4 and 8), each data

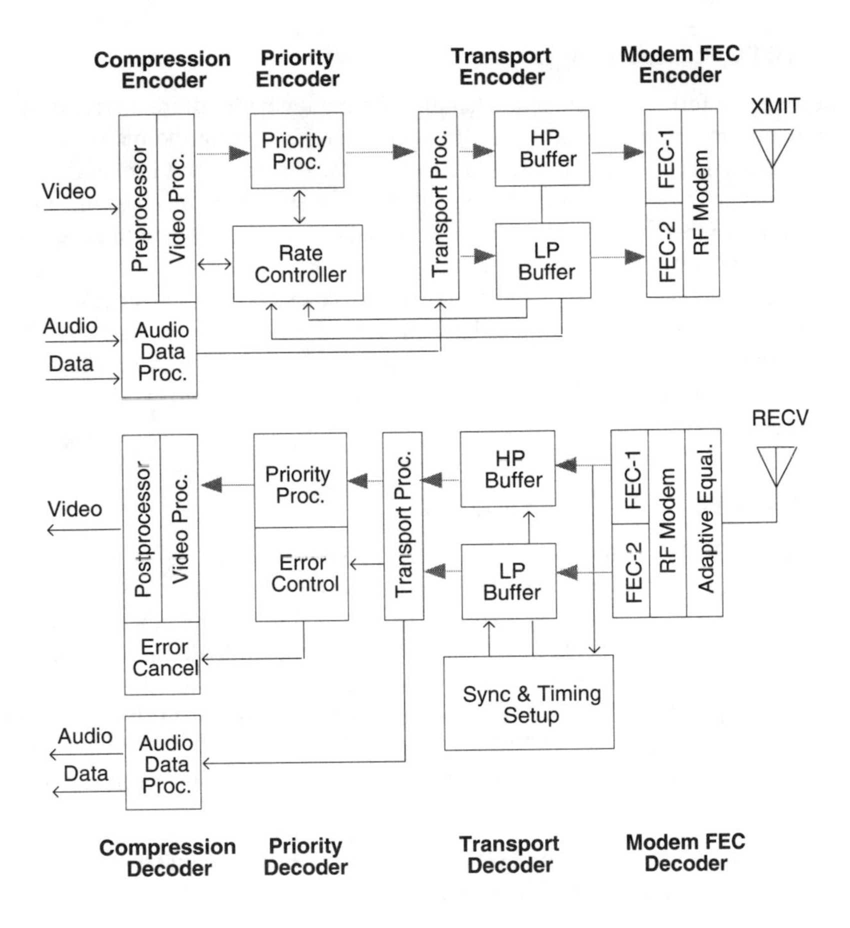

Figure 3.39 The block diagram of the ADTV system partitioned by the layers shown in Figure 3.38. (FEC = Forward Error Correcting.)

cell consists of a fixed size data, a header, and a trailer. The system is extremely versatile since, although the data word is of a fixed length, the video portion of the data words can vary in length. This provides a flexible way to multiplex auxiliary video, audio, and computer data, with an ISDN-type format. (Again, see Chapters 4 and 8.)

The Rate Controller is used to maintain and control the flow of the variable-rate data into the fixed-rate data channel. It communicates with the Compression Encoder to regulate the amount of video compression according to the output data rate and monitors the channel load, including the Transport Buffers, to instruct the appropriate module in the video encoder to increase or decrease the flow of data into the channel.

The ADTV Transport Layer

The transport format is designed to handle information with different error protection requirements. The data from different inputs, including the main video, is encapsulated in cells with different priority codes. The Transport Processor asynchronously multiplexes the complete payload in order to maximize channel utility and minimize the impact of gaps in the payload on system throughput efficiency. Cells are of fixed size and are prioritized according to their prescribed protection levels. Each cell has its own CRC (cyclic redundancy codes) error-checking bits. The HP and LP (High Priority and Low Priority) Buffers, in both the transmitted and received Transport Encoders, work together to ensure that the overall bit stream experiences a fixed delay despite the variable delays caused by the variable-length encodings.

The ADTV Physical Communication Layer

Before the data is modulated and transmitted, there is a final stage of error correction in the Reed-Solomon Forward Error Corrector (FEC). (See any advanced text on signal processing or the Sandbank reference for information about the use of this redundancy error-checking scheme, which is used in both digital audio and TV.) The transmission modem uses Quadrature Amplitude Modulation (QAM). The modulation algorithm provides a special spectral shaping that [1] assures a low sensitivity to cochannel NTSC-type interference in the ADTV receiver and [2] assures a strong rejection of any ADTV signals by the NTSC receiver.

Representative Models for an HDTV Standard—General Instrument Corporation's All-Digital DigiCipher System

In June1990, General Instruments stunned the FCC and the TV industry by proposing the first all-digital HDTV. Not only did this proposal have a profound psychological effect on the industry, it caused a delay in the testing of the proposed systems because the FCC had not acquired the facilities to evaluate digital submissions. In its pioneering effort, General Instruments had solved the compression problem of how to digitally fit a nominal 20-MHz plus bandwidth into the width of a normal 6-MHz NTSC channel.

NOTE: These HDTV systems are not presented chronologically in order to show a more logical transition from the analog to the digital systems.

An Introduction to the DigiCipher System

This first system somewhat resembles some of its analog predecessors. The video format is 1,050-line interlaced (1,050/2:1) with a 59.94-Hz field rate and an aspect ratio of 16:9. YUV signals are obtained from RGB signals in the analog do-

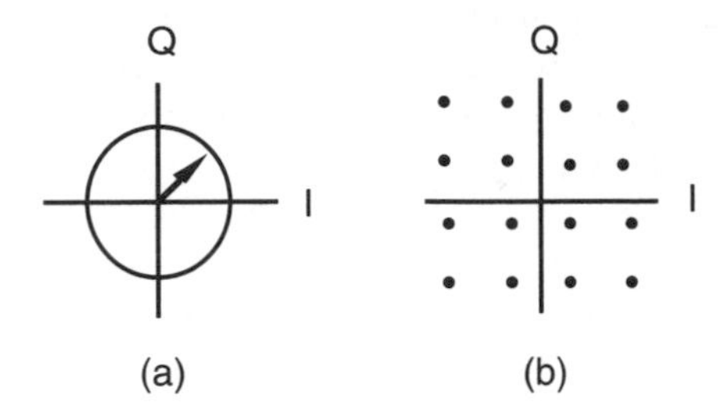

Figure 3.40 The I-Q (in-phase and quadrature) phase planes for analog and digital signals. (a) The position of vectors, which are created from plus or minus values of I and Q, may show up anywhere in the analog I-Q phase plane. (b) The positions of vectors created in the digital I-Q phase plane are limited to discrete locations. This figure shows the unique positions for a (4 × 4) 16-bit QAM signal.

main, are low-pass filtered, and converted to their respective digital counterparts with an A/D converter clocked at 51.80-MHz. At this stage, the video signals are compressed (which will be detailed in the next section) and formed into a serial data stream. Four channels of digital audio and four channels of digital data or digital text can also be multiplexed into this data stream. The data stream is further processed, including the addition of error correction, to become serial data at a rate of 19.42 Mbps. This serial data is changed into symbols and modulates the transmitter's picture carrier with 16-QAM (16-bit quadrature modulation) with a data rate of 4.86-MHz.

QAM is roughly the digital equivalent of general analog I-Q (in-phase and quadrature) modulation (see Figure 3.20 and 3.40). In the analog domain, ± I and ± Q signals can be combined to form a vector that can be placed at any point in the (continuous) 360° phase plane. For QAM, the positions of the vector combination of I and Q are discrete. For 16-QAM, there is a 4 × 4 matrix, uniquely clustered about the I and Q axes, that defines the only possible positions of the I-Q vector.

Compression in the DigiCipher System

The compression process in this or any HDTV system is much more complex than simply dropping out certain portions of information (which, hopefully, can be recreated by some interpolation process in the receiver). The following five steps are included in the overall DigiCipher compressing process.

1. Chroma Processing in the DigiCipher System

Because of the perception of color (chroma) by the human eye, the resolution of the chroma signal can be degraded, with the attendant reduction in bandwidth, without significant detection by the viewer. Therefore, as a bandwidth reduction technique, the U and V components are horizontally decimated by a factor of four and vertically by a factor of two. This decimation requires the application of pre-filtering prior to the decimation subsampling. The pixels are averaged in groups of four horizontally and in groups of two vertically. Since the vertical averaging is performed across two different fields, some degradation in motion rendition occurs. Because of the characteristic of the eye and the interpolation in the receiver, this is difficult to detect both in a static and moving picture.

2. The Discrete Cosine Transform (DCT) in the DigiCipher System

In optics, electronics, etc., there are a number of mathematical transforms that are used to make some processes more convenient, faster, or more understandable. One of the reasons to use the familiar Laplace transform is to provide equations that can be solved using algebra rather than having to solve those wonderful differential equations. The DCT transforms a picture, coded as a number of pixels (samples), into an equal number of transform coefficients. Since the coefficients usually represent differences, the digital codes of the coefficients are usually smaller than the codes of the original pixels. Thus, a serial digital stream of Cosine transform coefficient will most often contain far less total bits than the equivalent data stream of coded pixels. There are, of course, inverse Cosine transforms to return to the original Discrete Cosine transform Domain. The major problem in using the coefficients is that in the compression processing, rounding errors occur that make the inverse transformations or the interpolations inaccurate. (See any intermediate or advanced text on digital signal processing or, for TV, the Sandbank reference.)

3. Coefficient Quantization or Normalization in the DigiCipher System

Coefficient quantization is another process that is used to improve the coding efficiency and thus reduce the required bandwidth. The first quantizing step is to truncate the DCT coefficients (shift left and discard) to a lower fixed number of bits. Likewise, the number of bits can be further reduced or increased to maintain a constant bit rate. Again, if this process is carried to extremes, there is a chance for objectionable artifacts to appear in the received picture. The DigiCipher system has been specifically designed to prevent such artifacts from being visible at normal viewing distances (three times the picture height) and only occasionally visible at very close viewing distances.

4. Hoffman Variable-Length Coding in the DigiCipher System

Hoffman coding is the fourth processing step in the signal compression operation. It improves the efficiency of the compression coding by the use of *statistics*. It will be remembered that coefficient quantization or normalization improves the compressibility of an image by reducing the amplitude of the transform coefficients. However, depending upon the picture content, another algorithm is needed to vary, at certain times, the number of coefficient bits to maintain a constant bit rate.

Hoffman coding is used to provide the best statistical distribution of information in a data stream—given a priori knowledge of the probability of all possible events. (See the Fink and Christiansen reference.) The encoder can generate such probability distributions and send them to the decoder prior to the transmission of a given frame. A look-up table is used to derive Hoffman code words where relatively short code words are assigned to events with the highest probability of occurrence. In the receiver, an identical code book is used to match each code word with the actual event.

5. Motion Estimation in the DigiCipher System

Motion estimation algorithms can be divided into two general classes—those which focus primarily on extracting three-dimensional motion (temporal) parameters from a sequence of two-dimensional images or projections, and those which estimate velocities on a point-by-point or region-by-region basis without any consideration of how the object(s) is (are) moving as a whole. In the first case, a common formulation of the problem is to assume rigid three-dimensional bodies with movement patterns that can be described by a translation component, a rotation component, a zooming component, and a center about which the rotation or zooming is occurring.

It is clear that typical TV imagery is far too complex to be satisfactorily characterized by the first, sequential motion description. Therefore, although the common implementation of the second point-by-point algorithm is unintelligent (it has no understanding of the higher-level motion that is occurring), by using it to compare frame-to-frame motion in each superblock area (an area that is approximately 32 horizontal pixels by 16 vertical pixels), a satisfactory estimate of the motion can be obtained. Figure 3.41 shows the signal flow in the DisiCipher system.

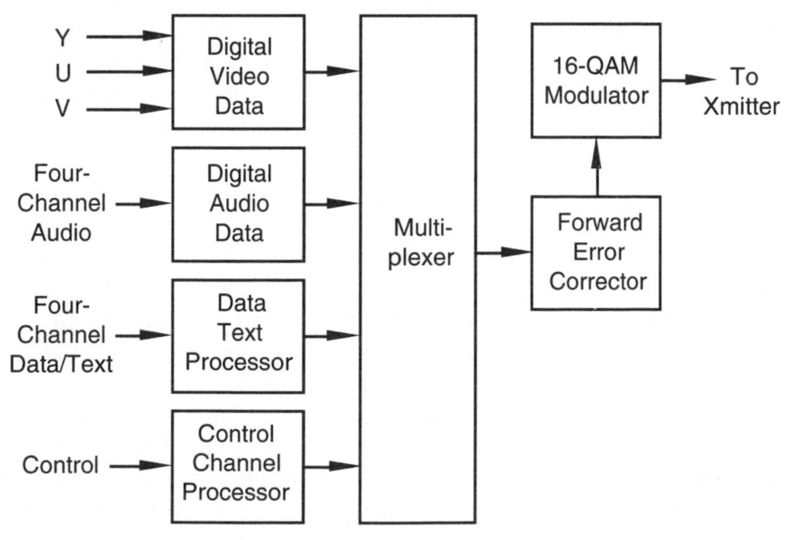

Figure 3.41 The signal flow diagram for General Instrument's DigiCipher HDTV transmission system.

Representative Models for an HDTV Standard—The Zenith/AT&T Digital Spectrum Compatible (DSC) All-Digital System

Like other HDTV proposals, the Zenith/AT&T digital proposal followed its original analog proposal. Also, like most of the other proposals, they relied heavily upon computer simulation. One of the major characteristics AT&T and Zenith claim for this 6-MHz, single-channel simulcast system is its channel compatibility with the present NTSC system. By especially computer shaping its DSC signal spectrum, it is claimed that their signal will significantly reject interference from any NTSC cochannel. This will permit an average 12-dB reduction in DSC radiating power to give a service area equal to a similar NTSC transmission. Likewise, the DSC transmission signal is more noise-like (a more even power spectrum) in character than an NTSC signal, which has further reduced the visibility of its interference into any NTSC channel. One of the other interesting characteristics of the DSC system is the input video's horizontal scanning rate. Although it uses a vertical rate of 59.94 fields per second, the horizontal rate is 787.5 lines per field. By using progressive scanning, the receiver presents a picture with 1,575 (unweighted) lines per field. The initial, noncompressed Y signal has a bandwidth of 34 MHz and the U and V signals occupy a band of 17 MHz.

The Zenith DSC Data Field and Frame

The DSC signal format (see Figure 3.42) somewhat resembles that of an NTSC signal. However, it is important to note that the details inside the digital format do not necessarily have the same names or interpretation as those in the NTSC format. For instance, a DSC Data Segment has a duration of 63.56 μs, just as a line in NTSC. In DSC terminology, the word "line" is reserved for the receiver the display parameter that has a duration of one-third of a Data Segment. Figure 3.43 shows how the audio, ancillary, (computer, text, etc.), and control digital data reside in the same data stream as the video. The Zenith system, like most others, does not require an aural transmitter, which therefore frees additional channel bandwidth and facilitates a less powerful transmitter and a smaller antenna.

The Zenith DSC Video Compression Encoder

Motion estimating is fundamentally a combination of analyzing both larger-image sequence differences and smaller differences in point-by-point, nonglobal areas. However, the complete motion analysis, including the production of the final transmitted vectors, is accomplished in three steps. In the first operation, the frame video signal is filtered and reduced in resolution decimated by a factor of two in both the vertical and horizontal directions. This reduces each block size, and thus the required search area, by a factor of four. Second, for these newly decimated blocks, course motion vectors are generated based on the frame-to-frame displacement differences and are sent to a second, finer estimator, which can predict the motion

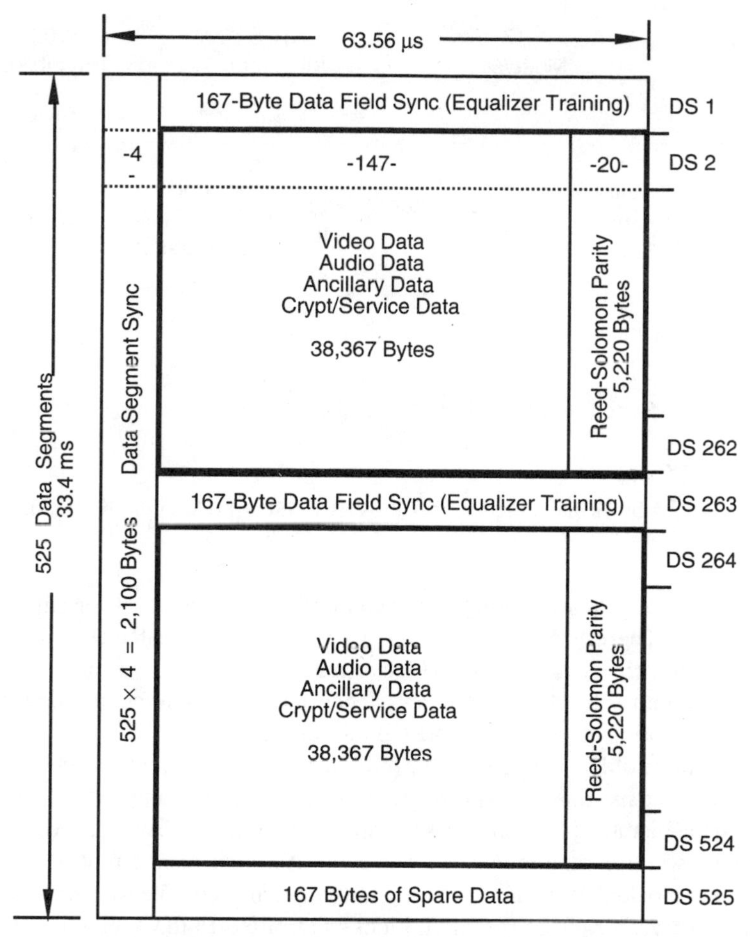

Figure 3.42 The DSC Data Frame (two fields) configuration. Each DS (Data Segment) is the same duration of an NTSC line. However, in the final receiver display, there are three TV lines for each Data Segment.

of small blocks to within subpixel accuracy. In this second step, each new frame is analyzed, prior to coding, to determine its rate of motion versus perceptional distortion and the dynamic range needed for each coefficient. Quantization of each transform coefficient is based on the importance of that coefficient based on a model of the human visual system.

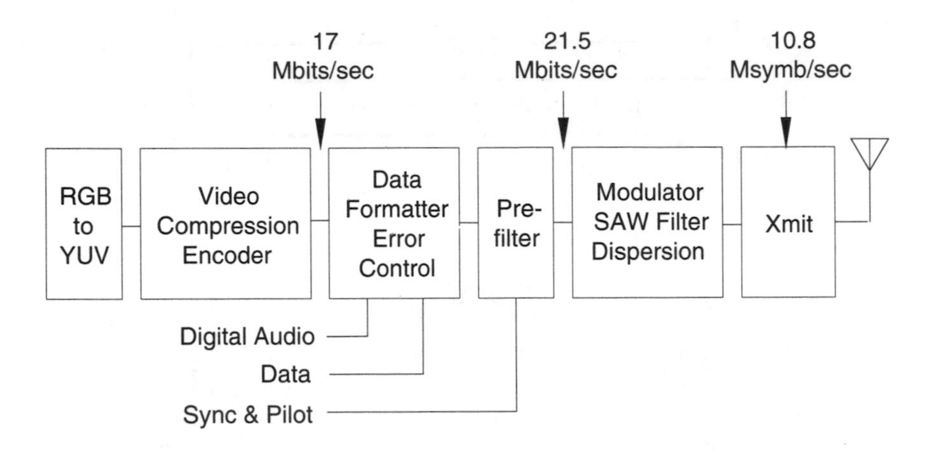

Figure 3.43 A block diagram of the Zenith/AT&T DSC HDTV studio-transmitter system.

The third motion-estimating step generates prediction errors for small blocks in the image. These errors are passed on to a prediction error calculator, which sums the prediction errors of these small blocks and, with the summed values, generates final motion error for the larger blocks. The final process in the motion estimator uses these motion errors to generate the motion vector.

Given the motion vectors from the motion estimator, the motion vector encoder must select the motion vectors that will give the best prediction of the frame while limiting the bit rate of the compressed motion vector data. This is achieved by sending two resolution motion vectors. The first set represents the motion vectors of the course blocks, which are unconditionally transmitted. The transmission of the other set of vectors from the small blocks is limited to the available remaining bandwidth.

The Zenith DSC Video Data Formatter and Error Controller

This section is used to combine digital audio and computer or text data to the compressed video data stream. It also adds error-checking codes compiled using a form of the Reed-Solomon redundancy error checking.

The Zenith DSC Prefilter

The major function of the prefilter is to preprocess the video so that, combined with further allied processing in the receiver, the effects of any NTSC cochannel signal will be minimized. Besides this function, the prefilter adds a special pilot signal that enables the receiver to recapture quickly the signal carrier in the presence of extreme interference or when the channel is switched. Also, encoded synchronization signals are added in this prefilter section. The data that is added in the prefilter section increases the overall bit rate from about 17 Mbits/sec to 21.5 Mbits/sec.

The Zenith DSC Modulator

The function of the overall modulator circuitry is to process the 21.5-Mbits/sec composite signal in a way that it can be squeezed into a normal U.S. 6-MHz TV channel. The method of modulation is single-sideband, suppressed carrier, where the position of the suppressed carrier would be located about 0.31 MHz up from the lower edge of the 6-MHz channel bandpass. The trick is to limit a pulse-amplitude modulation (PAM) signal to only four levels—a 2-bit symbol code—that can be accommodated in a 6-MHz channel.

Signal dispersion is one final nuance item that is added to achieve more closely the goal of a noise-like DSC signal character. Throughout the various stages of signal processing, special efforts are made to make the signal data steam as random as possible. The one glaring exception to the desired random signal is the necessary orderly occurrence of sync. By using dispersion, some of the normal random signals that either precede or follow the non-random sync codes are superimposed on the sync signals to give them a measure of randomness. This dispersion is achieved by using a special surface acoustic wave (SAW) filter to give the tailored "signal smearing" tilt. Complementary dispersion is implemented with a SAW filter in the receiver to restore the proper sync codes.

The Zenith DSC Receiver—The Tuner, IF, and SAW Filter

Figure 3.44 shows the DSC receiver system. The DSC receiver processing is essentially the reverse of the DSC studio-transmitter processing. The SAW filter provides complementary dispersion to restore a stable, noise-free sync signal. The synchronous demodulator uses the familial Frequency and Phase-Locked Loop (FPLL) circuit. The receiver functions of automatic carrier frequency control, automatic gain control, and automatic carrier phase control are all performed under control of the FPLL. The Synchronous Demodulator has in-phase and quadrature outputs.

The Zenith DSC Receiver—Data Clock and Data Field Synchronization, and the NTSC Interference Filter

The composite baseband data signal coming from the synchronous demodulator is sampled and analog-to-data converted. Next, the data clock and data fields are synchronized.

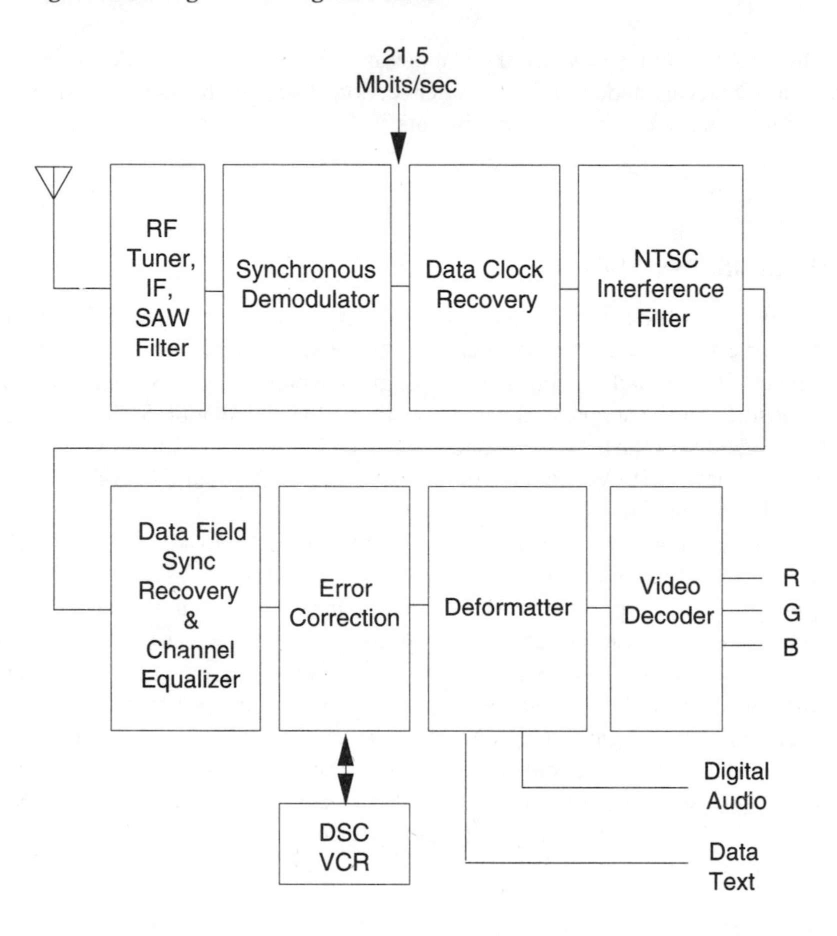

Figure 3.44 A block diagram of the Zenith/AT&T DSC HDTV receiver system.

Data clock synchronization requires the recognition of the precise instant of the zero crossing of the data segment sync in-phase component among all the other zero crossings from the constantly varying data patterns. This is accomplished in a correlator circuit, using the periodic recurrence of the identical sync quadrature component in every data segment. Random data patterns equal to the sync pattern could occur. However, they have such a low probability of repeating in a sufficient number of data segments in the same location that false detections are prevented. The circuit has an excellent signal-to-noise ratio. When the data segment sync is detected, the video data clock and the transmission data clock are synchronized.

After the clock timing is aligned, the data fields must be synchronized. Referring to the DS 1 and DS 263 lines in Figure 3.42, each data field sync consists of two 255-symbol pseudo-random sequences. Data field detection is achieved by comparing the incoming signal data to a local look-up table, data segment by data segment. Even under strong ghosted conditions of the incoming sequences, the correlation is high and the data field sync detection is secure. Ghost canceling for digital signals is usually more effective than the comparable processing for analog signals.

Since the transmitted sync is preceded and followed by random pulse signals, dispersion causes a portion of the preceding or following random signal to be superimposed on the nonrandom sync signal that gives it a measure of randomness. The complementary receiver dispersion restores the proper sync codes and spreads any sharp transitions in an interfering, cochannel NTSC signal—thus reducing cochannel interference.

The Zenith DSC Receiver—Error Correction

The Reed-Solomon code reduces the bit-error rate to negligible proportions. Moreover, it has been modified, based on a model of the human eye, to provide an optimized signal-to-noise and signal-to-interference ratio to the viewer.

3.8 TELEVISION SIGNAL PROCESSING—THE CHAPTER IN RETROSPECT

Television—The Keystone Chapter

The chapter on television is created as the keystone chapter. This chapter, coupled with the two introductory chapters on audio and radio and Chapter 4 on telecommunications, provides a firm foundation for signal and image processing and a springboard for the remainder of the book.

Television—The Process of Synchronization and Scanning

In its introduction, this chapter shows that television brought two new signal complications that had not been present in most radio transmissions. In TV, there has to be a direct and precise relationship between the time and position of the instantaneous camera pickup of a pinpoint of light from the studio scene and the time and position when the corresponding pinpoint of light is emitted by the receiver's picture tube. Also, because one part of the studio scene might be quite bright and an adjacent part might be quite dark, the television transmission and the receiver system have to routinely respond to very rapid changes in signal amplitude, again not usually encountered in radio transmissions. Thus, the dictates of picture timing and synchronizing, along with the rapid transient or pulse processing, required upgrades in both circuit theory and application. Because of these requirements, new relationships between picture resolution, transient response, and system bandwidth had to be developed and are presented. Likewise, special measurement techniques and waveforms are shown that are required to assure that a given TV processing component or system is performing to its design specifications.

The Additional Demands of Color Processing

The addition of color presented at least two major additional demands on the system and its signal processing. It is shown that the additional color information required more bandwidth and circuits with even better transient, phase, and delay performance. In the resulting NTSC system, the additional band space was provided, not by extending the bandwidth above the standard 6 MHz, but by filling in some of the "holes" in the existing bandwidth. The transient and phase response of the affected circuits was improved by using both a better design theory and a more advanced technology.

Improved Color Processing

This chapter further illustrates that the improvement or introduction of more advanced components and techniques such as the microprocessor, the surface acoustic wave filter, and expanded memory devices that could be used for comb filters, made it possible to design improved studio and receiving equipment that utilized more of the potential of NTSC. In the studio, digital memories improved the performance of TV tape with better time-base correctors and improved the performance of signals from nonstudio sources with the digital synchronizer. The comb and SAW filters increased the receiver's usable bandwidth, and thus the performance of the color channels, while reducing their crosstalk. The microprocessor made it possible to improve such things as picture sharpness and color balance or "flesh tones" and gave the viewer the convenience of a much improved remote control.

Processing for High-Definition Television

Like many topics in contemporary technology, a precise definition of high-definition is somewhat controversial. The presenter must specify such things as how "definition" is to be stated; what bandwidth will be required for the system; will it be compatible with the present NTSC color receivers; and will it be compatible with present TV transmitters, cable systems, and satellite systems. Several definitions are presented to illustrate the compromises required in any high-definition television system.

The concluding part of this chapter describes five of the HDTV systems that were submissions to the FCC. In all cases it was the intent to present them without any bias. Some of the systems are analog and some are all-digital, including the sound channels. Some are not even remotely NTSC compatible, some could be made to work (at a lower definition) with present receivers, and most are based on many of the principles, such as scan rates, of NTSC, but would require new receivers to present the picture in high definition.

Since it was the hope, if not the intent, of many in the U.S. TV industry to have a high-definition system in a restricted bandwidth—hopefully the present 6 MHz —the thread that runs through most of the proposals is the presentation of their techniques of bandwidth allotment and compression. In addition, with compression, each submitter had to use special algorithms to produce the proper picture motion in the compressed bandwidth. It is especially interesting to note that the data stream of many of the systems resembles the high-level communication data and protocol structures, such as ISDN, that are presented in Chapter 8.

Archival and Cardinal References

Anner, George E. *Elements of Television Systems.* New York: Prentice Hall, 1951.

Cochran, L. A. "The XL-100 ColorTrack System" *The RCA Engineer,* July 1976.

Corrington, M. S., and Murakami, T. "Applications of the Fourier Integral in the Analysis of Color TV Systems." *IRE Transactions on Circuit Theory,* vol. CT-2, no. 3, September 1955.

Ennes, Harold E. *Television Broadcasting: Equipment, Systems, and Operating Fundamentals.* Indianapolis IN: Howard W. Sams & Co., 1971.

Fink, Donald G. *Television Engineering.* New York: McGraw-Hil, 1952.

Harwood, L. A. "A Chrominance Demodulator IC with Dynamic Fleshtone Correction" *IEEE Transactions. on Consumer Electronics,* February 1976.

Horowitz, Paul, and Hill, Winfield. *The Art of Electronics.* Cambridge and New York: Cambridge University Press, 1983.

IEEE, *IEEE Trial-Use Standard on Video Signal Transmission Measurement of Linear Waveform Distortion.* New York: The Institute of Electrical and Electronics Engineers, 1974.

Rider, John F. *The Cathode-Ray Tube at Work.* New York: John F. Rider Publishing, 1945. A delightful early book, first printed in 1935 by one of the early publishers of radio and TV books.

Schade, Otto H. Sr. *Image Quality—A Comparison of Photographic and Television Systems.* Princeton NJ: RCA Laboratories, 1975.

Contemporary References

Benson, K. Blair, and Fink, Donald G. *HDTV Advanced Television of the 1990s.* New York: McGraw-Hill, 1991.

Benson, K. Blair, and Whitaker, J. (eds.) *Television and Audio Handbook for Technicians and Engineers.* New York: McGraw-Hill, 1990.

Benson, K. Blair, and Whitaker, J. (eds.) *Television Engineering Handbook*, rev. ed., New York: McGraw-Hill, 1991.

Dorf, Richard C. (ed.) *The Electrical Engineering Handbook.* Boca Raton, FL: CRC Press, 1993. (See especially the sections on the Discrete Cosine Transformon and SAW filters.)

FCC et al. The submissions of the various laboratories and organizations to be considered as an HDTV standard. 1990–1991. (These submitted documents became part of the public record and are available from the Federal Communications Commission, Washington, DC.)

Fink, Donald G. and Christiansen, Donald (eds.) *Electronics Engineers' Handbook 3rd ed.* New York: McGraw-Hill, 1989.

Gersho, A and Gray, R. M. *Vector Quantization and Signal Compression.* Norwell, MA: Kluwer Academic Publishers, 1991.

IEEE Transactions on Microwave Theory vol. 21, no. 4, 1973, special issue on SAW filters.

Isnardi, Michael A. "Multidimensional Interpolation of NTSC Encoding and Decoding" IEEE Transactions on Consumer Electronics, Vol. 34, No. 1, 1988.

Jayant, N. "Signal Compression: Technology Targets and Research Directions." *IEEE Journal on Selected Areas in Communications*, vol. 10, no. 5, 1992.

Lynn, Paul A., and Fuerst, Wolfgang. *Introductory Signal Processing with Computer Applications.* Chichester and New York: John Wiley & Sons Ltd., 1989.

Netravali, A. N., and Haskell, B. G. *Digital Pictures: Representations and Compression.* New York: Plenum Press, 1988.

Nelson, M. *The Data Compression Book*. San Mateo, CA: M & T Books, 1991.

Nilsson, James W. *Electric Circuits.* Reading, MA: Addison-Wesley, 1986.

Prentiss, Stan. *HDTV High-Definition Television.* Blue Ridge Summit, PA: Tab Books, 1990.

Roody, Dennis, and Coolen, John. *Electronic Communication.* Reston, VA: Reston Publishing Co., 1984.

Sandbank, C. P. (ed.). *Digital Television.* Chichester: John Wiley & Sons Ltd., 1990.

4

Telephony

From "Mr. Watson, Come Here; I Want You," to Satellites, Cellular Telephones, and Global Telecommunication Standards

Analog Telephones and Transmissions

4.1 AN INTRODUCTION TO THE TELEPHONE AND FUNDAMENTAL TELEPHONE TECHNOLOGY

Alexander Graham Bell (1847–1922) was the son of the internationally known inventor of visible speech, a system of visual symbols that indicated the positions of the human speech organs in the production of any vocal sound. As a professor of vocal physiology, the younger Bell not only made some improvements on his father's techniques and invention, but was also involved in speech training for the deaf. With the help of a professional electrician, Thomas Watson, he first worked on what he called the harmonic telegraph as an aid to send aural messages to his deaf students. These experiments, with the help of the inventor's serendipity and the legendary fortunate accident with acid, led to the creation of a revolutionary device that would transmit and receive the spoken word over electric wires. Although the first few public demonstrations were perceived as some sort of trickery, within about five years, there were telephone systems in several eastern U.S. cities.

The Principles of the Very Early Telephones

The very early telephones used a single instrument for both the transmitter and the receiver. The single transmitter-receiver can be analyzed both as an elementary electric generator and an elementary electric motor. As a transmitter, the voice (sound) waves would slightly vibrate the diaphragm, placed in an electromagnetic field, which, in turn, would produce a miniature electromagnetic force (EMF) or voltage. As a receiver, the incoming pulsating EMF would produce a changing magnetic field that would vibrate the diaphragm and reproduce the original sound waves. Later, a more effective system used a separate and more efficient transmitter, usually with a carbon element as well as the electromagnetic receiver.

Very Early Telephones

For an electromagnetic (moving-coil) transmitter-motor or receiver-generator:

1. When used as a simple electrical (transmitter) generator, the output voltage of an elementary electromagnetic telephone, caused by the sound-activated motion of the diaphragm, is:

$$emf = Bl\,(vol) \tag{E 4.1}$$

where:

B = the flux density in the air gap, in gauss
l = the *equivalent* length of the diaphragm, in centimeters
vol = the velocity of the diaphragm, produced by the input sound, in centimeters per second. Note that, because the diaphragm acts as a single conductor cutting the magnetic field, the efficiency of this type of telephone transmitter-generator is very low.

2. When used as a simple (receiver) motor, the force moving the sound-producing diaphragm of an elementary electromagnetic telephone is:

$$f = B\,l\,i \tag{E 4.2}$$

where:

B = the flux density in the air gap, in gauss
l = the length of the conductor (the diaphragm) in meters
i = the coil current in amperes

(The above two equations are the elementary generator and motor equations. They do not include the complex actions of the sound waves entering or leaving the diaphragm with its involved acoustical and mechanical parameters. See the microphone and loudspeaker sections in Chapter 1.

3. For a carbon-element telephone transmitter, the AC-voltage output, assuming a quiescent DC current, is a form of Ohm's law as:

$$emf = \Delta e = I \Delta R \tag{E 4.3}$$

where:

Δe	=	the changing voltage caused by the changing resistance
I	=	the quiescent DC flowing in the carbon element
ΔR	=	the change in the resistance of the carbon element due to the motion of the diaphragm from the effect of the sound waves. Note that the carbon transmitter was much more efficient than the elemental moving-coil transmitter, i.e., there is much more voltage output for a given sound pressure input. Again, this simple equation does not include any acoustical or mechanical parameters. (See Chapter 1.)

It is interesting to note the care that was taken to design the mouthpiece on the carbon-button transmitter, as illustrated in Figure 4.2 (b). The flared (exponential) shape of the mouthpiece not only had a pleasing appearance, it served as a tapered transmission line to match the acoustical impedance of the air to the acoustical impedance of the cavity containing the diaphragm. The horns on loudspeakers, as well as the canals in the ears, use the same principle to improve their efficiency. (See the section on mechanical and acoustical resistance and impedance in microphones and loudspeakers in Chapter 1.)

4.2 EARLY TELEPHONES AND THE DEVICES AND CIRCUITS THAT MADE POSSIBLE TELEPHONE NETWORKS

Very Early Telephone Instruments and Circuits

The first telephone instruments did not use a separate transmitter and receiver but only one combined transmitter/receiver, as sketched in Figure 4.2 (a). In the first year or two, the exact shape of the instruments changed in rapid succession. For instance, the units that were demonstrated at the 1876 Philadelphia Centennial resembled a large, horizontal flashlight with an invented megaphone on one end.

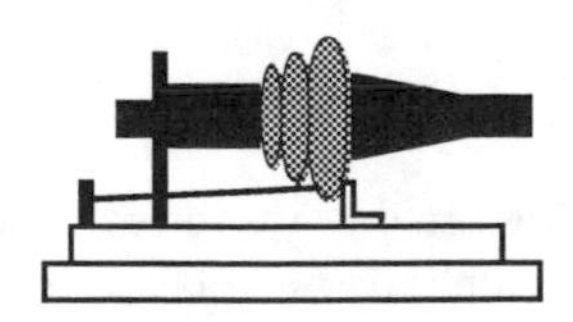

Figure 4.1 A sketch of the early telephone configuration that was demonstrated at the 1876 Philadelphia Centennial.

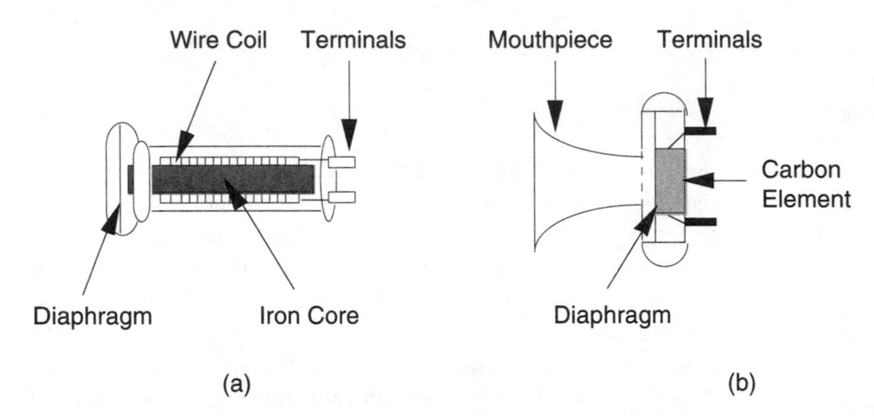

Figure 4.2 (a) A cutaway sketch of a simple telephone receiver/transmitter. As a transmitter, the sound-actuated motion of the diaphragm, in the electromagnetic field, will generate a small EMF, which is the analog of the input voice-wave pressure. As a receiver, the changing magnetic field, caused by the incoming line current in the coil, will move the diaphragm to produce sound waves. In later designs, permanent magnets were added to improve the sensitivity. (b) A sketch of a simple carbon-button transmitter. The changes in air pressure on the diaphragm caused by the voice waves compress and release the carbon particles that, in turn, vary the resistance of the carbon-button element.

The first demonstrations were simple, sometimes only from room-to-room, using only elementary two-wire circuits (Figure 4.3). It must be remembered that in these early days of the telephone, there were no vacuum tubes or transistors to amplify the sound-produced current as it traveled through the inherent resistance of the connecting lines. The power to excite the electromagnets, and later the carbon transmitters, was supplied from carbon-zinc or lead-acid storage batteries. Once the telephone started to proliferate there were several basic problems, including connections, signaling, and switching, that had to be solved as the number of users continued to increase. A few telephone circuits used one wire and a ground or earth return. However, the earth is usually a very noisy conduction path as well as having a varying impedance. For these reasons, the use of a ground-return system is generally considered unsatisfactory. Some of the very interesting early developments that helped expand telephone technology were the magneto (bell) ringer, the telephone dialer, and the three-for-two party line phantom circuit.

The Magneto Ringer

The magneto ringing generator (Figure 4.4) was an early reliable signaling device used to actuate a called party's telephone bell. The signaling bell required more current than could be reliably and economically provided by the batteries that were normally used to power the telephone conversations. A special hand-cranked generator would supply this current and at a frequency especially tailored to effectively ring the bells. It would automatically disconnect from the line when not in

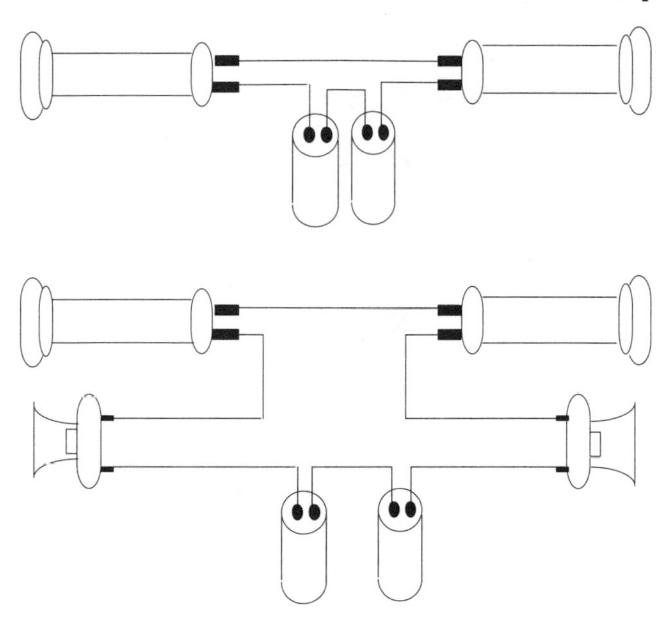

Figure 4.3 The connection diagrams of two early telephone circuits. These circuits were not efficient. It took several years of research by AT&T and Western Electric, Automatic Electric, Kellogg, Stromberg-Carlson, and many others, to develop the ideas of correct impedance matching. (See the Libbey reference or any text on circuit design and analysis.)

use. The magneto generator was used in several system configurations. Often, it was simply used to alert "central," the switchboard operator, who in turn would ask for the number you wished and ring that telephone. The magneto ringer was also used as a convenient signaling device for multiuser or party lines—one ring for the Joneses, two rings for the Smiths, etc. (It is interesting to note that many times a party line would enable all involved the opportunity to listen in. More often than not, this was considered a convenience and part of the neighborhood security system rather than an unwanted invasion of privacy.)

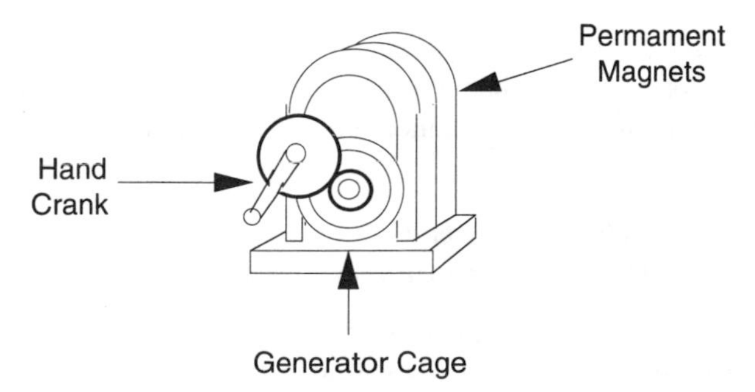

Figure 4.4 An early magneto ringing generator. Such a magneto was typically housed in the highly polished oak telephone set cabinet with only the hand crank visible.

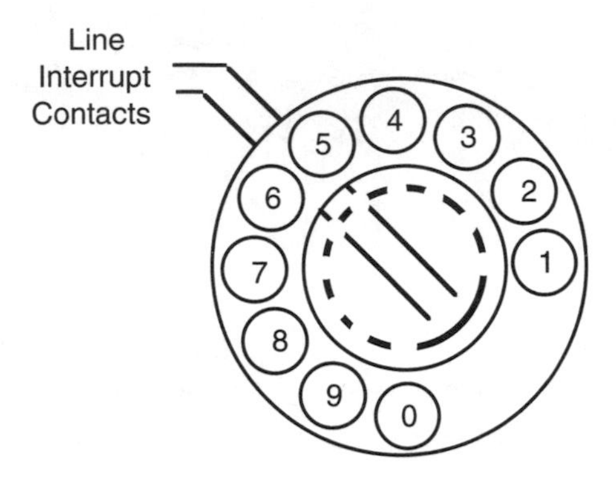

Figure 4.5 A graphic of a telephone dial with a schematic representation of the internal pulser or interrupt mechanism. The disconnect that prevents any pulses resulting from the clockwise motion of the dial is not shown.

The Telephone Dialing Mechanism

As telephones began to proliferate, it was necessary to invent some signaling and addressing device to provide a means to automatically call the desired party. The dial or dialer, shown in Figure 4.5, proved to be a very clever and reliable device to fulfill this requirement. The unique feature of the dialing mechanism was that it was not connected as the caller used his or her finger to rotate the selected number clockwise. Thus, a rapid or a slow dialing action did not matter. Only when the dial was allowed to return at an unattended, constant rate did the dialer send out the selected number code. Each time, in its counterclockwise return, the dial would then send a constant-rate pulsed code that would be received by the central switcher and used to help select the number or switcher address of the called party.

The Three-for-Two or Phantom Circuit

As telephone use expanded, several clever circuit configurations were developed to provide more users per two-wire circuit—often termed a line pair. One approach was the three-for-two or circuit (see Figure 4.6). By using transformers (telephone companies called them repeating coils) or inductors in a special bridge circuit, a third circuit could be obtained while still using just two line pairs.

Higher-Capacity Telephone Circuits—
Carrier Current Modulated Transmission Systems

The Figure 4.6 phantom technique produced a 50 percent (three-for-two) gain in line-pair utilization. However, in the first half of the twentieth century, the demand for line capacity and utilization was as great as, if not greater than, it is

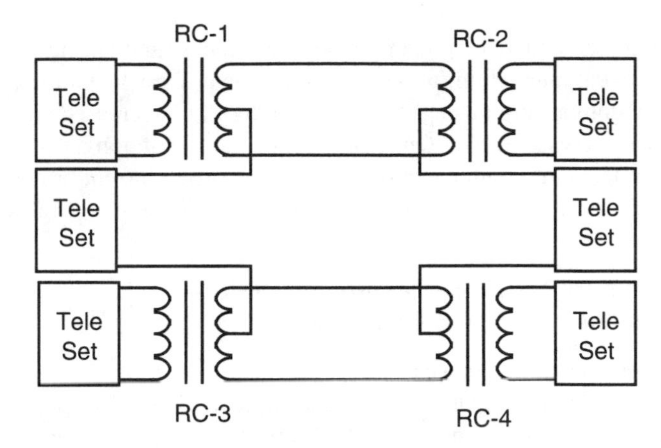

Figure 4.6 The fundamental two-line-pair phantom circuit. The two telephone sets in the top row use RC-1 (Repeat Coil-1) and RC-2 to complete their circuits. Likewise, the bottom two sets exclusively use the normal transformer action of RC-3 and RC-4 for their circuits. However, the middle two phantom set pairs use a combination of all the repeat coils for their circuits. Since the phantom circuit uses the center taps of the repeat coils, their currents are balanced (nulled) out and do not affect the top and bottom normal set pairs.

today. Another approach to higher line-pair utilization was a radio-related scheme termed carrier current modulation. Using a device called a balanced modulator, the voice signal from one telephone input could be scaled up in frequency and made to ride on a second higher-frequency signal termed a carrier. Using this idea, the modulated carrier would be sent over the line pair rather than the original unmodulated signal. Likewise, a second and third voice input could also be modulated with a different carrier and, in turn, the other modulated carriers could also be sent over the same line pair.

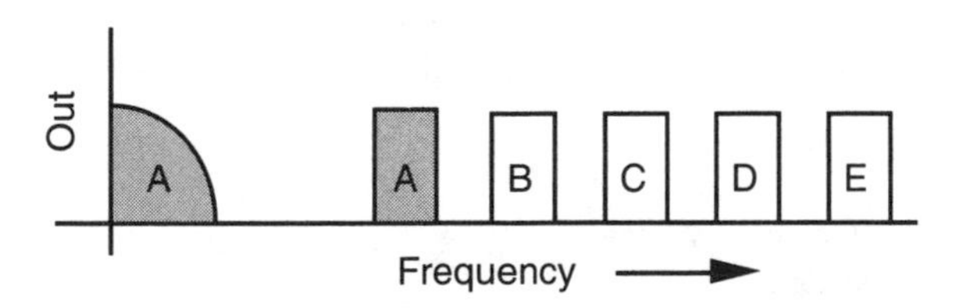

Figure 4.7 The frequency allotment of different voice-channel inputs in a carrier-current modulation system. The voiceband A is the normal low-frequency or baseband spectrum of a simple telephone voice output. The spectrum A' represents the modulated band of A, which is now modulated on a higher-frequency carrier. The modulated bands B', C', D', and E' represent other modulated telephone inputs that can also be transmitted on the same line pair. Both the telephone baseband and the modulated-band spectrum were, and still are, typically 3 to 5 kHz.

By the judicious use of proper carrier frequencies and special band-limiting filters to prevent interaction from one modulation process to another, the number of two-way conversation using one wire pair could be increased significantly. A per-line-pair increase of over ten to one in the number of conversations was possible. It is interesting to note that the early term carrier current has been modernized and this technique is now known as frequency division multiplex or FDM.

Higher-Capacity Telephone Circuits and Devices— The Two-to-Four-Wire Terminating Set

The complete integration of a carrier-current system incorporates several building blocks not yet described. One of the most important is the four-wire terminating set given in Figure 4.8. This or similar circuits make possible two-way (sometimes termed east-west) communication using a modulated two-wire system. This particular transformer, with its scheme of carefully balanced windings, is often termed a hybrid coil.

Higher- Capacity Telephone Circuits and Devices— The Diode-Bridge Balanced Modulator

To facilitate the carrier current modulated transmission systems, illustrated in Figure 4.7, required a simple and inexpensive modulator and demodulator for each voice channel transmitted. Actually it required two modulators and two demodulators—one modulator-demodulator pair was required for sending and one for receiving. The simple copper-oxide diode bridge modulator was small, rugged, and inexpensive. Like the very early crystal or "cat's whisker" detector, the copper-oxide rectifier and the selenium rectifier were semiconductor-type devices before the age of semiconductors. The copper-oxide bridge shown in Figure 4.9 provided an extremely simple type of balanced amplitude modulator.

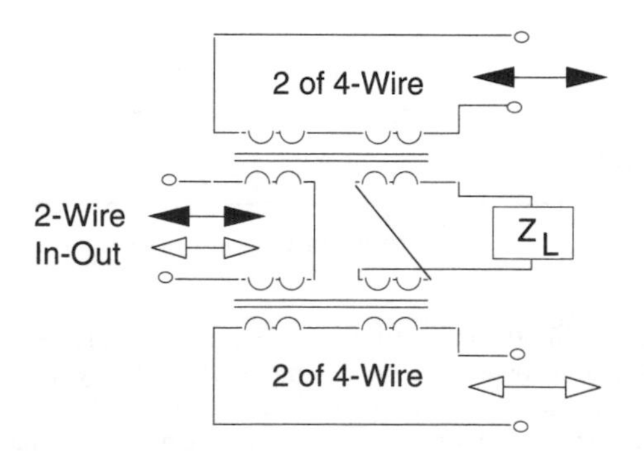

Figure 4.8 The two-to-four-wire termination that was used in analog repeater amplifiers and carrier systems. The impedance Z_L was the characteristic impedance of the (two-wire) transmission lines—typically 600 ohms.

Conventional amplitude modulation, such as is used for standard AM broadcast, results in a modulated carrier that contains three frequency components—the original carrier, the carrier plus the baseband voice frequencies (the upper sideband), and the carrier minus the voice frequencies (the lower sideband). These three frequencies must be transmitted for proper reception with normal AM radio receivers. In a balanced-modulating process, the carrier is balanced or nulled out, leaving only the two sidebands. Likewise, before transmission, either one of the sidebands can be selected with a bandpass filter (BPF), which will automatically eliminate the other one. Thus a carrier-current system, as shown in Figures 4.9 and 4.10, uses a minimum amount of bandwidth for each voice transmission.

For broadcast-type AM with maximum modulation:

$$e_{mod} = \sin\omega_c + \frac{1}{2}\left[\cos(\omega_c-\omega_m) - \cos(\omega_c+\omega_m)t\right] \quad \text{(E 4.4)}$$

where:

e_{mod} or f_{mod} – the final modulated wave

f_c or ω_c = the carrier frequency

f_m or ω_m = the modulating frequency

For a balanced modulator for carrier-current AM:

$$e_{mod} = \cos(\omega_c-\omega_m) - \cos(\omega_c+\omega_m)t \quad \text{(E 4.5)}$$

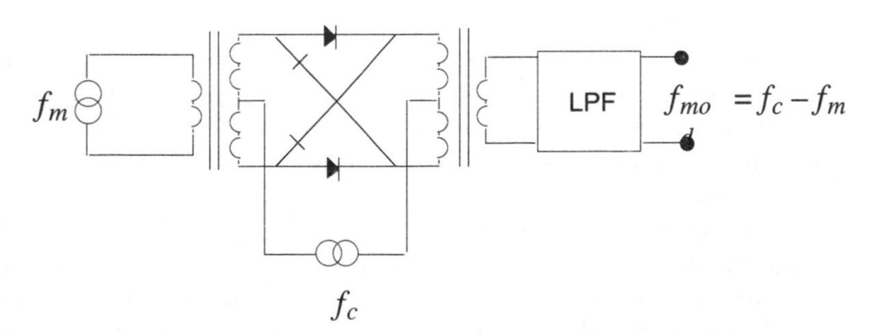

Figure 4.9 The diode-bridge balanced modulator*:* the circuit for a copper-oxide balanced modulator. By changing the input frequency from f_{in} to f_{mod}, the device can also act as a demodulator.

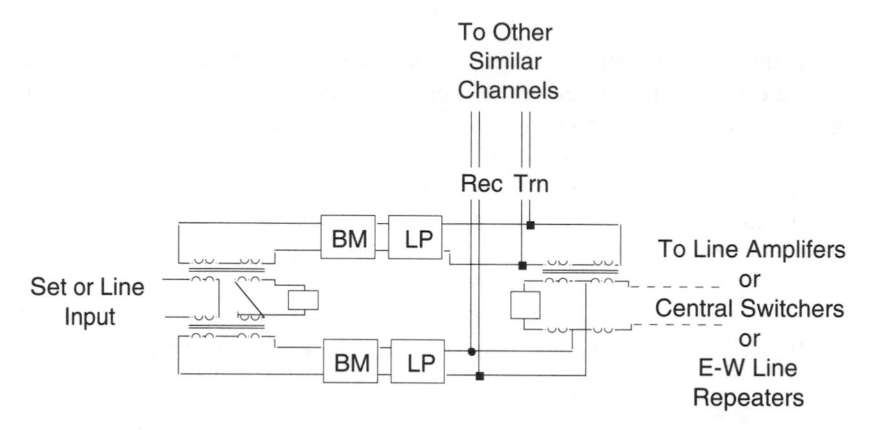

Figure 4.10 A complete balanced-modulator carrier system: a simplified schematic of a composite carrier system. Note the special uses of the hybrid coil to differentiate between sending and receiving signals and east-west signals. There are numerous variations and refinements to this elementary illustrative circuit. The small unmarked boxes represent the impedance-matching loads.

Higher-Capacity Telephone Circuits and Devices—A Complete Carrier System

By using the building blocks of the two-to-four-wire terminating set, the balanced modulator and bandpass filter, and some additional amplifiers and equalizing filters, a complete two-direction carrier current telephone system can be created. One of the very convenient features of the two-to-four circuit is its ability to differentiate between incoming and outgoing signals.

4.3 SIGNAL ROUTING AND MECHANICAL SWITCHING SYSTEMS

Electromechanical Switches

The concurrent introduction of the circular telephone dialer and the pulse-selectable electromechanical switch made possible a completely automatic routing system. The original pulse-selectable switch, the Strower step-by-step switch (see the 1981 Freeman reference), resembled a cascade of ten half-circle rotary switches. The incoming pulses could choose any of the ten decks and then select one of the ten switch-pair contacts on any given deck—ten decks with ten contact pairs allowed for 100 telephone line connections. A later and more universal switch, the crossbar switch, was constructed in the form of a grid with ten horizontal rows and twenty vertical columns to create a 200 switch-pair matrix. Although there were several versions of both the step-by-step switch and the crossbar switch, the former would usually permit only one line connection at a time, whereas the latter would support twenty simultaneous connections. The selection of the decks or columns and the contacts in a row was done by mechanisms actuated by electromagnets. Once the connection was made, it would be latched and the mechanism was available to respond to another caller and choose another switch contact (termed a cross-point in the crossbar switcher). Even in the early part of the twentieth century, each central exchange would require hundreds, if not thousands, of these high-maintenance electromechanical switches. There was also a "preswitch" or line-finder switch used to assure the caller a clear line to the bank of switches (see the Roddy and Coolen reference).

Telephone Circuit Routing and User-Number Codes

A much better understanding of the switching routing and hierarchy can bc obtained by investigating the evolvement of the United States telephone numbering/ number system. The model number will be 1-123-456-7890 where the 1-123 can represent the present-day area access and code, the 456 the local exchange code, and the 7890 can represent a typical individual user code. Historically and logically, the use of this model number should bc cxamincd starting on the right rather than the left.

As has been inferred, the local exchange is the foundation switching unit. The development of first local, and then regional, telephone networks were built around the local exchange. Although some exchanges, be they in a small town or in a small section of a major city, may serve less than 10,000 customers, the four-digit user code will permit 10^4 or 10K lines. Besides the four-digit user code, each local 10K line-switch group also requires its particular identifying three-digit exchange code. If a given locality continues to grow and needs more phone access, an additional switch bank must be added. This usually contains an additional 10,000 (10^4) switches (user codes) that can be accessed with a new three-digit exchange code.

The addition of the three-digit area code (and later a prefixed "1-" access code) made possible direct short- or long-distance dialing to almost any telephone in the United States. It is important to note that not only does this area-coding system put an additional burden on the switching and routing capabilities of the local and regional telephone networks, it also necessitates a national system of customer toll changing—a rather nontrivial task. (After the breakup of AT&T in 1984, when other carriers were added, the billing overhead was greatly expanded.)

Dial Coding for International Calling

In theory, all national telephone number codes (including those of the United States) should adhere to the 12-digit limit set by the on International Telephone and Telegraph Consultative Committee (CCITT) or Comité Consultatif Internationale de Téléphonique et Télégraphique. Most of the intercountry telephone numbers in upper North America do not exceed this 12-digit limit. However, some European intercountry numbers, including the international prefix, the called country code, the city code, and the subscriber number, add up to 13 or more digits. European number code systems vary widely from country to country. Sometimes the number of combined digits in the exchange and subscriber codes will change depending on whether they are being called locally, in the general region, or to another country.

Tariffs in the International Community

There are two major methods of automatic billing—bulk billing and detailed billing. Bulk billing is a simplified method that uses a timer on each user line. The number of billing "ticks" per minute is regulated by the exchange or area code that is dialed. Thus, a regional call might produce one billing-rate tick per minute while a long-distance call might produce three or five ticks per minute. This is a simple and inexpensive way to produce a bill but, because it tends to be quite imprecise, there are numerous subscriber complaints. In North America the common method is a more detailed billing. Whereas the installation and administration costs are relatively high, it does provide the customer and the company with an accurate, detailed account of each toll call.

4.4 TWO-WIRE TRANSMISSION LINE SIGNAL PROCESSING

The care and feeding of telephone transmission lines, and their amplifiers, represented a major portion of the communication-circuits literature in most of the first half of the twentieth century. Using two wires to transmit and receive telephone calls for a few blocks, or even a few miles, is usually quite simple. However, as the distances and complexity of the two-wire and four-wire network materially increased, it was necessary to both understand and implement more efficient line-usage techniques.

One of the first improvements resulted from understanding the wave nature of long transmission lines. Besides carrying the voice-actuated electric current, a transmission line can be visualized as a conduit carrying a fluid that is prone to wave motion. If the line characteristics and the characteristics of the sending generator and the receiving load are not properly controlled, the line will have standing waves—reminiscent of ocean waves—that will cause an unnecessary loss of signal energy. To reduce or eliminate the standing waves, the impedance of the generator and the load should equal the characteristic impedance of the line. (See any text on communication circuits.)

Even if a transmission line is properly source and load terminated, there are still losses at the higher-frequency end of the voice spectrum. Above 1 kHz, these losses will cause frequency distortions and result in a general lack of signal fidelity. The classic way to reduce these losses is with loading coils (inductors). By using special added inductors at intervals along the lines, the amplitude of the mid- to higher-range voice frequencies is increased at the expense of the much higher, and usually nonessential, frequencies—see Figure 4.11.

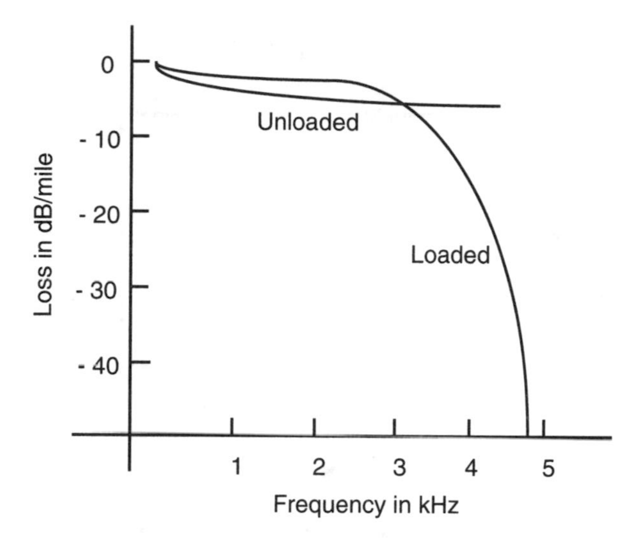

Figure 4.11 The typical effect of loading coils.

Signal Processing to Reduce or Cancel Line Echoes

Most user sets (the transmitter and receiver) incorporate special circuitry that will allow the user to hear, at a reduced level, what he or she is saying as well as what the other connected party is saying. This feedback of the user's voice is termed a sidetone and is psychologically necessary to assure the user that the system is working properly. However, if this sidetone, or any other replica of the talker's voice, is delayed, its presence can be very disconcerting. When a long-distance telephone call travels over several hundred or even several thousand miles, there is always a distinct possibility that one of the talkers will hear an annoying reflection (echo) of his or her voice.

Any electrical transmission medium, such as television twin-lead, a coaxial cable, or a two-wire telephone line, requires a finite time for the signal to travel through from end to end. For very short two-wire telephone lines the delay may be in nanoseconds (10^{-9} seconds). However, for a very long line with loading coils, other equalizers, and repeater amplifiers, the total delay may be in the order of tens of milliseconds (10^{-3} seconds). If the circuit uses two-to-four wire hybrids, there is a high probability that there will be delayed line pair-to-line pair crosstalk and a disconcerting echo will be created. In addition to the echo problem, there may also the so-called singing problem. The singing problem is analogous to the whistle that is often heard in public-address systems. Both of these problems are created when the gain of the amplifier is too high or something has caused an unexpected change in the amplifier's or system's delay or bandwidth. In telephone circuits, these unexpected changes often can be caused by switching to a new line with an improper gain, a different delay, or improper equalization.

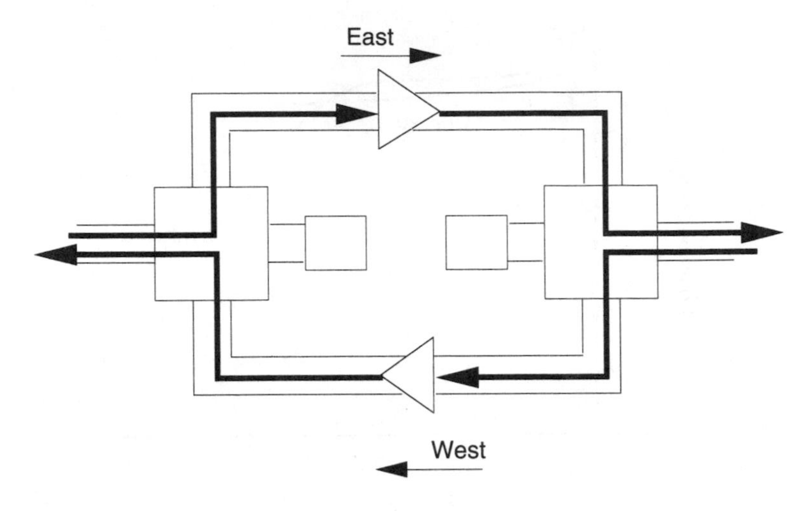

Figure 4.12 An elementary closed-loop two-to-four-wire circuit with repeater amplifiers in both directions. Although this graphic shows the east and west lines completely isolated, in practice this might not be true. There are many opportunities, such as mismatch, for echo-producing crosstalk.

There are two common techniques for solving the echo and the singing problems. In simple terms, one is known as the reduction technique and the other the elimination technique. The reduction circuit samples the energy in both the talk and listen sides of the network. If an energy detector (ED1) senses the energy

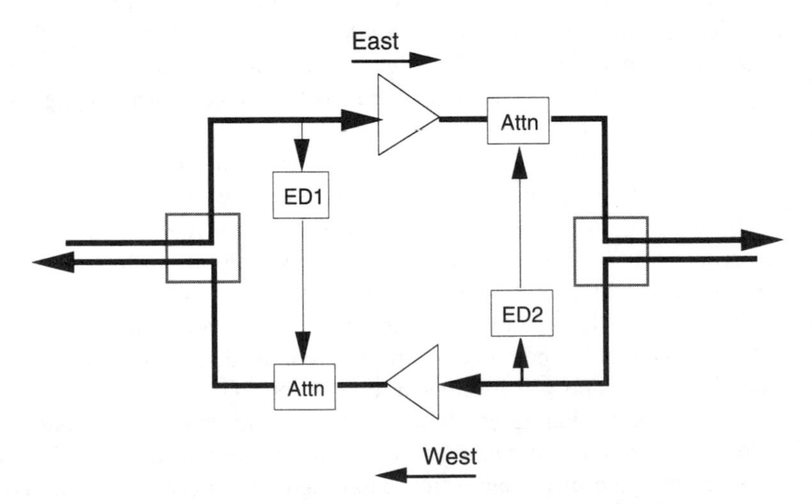

Figure 4.13 An elementary circuit for echo suppression adapted from Figure 4.12. The amount of attenuation in either "Attn" block is controlled by the output of the "ED" energy detector. If the power level exceeds a preset threshold in the talking (annoying) side, a high amount of attenuation is inserted to greatly reduce the listen-side energy.

above a preset level in the talker's side from an echo, it places a high attenuation in the incoming listener's echoing side to prevent its reception by the talker. However, when the roles are reversed and the listener becomes the talker and the old talker become the new listener, there may be some time delay until the energy sensor removes the line attenuation. If there is a delay, the first few syllables of the new talker's conversation may be very weak or missing. This general type of echo suppressor also changes the circuit from a full-duplex (two-way) to a half-duplex (one-way) communication circuit—see Figures 4.12 and 4.13.

The echo canceling circuit shown in Figure 4.14 is a much more complex circuit but it almost eliminates echoes. It is especially interesting because this is an early (and now a classic) example of a neural network. Neural networks require careful design, based upon the specific application, and they also require real-circuit training for precise operation with a wide variety of changing input conditions. (See the section on the Adaline network in Chapter 10.)

Digital Telecommunications Circuitry

4.5 WHY DIGITS?

An Introduction to Digital Telephony—Why Digits?

It is interesting to speculate on the mindset of the technical personnel and the planners in the world's telephone community at the dawn of the digital age—circa 1955. Could any of their dreamers have anticipated the revolution that the industry has seen? For background, and admittedly with more than a little hindsight, what were the major reasons for the gradual shift to the nearly all-digital systems we use today?

1. *Channel capacity and cost:* Even with the advent of larger and more efficient coaxial cable and microwave links, analog multiplexing technology might have been inadequate, and certainly increasingly more expensive, to meet the demands of the growing telecommunications industry.

2. *A much smaller, more versatile technology:* Replacing the electromechanical switches with their included electromechanical logic not only saved millions of dollars in building and maintenance costs, but also provided a much more reliable system that could be more conveniently updated because of expansion and technology updates. A generalized digital format makes quality monitoring and billing easier and more reliable.

3. *Ease of signaling:* Special nonmessage control codes for information such as on or off the hook, coin deposits, and special addresses can be inserted without disturbing the content or sequence of the main message. Both the digital control and digital message formats made possible a wide variety of additional services. These services, such as call waiting, three-party calls, etc., constitute the Integrated Services Digital Network—ISDN.

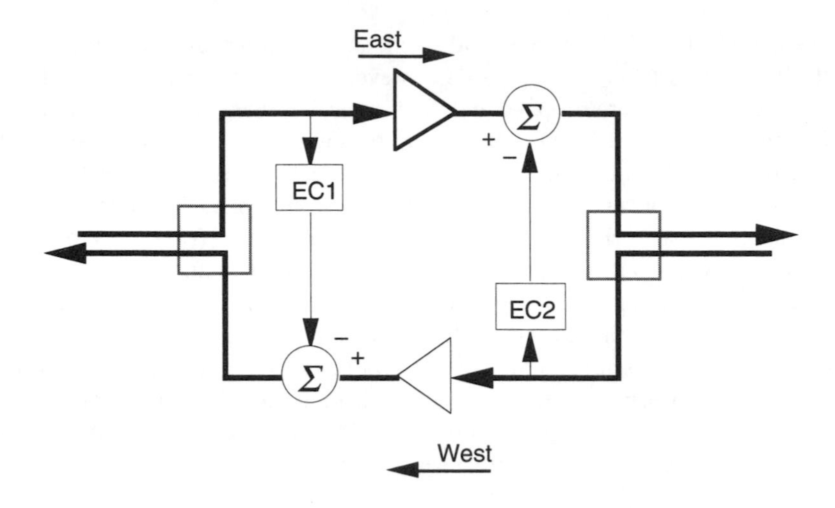

Figure 4.14 The scheme of an echo-canceling circuit. This early neural network-type circuit was invented by Bernard Widrow of Stanford University (circa 1960) and updated versions are still widely used today. Twenty-five years after his original invention, Widrow again demonstrated it to the delighted attendees at a neural networks computer conference as a very old example of the new technology of neural-network processing. The "EC" echo-canceling circuits require initial training. (See the Nelson and Illingworth reference.)

An Introduction to Digital Telephony—A Comparison of Analog and Digital Modulation Techniques

It is somewhat difficult to exactly equate apples to apples when comparing these two modulation techniques. For instance, the cost of the analog FDM technique is relatively inexpensive when it uses frequencies under about 500 kHz. However, above this general range of frequencies, the cost of analog components for the modulators and demodulators and line-repeating amplifiers increases. Likewise, higher-frequency switchers must include designs that prevent undesirable losses. Also, there probably will be some digital signals (computers, etc.) on any analog system and usually many analog (A-to-D converted) signals on any digital system. Nevertheless, for perspective, it is useful to review and compare both the analog FDM and the digital time division multiplex modulation (TDM) methods as they were and are used in telephony.

Telephone FDM and TDM Two-Wire Circuits

Earlier sections showed the use of both the nonmodulated phantom circuit and the FDM technique, including Figures 4.6 and 4.7, as methods for increasing the number of coexisting channels in a two-wire line. It will be remembered that because of the telephone transmitter and receiver and loading coils, the elemental two-wire base bandwidth was limited to under 4 kHz. However, the response of an unencumbered, unloaded two-wire line, while not flat (see Figure 4.11), will extend well beyond 4 kHz and only be down in the order of 15 dB at 1 MHz. By using equalized repeater amplifiers, the usable (nominally flat) bandwidth of such a two-wire line can be in the order of 1 MHz. On the basis of this information, AT&T designed an early benchmark analog FDM system, the A5, with twelve channels that occupied a frequency band of about 60 to 108 kHz. Likewise, five of these A5 groups could be upscaled to form a super group of sixty channels that occupied a band of 312 to 552 kHz—still well under the nominal 1 MHz top limit.

One of the first (1962) AT&T TDM digital systems, the T1 (sometimes also referred to as the DS1), had twenty-four voice circuits and required a bandwidth that would allow a bit rate of 1.544 Mbps. Several DS1 systems could be combined to form 48 and 96 channels that required two or four times the band rate and the bandwidth. (See the relationships between the pulse width and the bandwidth in Chapter 3.) The digital repeater amplifiers, also called regenerative repeaters, are used to restore the attenuated pulse signals to their proper digital form and amplitude. This digital regeneration process is usually much easier and cheaper than restoring analog signals. Even if the A/D and D/A converters are included, the overall cost of a digital channel was and is usually cheaper than an analog channel. Likewise, if the analog signals are overly attenuated, the amplification (repeater) process will often produce distortion—this is not the usual result in digital pulse regeneration.

The generic T1 resulted in a number of multiple-channel standards in both North America and Japan. The technology of the DS1 was expanded to include 48-, 96-, 672- and 4,032-channel systems. In Europe, CCITT set standards for five similar TDM systems ranging from 30 to 7,680 channel systems. (See the Bellamy reference.

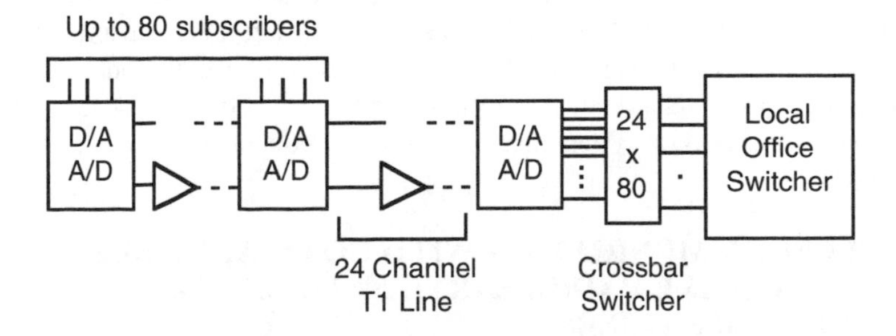

Figure 4.15 A prototype pair-gain system that can be used for a small group of subscribers. Later models did not need the crossbar switcher and used a T1-type system for 40- and 96-channel systems. (See the Bellamy reference.)

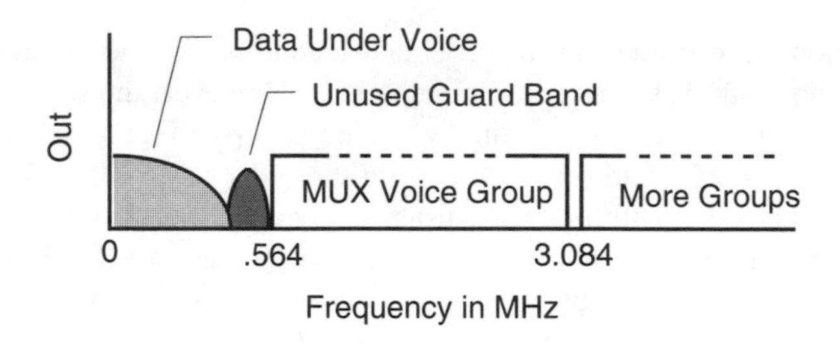

Figure 4.16 The spectrum of data under voice. (See the Bellamy reference.)

Other Two-Wire Digital Applications—Pair Gain and Data under Voice

Digital pair-gain systems such as Figure 4.15 were developed to increase the channels per line pair for small communities or for areas that developed overnight, such as shopping centers or housing and trailer-park developments. In small communities, as the older, unattended but very high maintenance analog switchers became obsolete, it was economically wise to replace them with some alternative higher-capacity digital system.

These digital pair-gain systems resemble the T1 systems but in the early versions replaced the pulse-code modulation with the less expensive delta modulation. Even though these pair-gain systems included A/D and D/A converters for each subscriber, the cost savings of the digital circuitry justified the additional technical complication.

The data under voice technology was developed to add digital data capability to the T1-type voice carrier system (Figure 4.16). Some of the later T1-type systems, used in office-to-office trunk or long-distance circuits, extended the analog multiplexing capabilities to systems that could be combined to carry from 600 to over 3,600 voice channels with coaxial cable. In these more complex, higher-capacity FDM systems, the first group of 600 voice channels would cover the frequency range of 0.564 to 3.084 MHz. This originally left the spectrum below the 0.564 MHz unused. By adding a T1 system, modified to accommodate a digital-only format, digital data (known as Dataphone Digital Service or DDS) could be sent over a transmission system that was originally conceived for voice only.

4.6 TELECOMMUNICATION WITH COAXIAL CABLE, MICROWAVE RADIO, EARTH SATELLITES, AND GLASS FIBERS

Telecommunication with Coaxial Cable

Following World War II, there was a dual pressure on the telephone companies (both the Bell System and the independents) to expand service. Not only were there the expected requests for expanded telephone routes and service but TV began to expand and needed city-to-city links by a common carrier. Coaxial cable (coax) not only provided many more voice channels but, in some situations, provided a relatively simple and reliable method of providing network TV service. (Its competitor, microwave radio, is covered in the next section.) Since coax is by definition surrounded by a braided electrostatic shield, it provides a higher degree of noise immunity. It is usually much better than two-pair lines, whether they are using baseband or carrier transmissions. Likewise, the nominal 3 to 6 MHz bandwidth of TV (the first signals were black and white) precluded the use of wire pairs. Like open-wire lines, numerous repeating amplifiers were needed for both telephone and video transmissions. (A pair of cables, with directional repeating amplifiers in each side, was needed for two-way telephone communication.)

Coax systems that use a cable bandwidth of 12 MHz and 60 MHz are typical, although not exclusive, in the industry. As both the telephone and TV industries expanded after World War II, both the Bell System and CCITT created standards for coaxial cable systems. In representative cases, a 12-MHz cable will use FDM carrier systems to transmit about 2,700 voice channels. A 60-MHz cable can accommodate 10,800 voice channels. Although there are some schemes that allow both voice and video in one cable, this is not common practice.

The design of gain and equalizing amplifiers for cable is somewhat different from using them for wire-pair repeater amplifiers. With the wider bandwidths, many more carrier channels, and the very critical color TV signals, the amplifier's phase and transient response, as well as frequency response, became increasingly critical. With the multiplicity of carriers and sidebands, special care had to be taken to minimize intermodulation distortion as well as simple harmonic distortion.

Intermodulation distortion results when one carrier or sideband unintentionally modulates another unrelated carrier or sideband. Just as the original desired carrier modulation was caused by a special nonlinear process, so can this undesired intermodulation be caused by undesired nonlinearities if the repeater amplifiers are not properly designed.

Beside selecting repeating amplifiers that produce very low amplitude and intermodulation distortion, it is interesting to understand some of the factors that influence the number of amplifiers that are required for a given cable run. For an example, assume a coaxial cable with an average attenuation of 15 dB per mile or 1,500 dB per 100 miles. One completely ridiculous approach would be to use one 1,500 dB repeating amplifier—an idea that would be rejected immediately. Likewise, two 750 or four 375-dB amplifiers would be unworkable. In designing any wide-band electronic amplifier, experience has shown that it is difficult to produce a workable, stable, and reliable amplifier that has a gain much over 40 dB. If the amplifier has a bandwidth of 12 or 60 MHz, as is typical in telephone carrier systems, it is usually wise to design for even a smaller maximum gain. A typical cable line amplifier is designed for a gain of about 30 dB. This would require fifty 30-dB amplifiers for a 100-mile run of cable. Note that this means that the cable is in fifty lengths, that there are thus 100 connectors that might become loose and corroded, and each of these amplifiers requires a power supply. The power supply problems can be quite dominant when the coaxial cables (or even wire-pair cables) are in remote areas or on the ocean floor. (See Chapter 7 of the Freeman 1989a reference for a more detailed example.)

Telecommunication with Microwave Radio

One major advantage of microwave radio towers over wire pairs or coaxial cable is their limited requirement for land and the accompanying right-of-way acquisition. The microwave radio links (Figure 4.17) have a relatively wide bandwidth, since their fundamental carrier is usually 1 to 10 GHz (GHz = 10^9 hertz). Voice channels and/or a video signal are multiplexed and up converted to modulate (usually frequency modulate) a 70-MHz subcarrier on the microwave signal. This makes possible the transmission of a TV signal and/or up to 2,700 voice channels.

While the microwave links provide relatively unencumbered signal relays, there are several signal propagation problems that make them far from the most ideal or economical choice for signal transmission. For instance, at some microwave wave lengths, rain drops will materially attenuate the transmissions. The dominant factors involving the quality of the microwave signal include [1] the height of the antenna tower, [2] reflections from natural object such as trees or mountains and diversity or multi-path propagation, and [3] noise interference.

Telecommunication with Microwave Radio—Tower-Height Requirements

The antenna must be elevated because of potential interfering objects and to obtain the best line of sight because of the curvature of the earth. Besides getting the

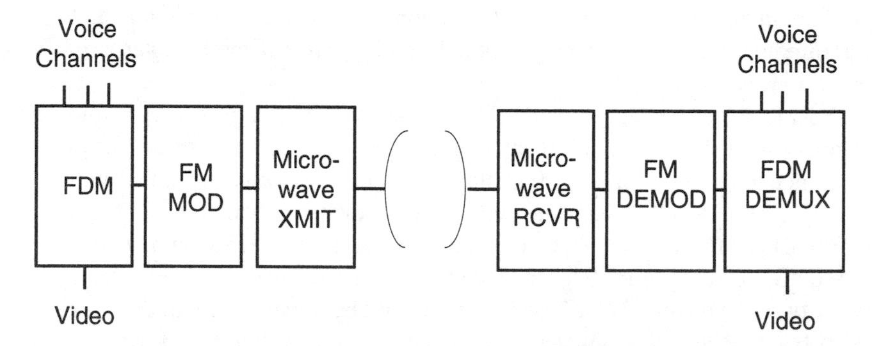

Figure 4.17 An elementary microwave-radio link. The FDM signals, whether voice or video, were originally FM modulated on the microwave carrier. However, more recent systems use digital modulation techniques. A microwave repeater would include the Microwave XMIT and RCVR blocks, an additional repeater amplifier with the receiver and an antenna dish for each direction. For clarity, this graphic is simplified and does not show the normal two-way traffic that uses a single antenna to accommodate both transmitted and received signals.

towers high enough to obtain obstacle-free transmission, the tower-to-tower path, termed the hop, must always be at least 40 to 50 feet above any trees. If there is a rise or hill along the path, a complicated triangulation formula, involving the distances from the hill to each antenna, must be used to calculate an additional tower-height factor. There is also another factor, involving the property of radio waves and termed the Fresnel zone factor, that must be included in the overall tower-height calculations. If there are one or more buildings or a water tower near the hop route, there must be an additional tower height to prevent multipath reflections. These reflections could produce echoes or drop-outs in the voice communications and ghosts in the TV pictures.

Telecommunication with Microwave Radio—Transmitter Power Calculations

To obtain the lowest practical installation and maintenance cost, consistent with good operation, the output power from each tower transmitter is kept as low as possible. The exact determination of this power is very complex, involving the hop distance, the microwave frequency, the method of modulation, the various losses in the accessory cables and connectors, and especially the noise performance of the repeating receiver. If the frequency is above a few gigahertz, the connecting feed to and from the tower as well as from the transmitters and receivers must be wave guides rather than coaxial cable. The following section illustrates some of the factors involved in the calculations required to determine the power output for a microwave-link transmitter. The problem is simplified for clarity.

Microwave Link Problem Assumptions

- The link distance is 30 miles.
- The modulation is FM.
- The microwave frequency is 1 GHz—this makes calculation and scaling easy and also assumes coaxial feed lines.
- A 10-dB FM carrier signal-to-noise ratio at the receiver input is to be maintained.
- The gain of each parabolic dish antenna will be assumed to be 23 dB—this number will be derived.

Path Loss

Whether it be AM or FM radio, TV, or microwave transmission, most calculations of radiated power start with the assumption that the antenna is a point source and is radiating equally in all directions as it would if it were at the center of a spher—this is known as free-space, isotropic radiation. The area of a sphere can be calculated by:

$$A = 4\pi r^2 \qquad \text{(E 4.6)}$$

Thus, if one watt of power is radiated from the center point source "c," that power is distributed uniformly over the total surface area of the sphere with a radius "r." In any given section of the spherical surface, termed the solid angle, the portion of the total radiated power may be quite small. If a receiving point is placed "d" distance from the center radiating point c, the path loss can be calculated by (see the Freeman 1989a reference):

$$L_{dB} = 96.6 + 20 \log_{10} F + 20 \log_{10} D \qquad \text{(E 4.7)}$$

where:

L_{dB} = the free-space path (power) loss in decibels or dB W where the "W" denotes a reference of one watt
F = the microwave frequency in GHz
D = the path distance in miles

As an example, if F is 3 GHz and D is 30 miles, the loss is:

$$L_{dB} = 96.6 + 9.54 + 29.5 \approx 135.5 \text{ dB.} \qquad \text{(E 4.8)}$$

Receiver Thermal Noise Threshold

Designers from the beginning of the electronics age (starting with vacuum-tube amplifiers) have wished for and sweated to perfect an amplifier that did not add noise to the incoming signal. However, due to the physical principles involved, there is always some thermal (atomic) agitation that produces an additional unwanted noise signal. Like many equations in electronic communications, the receiver noise formulas can get more complicated when they are written in the context of a specific problem. The formulas for the receiver *thermal noise* level include:

$$N_{dB\,W} = 10 \log_{10} kTB \qquad \text{(E 4.9)}$$

The generic thermal noise formula, where:

k = Boltzmann's constant = 1.3803×10^{-23} J/°K
T = receiver noise temperature in degrees Kelvin
B = the noise bandwidth in hertz

For telecommunications radio links using FM, E 4.9 can be transformed, to give:

$$N_{\mathrm{dB\,W}} = -228.6\ \mathrm{dB\,W/Hz} + 10 \log_{10} T + 10 \log_{10} B_{IF} \tag{E 4.10}$$

where:

228.6 dB W/Hz	= the thermal noise of a receiver with a 1-Hz bandwidth, unadjusted for noise temperature
10 log T	= will adjust the receiver to the real noise temperature
B_{IF}	= the calculation for the proper (FM) IF bandwidth

The equation for the more common noise figure (NF), that converts the effective temperature from the 290°K room temperature can be written as:

$$NF_{\mathrm{dB\,W}} = 10 \log_{10} \left(1 + \frac{T_{eff}}{290\,°K_{Room}} \right) \tag{E 4.11}$$

Likewise, the noise figure, which is a measure of the noise a practical network produces compared to an ideal, noiseless one, can be written:

$$NF = \frac{S/N_{in}}{S/N_{out}} \tag{E 4.12}$$

Expressed in decibels:

$$NF_{\mathrm{dB}} = 10 \log_{10} NF \tag{E 4.13}$$

If room temperature (+290°K) is assumed, E 4.9 can be simplified as:

$$N_{\mathrm{dB\,W}} = -228.6\ \mathrm{dB\,W/Hz} + 10 \log_{10} 290\,°K + NF_{\mathrm{dB}} + 10 \log_{10} B_{IF}$$

$$= -204\ \mathrm{dB\,W/Hz} + NF_{\mathrm{dB}} + 10 \log_{10} B_{IF} \tag{E 4.14}$$

If a receiver has a 10-MHz bandwidth and a noise figure of 10 dB:

$$N_{\mathrm{dB\,W}} = -204\ \mathrm{dB\,W/Hz} + 10\ \mathrm{dB} + 10 \log_{10} 10^7$$

$$N_{\mathrm{dB\,W}} = -124\ \mathrm{dB\,W} \tag{E 4.15}$$

The foergoing exercise on the math of noise, although dull and laborious, is typical of the thought processing that is often required to go from the equation of a fundamental physical principle, as given in E 4.8, to a practical application, using real-world numbers, such as was finally derived in E 4.14. This final receiver noise threshold is one-half of the proble—this is the constraint on the input signal. Now, using this information and then calculating the microwave path loss, it will be possible to calculate the output power required from the microwave transmitter.

The example associated with E 4.7 and E 4.8 showed a free-space path loss of 127 dB. In the real world there will also be some cable and connector losses that can be approximated as an additional 4 dB. The total loss is now 127 + 4 = 131 dB. It will be remembered that the incoming signal must have a power at least as great as the receiver noise threshold, that is, –124 dB W. (124 dB down from one watt.) If, as a starting point, a 1-watt power output is assumed, the result would be 131 dB down from 1 watt (–131 dB W) which is 7 dB (131 – 124) below the receiver noise threshold. Thus, the output of the amplifier would be required to be 7 dB above the assumed 1 watt (P_1), which would be (dB = 10 log P_2/P_1 or $P_2 = 10^{dB/10}$)—about 5 watts. If it is desired to have a carrier-to-noise ratio of 10 dB, a minimum requirement, it would require that the transmitter output be 7 + 10 or 17 dB above 1 watt or about 50 watts. This would be very difficult (expensive to purchase, operate, and maintain), if not prohibitive. However, all is not lost. The gain of the dish antennas has not been factored in.

Telecommunication with Microwave Radio—Parabolic Antenna Gain

The operation of the elementary dish (there are other, more sophisticated designs including exponential horns—see Chapter 5) is analogous to a glass lens; it concentrates a diverging beam into a narrow, almost parallel ray with very little divergence. For a properly shaped dish in the form of the parabola, with free-space isotropic (in all directions) radiation, the gain G increase over that of a point source in an imaginary sphere, calculated in dB, is:

$$G = 10 \log_{10} (4\pi A\eta / \lambda^2) \quad \text{(E 4.16)}$$

where:

A = the antenna aperture—the effective size
η = the aperture efficiency—in the order of 50 percent
λ = the wave length of the radiated (carrier) frequency

Again, as with the noise equations, this formula can be reduced to practical numbers as:

$$G = 20 \log D + 20 \log_{10} F + 7.5 \quad \text{(E 4.17)}$$

where:

D	=	the dish diameter in feet
F	=	the radiated frequency in gigahertz

For a six-foot dish radiating a 4-GHz signal:

$$G = 20 \log_{10} 6 + 20 \log_{10} 4 + 7.5$$

$$G = (20 \times .778) + (20 \times .6) + 7.5$$

$$G = 35 \text{ dB per dish} \qquad \text{(E 4.18)}$$

The reader will at once see the fallacy in the previous noise calculations. Without going through the arithmetic again, for the assumed distance of 35 miles, the two dishes will provide more than enough gain to run the transmitter(s) at a maximum power of 1 watt or less. In a practical design, the diameter of the dishes would probably be reduced.

Telecommunication with Satellite Transmissions

For a short period in the history of telecommunications, the AT&T Telestar defined satellite telephone communications to millions of people. To keep the subject matter consistent, the general subject of satellite communications is included in Chapter 6.

Telecommunication with Light Using Glass Fibers—An Introduction to Quantum Physics

Light has been used as a signaling medium for thousands of years. The use of signal fires or smoke-signal schemes represents some very early illustrations of modulating a source to provide coded intelligence over long distances. Alexander Graham Bell experimented with modulating a beam of sunlight for voice transmission. In the latter part of this century, several technologies, including lasers, light-emitting diodes (LEDs), and low-loss glass fibers have converged to make modulated light transmission one of the most attractive methods of transmitting telecommunication information.

It is interesting to deviate a few moments and review some of the esoteric properties of light. Out of sheer frustration, light has been called a waveform on Mondays, Wednesdays, and Fridays and a packet or quantum form on Tuesdays and Thursdays. The confusion comes from our lack of a complete understanding of the nature of light. The Scottish physicist James Clark Maxwell (circa 1864) produced the fundamental equations (Maxwell's equations) that have been used to define and explain the electromagnetic wave phenomenon. In many cases, light can be considered a very, very high-frequency electromagnetic wave. However, there are some instances where light can be best conceptualized as being a series of individual bullets of matter (just what type of matter is still a mystery) that form a stream or ray. Einstein, Bohr, and many others spent a significant portion of their lives working on the problem of how to characterize light. For a background on ideas on what light may be, see any introductory treatise on quantum physics. The Herbert reference is an excellent introduction to the wave and packet theories of light.

As a starting reference, the technique of using optical fibers for telecommunication can be compared with that of coaxial cable. Coax, with its center conductor and its outer shielding (circuit return) conductor, will permit the flow of *electric current* (whatever that is—again, if you're titillated, see references on quantum physics) from DC to several hundred megahertz. Telephone communication is most often accomplished by using one or more modulated carriers in the coax (recall the generic T1 systems). By analogy, elementary optical transmission can be also visualized as a carrier system. However, in optical fiber, the carrier is light with a frequency much higher than any carrier used for coaxial cable transmission—in the order of 10^{15} hertz (1,000 Terahertz). Converted to wavelength, fiber-optic transmissions usually use laser or LED transmitters that emit in the range of 800 to 1,600 nanometers (10^{-9} meters)—see Figure 4.18. The optical transmitters are modulated with some kind of amplitude modulation, including the on-off coding of PCM.

Telecommunication with Light Using Glass Fibers—Modes of Operation

The use of "light pipes" made of formed glass or clear plastic to transmit light around corners has been done for several centuries. In most cases, the path length was short and any attenuation of the light was not significant. The Pierce and Noll reference (page 130) shows that the light loss in glass has been reduced from 10^7 dB per km with early Egyptian glass, down to about 0.1 dB per km with contemporary optical fiber. In fact, the losses resulting from the connections to the light emitters and receptors, as well as the cable splices and other connections, present a significant portion of the attenuation in fiber-optic runs.

The light that is used as the transmission medium, the primary carrier in fiber-optic transmission, usually comes from one of two types of modulatable sources: LEDs or injection laser diodes (ILDs). LEDs are inexpensive and rugged while ILDs are more expensive and slightly less durable. The ILDs, being members of the laser family, emit a waveform that is almost monochromatic (a single frequency—sometimes termed coherent). This narrow, nearly monochromatic wave is more intense than the beam from LEDs and is thus easier to inject, transmit, and detect. When a ray is injected into a glass fiber, care must be taken to assure that it will stay within the boundaries of the fiber and not be refracted (bent) out of the

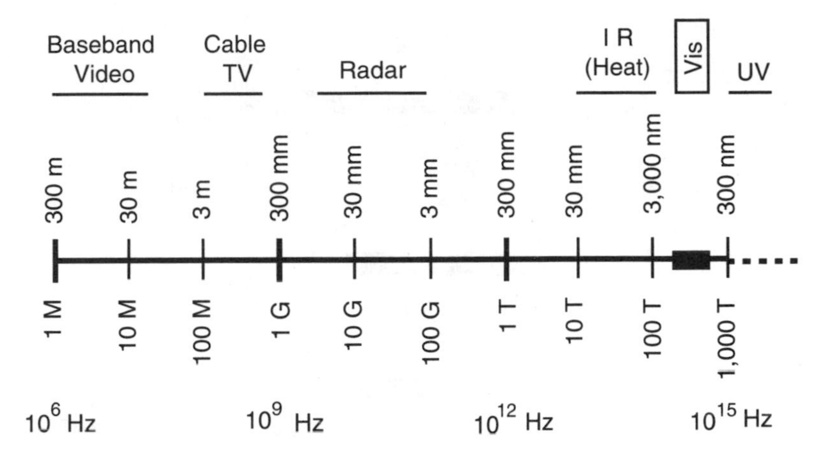

Figure 4.18 An abbreviated electromagnetic spectrum starting at 1 MHz and continuing through the Infrared (IR), Visible Light (Vis), and the Ultraviolet (UV). The area between 3,000 and 300 nanometers (nm) represents the general spectrum used for fiber-optic transmissions. Both the frequency, in hertz, and the wave length, in meters, are shown. The spectrum above ultraviolet, not shown, would represent x-rays and alpha and gamma radiation. In the upper portion of the graphic m stands for meter and in the lower portion M is used to abbreviate MHz.

path provided by the fiber. The physical principles that govern the way a beam of light reacts (bends) when it travels from a medium of one optical density to that of another were described by Willebrord Snell, a Dutch astronomer and mathematician, in 1621. The casual observer has seen the effects of reflection and refraction associated with the tricks that the water in a swimming pool plays on optical images. Snell's law states that when a beam of light passes from a material with a lower index of refraction, such as air equaling 1.0, to a material with a higher index of refraction, such as optical fiber with an index in the order of 1.4 to 1.5, the beam will be bent towards the normal or center line—see Figure 4.19.

The relative ratios of the glass fiber diameter to the (glass) cladding thickness change depending on the type and use of fiber-optic cable. For minimum attenuation undersea fiber-optic cable, the diameter of the fiber is extremely small—in the order of only 10 microns (10 one-millionths of a meter). The cladding will usually be several times that thickness and there will also be an outer plastic jacket for physical protection. If an optical cable is used for short runs, where minimum attenuation is not the domination factor, the glass fiber may be much larger, which will make it much less expensive.

Fiber-optic cables can take several configurations that are classified as modes, as indicated in Figure 4.20. A cable with a very small core compared to the cladding is termed single mode because the ray will propagate primarily in a straight line parallel to the normal. A single-mode cable will have a very narrow cone of acceptance (numerical aperture) but will have the smallest signal attenuation. Of

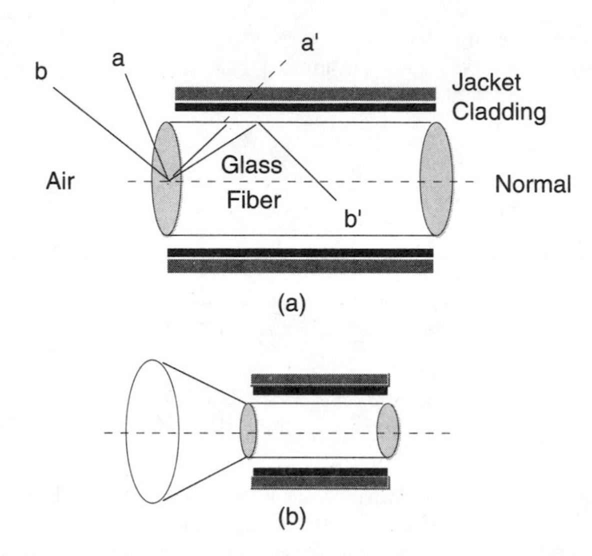

Figure 4.19 Ray paths from air to glass fiber. (a) Using the principles of Snell's law, the "a" ray will be refracted towards the dashed normal line. Even so, this ray may still leave the confine of the fiber and the surrounding cladding and not be propagated. The "b" ray arrives at a less oblique angle and thus the refraction will assure that it is transmitted within the glass-fiber medium. (b) A cone of acceptance (also termed the numerical aperture) can be postulated to illustrate the maximum solid angle that assures a beam will be propagated down the optical fiber inside the cladding. (For illustration, not to scale.)

greater importance, it will produce less pulse stretching (termed pulse or phase distortion) because different parts of the wave will not take longer or shorter paths. When the diameter of the fiber is enlarged, there is a high probability for multipath transmission that canlead to pulse distortion. By specially fabricating the fiber

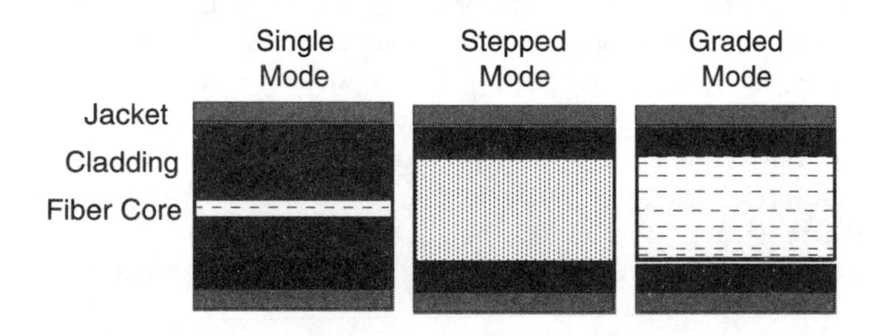

Figure 4.20 The different types or modes of fiber-optic cable. The single-mode cable, with the small, expensive core, has the widest signal bandwidth and the least pulse distortion. The stepped mode, with three abrupt refraction indexes, will produce the most pulse distortion. The graded mode, with better pulse response than the stepped fabrication, is a compromise between cost and performance. (Not to scale.)

core in a graded or tapered manner from the center, a compromise index of refraction can be created that will tend to keep a more uniform signal path but also allow a larger, easier to fabricate fiber core.

Telecommunication with Light Using Glass Fibers—An Introduction to Systems That Can Contain Individual Features and Still Talk to Each Other

The change in transmission practice from analog to digital was, at first, quite subtle and low-key. Most of the early (circa 1970) digital transmissions were each based on proprietary protocols that were limited to that particular vendor's product. Any techniques of interfacing to another system with another protocol were either not very well thought through or purposely ignored in an effort to promote the sale of a given vendor's product line. As digital transmission systems continued to proliferate and expand, the ability to interact and interchange information between competitive systems soon outweighed the merits of exclusivity. Chapter 8 describes two master models or master architectures for the transmission of primarily digital data—the OSI model and the ISDN model. Both of these transmission media models attempt (really, quite successfully) to provide a universal architecture that will allow each single system designer to fill in the blanks for their particular use and still produce a semiuniversal product that can interface with other individual products that, likewise, adhere to the rule of the universal models.

Telecommunication with Light Using Glass Fibers—SONET, the Synchronous Optical Network

SONET is a system of multiplexing that will accept several types or combinations of electrical signals and convert them to (fiber) optical signals.

1. SONET uses the 24-channel, 1.544-Mbps DS1 digital signal as its fundamental multiplexing building block. Each frame of the DS1 (earlier termed the T1) signal is sampled at 8,000-Hz (or bps) for a configuration equal to: (8 bits $\times$ 24 voice channels) + 1 framing bit = 193 bits and, multiplying each frame times the sampling rate, gives a data rate of 193 bits $\times$ 8,000 = 1.544 Mbps.

2. A composite SONET signal can be made up of several basic digital formats such as the DS1 (24 channels), the DS2 (96 channels), the DS3 (672 channels), or the DS4 (4,032 voice channels). (See the Bellamy reference.)

3. These basic nonsynchronous electrical DS channels are usually synchronized and combined with overhead information to produce the Synchronous Transport Signal (STS). For example, the elemental DS3 signal contains twenty-eight of the 24-channel DS1 signals for a combined capacity of 672 voice channels. However, the combined data rate of a DS3 channel is 51.840 Mbps, not 1.544 $\times$ 28 or 43.232 Mbps. The additional 8,608 Mbps

are used for special overhead data. This overhead includes parity, frame addressing and general operations, and administration and maintenance (OAM) information. It is this rich OAM structure that provides SONET the ability to service multiple vendors with their varied data structures and protocols.

4. These various STS electrical carrier signals are multiplexed, scrambled for security, and a final, overall parity calculation is made before the signals are transferred into a PCM optical carrier (OC) format for the fiber cable. It should be noted that except for the scrambling the optical data format is exactly the same as the electrical data format.

5. Besides the original U.S. correspondence between the STS electrical and OC optical signals, there are also corresponding CCITT STM designations.

It should be remembered that the all-fiber network, including one or more fiber-optic cables to every customer's home, is still in the future. Besides the obvious expense involved, the technology is constantly changing. For instance, the fantastic advances in data compression, including picture compression, force us almost daily to rethink the ways to accomplish telecommunications in the future.

4.7 THE SIGNAL PROCESSING USED IN THE CONVERSION AND MODULATION OF TELECOMMUNICATIONS SIGNALS

Special Digital Conversion and Modulation Techniques for Voice — An Introduction

Chapters 1 and 3 presented a variety of methods for coding an analog signal in the digital domain and gave examples of applying these different techniques to au-

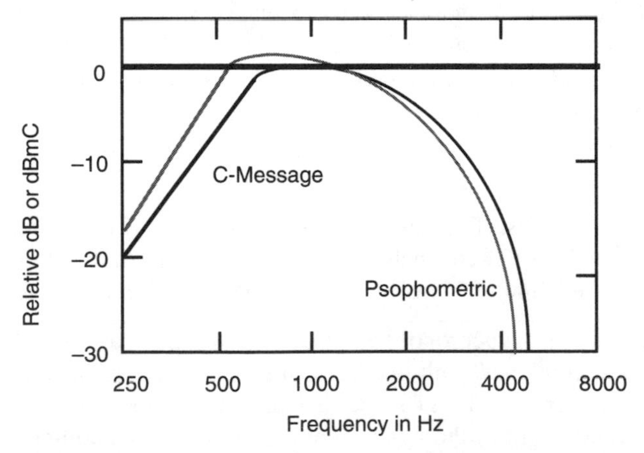

Figure 4.21 Approximate weighting responses based on standard reference telephone sets and measurement procedures.

dio, radio, and video signals. Because of the unique characteristics required for both high-quality and efficient voice communications, special developments peculiar to voice transmissions were (and still are) required. For example, the average power in human speech is reasonably constant—rapid, loud crescendos are usually not a problem. Therefore, in some cases, the headroom required for music and some video signals is not required. Likewise, there are numerous gaps or pauses in human speech. This dead time can, with one general compression technique, be eliminated during the transmission or, using a different approach, other communications can be judiciously squeezed in the gaps. Speech processing in telephony represents one of the most highly developed areas of signal processing and its advances have served as models for progress in many other fields.

Telephone Set Frequency Response, Noise, and Weighting

Starting from the early telephone instruments (see Figures 4.1 and 4.2), telephone designers have attempted to improve the overall intelligibility of the complete transmission system. The original transmitters and receivers had a very limited, ragged response that somewhat limited the quality of reception. Throughout the years, the combined electrical and acoustical response of the transmitters and receivers, the set, has been periodically altered to improve its intelligibility, i.e., the perceived quality. These early developments culminated in a response standard based on the Western Electric type 500 telephone handset. This standard is known as the C-message weighting. There is a similar European standard known as Psophometric weighting. The significance of the weighting is that all response and distortion measurements in the voice band must be scaled to comply with the C-message or the Psophometric curve—see Figure 4.21.

Distortion and Noise in the Voice Channel—Simple Quantizing Noise

The C-message roll-off of the low frequencies not only adds to the intelligibility but eliminates the interference that would result if excessive 60- or 120-Hz hum was reproduced. Likewise, any high-frequency noise, especially the potential hiss during the normal gaps in the voice communication, is attenuated by the high-frequency roll-off. There are other noise sources that must be considered when an audio signal-to-noise ratio is calculated for telephone transmissions. When the analog voice is digitized, the quantizing noise must also be considered (see Chapter 3 for background). For sine-wave signals, the signal-to-quantizing noise ratio is:

$$SQR = 7.78 + 20 \log_{10} \frac{A}{q} \qquad \text{(E 4.19)}$$

where:

A = the peak amplitude of the input sine wave and
q = is the magnitude of a quantizing interval

For example: If A is 1.0 volt (peak) and, for 4-bit quantizing, with $q = 0.0625$ volts $= 62.5$ mv,

$$SQR = 7.78 + 20 \log_{10} \frac{1{,}000}{62.5} = 31.86 \text{ dB} \qquad \text{(E 4.20)}$$

Likewise: If A is 1.0 volt (peak) and, for 8-bit quantizing, with $q = 0.003906 \approx$ 0.004 volts,

$$SQR = 7.78 + 20 \log_{10} \frac{1{,}000}{4}$$
$$= 7.78 + (20 \times 2.4) = 55.74 \text{ dB} \qquad \text{(E 4.21)}$$

Uniform and Nonuniform Quantizing—Companding

Earlier in this chapter it was mentioned that normal human speech is quite uniform in its average power and most often does not require the same amount of headroom as music. While this is certainly true, for realistic telephone conversations there must be some dynamic range in addition to the minimum generally recommended SQR of about 30 dB—about sixteen quantizing levels or four bits. If, for example, another 30 dB for range of headroom is desired, it would require a total quantizing range of 60 dB that, using E 4.19, is about 408 quantizing levels or approaching 9 bits. (It should be remembered that the audio on a CD uses 16 bits or 65,536 levels.)

Since the additional five bits (above the basic 4-bit signal) is not often required, telephone engineers developed an alternate way of having the reserve dynamic range without going to nine or more bits. The technique is known as *companding*. Companding is akin to the analog compression techniques that were described in Chapter 1. Digital companding (compression and expanding) is accomplished with nonuniform quantizing levels—there is more gain for the lower-level signals. The amount of compression depends on the circuit technology—discrete, simple integrated circuits or very highly integrated circuits. Two general standards have been used throughout the world: μ-law and A-law companding. The development of the U.S. μ-255 characteristic came about when, after the very successful experiences with the T1 system in interoffice exchange transmissions, AT&T wished to build a similar but more versatile and robust PCM system for toll networking. That new D2 channel standard included:

1. Eight bits (rather than seven in the older T1/D1 system) per PCM code word.
2. The companding and decompanding functions were included in the same circuits as the A/D and the D/A converters.
3. The companding characteristic was the new μ-255.

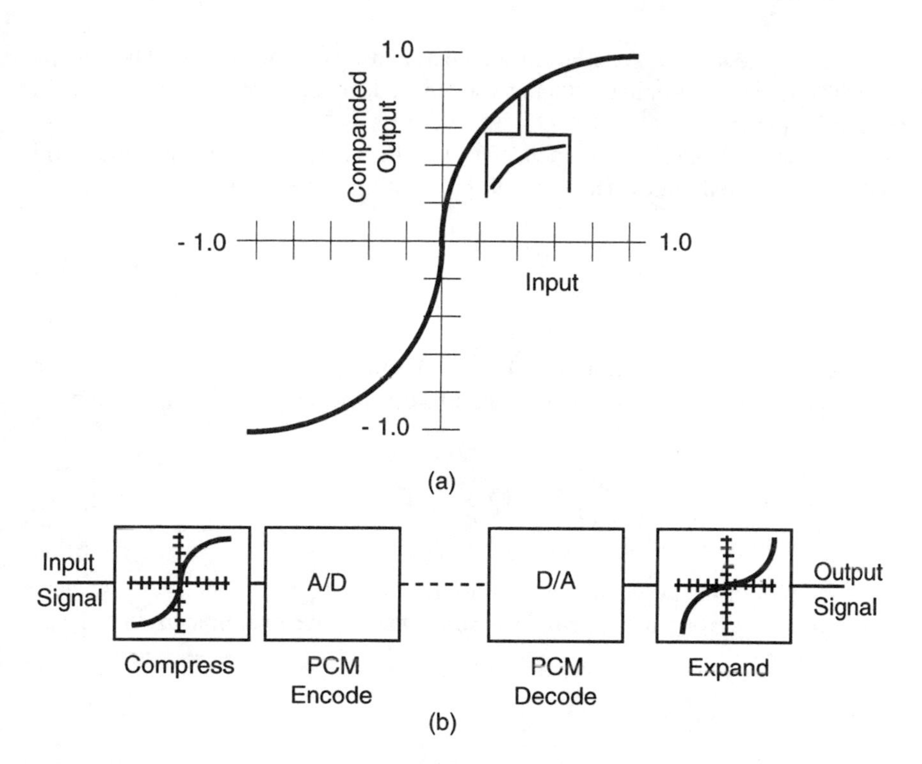

Figure 4.22 (a) A sketch representing the principle of digital compression or companding. When the analog input is small, the quantized digital units are small. However, as the input increases, it takes more input to produce a quantized unit output. The process is reversed when the PCM is reconverted to an output analog audio signal. The insert shows that the quantizing is actually piecewise rather than a smooth curve. (b) The overall PCM compress and expand system including the (dashed) transmission link. The actual circuits can be combined in one IC function or some of the companding may be accomplished in several other ways, such as software look-up tables.

It is interesting to investigate the derivation of the terms μ-law and A-law. Most nonlinear functions (curves) can be defined mathematically by using some form of n^x or, in its converted form, by using common or natural logarithms. The general formula for the μ-law function is:

$$F_\mu = \frac{\ln(1+\mu|x|)}{\ln(1+\mu)} \qquad \text{(E 4.22)}$$

Practical circuits do not exactly duplicate logarithmic functions but, if all the compression and expansion circuits are reciprocals, the overall effect is close enough to obtain the desired result. One of the earlier AT&T companding curves, as illustrated in Figure 4.22 (A), had seven segments and was termed μ-100. The later μ-255 D2 circuits, while still producing a segmented curve, more closely approached a logarithmic curve.

A-law companding is recommended by CCITT and is used in much of the world outside the United States. The form of the equation it follows is:

$$F_A(x) = \text{sgn}(x)\left[\frac{A|x|}{1+\ln(A)}\right] \quad 0 \le |x| \le \frac{1}{A} \qquad \text{(E 4.23A)}$$

where:

x is the input signal amplitude ($-1 \le x \le 1$), sgn(x) is the polarity (+ or –) of the signal x, and μ is the parameter used to define the amount of compression.

$$F_A(x) = \text{sgn}(x)\left[\frac{1+\ln A|x|}{1+\ln(A)}\right] \quad \frac{1}{A} \le |x| \le 1 \qquad \text{(E 4.23B)}$$

The first part of the curve, represented by E 4.23A, shows the linear A-law characteristic. Thus, for small signals it is sometimes easier to compand with the μ-law functions.

Special Signal Processing for Telephone Communications

4.8 SIGNAL PROCESSING IN TELEPHONY—A REVIEW

Intelligibility, Bandwidth, and Bits

After the first few years in the life of the telephone, the dominant thrust was providing more circuits together with more quality and reliability. Both of these items really involved the intelligent use of circuit and component bandwidth. The beginning of Section 4.7 described the C-message weighting as a technique for integrating the electrical and acoustical characteristics of the telephone transmitter, the receiver, and the line-pair bandwidth into an overall bandpass that would provide the best composite signal quality. Likewise, the various sections on phantom and bridge circuits, carrier modulation systems, coax, microwaves, and fiber optics, showed the diligent quest for increased signal-band capacity. In addition to these advances, there are a number of processing techniques that have either again extended the bandwidth or, more often, significantly improved the utilization of the existing band capacity.

For digital transmission, the use of the band capacity results in a trade-off between the bandwidth and the number of required quantizing levels—including any required overhead. The utilization of any transmission medium can best be analyzed by starting with Harry Nyquist's classic sampling theorem:

$$f_s \geq 2\,BW \tag{E 4.24}$$

where:

f_s = the sampling frequency and
BW = the bandwidth for the input signal

Thus, with a sampling frequency of 8 kHz, the maximum undistorted signal bandwidth will be equal to or slightly less than 4 kHz. In the digital domain, this means a maximum of 4 Kbps. If the data stream is divided into words, there can be 4,000 1-bit words, 2.000 two-bit words, 500 8-bit words, etc. For illustration, assume the 500 eight-bit words. If some device or technique can be created that will compress (and after transmission expand, i.e., compand) the data into seven bits instead of eight, the added one-eighth of the bandwidth can be used for additional data. Because this principle of finding methods to improve the data-handling capacity for a given bandwidth is so important in telephone and most other communication, the information in the next section will present alternate methods of limiting the amount of bandwidth that data requires. It should be noted that not all of these techniques are used in contemporary telecommunications but the principles apply to the practice of general signal processing.

4.9 TECHNIQUES TO REDUCE THE REQUIRED BANDWIDTH AND/OR REDUCE THE NOISE AND DISTORTION

Syllabic Companding

A-law and μ-law companding are based on the principle of reducing in dynamic range (and thus the required bandwidth) by creating nonuniform quantization intervals. Syllabic companding (SC) reduces the required bandwidth by reducing the dynamic range of the input signal. SC is based on the observation that the average power in speech is almost constant. However, if there is an instantaneous change in speech level, it will occur at the approximate rate that word syllables are spoken—about one every 30 milliseconds. SC is accomplished by continually monitoring the average power and, if there is a momentary increase, the input gain is reduced to again provide the same approximate speech power. At the receiving end, the instantaneous gain is increased by the same amount to recover the original dynamics of the signal. This system requires power sensors, as well as variable attenuators and amplifiers at both the input and output of the transmission system. An additional signal must be added to indicate any system gain change. Even so, since the sensors can be made synchronous at about every 30 ms, even with the additional hardware and signaling, there is still a worthwhile saving in bandwidth.

Adaptive Gain Encoding

One of the most conceptually straightforward methods of reducing the dynamic range of any signal, including voice, is automatic gain control. This technique has been used in radio for 60 years. The use of AGC for speech involves [1] a periodic level sampling scheme and [2] a gain-controlled amplifier. A workable AGC transmission system involves sensing amplifiers coupled to adjustable amplifiers at both the sending and receiving ends, as well as a method of encoding the periodic gain information in the bit stream. The encoded gain information is inserted in the data stream as blocks or frames. The AGC technique can produce a bit rate reduction of 30 percent and is one of the several tricks that is often used in implementing cellular telephones. (See the Bellamy reference.)

Speech Redundancies

The time domain waveform of most speech displays much repetition or redundancy. Speech patterns usually repeat at a rate based upon the characteristic pitch of a given voice and the particular sound that is being spoken. In average sampled speech, there is a correlation coefficient (a measure of predictability) of about 85 percent between adjacent samples. With this high degree of sample-to-sample correlation, i.e., repeatability, there is no need to continually send a complete, identical code word. One of the more efficient ways to reduce the code requirements, and thus the required bandwidth, is to encode only changes or differences, if they exist, between adjacent samples. (It will be shown later that this is in essence only encoding the slope or derivative of the signal.)

In addition to the sample-to-sample repeating, there are several other redundancies that may be used for bandwidth reduction. Inside the pitch interval there are often cycle-to-cycle correlations as shown in Figure 4.23. These individual repeating waves are at a higher frequency than the complete pitch-interval wave and any scheme including differences of encoding and decoding them is quite difficult. Likewise, because these higher-frequency waves produce the lifelike quality in the speech, errors in the encode-decode process will make the voice sound synthesized.

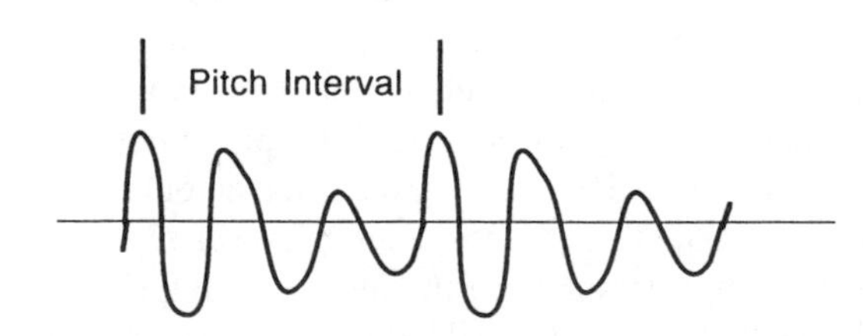

Figure 4.23 A time domain waveform of the human voice. This is the wave of a "voiced" sound and exhibits a high degree of redundancy. The pitch interval is the period between the puffs of air from the lungs that are used by the human vocal tract to create the sounds of speech.

Idle Channel Coding

One of the very productive methods of band compression is to encode and use the inactivity in a typical telephone conversation. For instance, in a model two-way (full-duplex) conversation, each party will only talk (send) 50 percent of the time. Measurements have shown that, including the normal pauses in human speech, the usage time of each party is closer to 40 percent. The utilization of this inactive time, and thus inactive bandwidth, dates back to the days of all-analog telephone networks and their Time Assignment Speech Interpolation (TASI). In the digital world this same general idea can be executed with more ease and accuracy. The contemporary Digital Speech Interpolation (DSI) senses the activity on a line and, if it is idle, momentarily uses that line to transmit other voice or data. Of course, the line must have an anticipatory monitor and switch so that, once the original voice transmission is resumed, the line will be released without any loss of continuity.

Differential Pulse Code Modulation

The fundamental technique for digitizing the baseband analog voice waveform is pulse code modulation, shown in Figure 4.24 (a), which includes the processes of filtering, clocking, A/D, and D/A. This fundamental system can be made more efficient if it transmits codes that represent the difference from sample to sample rather than the complete code. Figure 4.24 (b) shows an elementary conceptual diagram of how this sample-to-sample difference can be derived. Difference or differential PCM is based on the principle that the sample-to-sample difference will always require fewer bits, and thus less bandwidth, than the complete sample-to-sample code. This system includes a delay (of one clock pulse in a digital implementation) and store (accumulate) so that the first or N sample can be compared with the next N–1 sample. If there is no difference, nothing (or a 1-bit zero) is sent down the line. (It is tacitly assumed that there is also an accumulator in the receiver circuitry and, until a change is sent, that circuit will continue to use the last unmodified stored sample.) The conceptual system would suffer from the accumulation of quantizing error and cannot be implemented as a reliable working system. Figure 4.24 (c) illustrates a feedback version that overcomes this problem. The block in the lower part of the figure represents a combined accumulator and integrator. This circuit feeds back the differential of the delayed N–1 signal and, because of the continuous feedback update, does not accumulate overflow errors. It should be noted that although a clocking input is shown only as a reference in 4.24 (a), any of the digital circuit implementations must including clocking. (See the Bellamy reference.)

Delta Modulation

Analog voice signals can be digitized by a technique similar to differential PCM but limited to one bit in the difference signal. Such a system, termed delta modula-

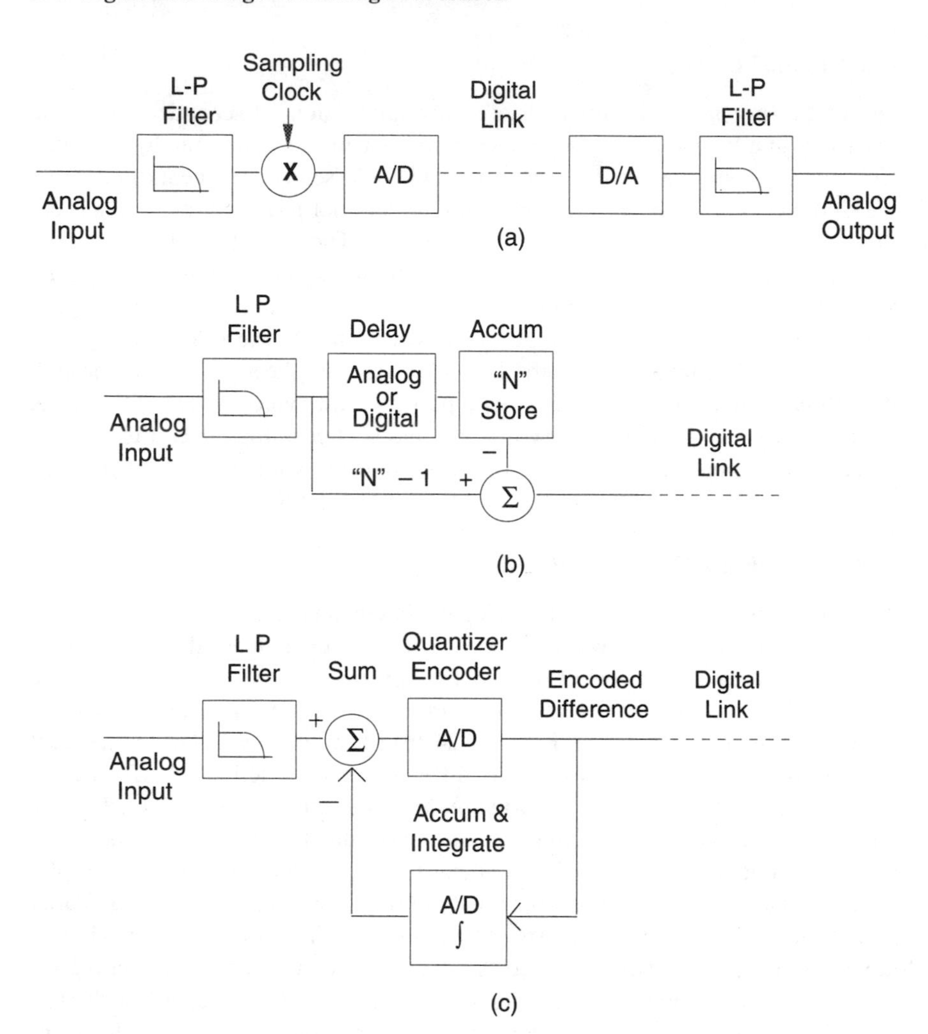

Figure 4.24 (a) The elemental pulse code modulation system. (b) A conceptual difference PCM system. (The receiver decoder would include a D/A and filter with a sample-and-hold or an integrator circuit.) (c) A differential PCM system.

tion (DM), again relies on an integrator (analog or digital) in a feedback loop to sense the changes in slope of an analog input signal. Figure 4.25 (a) shows the simplest conceptual delta modulator and demodulator. The comparator will sense whether or not the instantaneous level of the input signal is different from the feedback signal and so actuate the flip-flop. A digital 1 or 0 from the flip-flop tells the integrator to go up or go down. Likewise, the digital outputs of the flip- flop can be integrated at the receiving end and, after passing through a reconstructive low-pass filter, will simulate the original input voice signal. Both the quality

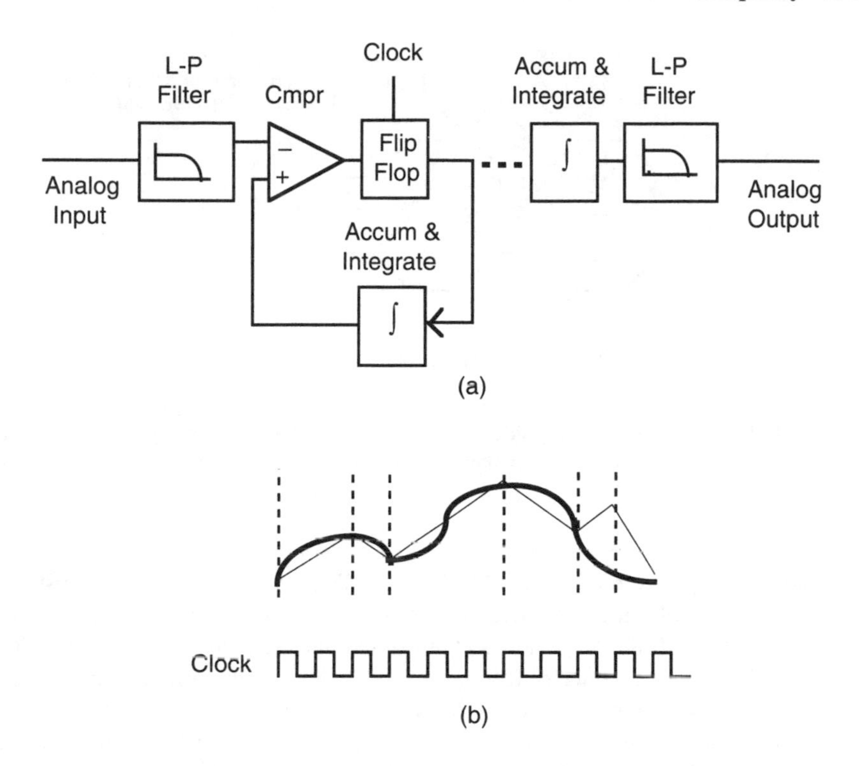

Figure 4.25 (a) An elemental delta modulator and receiver. (b) A simulated input waveform. The directions of the diagonal lines indicate how the output of the integrator is affected by the changing slope of the input signal.

and the dynamic range of a DM are dependent on the clock rate. An oversampling rate of at least four times (four times the Nyquist f_s) is usually recommended. A more complex form of DM, known as continuously variable slope delta modulation (CVSD) includes circuits similar to those used in syllabic companding. CVSD can reduce the required data rate and bandwidth as much as two to one if the user is willing to accept slightly less quality (realism) in the voice. It should also be noted that most forms of delta modulation do not accommodate tone signals or signals that depend on good phase characteristics such as those that use forms of phase encoding. (See the Bellamy and the Jones references.)

4.10 TELEPHONE SYSTEM PERIPHERALS

Cellular Radio Telephones

Vehicles, including cars, trucks, boats, etc., with mobile two-way radios that could be connected as part of the common-carrier telephone system have been available for many years. Before 1983, the service was limited to two very congested bands

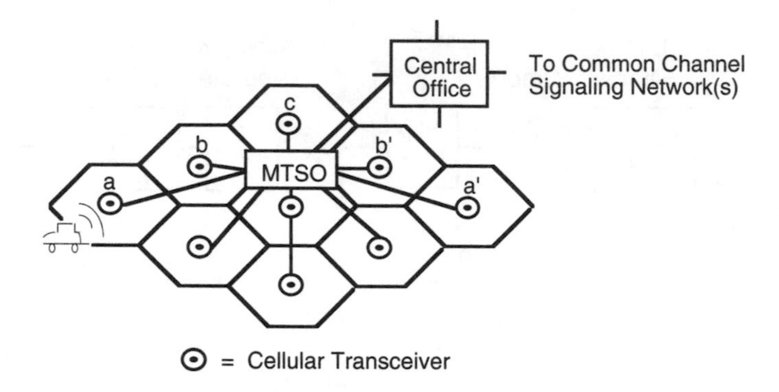

Figure 4.26 The rudimentary topology structure of a cellular telephone system. The original concept included provisions for expanding the system by subdividing any or all of the initial cells.

located in the spectrum at about 150 and 450 MHz. Most of these older systems required the use of a mobile operator to provide an interface to the normal telephone system. In 1983, a the new system of cellular transmission and switching was introduced in the Chicago, Illinois, area. The underlying principle of the cellular system, as shown in Figure 4.26, was a multiplicity of small, low-power radio transceivers that were all connected, originally by land lines, to a mobile telephone switching office (MTSO). The MTSO is in turn connected to the a common carrier's central office.

If a mobile unit is in the area of a given cell, such as "a," that transceiver will service the unit. If the mobile unit proceeds into an adjacent cell, the MTSO will automatically change to the new cell's transceiver. There are protections built in the system, such as in-cell, as well as cell-to-cell, channel switching to prevent call disruptions. When the subscriber crosses a given MTSO boundary, a common-channel signaling network transfers the call to the cells in the new MTSO but the connection to the common carrier is unchanged. When a contract is purchased from a cellular company, the subscriber can choose how much call-reception area (up to essentially worldwide coverage) is required. The cellular provider will then encode that address and search information with the subscriber's cellular telephone number. This information becomes an address header in the subscriber's message block. (See the Bellamy reference.)

Facsimile Transmission

It is interesting to note that the fax machine is older than the telephone. The first fax machines were based on synchronized pendulum clocks and were patented in 1843 by Scottish inventor Alexander Bain. The movable pendulum of the sending machine had a wire spring attached that would wipe across a block of metal type and, when it made contact with the extremities of the letters, it would produce an

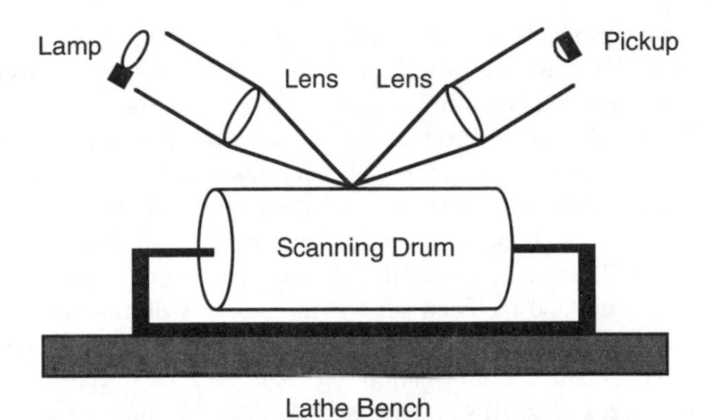

Figure 4.27 The rudimentary fax machine as used in the period of 1920 to 1960. In all models the drum would rotate. Some models moved the drum lathe assembly horizontally under a fixed light spot while others moved the lamp and pickup mechanism over the fixed rotating drum. The transmitting fax would convert the document reflections from the drum to a varying output analog voltage. The receiving fax would use the changing light intensity to expose a photographic film.

instantaneous electric current, which was transmitted to the receiving machine. The pendulum in the receiving fax had a small stylus attached that was synchronously scanning, but not touching, a sensitive recording medium. The currents that were sent to the receiver would cause minute arcs and thus burn an image of the original type in the medium. After each swing, the pendulums would be lowered a fraction of an inch to eventually scan the entire surface. It is also interesting to note that the Italian inventor Giovanni Casselli improved upon Bain's device and used it to start a commercial fax network in 1865.

The modern facsimile machine dates from the 1920s when newspapers, police departments, and industry used a spinning drum fax to send photographs. In fact, the AP Wirephoto machine became a legend in its time. This drum-type fax (Figure 4.27) used various forms of exciter lamps along with the appropriate optics to produce a small spot of light on the text or graphic to be transmitted. This spot of light was reflected through more optics and focused on a pickup photo tube. Since the amount of light that was reflected depended upon the reflectivity of the subject document, a varying, analog gray scale could be produced. Depending on the model of the machine and the content of the document, a contrast ratio to 15 to 30 dB (or about 5.6 to 32 times) could be obtained. Using a more contemporary scale of measurement, this would convert to a ratio of slightly over two bits to slightly under five bits.

In the 1980s, a new generation of facsimile machines was introduced based on technology similar to laser printers. The lamp and analog pickup were replaced by lasers or LEDs and the pickup and the photographic paper were replaced by a new technology electrostatic system.

As with many products that are rapidly accepted by the public (such as B&W and then color TV), contemporary fax machines adhere to specific, well-defined standards. The older rotating-drum machines were not designed to any universal standard. If brand A was used to pick up and transmit, it was mandatory also to use brand A as a receiver and copier. The first fax standards were developed in 1966 in the United States by the Electronics Industries Association (EIA) and shortly the CCITT followed with the foundation of the standards that are used today.

The first CCITT fax standards included Group I and Group II. These standards were quite complete and included such items as line scan time as well as total document scan time, scan density (lines/inch), modulation frequency, answer-tone frequency, end-of-message information, etc. However, they were all analog parameters. The later Group III standards specify digital techniques and still allow high bit rate transmission, up to 9,600 bps, over standard voice-grade telephone lines. The chronology of a typical fax message, using the Group III standards, is an example of how a very complex, but well-conceived digital transmission protocol can be used to produce a system that gives excellent and workable day-to-day results.

Before proceeding to an outline of the fax message phases, it is necessary to review the CCITT standards for modems, since they format and manage the overall fax message. One of the intriguing parts of the Group III standard is its ability to first attempt the transmission of the fax message at 9,600 bps. If the receiver finds this high rate unsatisfactory, the protocol, with appropriate handshaking, will automatically reduce the rate until a satisfactory message is obtained. The following is a list of the CCITT modem standards that are used in Group III fax modems. It should be noted that in most cases some original Bell standards predated these CCITT standards. Both standards are similar but many times the two frequencies, used to denote a MARK and SPACE in the FSK modulation, differed slightly. (See the Campbell reference.) Note also that in some cases, the CCITT standards may also include other data rates, such as 200 bps, that are not applicable to fax modems.

CCITT V.21	300 bps FSK	Full Duplex
CCITT V.22	600/1200 bps DPSK	Full Duplex
CCITT V.22b	2,400 bps QAM	Full Duplex
CCITT V.27ter	2,400/4,800 bps PSK	Half Duplex
CCITT V.29	7,200/9,600 bps QAM	Half Duplex

T.2 A Protocol and Modulation Standard Primarily associated with Group I.

T.3 A Protocol and Modulation Standard Primarily associated with Group II.

T.4 A Protocol and Modulation Standard primarily associated with Group III.

Establish Call –Phase A–	Premessage Procedure –Phase B–	Message Transmission –Phases C1, C2–	Postmessage Procedure –Phase D–	Call Release –Phase E–

Figure 4.28 The five phases of a fax call using CCITT T.30.

The T.3 specification, although primarily associated with Group II, is also used as the umbrella specification for the digital Group III. As the call progresses through its five phases, the protocol uses both T.3 and T.4.

Figure 4.28 shows the protocols for a contemporary fax transmission. In Phase A, the calling device dials the number of the device to be called, usually using V.21. After the initial contact is established, there is a period of hand-shaking with ID and machine-specific information exchanged in Phase B. This message identifies both machines and provides the sender with information about any nonstandard features of the called device. The information is framed using the HDLC standard (High-Level Data Link—see Chapter 8), which includes error-checking procedures. The CCITT T.30 specification also allows for the V.27ter rate. One of the information data frames that the called fax can return is termed NFS and it can give the caller special information about encryption and fax store and forward.

Once these preliminaries are completed, the transmission rate is changed and the fax can be transmitted at a higher bit rate. The V.29 9,600 bps rate is first used to test the line conditions. If there are excessive errors, the rate is reduced in steps until a satisfactory message can be transmitted. It should be remembered that the receiving fax may be a slow thermal printer or a very fast computer. The T.30 specification includes this initial C1 start-up phase and the C1 message transmission phase. After a fax page is sent, the communication goes to the slower postmessage D-phase. Here, if the sender has more pages to send, he or she will send a Multipage Signal (MPS) and, after more handshaking, revert back to the higher-speed C2-phase for more fax transmission. When the transmission is finally completed, the D-phase, End of Message (EOM) is sent to prepare for the final call-release phase. This last part may include some instructions for manual shutdown.

Although the transmission of the actual fax in phase C3 usually is sent at 9,600 bps, the time to send a complete page of information could exceed the three minutes that was the page rate for the older Group II process. To compete with Group II required signal compression, there are two compression algorithms normally used in Group III systems. These two types of coding, termed one-dimensional (1D) and two-dimensional (2D), are adaptations of the idea of sending the change in code rather than the complete code itself. One-dimensional coding sends the changes in codes in a given line while two-dimensional coding relies on line-to-line coding differences. With 2D coding, the first line that is sent is coded 1D and is used for a reference. From then on, the differences between this reference line and the current line are sent as the fax message.

4.11 TELEPHONY—THE CHAPTER IN RETROSPECT

The Telephone—Its Romance and High Technology

The telephone has been part of Americana for more than 100 years. Much of its original theoretical and applied work has been borrowed or adapted by most other parts of the electronics art and industry. The technical and patent literature is replete with innovations derived from the work of telephone pioneers.

Early Telephone Devices and Signal Processing

The early telephones both looked and performed like side-show electrical gadgets. However, within five years after its invention (circa 1875) there were complete, working telephone systems in several U.S. cities. At first one apparatus was used to both send and receive but soon its inefficiencies were recognized and it was replaced with the carbon-element transmitter and the electromagnetic receiver that served as prototypes for devices used for about a century. As the telephone proliferated, first a ringer and then the familiar dials were added to signal the called party. In early systems, all central switching was done by human operators, but with the addition of the dial, automatic, coded mechanical switching could be used.

Several developments were incorporated in the early part of the twentieth century to increase the two-wire line usage. First, by giving several users a specific number of rings for their individual code, several users could be grouped to use what was called a party line. Second, by using special transformers or "phantom" repeat coils, three users could be served with only two line pairs. Later, by using the repeat coils in a semibridge arrangement, two users could be served with only one line pair. This was especially efficient for long-distance service. Finally, inexpensive copper-oxide bridge rectifiers were introduced as modulators and demodulators in the first multiparty frequency division multiplexed carrier current modulation systems.

More Advanced Switching and Line Processing

A prelude to the age of digital technology came with the deliberate progression of telephone number coding. With the continued expansion of telephone usage, it became necessary to use digital coding for switching. First came the 4-bit switchbank code that provided an access to 10^4 switcher cross points. A second 3-bit code was added to designate each separate exchange. Later the now familiar 3-bit area codes and the 1-bit access codes were also added. Similar additional codes were also added in most parts of the industrial world with the guidance of the international committee CCITT.

Along with the developments in the switching and access coding systems came more advanced processing to improve the quality and reliability of telephone service. Improvements in the listener's sidetone were developed as well as processing techniques to reduce or eliminate echoes on long distance lines. Improvements were also periodically made in the quality of both the transmitter and receiver and the frequency response of the lines.

Coaxial Cable and Microwave Transmissions

The increased bandwidth of first coaxial cable and later microwaves gave the telephone companies relief from the ever-growing demands on their line-pair circuits. Likewise, the growth of the television networks, with their requirements for wide band circuits with excellent transient and phase response, produced additional demands of transmission capacity. This chapter provides both background and example problems to illustrate the principles and use of both coax cable and microwaves for telecommunications.

True Digital Telephony—The Introduction of Digital Transmission

AT&T introduced is first digital transmission system in 1962. It had 24 TDM voice channels and was designed to augment or replace their existing analog FDM systems. This new T1 system could, at first, be cascaded to provide up to 96 voice circuits but later developments provided for up to 4,032 circuits. Other early digital gain-pair systems were developed to interface with analog long lines and serve units of 80 subscribers in rapidly growing communities. Both the digital techniques and the equipment rapidly expanded with the introduction of microwaves and, especially, the use of glass fiber for transmission. The subjects of glass fibers and fiber optical transmission is covered in a section of this chapter.

SONET—A High-Level Telecommunications Network Processing System

The introduction of wide-band fiber optics not only gave network designers much more latitude on how to transmit and distribute signals, it also made it possible to rethink the entire idea of what a telecommunications system could and should do. SONET, the Synchronous Optical Network, is a model that permits not only very expanded signal capacity but provides for the interface of both existing analog and digital telephone equipment as well as present and yet-to-be-designed peripheral equipment. This model also provides for additional telephone services such as call waiting, call forwarding, etc.

Special Signal Processing for Telephony

A special section, near the end of the chapter, reviews the unique needs for telephone processing. Such topics as companding, adaptive gain, the use of redundancy, as will as the coding and modulation techniques for telephone signal proccessing are included.

Telephone Peripherals

The final section of the chapter discusses cellular telephones and fax machines and the techniques and the national and international protocols that are required to operate such devices on a worldwide basis.

Archival and Cardinal References

Brooks, John. *Telephone: The First Hundred Years.* New York: Harper & Row, 1976.

Gilbert, E. N. "Information Theory after Eighteen Years." Bell Telephone Monograph, Holmdel, NJ: Bell Telephone Laboratories, 1965.

Nyquist, H. "Certain Topics in Telegraph Transmission Theory." *Bell System Technical Journal*, April 1928.

Pahlavan, K., and Holsinger, J. L. "Voice-Band Communication Modems: A Historical Review." *IEEE Communications Magazine*, January 1988.

Shannon, C. E. "A Mathematical Theory of Communication." *Bell System Technical Journal* July and October 1948.

Contemporary References

Adams, R. W. "Companded Predictive Delta Modulation: A Low-Cost Conversion Technique for Digital Recording." *J. Audio Eng. Soc.*, March 1984.

Banks, Michael A. *The Modem Reference: The Complete Guide to Selection, Installation, and Applications.* New York: Brady/Simon & Schuster, 1991.

Basch, E. E. (ed.) *Optical Fiber Transmission.* Indianapolis IN: Howard W. Sams & Co., 1987.

Bellamy, John. *Digital Telephony.* New York: John Wiley & Sons, 1991.

Campbell, Joe. *C Programmer's Guide to Serial Communications.* Carmel IN: Howard W. Sams & Co., 1989.

CCITT. *Red Book(s).* (Recommendations) Malaga-Torremolinos.

Freeman, Roger L. *Telecommunication System Engineering* 2d ed. New York: John Wiley & Sons, 1989.

———. *Radio Systems Design for Telecommunication.* New York: John Wiley & Sons, 1987.

——— . *Telecommunication Transmission Handbook* 2d ed. New York: John Wiley & Sons, 1989 b.

Gage, S. et al. *Optoelectronics/Fiber-Optics Applications Manual* 2 d ed. New York: Hewlett-Packard/McGraw-Hill, 1981.

Herbert, Nick. *Quantum Reality.* Garden City, New York NY: Anchor Press/Doubleday, 1985.

IEEE Transactions on Microwave Theory, vol. 21, no. 4, 1973, special issue on SAW filters.

Jones, Don. "Delta Modulation for Voice Transmission." Application Note 607. Melbourne, FL: Harris Semiconductor Div., 1980.

Lee, William C. Y. *Mobile Cellular Telecommunications Systems.* New York: McGraw-Hill, 1989.

Libbey, R.L. *Handbook of Circuit Mathematics for Technical Engineers*, Boca Raton FL: CRC Press, 1991.

Nelson, Marilyn McCord and Illingworth, W.T. *A Practical Guide to Neural Nets.* Reading MA: Addison-Wesley Publishing Co., 1990.

Pierce, John R., and Noll, A. Michael. *SIGNALS: The Science of Telecommunications.* New York: Scientific American Library, 1990.

Roddy, Dennis, and Coolen, John. *Electronic Communications.* Reston, Virginia VA: Reston Publishing Co., 1984.

——— . "Telecommunications Quality."*IEEE Communications Magazine*, October 1988 (special issue).

5

Radar Signal Processing

An Introduction to Radar—Radio Detection and Ranging

5.1 FUNDAMENTAL RADAR MODELS

Surveillance and Tracking Radars

There are at least a dozen examples of the use of specially transmitted or specially received and processed radio signals that can be gathered under a general umbrella termed radar. As generic touchstones, *surveillance* and *tracking* can be considered the root radar configurations. Surveillance radars usually combine observations to detect the presence of targets (or weather patterns, etc.). They usually have the difficult task of watching large portions of the sky. Tracking radars are used to find (acquire) targets and follow them while continually returning information that can be processed to find their *azimuth* or magnetic compass bearing angle (θ) and their *elevation* angle (ϕ) above the horizon. Most often, the numbers representing these angles can be used to calculate the instantaneous distance and the speed and direction of flight.

The Elementary Graphical Model of a Radar System

It is prudent to return to the general forms of the text model, the graphical model, and the mathematical model to describe the fundamentals of radar.

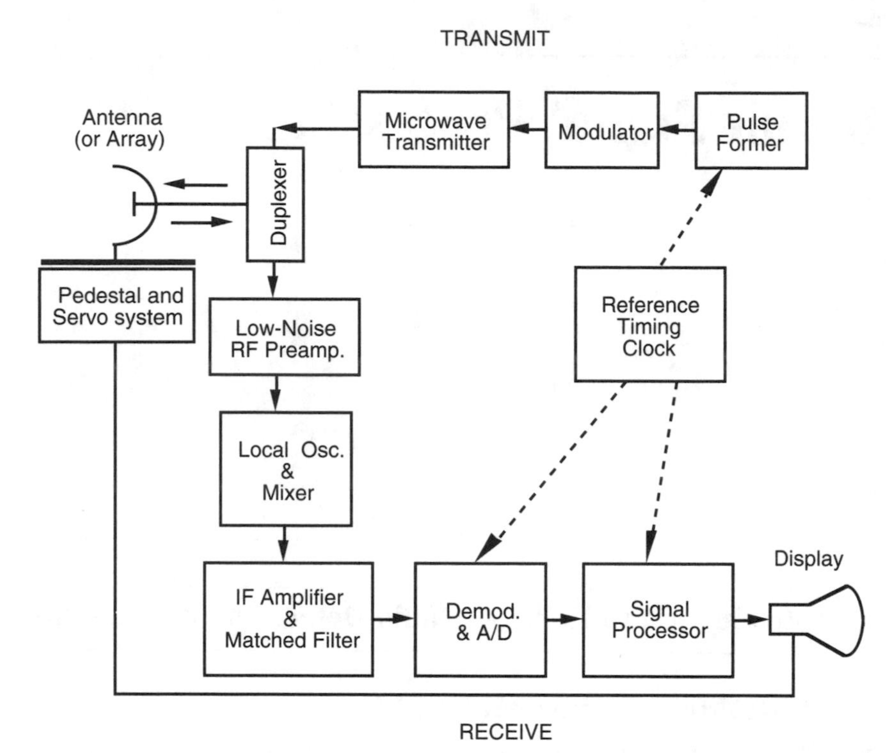

Figure 5.1 The elementary graphic model of a radar system. Like all beginning models, this graphic includes several universal and simplified assumptions. For instance, the transmitted signal may contain other waveforms than pulses and the antenna may be a cellular array rather than a rotating dish.

The Elementary Text Model of a Radar System

In its simplest form, a radar system will simultaneously transmit a pulse and start a synchronized and calibrated clock. When the pulse hits a target, an echo is sent back to the receiving antenna and, after the pulse is detected and processed, the clock is interrogated and the elapsed round trip time is calculated. From this time information, the distance can be calculated. The next level of sophistication in this elementary model would be the derivation of azimuth and elevation information about the way the antenna was pointed, which can be used to calculate where the target was at the instant it was acquired. When the repeating pulse echoes are received and processed, the target's direction of travel may be obtained.

A slight reflection upon the brief preceding material will reveal several of the voluminous calculations that are required to produce the desired information about the position, the vector motion, and speed of the target. The advances of the speed and accuracy of both military and commercial radars has paralleled not only the advances in the speed and precision of computer hardware but especially the advances in the computational algorithms and their efficient use in software. In fact, most of the sections to follow will detail the calculations necessary to obtain the information about location and motion from an environment that is quite hostile, not only from a military perspective but also dominated by problems from nature.

The Elementary Mathematical Model of a Radar System

Many textbooks on radar prefer to introduce the subject with the following well-known radar equations. For a radar with isotropic radiation, i.e., radiating in all directions as from the surface of a minute sphere, the power density per unit area is:

For an isotropic radar antenna:

$$\mathbf{P} = \frac{P_t}{4\pi r^2} \qquad \text{(E 5.1)}$$

where:

P	=	radiated power density in watts per unit area
P_t	=	the peak power in the transmitted pulse
R	=	the distance from the antenna to the target
$4\pi r^2$	=	the surface area of an imaginary sphere with radius r

For a directional radar antenna:

$$P = \frac{P_t G}{4\pi r^2} \quad \text{or} \quad P = \frac{P_t G A_1}{4\pi r^2} \qquad \text{(E 5.2)}$$

where:

G	=	the directional gain of the antenna measured in the direction of the target
A_1	=	the radar cross section of the target. Most targets will have a radar cross section smaller than the physical cross section and stealth targets will be much smaller.

The above equations describe the propagation of radar signals from the transmitting antenna to the target. More mathematics are required to complete the round-trip path of the signal back to the radar receiving antenna, which is usually (but not always) the same physical antenna. Again, the power dimensions are in watts per unit area—most often in square meters.

For the round trip, including the returning echo:

$$P = \frac{P_t G A_1}{4\pi r^2} \frac{A_2}{4\pi r^2} = \frac{P_t G A_1 A_2}{(4\pi)^2 r^4} \tag{E 5.3}$$

where:

A_1 = is the target cross section
A_2 = is the effective cross section of the receiving antenna

Finally, these equations can be transformed into a more workable and understandable form—the predicted range of an elementary radar system.

$$R_{max} = \left[\frac{P_t G A_1 A_2}{(4\pi)^2 S_{min}} \right]^{1/4} \tag{E 5.4}$$

where:

R_{max} = the maximum range of the somewhat simplified radar system
S_{min} = the minimum detectable signal at the receiver, to be described later

This E 5.4 can further be put in a more conventional and convenient form by using a relationship from fundamental antenna theory where $G = 4\pi A_2/\lambda^2$ where λ is the wave length of the antenna. (See the Terman, Kennedy, or Skolnik references.)

$$R_{max} = \left[\frac{P_t G^2 \lambda^2 A_1}{(4\pi)^3 S_{min}} \right]^{1/4} = \left[\frac{P_t A_2^2 A_1}{(4\pi) \lambda^2 S_{min}} \right]^{1/4} \tag{E 5.5}$$

As with all complicated equations, it is sometimes difficult to realize the proper interpretation. The following guidelines can aid in the interpretation and use of the radar equations.

1. All these equations oversimplify the statement of the problems in designing and implementing radar systems. Although they provide convenient packets of starting information, they are just that—starting information.

2. The equations do point out the significance of the directivity gain G of the transmitting antenna and the area A_2 of the receiving antenna. Of course, the effective area (or stealth area) A_1 of the target is dominant in the radar process.

3. The radar range is proportional to the fourth root of the peak radiated power. This means that to double the maximum range would require a 16-fold increase in transmitted power.

4. E 5.5 can give the false impression that the range of radar varies as either $\lambda^{1/2}$ or $\lambda^{-1/2}$. Both assumptions are incorrect, as E 5.4 shows the range of the radar is independent of λ. (See the 1980 Skolnik reference.)

5. Most often the calculated range values are too large by a factor of about two to one. For example, the equations do not consider the problems of receiver noise or sky clutter.

The following sections will build upon the graphic, text, and mathematical models. First, they expand on the general radar processes represented by the blocks in Figure 5.1, and then they include some special radar processing problems and specific implementations.

Radar Signals

5.2 SINE WAVES, PULSES, AND MODULATION

Some of the most intriguing facts about the modulation process in high-power radar and TV transmitters are the magnitudes of voltage, current, and their rapid change with time. These high-power transmitters often emit peak powers in the order of 1 to 10 megawatts. That means modulation voltages of 5 to 20 kilovolts, with pulse currents from 200 to 1,000 amperes. The power levels and the rate of change of the pulses are so high it almost gives a new meaning to the familiar equation "di/dt". A detailed description of the devices used in high-power modulation and output stages is beyond the scope of this book, but they include magnetrons, klystrons, and traveling-wave tubes, sometimes working together in unique configurations. It should also be noted that the power supplies for high-power radar and TV transmitters, with their required high reliability and protective circuits, represent some of the most creative of all circuit designs.

CW Doppler Radar

The most elementary radar signal is a continuous, unmodulated sine wave carrier. A moving target will return an echo that is slightly different in frequency than the transmitted frequency. With proper processing, the receiver can convert this frequency shift into target velocity. One of the disadvantages of CW radar is its required high average power. Since CW Doppler radar transmits a continuous signal of maximum amplitude, for comparable penetration, it usually requires much more input and radiated power than the prototype low-duty, pulsed radar. For this reason, CW Doppler radar is most often used in short- range applications, such as automobile speed-control radar, since they require only low operating power levels. Doppler radar is further detailed in the section on radar receivers.

Linear FM Pulse-Compression Radar

A modified form of pulse radar uses a short burst of the sine wave-type signals rather than a rectangular pulse. However, during the time of this burst, the transmitted signal is linearly frequency modulated. One of several ways to frequency modulate the carrier is to use a positively or negatively going ramp during the equivalent pulse-width time interval—see Figure 5.2. Where carriers are frequency modulated by ramps or similar functions, the detected output is usually audible and thus this type of modulation is often termed a chirp. Frequency modulation provides a very common type of radar termed pulse compression radar that will be further detailed in the section on radar receivers.

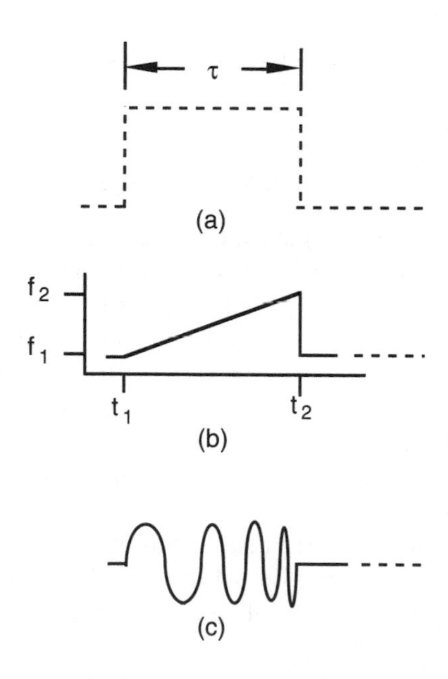

Figure 5.2 The derivation of a chirp or frequency-modulated tone burst in the time τ normally allotted for the prototype rectangular radar pulse: (a) shows a prototypical pulse lasting τ-seconds from time t_1 to t_2, (b) shows a linear ramp that will frequency modulate the transmitter carrier producing the modulated output burst signal shown in (c).

Pulsed Radar and the Pause Formers

In theory, the simplest pulse former and modulator would be a switch that would periodically switch the radar carrier on and, usually in about 1 μs, switch it off again. Although this crude approach, if practical, would produce a carrier "burst" rather than a rectangular pulse, there is a variation on this switching idea that is used in some pulse radars—the switched pulse-forming network.

Figure 5.3 shows one way to produce a nearly rectangular pulse signal for high-powered radar transmission. It is well known that an unterminated transmission line can be configured to reflect a voltage equal (not counting losses) to a voltage impressed on the input of the line. By slowly charging a special transmission line through a series impedance and rapidly discharging the line into a load, which is usually made the same value as the characteristic impedance of the line, the line will produce a voltage pulse with a width that is a function of the length of the transmission line. For instance, a transmission line of about 250 feet will produce a 1 μs pulse. A much simpler solution is to create an artificial L/C transmission line network, as shown in Figure 5.3. (See the Terman or Skolnik references.)

A second method of pulse modulation uses components that are a variation on the more common practice of using a preformed pulse from a function generator to drive the grid or base of an output modulating device. One must realize that the concept of a simple pulsed output device often becomes very complex when implemented in practice. Because of the high peak powers and the microwave frequencies, the modulator and output devices often become some combination of magnetrons, klystrons, discharge, and traveling-wave tubes. Likewise, it should be remembered that much of the circuitry includes bulky wave guides or components configured from wave guide-like devices.

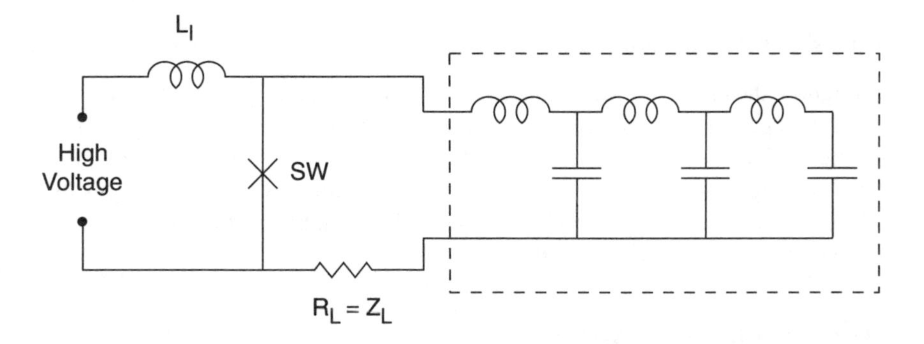

Figure 5.3 An elementary pulse modulator that includes an artificial transmission line network. The high voltage will slowly charge the network through a constant-current device such as the inductor L_1 and then the shorting switch SW will be closed to rapidly discharge it into the load R_L. The shorting switch can be a triggered gas-discharge tube and the load R_L can be the combined circuits of the output (tube or transistor) devices coupled to the load of the radar antenna.

5.3 RADAR ANTENNAS

Processing with Dishes

Radar antennas can normally be divided into the classes of parabolic dishes, horns, lenses, and arrays. A parabola dish has a focus point or vertex that provides a single point to inject the signal for radiation transmission or for receiver pickup. If the signal is injected at the focus point, the resulting transmitted beam is almost parallel, i.e., pencil-like. Likewise, by mounting the dish on a swiveling cradle and rotating pedestal, the narrow radar beam can be moved over a wide solid angle. The major disadvantages of dish antennas are their inability to easily change the beam shape and their somewhat limited power-handling capability. This latter problem can be circumvented by changing from a solid to a dish-type formed grid. The formed-grid design can be built into other shapes to provide different fan-beam patterns.

Processing with Horns

Horns are also used for radar and other uses of microwave antennas. Whether it be a musical instrument, a loudspeaker enclosure, or a radar antenna, the tapered horn shape is used as an impedance matching device. In a cornet or French horn, the length of the tapered tube furnishes the pitch and quality of the note (to a first approximation), however, it is the long, convoluted taper that, in a sense, amplifies the sound by matching the confined mouth-mouthpiece opening to the open air at the end of the instrument's bell. (The taper acts in a manner similar to the step-up or step-down windings in an impedance matching transformer.) The horn-type loudspeaker cabinet performs in a similar way.

Processing with Lens Antennas

Throughout the history of radar, just as with optics, a variety of lens designs have been developed to give a particular radiating pattern that is based on staggering or altering the index of refraction (h) of parts of the lens. (See the section on optical fibers in Chapter 4.) The 1980 Skolnik reference details several lens designs that include metal-plate zoned lenses and those with tapered refraction indexes based on the early work of R. K. Luneburg.

Processing with Phased Arrays

One of the major advancements in radar antenna design was the phased array. This is usually a grid of many individual antenna elements (usually elementary dipoles) that can either be separately fed or fed (driven) in selected groups. By carefully choosing the amplitude and/or phase (usually with a dedicated computer) of the signals going to individual elements or groups of elements, the shape and direc-

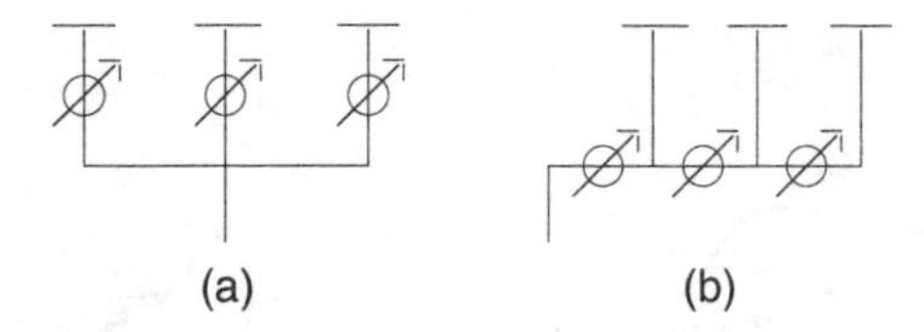

Figure 5.4 Two of several ways to feed the elements, usually dipoles, in phased arrays. The computer-controlled phase shifters are shown as circles with arrows.

tion of the composite transmitted beam can be precisely controlled—see Figure 5.4. It is not uncommon to have as many as 4,096 individual radiators that can have the relative phase of the emitted signal adjusted up to 180°, i.e., a carefully designed array can produce a beam that can be placed almost anywhere in a hemisphere. Common controllable phase shifters include PIN diodes, which can only shift in discrete amounts or ferrite rods that will continuously shift the phase of microwaves if the ferrite is placed in a controllable magnetic field. See the Skolnik reference. (The effects of phased arrays and other antenna configurations are discussed in more detail in the sections on receiver signal processing.)

5.4 RADAR SIGNAL PROCESSING WITH ANTENNA ARRAYS

Conical Scan Radar

The beam from a prototype dish, as shown schematically in Figure 5.1, can be visualized as a beam similar to that of a searchlight or in the form of an elongated petal or a lobe. Since any of the idealized beam configurations will be most intense along the central axis of the transmitted beam or, conversely, most sensitive along the central axis of the received beam, it is desirable to have the target in the center of the beam, i.e., on the central axis. There are several methods used to adjust the precise scanning motion of the antenna to both acquire the target in the center of the beam and then to keep it in the center of the beam while the antenna is following (tracking) the movement of the target.

Figure 5.5 illustrates the original processing required to place the target in the exact center of the beam. Note that when first acquired, the target is not on the axis of the antenna. Also, remember that this figure only shows the target in one plane. Since the radar needs to know both the elevation and the azimuth of the target, two sets of servo drivers, sensors, and controllers are required to alter the motion of the antenna scan. Once the target is roughly acquired, the antenna and thus the beam will rotate in a cone-shaped or conical scan. For some antenna dish or grid systems, where the weight is in the order of tons, it is no small achievement to control the position of the target relative to the axis of the antenna down to very small angles.

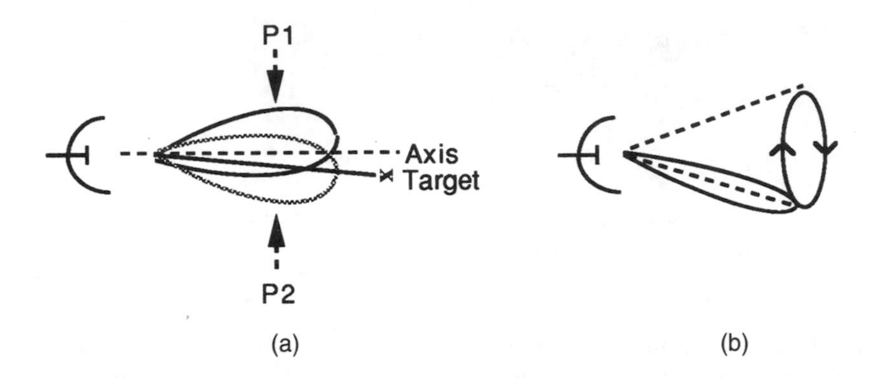

Figure 5.5 The processing required to place the target on the axis of the radar antenna. In (a) it is assumed that the system is reporting the target off axis with sweeps at positions P1, P2, etc. After a few scans, the servo controllers for the antenna will adjust the (phase) position of rotation so that the target will appear on the antenna axis. In (b) the same principle is expanded to three dimensions for both azimuth and elevation information.

Monopulse Radar Processing

Monopulse is one of the very intriguing radar antenna processing systems. In its simplest form, monopulse systems will duplicate the actions of conical scanning techniques without the need for the electro-mechanical servo systems required for elevation and azimuth control. These two servos are replaced by a four-element antenna matrix, combined with an all-electronic signal processor that uses the sum (Σ) and difference (Δ) signals from the antenna elements to derive the target direction and the elevation and angle estimates. The processor uses the output of two of the antenna array elements to compute the elevation angle estimate and two of the array elements to compute the azimuth angle estimates. The antenna elements can be hyperbolic dishes, horns, or a combination of the two. A common configuration uses a reflector dish with four horns clustered at the vertex to facilitate both transmission and reception, as shown in Figure 5.6.

Radar Receivers and Their Signal Processing

5.5 RADAR NOISE PROBLEMS

TV and Radar Signals in Noise

In some ways the problems of identifying a small object on a very noisy TV screen and of positively acquiring and following a radar target are similar. Both involve several types of noise reduction and signal processing techniques to be certain that

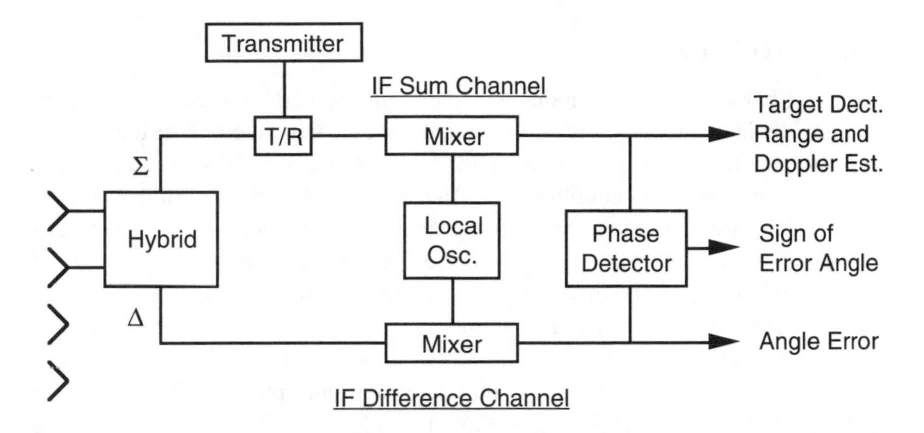

Figure 5.6 The elemental amplitude-comparison monopulse radar. Note that only two of the four antenna elements are shown. A complete system will generally use four Σ/Δ hybrids and three channels rather than two. (See the Fink and Christiansen reference.)

the representation of an object is truly a target or, say, a person's left ear. TV has several advantages. The most telling is its inherent redundancy. Sixty times a second one has several clues that help the eye and the brain decide if that is, to be sure, the outline of an ear. Certainly, it is more difficult to identify an ear (or the ball) in a rapidly moving basketball game than in a slower moving dramatic program. Still redundancy and usually some knowledge of the program content is very much on your eye's side. There are, however, some tricks that can be used in radar that are usually not applicable for TV, and these will be the subject of the following sections on radar signal amplification and processing.

5.6 RADAR RECEIVER INPUT STAGES

The Radar Input Noise Floor

Most, if not all, radar receivers follow the common superheterodyne configuration. As with TV and most types of radio, the noise of the input RF preamplifier may set the noise floor for the rest of the receiving system. Radar receivers can afford to pay more for their input components than can most radio and TV sets. A gallium arsenide FET is a very common microwave input RF component. A noise figure of about 1.5 dB is typical for a radar receiver, whereas a figure of 3.5 dB is more common for TV. In a few radar installations, the input components are cooled to reduce the equivalent noise temperature. (See the appendix.) Most often, the design of the radar will include a noise list or noise budget that tabulates all input noise sources from the antenna through the mixer stage.

Radar Mixer Noise

The mixer (which is really a modulator) stage can itself add to the noise in the radar signal by injecting unwanted noise signals that are generated in the local oscillator. Two of the methods that reduce this local-oscillator (LO) noise are the use of a special narrow-band coupling amplifier between the local oscillator and the mixer and the use of a special balanced mixer. There are several circuits and techniques that produce the balanced mixer action. All are based upon the principle of injecting only the sum and difference signals from the RF and the LO.

The mathematical analysis of a two-input signal mixer (modulator) varies depending on what assumptions are made about the linearity function of the mixing device. An ideal or balanced mixer/multiplier will only produce the two output terms of $(\omega_1 + \omega_2)$ and $(\omega_1 - \omega_2)$, as shown in E 5.6 (see Chapter 2):

For a balanced mixer:

$$\cos \omega_1 t \cos \omega_2 t = \frac{1}{2} \cos (\omega_1 + \omega_2) t + \frac{1}{2} \cos (\omega_1 - \omega_2) t \qquad \text{(E 5.6)}$$

If some other nonlinear function, such as a square function, is used to mix $\omega_1 + \omega_2$, additional noise terms are added that may partially corrupt the mixer output. For instance, if a square-law mixer is used, the appropriate trigonometric manipulations will produce (see the Horowitz and Hill reference):

For a nonlinear square-law mixer:

$$(\cos \omega_1 t + \omega_2 t)^2 = 1 + \frac{1}{2} \cos (\omega_1 + \omega_2) t + \frac{1}{2} \cos (\omega_1 - \omega_2) t +$$

$$\frac{1}{2} \cos 2\omega_1 t + \frac{1}{2} \cos 2\omega_2 t \qquad \text{(E 5.7)}$$

There are other noise items that may appear in the radar signal and require special processing if the aforementioned target is to be distinguished from the other confusing signals. Since the amplitude of the radar echo may be very small, the noises of the sky or the galaxy—that some contemporary theories predict originated with the big bang—are a dominant factor in the input signal. In addition, military radar signals will most often include clutter and jamming. Clutter is caused by echoes from either manmade diversionary objects, such as strips of metal foil, or from rain, clouds, trees, etc. Jamming is signals that are transmitted, using several clever techniques, to confuse the radar receiver processing.

Doppler Processing

Radar, like other RF or audio signals, is subject to the phenomenon discovered by the Austrian physicist Christian Doppler in 1842. It must be emphasized that any radar signal is affected by this physical occurrence of a frequency shift, including the special branch termed Doppler radar. Very simply, when a device such as a train whistle, a fire siren, or a radar target is emitting a constant frequency signal, if the source and the observer are moving closer together (either or both can be moving), the observer will hear more pulses or cycles of the signal per unit of time and, therefore, the sound will be observed as rising in pitch. If the source and the observer are moving away from each other, the observer will hear fewer cycles per unit time and the signal will be observed as going lower in pitch.

The Doppler shift f_d from a target is:

$$f_d = v_r \, 2/\lambda \text{ in Hz} \tag{E 5.8}$$

where:

v_r = the speed of the target toward the receiver
$2/1$ = the phase-shift distance to produce a 1-Hz change in the received signal frequency

Since the target is moving away from or toward the target (motion parallel to the receiver or angular motion requires more complex calculations), some time is required to calculate and resolve the continually changing Doppler frequencies. The time to resolve each two successive changing Doppler frequencies is approximately:

$$t \approx 1/\Delta f_d \text{ in seconds} \tag{E 5.9}$$

where:

Δf_d = the resolution between two successive Doppler frequency echoes

Likewise, there will be an average error in these measurements that can be calculated as:

$$\delta f_d = \frac{1}{t\sqrt{(2 \, x \, SNR)}} \text{ in Hz} \tag{E 5.10}$$

(The Kingsley and Quegan reference has a nice section on Doppler radar and the Doppler effects.)

Radar Signal Processing for Minimum Noise—The Use of Statistics

In former sections, the problems of noise in the receiver were discussed and several circuit techniques were discussed that will minimize the effects of noise in detecting a target. The reader might also review the techniques of noise reduction discussed in Chapter 3. In radar, since the received echo signal may be in the order of 10^{-14} watts (and often less) and is not necessarily redundant, noise reduction requires additional approaches. One of the techniques commonly used to understand

and reduce the galactic noise involves the general discipline known as probability and statistics. (Many other communications systems also use probability theory to improve the signal-to-noise and data rates. For the landmark paper, see the Shannon reference.)

Galactic noise can be defined as billions and billions of extremely minute explosions producing electromagnetic radiation. Since the number of these emissions is so vast, some scheme is required to predict data about the number, intensity, and spectrum of these emissions—this omnipresent galactic noise. One of the most appropriate statistical techniques is termed the probability density function. In simplistic terms, the probability density is the guess of how many events of one type will occur out of a large number of possibilities. For instance, out of an elementary school of 200 students, how many will have a height of 3' 6", 3' 8", or 4', etc.? What will be the density of the students with a height of 3' 8"? On average, the heights of the students will lie somewhere on the well-recognized bell-shaped or Gaussian standard deviation curve. By using some rather advanced and complicated mathematics, it can be shown that the root-mean-squared (RMS) spectrum of noise is usually but not necessarily Gaussian.

Signal Processing with a Matched Filter

Armed with a knowledge of the galactic RMS noise characteristic, it should be possible to design a filter that would produce the maximum possible pulse signal in the presence of this galactic ambient noise. This is indeed possible if it is kept in mind that the goal of the filter is to preserve, pass the maximum peak amplitude of the signal, and to inhibit as much as possible of the RMS noise.

Before specifying the details of this magical matched filter, it is important to again note that the goal called for the maximum peak amplitude of the signal but did not include any requirement to maintain its detailed pulse shape. Mathematically, a pulse can be specified several ways. In the time domain, [1] it is usually specified by its height, which can be some amplitude A or unity, [2] by its width in τ seconds, and [3] if it is repeating, the repetition time in T seconds. If the pulse is specified in or transformed to the frequency domain, its spectrum components (its frequency band) can be specified by a Fourier series or a Fourier integral (transform). A less familiar form to specify the bandwidth is:

For the 3-dB bandwidth B:

$$B = \frac{1}{\tau} \qquad \text{(E 5.11)}$$

where:

τ is the pulse width

To be sure, a filter with a bandwidth specified by this formula certainly would not pass a pulse with any semblance of its original shape. However, it will closely maintain the peak-pulse amplitude.

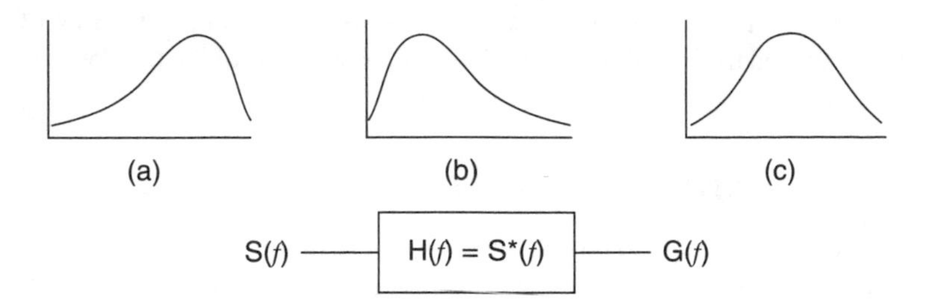

Figure 5.7 The action of a conjugate or matched filter. The frequency-domain response of the filter (b) is the conjugate of the spectrum of in input signal (a). The output (c) is obtained by combining (a) and (b) using the mathematical techniques of convolution for time-domain processing or multiplication for frequency-domain processing . (The signal sketches were not generated from specific signal data and are for illustration only.)

A network whose frequency response function maximizes the ratio of the output peak signal power to the RMS noise power is termed a matched filter. (This has nothing to do with the technique of impedance matching.) A matched filter is a network whose frequency spectrum or time function—$S^*(f)$ or $s^*(t)$—is the complex conjugate of the frequency spectrum or time function ($S(f)$ or $s(t)$) of the input signal. (The raised asterisk denotes the conjugate of.) The literature indicates many mathematical methods with many different symbols to derive and present the equations for, and the action of, a matched filter. Figure 5.7 indicated the rudimentary action of the most common or conjugate matched filter. (See the Couch reference.)

5.7 MTI—MOVING TARGET INDICATION

An Introduction

The processing for Moving Target Indication (MTI) is one of the most interesting topics in the field of radar, if not in the general field of signal processing. The premise of MTI is quite simple. Consecutive Doppler echoes that return from fixed or slowly moving bodies (targets) do not change their signal phase from echo to echo. However, if a target is moving, there will be a continuing phase shift from echo to echo. A phase detector can sample the phase of the reoccurring signals and, if there is no phase difference, these signals do not appear on the radar display. Thus, signals from fixed land masses, sometimes clouds, rain, etc., will not override or confuse the displayed target of interest signals. MTI radar is designed to eliminate the chaff from the radar display.

Besides its effective use in radar apparatus, the historical development of MTI can serve to chronicle the development of signal cancelers as filters, from the relatively simple analog comb filters, introduced in Chapter 3, through the general theory of transversal (lining across or intersecting) filters. This provides a background for the development and understanding of both recursive feedback and nonrecursive digital filters. Along with this will come a better understanding of both the frequency and the time domains and the widespread use of the fast fourier transform in DSP.

The Elementary MTI Receiver Circuits

MTI is not used in every radar or in every radar operation. It is designed for use in target tracking and is usually tuned off in the search and surveillance mode. An elementary radar system must include at least one, and usually two, reference oscillators that not only have very stable frequency characteristics but also very stable phase characteristics. Usually there is a transmitter stable local oscillator (stalo) that also works with the receiver mixer and a coherent local oscillator (coho) at the precise frequency of the receiver IF that is used as a phase reference for both the transmitter and the receiver—see Figure 5.8.

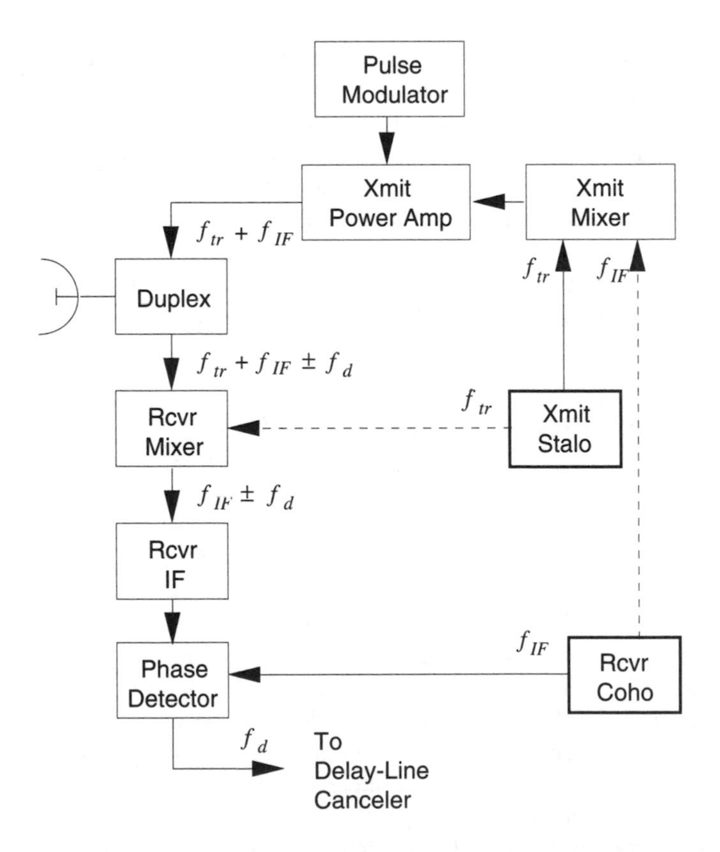

Figure 5.8 One version of an elementary MTI radar. The transmitted pulsed signal includes the mixed sum of the Xmit Stalo (the carrier oscillator) and the Rcvr Coho. The received signal is this transmitted signal $\pm f_d$, the Doppler signal. The phase detector will extract this Doppler frequency signal. The delay canceler, to follow, will eliminate any Doppler signals that do not show a phase shift.

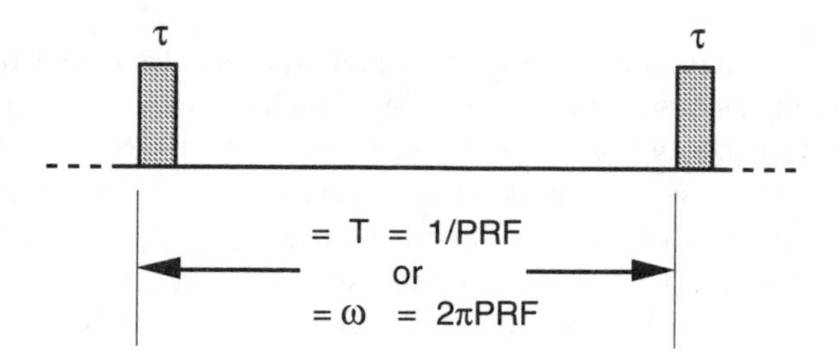

Figure 5.9 A timing diagram of the pulses that might be used in MTI radar. For example, the pulse width τ might be in the order of 1 μs and the time T between pulses could be in the order of 1 ms. This would give a pulse rate of 1,000 pps and a pulse repetition frequency (PRF) of 1,000 Hz. ω would be equal to 6,280 radians. The shading indicates that the modulation might be a simple pulse, a chirp, or some other coding. Note particularly that T is a time-domain dimension and ω is a frequency-domain dimension.

Signal Canceling (Filtering) With an Analog Delay Line

Chapter 3 introduced the idea of canceling signals using a comb filter. In television, the chrominance and luminance signals have staggered bands that are centered exactly 7,875 Hz apart. By using delay lines, these signals can be effectively separated or combed to eliminate intersignal interference. (See Figure 3.27.) Likewise, in MTI radar, the echo from a nonmoving target or land mass can be canceled. The most elementary delay-line canceler is illustrated by Figures 5.9 and 5.10.

It is important to go beyond the simple canceling action of the prototype delay line/subtracter combination to investigate the filtering action of the circuit. Figure 5.11 gives an analytical idea of multiple delay-line filters. Curve (a) is the darkened reference response where the T delay is equal to π radians (180°). (Remember there is a reciprical relationship between the Doppler repetition frequency and its wavelength T.) Note that for the dotted (b) curve, the nominal maximum response is down to less than 1.5 because the phase shift has changed to 0.5 π. Curve (c) shows a maximum relative response of about 0.75 because of the phase shift of 0.25 π and (d), with a phase shift of only 0.1 π radian, is down to about 0.3 and, of course, the (e) response (which is slightly skewed to the left with this computer plot) is down to zero because cos x – cos x = 0. Thus, using the circuits shown in Figure 5.10, a return echo with no echo-to-echo phase shift would be canceled.

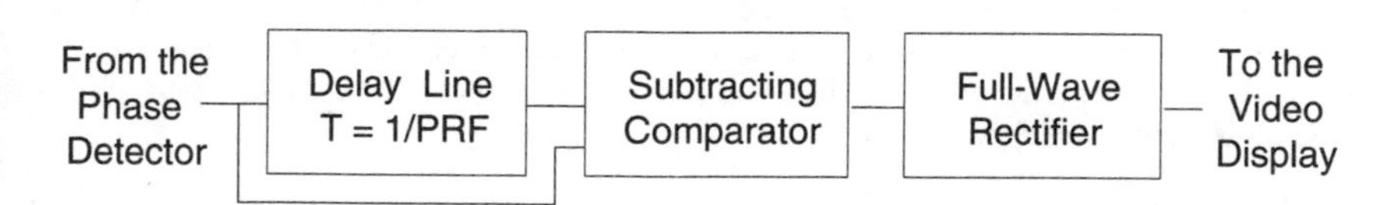

Figure 5.10 An elementary delay line canceler. The length of the delay-line in time is equal to the reciprocal of the pulse repetition frequency. This circuit will solve the equation $v_{out} = |\sin(x) - \sin(x \pm T)|$ where sin x (or cos x) is the input Doppler signal and T is the delay of the line. The full-wave rectifying circuit is used to derive the absolute nonnegative value of the resulting signals.

The minimums in Figure 5.11 all indicate DC values of frequency response and the slightly rising curves to the left or right of each minimum indicate the delay-line canceler's frequency-domain response at low frequencies. Since this DC and low-frequency response represents fixed or very slowly moving targets, such as clouds, it is sometimes desirable to broaden these valleys to provide higher low-frequency attenuation. This can be accomplished by cascading the canceling circuits. Figure 5.12 shows how the attenuation band can be widened by the addition of a second and also a third delay-line canceler. The one delay-line canceler produces an absolute-value sin (or cosine) response, a two-line canceler a $\sin^2$ response, and a three-line canceler a $\sin^3$ response.

The addition of more than one canceler circuit introduces two topics that are important background for the further understanding of analog and digital filters. These include [1] the type of configuration used in the canceler circuitry and [2] the use of weighting factors for the input signals to the summing and/or subtraction (Σ) circuits. Figure 5.13 shows two circuits that produce the same frequency response. Although more than two analog delay lines are unusual because of their high losses, many delays can be utilized in digital filters. This will be introduced in a later section of this chapter.

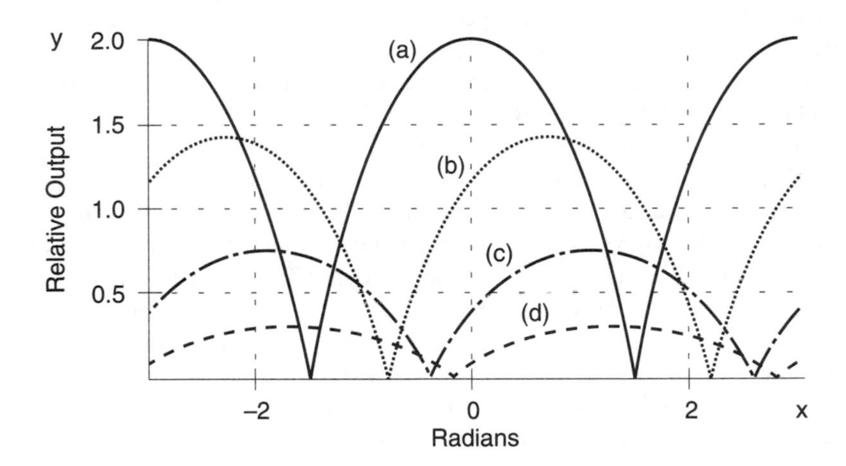

Figure 5.11 The operation of a delay-line canceler as a frequency-domain filter. Cosine terms, rather than sine terms, were used as a convenience to center the reference zero radian curve for this particular plot. All the outputs are absolute values. This graphic shows some plots skewed slightly to the left.

$(a) = y_{(out)} = |\cos(x) - \cos(x + \pi)|$
$(b) = y_{(out)} = |\cos(x) - \cos(x + 0.5\pi)|$
$(c) = y_{(out)} = |\cos(x) - \cos(x + 0.25\pi)|$
$(d) = y_{(out)} = |\cos(x) - \cos(x + 0.1\pi)|$
$(e) = y_{(out)} = |\cos(x) - \cos(x + 0\pi)|$

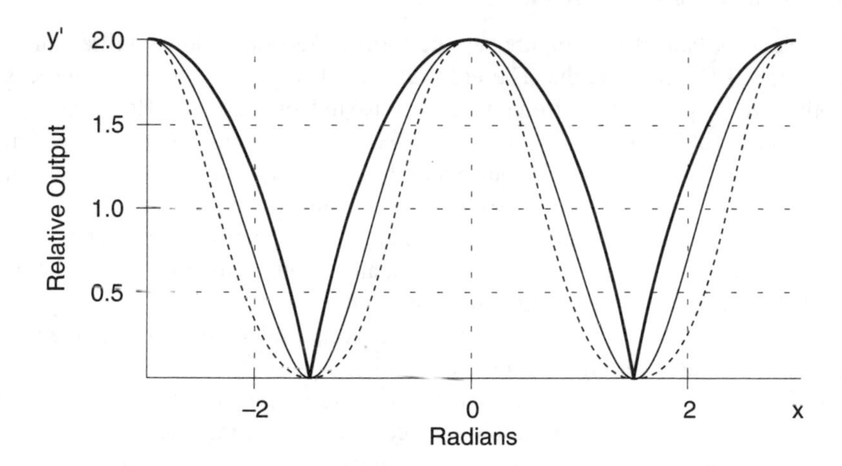

Figure 5.12 A normalized graph showing the advantages of using multiple delay-line cancelers to separate the fixed and slowly moving targets from the rapidly moving targets of interest. The heavy line is the reference one delay-line model, as shown in Figure 5.10. The light line indicates the frequency-domain response of two cascaded delay-line filters and the dashed line indicates three cascaded filters.

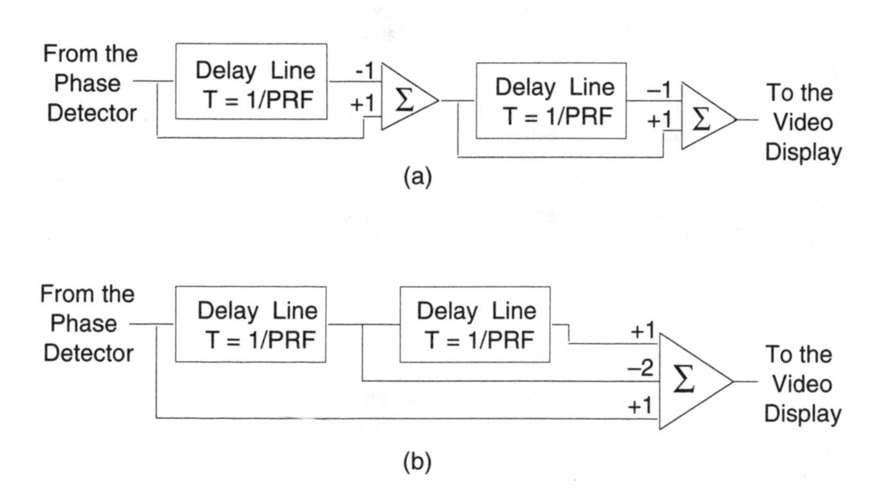

Figure 5.13 Two delay-line canceler configurations that produce the same response. Note the weighting coefficients shown as +1, –1, and –2.

Range-Gated Doppler Filters

The delay-line cancelers in Figure 5.13 perform as frequency-domain filters much as special RLC tuned circuits. Therefore, it should be possible to substitute RLC or other types of analog electronic filters for use in Doppler radar. However, such a component filter, if its bandwidth is narrow enough to filter the Doppler frequencies, will ring from the Doppler pulse and thus give false information. One solution to this problem is to use a series or block of filters to divide the required spectrum. In electronic communication, the meaning of a spectrum can mean a series of frequencies—thus, the use of spectrum usually connotes the frequency domain. If a waveform is given in the time domain, as the T or τ in Figure 5.9, a special mathematical manipulation, such as the FFT, is required to convert from the time domain to the more familiar frequency or ω domain.

Before a block of RLC or bandpass Doppler filters can be used, the PRF must be broken up into a series of samples. Since—again jumping back to the idea of the time domain—time intervals in radar can be equated to target range, the device to break the signal into specific time chunks is termed a range gate. The time width of each range gate is usually equal to the pulse width τ and its 1/pulse bandwidth as given in E 5.11. Thus, the pulse echo return of a moving target would progress from one range gate to another.

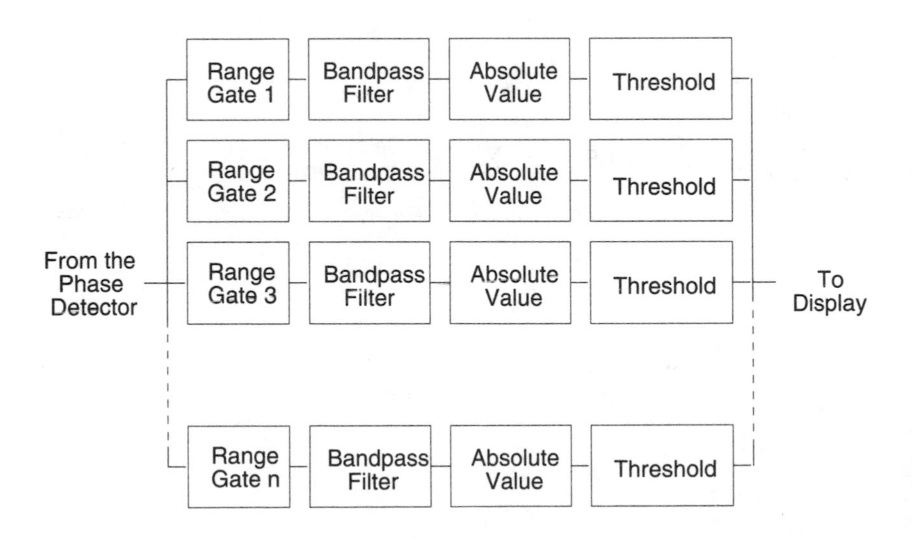

Figure 5.14 A block diagram of analog MTI radar processing using range gates and bandpass Doppler filters.

The threshold circuits are used to set a bias level for signal display. By adjusting this threshold either manually or automatically, it is possible to eliminate some of the unwanted grass in the picture. Although this threshold circuit is very helpful in producing a display without unwanted background signals, it must be carefully set so that weak targets are not eliminated.

Multipulse or Transversal Filters in the Frequency and Time Domains

The use of multiple delay-line filters, introduced in Figure 5.13, can be expanded in two ways.

1. Additional delay lines and adders can be used to give a three-pulse canceler, a four-pulse canceler, etc. There is usually a problem with noise if too many delay-line circuits are added as analog circuits. However, using similar principles, multiple delay cancelers work very well using the digital circuits, to be described later.
2. A second class of feedback or recursive circuits can be added that will aid in shaping the resulting filter's bandwidth to the desired shape. Both the feed-forward signals and the feedback signals can use the weighting factors to modify the signal magnitudes as they progress through the filter. Analysis in the analog domain is based upon the Fourier transform and in the digital domain on the Z-transform. The choice of the feed-forward or feedback weighting factors is usually dependent upon specific mathematical functions or factors that were derived based on the work of Butterworth, Bessel, Chebyshev, etc. (See any textbook on digital filter design.)

In the digital domain, the physical delay line is replaced by clocked storage registers (for example, J-K or D-type flip-flops) that can be used individually or cascaded Z^{-1} units to produce more delay—see Figures 5.15 and 5.16.

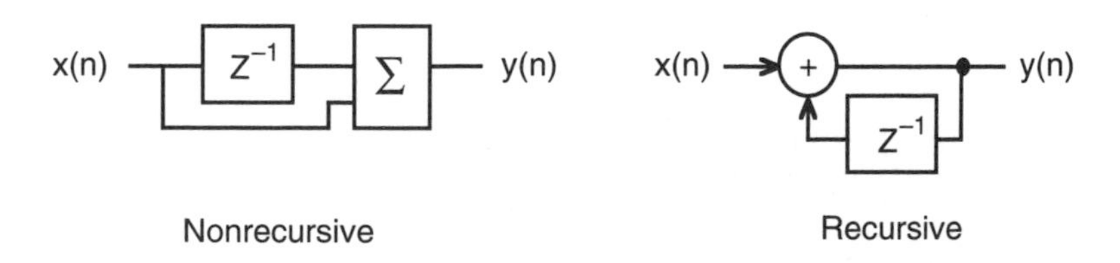

Figure 5.15 The building block nonrecursive (nonfeedback) and recursive (feedback) digital filters. If these building elements are ganged to give the digital equivalent of multiple delay-line analog filters, additional weighting summers, with the a_n and the b_n coefficients, are inserted in the lines going to the Σ and + blocks.

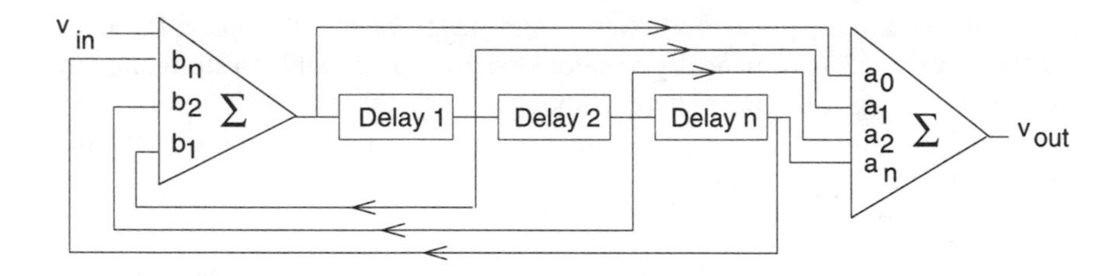

Figure 5.16 Multiple delay filters using both nonrecursive and recursive filters: the canonical (fundamental) configuration of analog or digital filters using both feed-forward and feedback signals to shape the bandpass characteristic. Older practice shows the a_n and b_n weighting factors inside the summing blocks, whereas small circles are usually used in contemporary drawings.

5.8 THE ELEMENTAL FUNCTIONS OF RADAR USING DIGITAL SIGNAL PROCESSING

An Introduction to Radar DSP

The proliferation, convenience, and reliability of highly integrated digital circuits starting in the late 1970s, gave radar designers much more versatility in their circuit designs. Like television engineers, radar engineers can use digital circuits to realize much more closely the predicted performance of their circuits that was often compromised by analog circuits. Specifically, the development of reliable integrated circuit analog-to-digital converters that could operate with clock speeds over 10 MHz made many real-time digital-processing functions possible. The improvements in general software techniques, as well as the development of processing

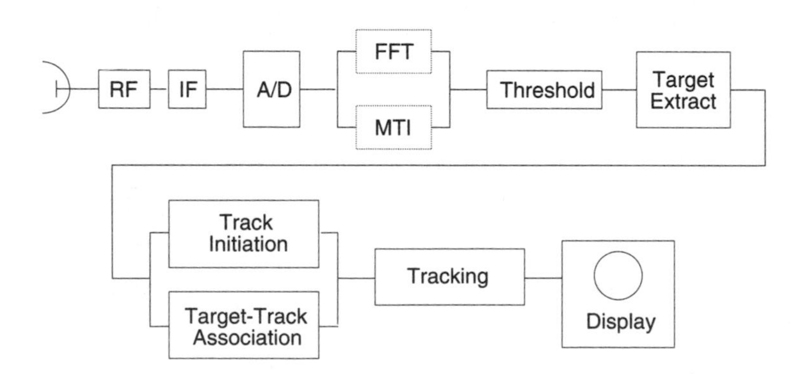

Figure 5.17 The elemental procedures in radar digital signal processing. This is a composite block diagram and not all of its elements will necessarily be used in one radar system.

algorithms such as the fast Fourier or Hartley transforms not only allowed innovation but made it possible to change procedures on the fly that had been impossible with the all-hardware analog designs.

Digital Circuits for MTI and Clutter Reduction

For review, let it be remembered that the MTI signal processing is used to filter out the redundant and often objectionable signals from fixed or slowly moving targets, clouds, land masses, clutter, etc. However, these filters must not destroy the information in the echoes of the rapidly moving targets. The filters usually take one of the aforementioned three forms including:

1. The return echoes can be filtered using analog delay lines in a comb-filter configuration. Since the echoes of fixed or slowly moving targets will return at a fixed repetition rate, these comb filters or delay-line cancelers can be used to eliminate these constant, unwanted target return signals.
2. The pulse repetition time can be divided into a series of much shorter snapshots, called range gates, with their bandpass frequency response approximately equal to the reciprocal of the radar pulse time. The circuitry for these bandpass filters can, in theory, be either analog or digital.
3. The use of digital delays (active circuit delays or memory storage) can be used both as delay-line cancelers and as digital bandpass circuits in the range gates.

MTI Processing—FFTs and Digital Range-Gate Filters

The FFT is a mathematical technique to convert a function that is displayed or operating in the time domain to a display or operation in the frequency domain. The resulting frequency-domain representation will be the spectrum of the signal. A practical example of a spectrum analyzer is the idea and use of the range gates. An elementary laboratory spectrum analyzer can be constructed using a group of tuned circuits with their center frequencies tuned to progressively higher frequencies (see Figure 5.18). (Early analog band analyzers and equalizers in hi-fi and stereo receivers used this principle.) The range-gate filters, analog or digital, follow the same pattern.

Figure 5.18 The bandpass of adjacent filters that can be used to analyze the frequency spectrum of a signal. Note that the two on the left illustrate passbands that do not join or overlap while the two on the right show overlapping passbands. Adjacent range-gate bandpass filters are shown in Figure 5.14.

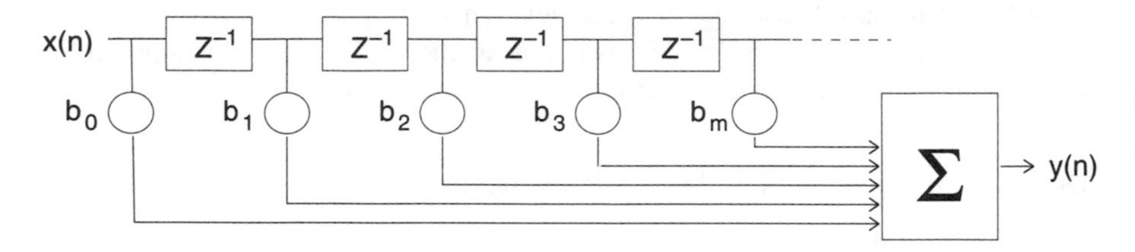

Figure 5.19 A convenient and understandable form of a nonrecursive digital filter. In radar, the weighting factors b_0, b_1, b_2, etc., often are derived from the mathematical series termed Bessel functions and thus produce a Bessel filter. Often Butterworth and Chebyshev weights are also used. Compare this with its prototype shown in Figure 5.15.

The basic or canonical form (derived from the same root as canon law) of a digital filter is shown in Figure 5.16. However, most radar filters are nonrecursive, as shown in Figure 5.19.

Most of the model filters that have been presented assumed a low-pass configuration. However, both high-pass and bandpass delay-line and digital-delay filters can be and (especially the digital implementations) are used. The design of these filters employs another mathematical tool called the z-transform and is beyond the scope of this work. However, the fundamental idea for a digital bandpass filter is illustrated in Chapter 3, Figure 3.27. Note in Figure 3.27 that the spectrum of the top comb filter is maximum at zero frequency, zero at a frequency represented by 0.5 the line rate, and maximum again at the one-line point. Now observe the behavior of the lower comb-filter spectrum. Its response is zero at zero frequency and maximum at a frequency represented by 0.5 the delay, etc. Thus, the lower spectrum and its circuit is a bandpass filter around the frequencies represented by one-half of the line delay.

In most cases, it is assumed that the bandwidth of the range-gate bandpass filter will be $1/\tau$, as given in E 5.11. Figure 5.20 illustrates the inverse relationship between the width of the time-domain pulse and its frequency-domain spectrum. In fact, if the pulse width in the limit goes to zero (which is termed a delta or δ function), the spectrum will, again in the limit, become infinite, i.e., all frequencies will be represented. Note that in the spectrum represented by the wide (top) pulse, the major frequency component has an amplitude of about 0.5. However, the very narrow pulses, although there are many frequencies in their spectrum, all have a very low magnitude. Thus, it can be seen that the spectral density, i.e., the number of components times their magnitudes, is about constant.

MTI Processing—Digital Delay-Line Cancelers

The processing system shown in Figure 5.17 is designed to show a few of the many different configurations that can be used in search and/or tracking radar. Besides the use of digital filters as range-gate spectrum analyzers, they are often used as

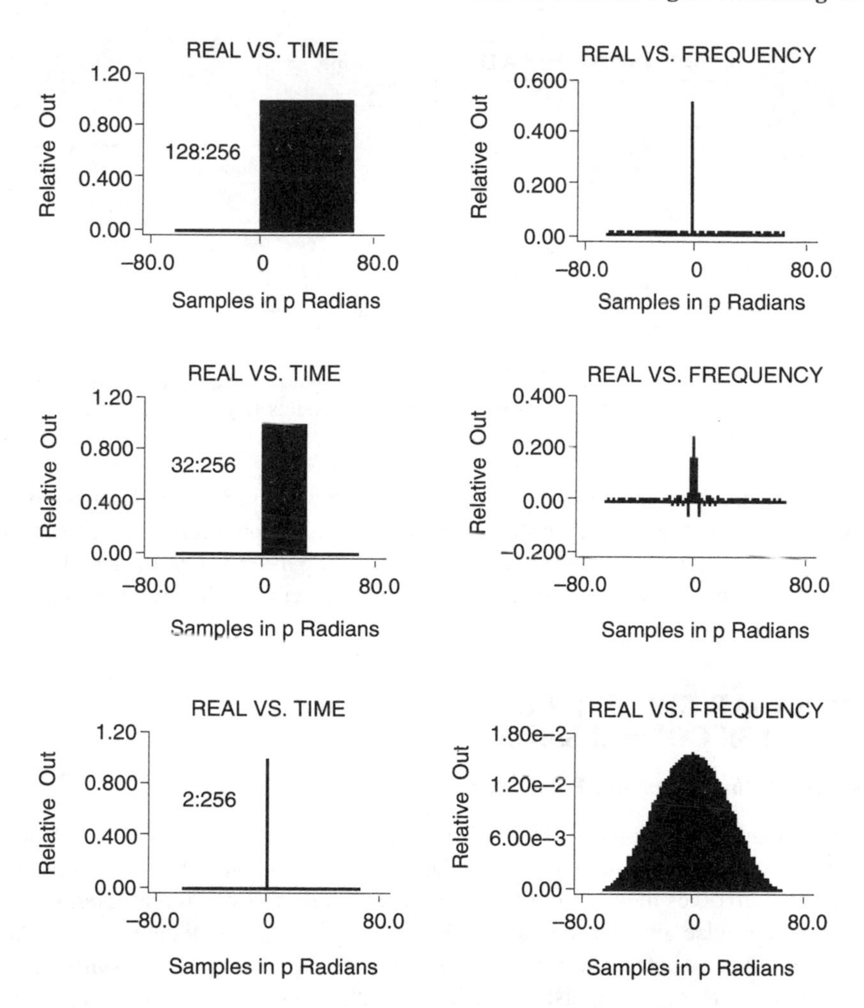

Figure 5.20 The relative spectrums for three pulses. The pulse in the upper left has a τ of 64 samples and a T width of 128 samples. The middle pulse is 32 of the 128 samples and the lower left shows pulse uses 2 of the 128 samples. A computer program calculated their FFTs which are displayed to the right of each pulse. The caption samples can be converted to time for the pulses or frequencies of the spectrum.

delay-line cancelers based on the original comb filter. One of the more sophisticated forms of MTI processing includes a more complex phase detector than is shown in Figure 5.8. This alternate detector, shown in Figure 5.21, is reminiscent of TV color demodulators that separate the signal into I (in phase) and Q (quadrature) components. Forming the I and Q radar signals makes it possible to solve a problem in MTI radar known as blind phases. Blind phases can occur when only the normal, or I signal, is sampled and, after the delay, is compared with the next incoming echo signal. At times, even though the echoes are from fixed targets, a

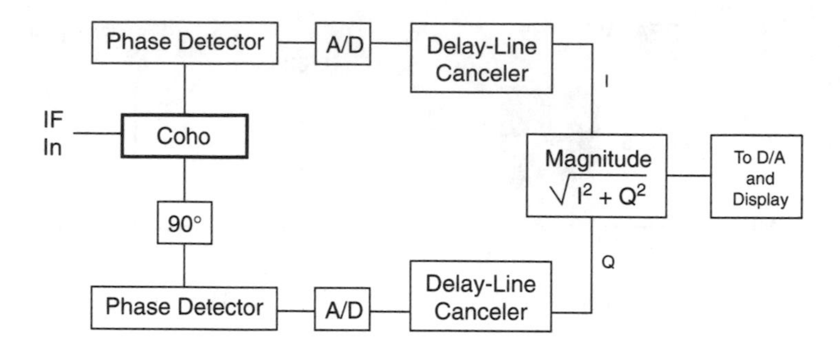

Figure 5.21 An MTI phase detector that creates both in-phase and quadrature signal components. This is only one example of several variations on this I-Q detector. See any radar text.

delayed signal may not be sampled exactly the same as the next incoming signal. The subtraction process, which should produce a zero output for fixed targets, will sometimes incorrectly produce a residue. By breaking the signal into I and Q components, the processing will always produce a correct answer. (See the Skolnik reference.)

5.9 RADAR RECEIVER CONFIGURATIONS—PULSE COMPRESSION

Receiver Variations and Pulse Compression

The previous sections have described a variety of techniques for recovering and accurately identifying the radar echo. One of the major radar problems not yet addressed involves the trade-off between the amount of power represented in the transmitted pulse and the accuracy of identifying and properly categorizing the received echo. To detect a target at a large distance or to burn through disturbances, such as some forms of sophisticated chaff, weather, and jamming, often requires tremendous amounts of power. Some military radars use peak powers in the megawatt range. However, using this power effectively might well require a relatively wide pulse. Again, pulse may mean the generic on-off waveform or may mean the more common modulated signal burst such as the FM chirp signal described in Figure 5.2.

Since wide pulses contain narrow bandwidths (see E 5.11), and their information and the information in their echoes is relatively limited, the exact range or position of the target is difficult to predict accurately. If the pulse is narrowed, which gives it and its echo more information, the peak power may become unwieldy, if not impossible. A very innovative solution to this quandary is to use the technique termed pulse compression. Pulse compression is a substitute for short-pulse radar. It allows the radar to utilize a long pulse to achieve a large total radiated energy, but simultaneously to obtain the resolution of a short pulse. The technique of pulse compression involves both the transmitter and various processing configurations in the receiver. Figure 5.22 illustrates three methods of implementing pulse-compression radar.

Pulse Compression Using a Conjugate Matched Filter

Figure 5.22 (a) indicates one method of radar pulse compression. Note that before the transmitter is modulated, the input pulse is passed through a matched filter to optimize its signal-to-noise ratio (SNR). When the return echo is received and converted to an intermediate frequency, it is then passed through a conjugate matched filter. The response of a conjugate filter (specified with the superscripted

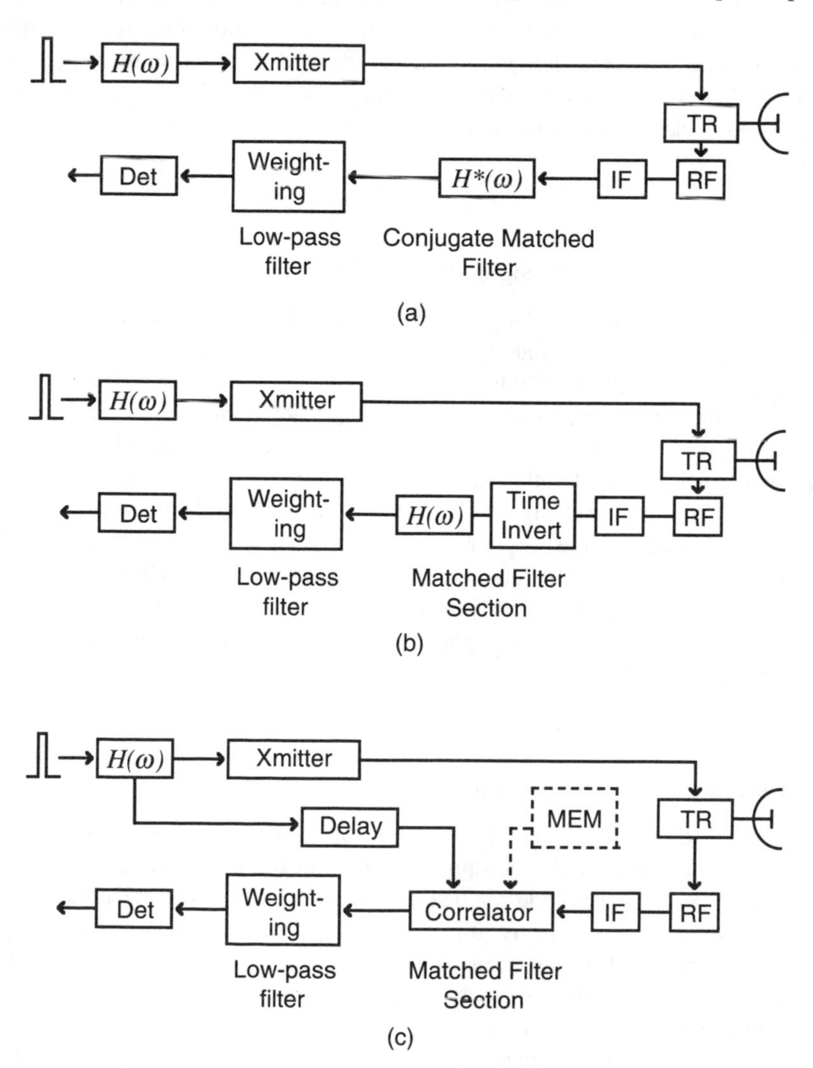

Figure 5.22 Three methods of pulse compression including (a) conjugate filters, (b) time inversion, and (c) correlation; (c) is also termed the correlation receiver.

asterisk) has the sign of the imaginary term reversed, i.e., if the H(ω) is R+jX, the conjugate H*(ω) will be R–jX. The output of the conjugate matched filter, together with its weighting filter, will be a waveform that has the characteristics of a narrow pulse with its inherent detailed information.

Pulse Compression Using a Matched Filter with a Time-Signal Inverter

The circuits in Figure 5.22 (b) give similar results to the circuits in Figure 5.22 (a). The reason for inverting the time sequence of the input signal involves a time-domain process termed convolution. Convolution in the time domain is similar to multiplication in the frequency domain; however, depending on the functions involved, there may be an inversion of the signal sequence. It is suggested that the interested reader consult specific texts on signal processing such as the Lynn and Fuerst reference.

Pulse Compression Using Signal Correlation

One of the most widely used signal processing techniques is correlation, which is the comparison of two similar signals. Often in signal processes, it is important to compare a signal with itself, which is termed autocorrelation. Whether it be correlation or autocorrelation, the result of the comparisons can be used to actuate decision processes. Figure 5.22 (c) shows the IF signal being compared with the H(ω) signal that was the input to the transmitter. If the delayed transmitted signal correlates with the received IF signal, the target is not moving. If, however, the correlated signals do differ, the output represents a moving target. Note that Figure 5.22 (c) also shows a dashed box labeled MEM for memory. In some radar systems it is prudent to store a digital replica of the output signal from the transmitter's matched filter H(ω) and use this stored signal in the correlation processing rather than to continually feed a signal from the transmitter.

Phase-Coded Pulse Compression

In addition to the FM chirp and other types of pulse modulation, it is common to use digital coding for pulse compression. One of the major advantages of phase coding is that the digital code can be easily changed in software, whereas a change in the chirp modulation may require exchanging (analog) delay lines or switching filter components. The transmitted pulse is divided into a number of equally divided subpulses. The phase of each of these subpulses is coded as being 0° or 180°. (The literature usually indicates a 0° subpulse as a "+" and a 180° phase pulse as a "–".) Either an analog tapped delay line-filter or an analog or digital matched filter can be used for pulse compression.

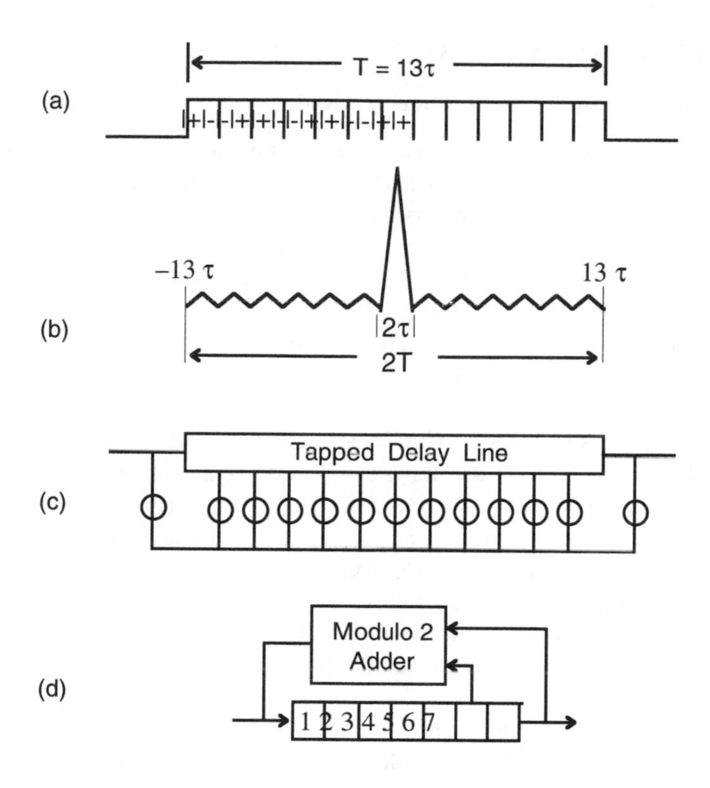

Figure 5.23 The digital phase-coded pulse for pulse compression and how it can be created. Part (a) is the wide pulse used for transmission that is composed of n (here 13) coded subpulses. The phases can be 0° or 180°. Part (b) shows the resulting pulse after it is compressed with circuits such as those shown in the receiver parts of Figure 5.22. The thumbtack spike is not to scale, as it should be $n\tau$ (from the pulse) or 13 units high. In (c) the pulse code can be created using a taped delay line in the analog domain. The circles indicate the multipliers for the + or – phase codes. Part (d) indicates that the pulse code can be created using a special software routine or a hardware pseudo-random generator in the digital domain.

If the originating pulse code is created with a software or hardware pseudo-random generator, the skirts or time-domain sidelobes of the received and compressed waveform will take on a noise-like characteristic. However, it is often more desirable to have the output (the autocorrelation function) produce more uniform, equal-time sidelobes. This can be accomplished by producing the original pulse coding from a digital function termed Barker coding. There are several Barker codes to choose from, including one with a code length of 13, as shown in Figure 5.23 (a). (See the 1980 Skolnik reference.) The digital compressed codes can also be broken into I and Q components using the digital counterparts of Figure 5.21 plus additional shift registers. (See the Fink and Christiansen reference.)

5.10 SYNTHETIC APERTURE RADAR

An Introduction to Synthetic Aperture Radar—SAR

Starting from a nontechnical perspective, SAR is similar to the previously described conventional radars—turned upside down. Whereas conventional radars usually have their transmitters and receivers on the earth and use them to search or track targets in the sky, SARs have their transmitters and receivers on airplanes or satellites and investigate targets on the face of the earth or other celestial bodies. By combining the information received from a given number of successive pulse echoes, a composite view of an area can be assembled. Thus, a new, composite radar aperture is synthesized by combining the results of the motion of the aircraft and the image-signal processing.

SAR, which was invented by Carl Wiley of B. F. Goodyear in 1957, is usually described as land-mapping radar. It has developed into much, much more. It is used extensively for ecological studies over both land and sea. There are extensive programs to study the ocean currents, not only as an aid to navigation, but to understand more about the make-up and movement of the oceans and the movements and migrations of the inhabitants of those oceans. Likewise, SAR has been widely used for space studies. Since it is usually not hampered by clouds or water vapor, it has permitted the observation of cloud-covered Venus where optical methods failed.

It can be seen that the principles of SAR are relatively simple but the implementation, to be described in the following sections, is particularly difficult and presents some interesting challenges even to contemporary signal processing. Three

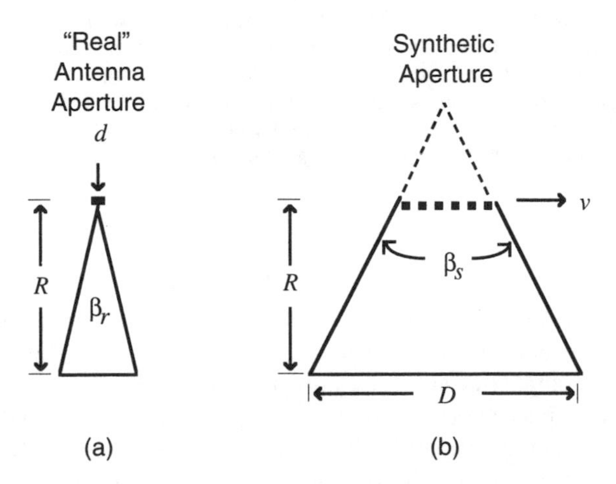

Figure 5.24 The elementary operation of SAR: (a) shows the beam from a single real, physical antenna such as a dipole, (b) indicates the broader beam created from the emission of a similar antenna as it is carried horizontally by a spacecraft at a velocity v. (See the text for symbol definitions.)

of the most important facets of SAR, as described in Figure 5.24, are the width (D) of the area scanned and its beam width β; the resolution of areas or objects in this beam width; and, of course, the processing required to combine the echo signals and finally produce the required topographical graphic.

SAR Compared to Conventional Radar—The Elementary Geometry

To provide a better understanding of the operation of SAR, most of the following text and mathematical models will include both the conventional or normal radar, as will as SAR. The subscript "*r*" will stand for a real or physical antenna, "*sa*" for synthetic aperture, "*d*" for the dimension of the real antenna, and "*D*" for the length of the synthetic antenna at a range "*R*," λ will stand for the transmitted signal wavelength and β, with the appropriate subscript, will indicate the beam width.

The antenna beam width, at range R, is:

$$\beta_r = \frac{\lambda}{d} \quad \text{and} \quad \beta_{sa} = \frac{\lambda}{2d} = \frac{d}{2R} \quad \text{(usually in radians)} \qquad \text{(E 5.12)}$$

The antenna length, at range R, is:

$$D_{r \text{ or } sa} = R\beta_{r \text{ or } sa} = R\frac{\lambda}{d} \quad \text{(usually in meters)} \qquad \text{(E 5.13)}$$

The minimum theoretical along-track resolution is:

$$\text{res}_r = R\beta_r = R\frac{\lambda}{d} \quad \text{and} \quad \text{res}_{sa} = R\beta_{sa} = \frac{d}{2} \quad \text{(in meters)} \qquad \text{(E 5.14)}$$

Thus, to a first approximation, to get better resolution, make the antenna SAR element shorter, which usually means make the carrier frequency higher.

A simple example can be used to compare the resolution of both ordinary or real radar resolution and SAR. Assume both types of radar are operating at 3 GHz, which has a wave length λ of 0.1 meters. They both are operating with a range R of 100 km (100,000 meters) and with real antennas of 10 meters.

The resolution of a normal radar at 100 km is:

$$\text{res}_r = R\frac{\lambda}{d} = 100{,}000 \times \frac{0.1}{10} = 1{,}000 \quad \text{meters} \qquad \text{(E 5.15)}$$

The resolution of an SAR at 100 km is:

$$\text{res}_{sa} = \frac{d}{2} = \frac{10}{2} = 5 \text{ meters} \qquad \text{(E 5.16)}$$

SAR Optical Processing

The original developments in SAR occurred before the perfection and common use of the digital computer. Some insight into the techniques and processes required for SAR can be obtained by a review of its early use of optical processing.

Each echo that is received as the aircraft flies over the target area must be stored for later integration into the final composite graphic. These analog time-domain signals were typically stored on optical disk or photographic film. One type of optical recorder consisted of [1] a cathode-ray tube that was modulated by the echo and served as a light source, [2] a condensing lens system, and [3] a moving reel of photographic film that recorded the return echoes, frame by frame. (See the Skolnik references.) The photos of the echoes, which were recorded in real time, were most often processed using a much slower, non real-time scale. Often the images on the film were later reconverted back to electrical signals and transmitted to a ground station for final processing.

Perhaps the most interesting early processing technique was the use of an optical system to perform a Fourier transform. Many will remember the classic physics demonstration of using a glass prism to decompose white light into a spectrum of rainbow colors. The prism was, in essence, taking the FFT of the white light. For SAR echo images on film, a series of conical, cylindrical, and spherical lenses can be used to transfer the image from the filmed horizontal plane and display it as a Fourier transformed image in the vertical plane. (Again, see the Skolnik and other references.)

The Digital Signaling Processing Techniques Required for SAR

The advent of DSP accelerated and perfected the processing required to bring SAR up to many of its theoretically predicted potentials. The processing requirements are varied and difficult. For instance, not only is there the forward (or circular or elliptical) motion of the plane or satellite, but there is also its rolling pitch and yaw that must be constantly factored in. If the SAR is on a satellite, compensation for the rotation of the earth must be added. Another complexity that must be factored in is the shape of the electromagnetic field, i.e., its wave front must be considered.

The general theory of antennas, which was developed from a combination of optical and electromagnetic wave theory, including both quantum mechanics and the famous Maxwell's equations, is very complex and beyond the scope of this work. However, one of the ideas of antenna theory will be mentioned briefly to show more of the processing that is required for SAR.

Digital Signaling Processing as a Function of Field Distance

The field emerging from an antenna can be divided into the near field, the Fresnel region or field, and the far or Fraunhofer region. The near field and the resulting shape of its radiation extends only a few antenna lengths (or diameters) away and most often is not a factor in SAR processing. In the next Fresnel region, the waveform is curved and the degree of curvature changes with the distance from the radiating source—the antenna. If the target is in this Fresnel region, a correction factor must be added to each returning echo depending upon the distance and the angle of the target. If the target is in the far region (the usual case), the wave front is considered parallel to the antenna (a plane wave) and no wave-shape correction is needed. The boundary between the Fresnel and the far or parallel wave field is somewhere between $(d_{sa})^2/\lambda$ and $(2d_{sa})^2/\lambda$. Stated in different terms, in the far field, with plane-wave fronts, the antenna is considered focused and the resolution is $d/2$ meters. If the target is in the Fresnel area, the antenna is considered not focused, and if the echoes are not weighted, the resolution will be a function of the distance R and equal to $\sqrt{R\lambda}$ meters.

Another technique related to the wave field is stereoscopic SAR used to provide three-dimensional images of the terrain. Stereoscopic images can be produced by multiple fly-overs or by alternating the position of the beam (multiplexing the beam position) during a single pass. Some of these stereo techniques include changing the shape of the beam as it is alternated back and forth. This, of course, adds a significant additional amount of signal processing.

SAR Digital Signaling Processing to Compensate for Undesired Surface-Related Echo Modulation

The introductory model of SAR assumes a nearly flat target terrain or, if not relatively flat, a surface that is not moving. In the real world of SAR, these idealized conditions, although they do happen, have many exceptions. Mountaintops may appear to be moving towards the radar compared to their valleys in two-dimensional SAR. In mapping snow-covered regions, these signal distortions tend to give false indications of the snow on the lee side of the mountains. Also, the waves of the sea tend to distort (modulate) the echo signals. (See the Kingsley and Quegan reference.)

A Final Overview of SAR Signal Processing

The two outstanding constituents of synthetic aperture are [1] the extremely imaginative ways that the engineers and scientists have used DSP and [2] the overwhelming volume of data that needs to be accurately processed to produce those essential images. (Millions of us can still remember the first TV pictures of the outstanding Magellan spacecraft high-resolution images of the surface of Venus.) The required data rates, in the tens and hundreds of megabytes per second, and the millions of multiplications per second, challenge even the most sophisticated contemporary digital signal processors. Likewise, although they were not repeated in this section, such procedures as pulse compression, FM chirp modulation, matched filters, as well as the workhorse of DSP, the fast Fourier transform, are used to their fullest to provide the many variations of mapping and surveillance that have become the trademarks of synthetic aperture radar.

5.11 RADAR SIGNAL PROCESSING—THE CHAPTER IN RETROSPECT

Radar and Radar Models

Radar serves as an excellent illustration of contemporary signal processing because of the breadth of technology and mathematical rigor that has been used to describe and develop it. It uses even more exotic pulse-forming processing than normal television and its signal and image processing is among the most advanced and complex in the art of electronics.

The introductory sections of the chapter present three different, but parallel, perspectives of a model radar system. The fundamental system components are first presented in a block diagram called a graphic model. Next the fundamental operations represented by these building blocks are detailed in text to form a corresponding descriptive text model. Finally, the elements presented in the graphic and text models are further clarified by giving their appropriate equations in mathematical models. With these models as background, details are presented on how these (usually) narrow, high-powered pulses are formed and modulated. A section on radar antennas shows how, in most cases, the antennas play a very integral part in radar signal processing. Of special interest are the antenna patterns that can be created with monopulse and phased-array antennas.

Special Problems in Radar Signal Processing

A typical incoming (return-path) radar signal is degraded or changed by noise, both from the natural noise in the sky and from manmade interference; by the (Doppler) frequency shifts due to the two-way travel time; by motion of most targets; and by unavoidable circuit delays and nonlinearities in the transmitter and the input of the receiver. The processing used to partially or completely solve these problems determines the effectiveness of the radar system.

Descriptions are given to show that receiver noise can be reduced by very special low-noise input amplifiers and the natural sky or galactic noise can be reduced by the application of probability theory using special circuits, called matched filters. Likewise, a special processing circuit, called MTI or moving target indication, is described that essentially discards the signals from stationary or slowly moving targets so that they do not confuse the display of the major targets of choice.

The description of these MTI circuits is quite thorough and built upon the analog delay-line comb filters of Chapter 3. The radar circuits show that similar circuits can be used to cancel the signals of fixed or slowly moving targets. With this background, corresponding digital delay-line cancelers are described and it is shown how these digital circuits, and their building blocks, correspond to the canonical nonrecursive and recursive digital filter configurations. Other MTI circuits are presented and compared to the I and Q phase detection circuits presented in Chapter 3.

Pulse Compression Processing for Minimum Power with Maximum Target Resolution

One of the very innovative radar processing circuits makes it possible to use wider pulses, with a corresponding decrease in instantaneous peak pulse power, and still obtain the precise target resolution associated with short, high-power, pulses. Several approaches are discussed, in both the analog and digital domain, together with the required coding.

Radar Signal Processing from a Different Perspective—Synthetic Aperture Radar

This "upside down" radar, which is used to investigate the earth from airplanes or satellites, is presented along with the signal processing that is unique to its operation. A theoretical, idealized comparison between conventional R and SA radar is presented along with a brief description of some of the early SAR processing including using optical techniques to perform Fourier transforms. Since the plane or satellite, as well as the earth, is continually moving, the processing for any received radar echoes must include compensations for these movements before meaningful, accurate images can be created. Of course, the processing load is increased when some of the images are created in stereo. Also covered is the distance compensating processing that may be required when the (earth area) target is in the optical near or far field of the radar's equivalent vision. Likewise, since the surface of the earth is curved, some additional processing may be required to produce an accurate two-dimensional or stereo image.

Archival and Cardinal References

MIT. *Radar System Engineering.* New York: McGraw-Hill, 1947.

Shannon, C. E. "A Mathematical Theory of Communication." *Bell System Technical Journal,* July 1948.

Terman, Fredrick Emmonds. *Radio Engineering.* New York: McGraw-Hill 1947.

(Take time to see the wealth of material in these early but still useful landmark texts. Especially, for THE background material on the origins of radar, particularly in the United States, see *Radar System Engineering,* which is volume 1 of the 28-volume (MIT) Radiation Laboratory Series.)

Contemporary References

Barton, D. K., and Ward, H. R. *Handbook of Radar Measurement.* Norwood, MA.: Artech House Inc., 1984.

Couch, L. W., *Digital and Analog Communication Systems.* New York: Macmillan and Co., 1990.

Curlander, John C., and McDonough, Robert N. *Synthetic Aperture Radar Systems and Signal Processing.* New York: John Wiley & Sons, 1991.

Fink, Donald G., and Christiansen, Donald (eds.). *Electronics Engineers' Handbook.* New York: McGraw-Hill , 1989.

Horowitz, Paul, and Hill, Winfield. *The Art of Electronics.* Cambridge: Cambridge University Press, 1983.

Hovanessian, S. A. *Radar Design and Analysis.* Norwood, Mass.: Artech House Inc., 1984.

Kennedy, George. *Electronic Communication Systems.* New York: McGraw-Hill, 1985.

Kingsley, Simon and Quegan, Shaun. *Understanding Radar Systems.* London and New York: McGraw-Hill, 1992.

Lynn, Paul A., and Fuerst, Wolfgang. *Digital Signal Processing with Computer Applications.* Chichester and New York: John Wiley & Sons, 1989.

Nathanson, F. E. et al. *Radar Design Principles.* New York: McGraw-Hill, 1991.

Skolnik, Merrill I. *Introduction to Radar 2d ed.* New York: McGraw-Hill, 1980.

———. (ed). *Radar Handbook 2d ed.* New York: McGraw-Hill, 1990.

6

Image Processing

With Examples from Medical and Space Technology

6.1 THE RUDIMENTS OF IMAGE PROCESSING

An Introduction to Image Processing

This chapter is primarily about the production and processing of a variety of physical images. However, it must always be remembered that these images are created for the use of human beings. Only when one or more humans can relate to or learn from these images can they and their processing be considered successful and worthwhile. The final physical image must, in some way, relate to its resulting mental image.

There are two broad classifications of image processing: those techniques that enhance or clarify and the techniques that transform from one domain, such as time or space, to another, such as frequency. An argument can be made that any of the mathematical transforms are a subset of the clarification process. However, as an aid to understanding, this work prefers the two classifications. Likewise, this chapter will concentrate only on the nonhuman production and processing of physical images. The physics and physiology of vision, along with the extremely complex and interesting processing of the human brain, is beyond the detail and scope of this chapter.

A Simple Model—What Is an Image?

As a reference perspective, an image can be assumed to be a single point or group of points that can be defined in a two-dimensional position by two variables, such as x and y, u and v, etc. This is in keeping with the fundamental propositions in geometry that a line can be made from a (an infinite) series of points and figures can be constructed with a group of straight or curved lines. (See also the ideas and figures relating to lines and objects in Chapter 10.) Besides the x, y position of a point, a line, or an object, an image model must contain brightness and contrast.

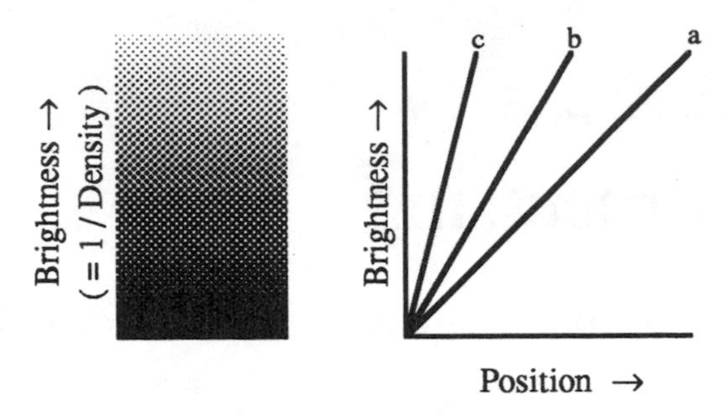

Figure 6.1 Two graphics showing a gradual change of brightness with position. The left pattern shows brightness increasing from the bottom to the top. In the right graph, a shows the same gradual increase in brightness as the left figure. The steeper slopes of b and c would represent pictures with higher contrast ratios than a.

The brightness of a scene or picture is the average amount of light that is emitted or reflected. If the brightness is linearly increased 20 percent, the light output of all parts of the picture (at least in theory), be they the lighter parts or darker parts, are increased 20 percent. Frequently the contrast of an image (such as in an x-ray film) is more important than the brightness. Whereas the brightness, temporarily discounting noise, is simply the gain of a system, the contrast is the ratio of low-brightness areas to high-brightness areas. The contrast represents the slope of the transfer function. In a high-contrast film, the change in brightness from a low-brightness area to a high-brightness area is much greater than in a low-contrast film. Contrast enhancement is one of the important processing functions—see Figure 6.1.

The ideas of position, brightness, and contrast are familiar in TV, motion pictures, and photography. Both photographic film and TV picture tubes exhibit similar input-output transfer relationships that are related to the Driffield and Hurter (D-H) or "gamma" curve. In photographic film the coordinates are usually log exposure (time) versus linear exposed density, whereas in TV an overall system D-H curve is often presented either as linear light into the transmitting camera tube versus light out of the receiver picture tube or as log light into the camera tube versus log light out of the picture tube.

Film Characteristics

Photographic (including x-ray) film must be considered an active component in image processing. The slope of the D-H transfer curve, as shown on the left of Figure 6.2, is commonly known as the speed of the film, i.e., how long the film must be exposed to produce a standard measured density. Unfortunately, often the price paid for high speed is miniature discontinuities or graininess in the film—

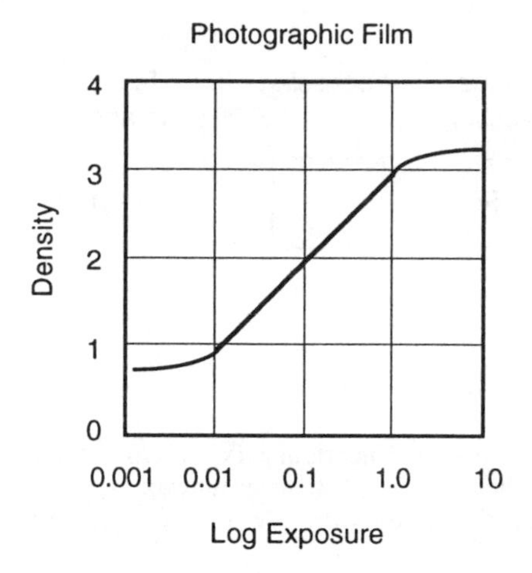

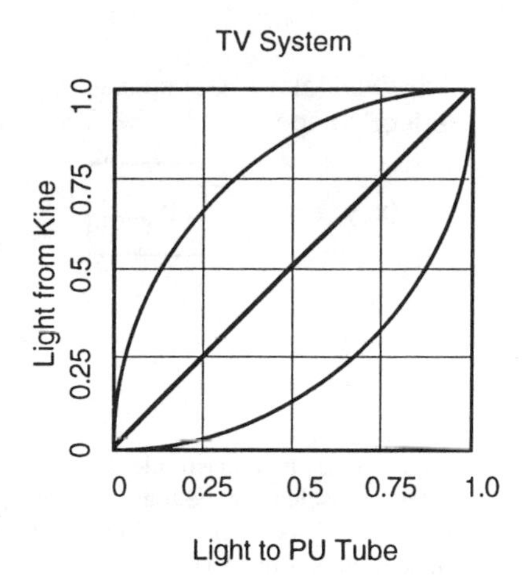

Figure 6.2 Two representative forms of the input-output or D-H curve. The transfer function on the left is used to relate film exposure time to resulting picture density. The graph on the right shows three different light-to-light transfer curves for a complete TV system. Whether it is the photographic or the TV medium, the first approximation ideal is to produce the linear (diagonal) characteristic in both cases.

thus the user must consider the trade-offs. The analog graininess is somewhat akin to the bit count or pixel density of a digital image. In addition to graininess, resolving power is another important image parameter. Resolving power includes graininess and the common but unwanted tendency of the film's emulsion to scatter the received light. Fine-grain films, which are inherently slower, provide less scattering and thus the best resolution. Fortunately, the quality of all film, including the type used for x-rays, is constantly being researched and improved.

6.2 IMAGE CORRECTION AND ENHANCEMENT

Pivotal Paradigms for Image Restoration or Enhancement

It is almost impossible to present a single coherent set of rules for the upgrad-ing or enhancement of images. The common techniques for improving the quality of an x-ray may not be the same techniques that are often used to enhance a signal received from, for example, a space probe. Perhaps the major problem is the non-universality for the definition of image goodness. In most cases, the bag of tricks that a given technology will use is based upon well-distilled heuristic procedures. Nevertheless, the three models to follow can serve as beginning thought models for image correction and enhancement.

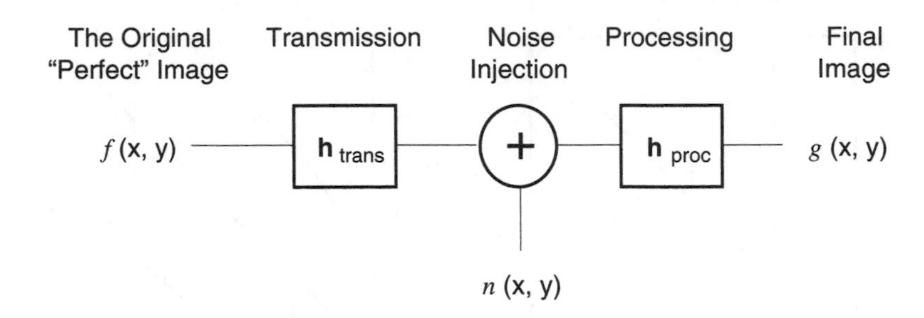

Figure 6.3 An elementary model for image processing. The original $f(x, y)$ is assumed to be a perfect, noiseless image that is altered by the transmission transfer function h_{trans} and noise $n(x, y)$. h_{proc} is the processing transfer function used to correct or reduce any defects to produce the resulting image $g(x,y)$.

Pivotal Image-Processing Models—The x, y Point

As an aid to thinking about how to transform or improve signals from the body and signals from space, image processing usually starts by assuming a two-dimensional image in some orthogonal, two-dimensional x, y plane. It is assumed that the starting or original image is nearly perfect and that additional noise and distortions are added by the transmission medium, be it an electrical network, the noise and debris of space, or such other transmission deterrents as body tissue or bone. In more technical terms, each transmission or correction medium presents an input-output transfer function H (often a lowercase h in the time or space domain or an uppercase H in the frequency domain), which modifies the original image in some way. Also, for analytical convenience, any added noise in the image is usually defined as some noise function such as n (x, y). Figure 6.3 presents a graph of such an image-processing model.

Pivotal Image-Processing Models—Local Neighborhoods

A second image-processing model that is often used when one is thinking of x, y-type images was presented in Figure 3.32 of Chapter 3. This figure shows TV lines broken down to individual pixels. By using one of many algorithms, a series of new pixels can be created that, in turn, form a new picture line. This idea can be expanded by using techniques to investigate and correct defects in the immediate neighborhood of a group of pixels. Neighborhood processing is shown in Figure 6.4.

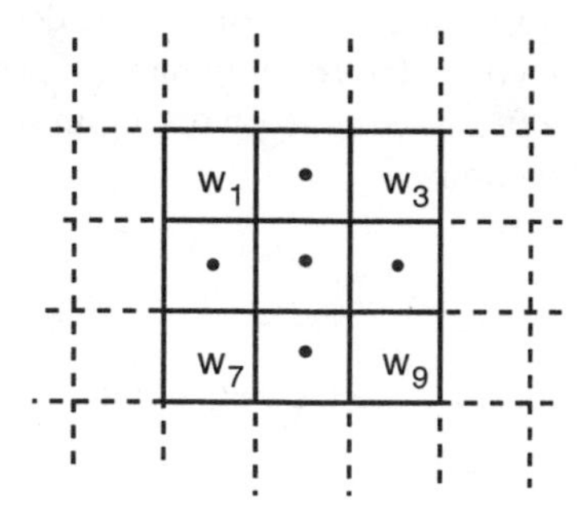

Figure 6.4 The general neighborhood or mask orientation for image processing. In each group, the center pixel is usually the reference point and the other pixels in the group are changed or weighted (w_1, w_2, etc.) by some predetermined processing algorithm. The dashed lines indicate adjacent neighborhoods that are also processed. The processing load, i.e., the number of multiples, is very great in such a scheme.

Pivotal Image-Processing Models—Global Areas or Frames

A third image-processing model that can be included when thinking about image processing was also introduced in Chapter 3. It was pointed out that some processing is best implemented in large or global areas rather than local or neighborhood areas. Illustrations of this include images that have smear caused by motion, such as TV pictures or frame-to-frame pictures that are successively taken by a fast-moving aircraft with synthetic aperture radar.

Pivotal Image-Processing Models—Compression

Many journals and periodicals have stated that, technically, the 1990s will be the decade of data compression. More than any other technique, compression has made HDTV possible. Most of us use compression every day with our computer disks, our faxes, and in the way we transmit and receive most picture and numeric data. When planning almost any data or image-processing-system, compression and its corollary expansion is usually considered.

6.3 TECHNIQUES FOR IMAGE CORRECTION AND ENHANCEMENT

Fundamental Image-Processing Techniques—Spatial-Domain Models

As stated in the last section, a good starting point for image processing is the assumption that local, not global, areas are composed of pixels. Using the values of these pixels, a processing transfer function can be derived based on the light output obtained from a given (transferred) light input. If $f(x, y)$ is the input, $g(x, y)$ is the output, and T (T is often used when working in the space domain) is assumed to be the transfer function:

$$g(x, y) = T[f(x, y)] \tag{E 6.1}$$

Likewise, if the variables are simplified to r equaling the input $f(x, y)$ and s equaling the output $g(x, y)$:

$$s = T(r) \tag{E 6.2}$$

Figure 6.5 illustrates the use of the transfer function T for simple contrast enhancement.

Fundamental Image-Processing Techniques—An Introduction to Histograms and Probability Pensity Functions

For any digitized gray-scale image, there exists n levels of gray, i.e., n changes from black (all gray) to white (no gray), as shown in Figure 6.1. Although pixels with a certain brightness, b, can be scattered throughout the image (the picture), the image can be analyzed to find how many pixels exist with this brightness b. If an image is quantized (digitized) into eight levels, these levels can be grouped (at least statistically) and represented by an eight-unit (zero plus seven) line graph termed a histogram. For instance, if a certain image is made up of a 32 × 32 pixel matrix, there are 1,024 pixels that can have a brightness ranging from level 0 to level 7 (8 bits). These 1,024 pixels will be distributed in eight groups. If most of the pixels have a brightness of six, seven, or eight levels, the overall picture will be relatively bright. Likewise, if most of the pixels are in the 1-, 2-, or 3-bit level, this is a dim picture.

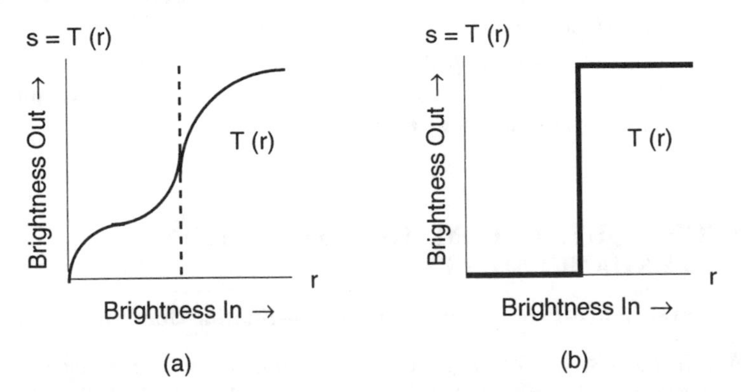

Figure 6.5 Two graphics showing the use of local (neighborhood) contrast enhancement following the elementary equation s = T(r). Referring to Figure 6.4, changing the coefficient algorithm will produce different contrast transformations.

In any related group of numbers, there are usually subgroups that occur quite close or equal one another. For instance, in a grammar school, although very few students are exactly six, or seven, or eight years of age, these ages can be used as density groupings. There may be one or two students that are five or perhaps thirteen or fourteen (assuming six grades) years of age but that probability is relatively low. Thus, some ages have high probabilities and some ages have a low probability of existing. The groupings are termed probability density functions.

NOTE: The literature on two-dimensional (x, y) functions and their transfer or transformation functions is not consistent in its symbols and terminology. The following sections use the symbols detailed here:

y = h x is the transfer function h operating on x to give y.

g(u, v) = h(u, v) f(u, v) is a similar two-variable transfer function.

s = T r can have the same or a very similar meaning, but s and r are abbreviated to denote the appropriate input or output (u, v) points and T is used as the abbreviation for the transfer or transformation function.

p(r) or p(s) = the probability density function of r or s.

Fundamental Image-Processing Techniques—Histogram Equalization

One of several ways to enhance a very high or very low contrast image and make its brightness more uniform is by [1] producing a brightness histogram of the original image, [2] producing a histogram equalization transform, and [3] processing the original digitized image with the derived equalizing transform. The calculations necessary to execute such a transformation are shown in Table 6.1. As a

Table 6.1 Histogram Equalization Calculations

A **b(k)**	B **n(k)**	C **PDF** n/1,024 (nn)	D **s_k or s_{bn}** n_0, n_0+n_1, $n_0+n_1+n_2$, etc.	E **b_n** Convert to Sevenths
b_0	5	0.005	0.005	0
b_1	10	0.010	0.015	0
b_2	25	0.024	0.039	0
b_3	35	0.034	0.073	1/7
b_4	75	0.073	0.146	1/7
b_5	124	0.121	0.267	2/7
b_6	250	0.244	0.511	4/7
b_7	500	0.488	0.999	7/7

hypothetical example, assume the above 32 × 32 matrix (1,024 pixels) with eight brightness levels. (b_k) has a distribution (n_k) of b_0 = 5 brightness levels, b_1 = 10, b_2 = 25, b_3 = 35, b_4 = 75, b_5 = 124, b_6 = 250, and b_7 = 500 pixels.

Column A is an index of the brightness levels b (k) and includes the eight brightness levels b_0 up to b_7. Column B shows the original number of pixels at each level. Column C, using the data in column B, shows the original probability density $p(r_k)$ versus the kth level using E 6.3.

$$PDF = p(r_k) = \frac{n^k}{n} \quad \text{(E 6.3)}$$

where n is the total number of pixels (here = 1,024) and n^k is the known probability of pixels at this k^{th} level.

Column D lists the calculations of the histogram equalization transformation function using E 6.4.

$$s_k = s_{b_n} = T(r_k) = \sum_{j=0}^{k} \frac{n^k}{n} = \sum_{j=0}^{k} p(r_k) \quad \text{(E 6.4)}$$

where s_{b0} = 0.005; s_{b1} = 0.005 + 0.015; s_{b3} = 0.005 + 0.015 + 0.039, etc.

Column E rounds off the numbers in column D to the nearest seventh because there are seven levels of brightness, plus all-black, to give the eight b_n levels. Figure 6.6 uses the data in Table 6.1 to produce histograms of (a) the original density distribution of the pixels as shown in column B; a histogram (b) of the (known) probability density as given in column C; the transformation function (c) as shown in column D; and the final discrete density distribution (d) taken from the quantized data in column E. This (d) histogram has collected all like data. Combining the information in columns B and E, there are 40 pixels with a brightness level at zero, 110 pixels grouped at the 1/7th range, 124 at the 2/7th range, 250 at the 4/7th range, and 500 at 7/7 or 1. Note that the final density distribution is still not uniform—the goal of this exercise. Since there are only seven density levels plus black, although this final image is somewhat more uniform, rounding errors made it impossible to obtain a completely uniform density. (Yes, the problem was rigged to show this.)

Fundamental Image-Processing Techniques—Using Special, Interactive Histograms to Equalize Contrast

In the preceding example, an image was processed using histogram equalization. There are many instances when another type of transformation might produce a more useful resulting image. In such a case, after a histogram equalization

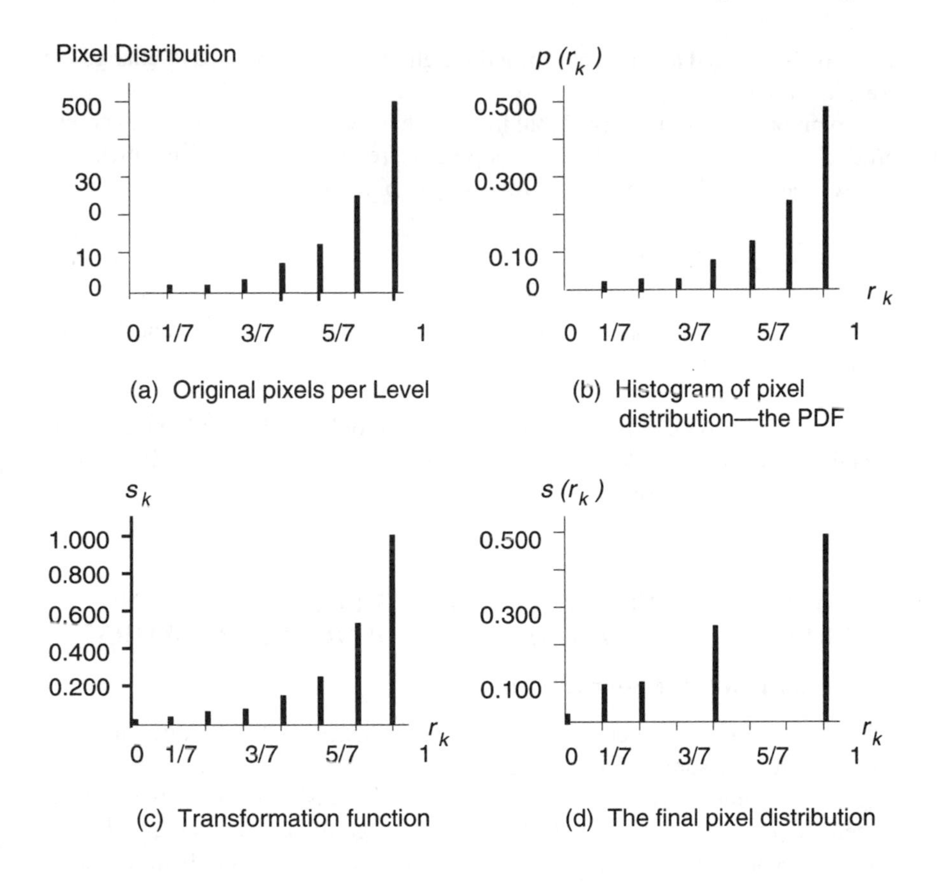

Figure 6.6 Histograms showing the use of equalization enhancement to obtain a more uniform density versus brightness level distribution.

has been completed and evaluated, it is most desirable to investigate the use of one or more other enhancements, i.e., use other probability density functions. For instance, the Gaussian distribution would be the most-used first choice. Also, a Rayleigh or some well-known log distribution might prove useful. Whereas procedures for calculating these new images are similar to the computations in the original histogram method, it is most desirable now to be able to interactively choose a succession of these particular algorithms and then to evaluate their effectiveness one at a time.

Fundamental Image-Processing Techniques—Local Enhancements

The advent of computer processing and the ability to easily select portions of a graphic or image has made it much more practical to use local area processing. The technique follows the scheme presented in Figure 6.4 where local areas are successively selected and predetermined weighting factors are used to change the pixels in the image.

A common local sometimes global image enhancement process involves neighborhood averaging for smoothing. This process, referring back to fundamental E 6.1, uses the following relationship for an by m pixel image:

$$g(x, y) = \frac{1}{M} \sum_{(n, m) \in s} f(n, m) \tag{E 6.5}$$

where the $\in$ means n and m relate to or are a part of the set of points s and M is the total number of points.

If the averaging produces blurring, a threshold factor T can be used to inhibit the averaging with large gray-level variations. This will still permit crispening of low-variation areas such as edges.

6.4 IMAGE ENHANCEMENT IN BOTH THE SPATIAL AND FREQUENCY DOMAINS—THE FOURIER TRANSFORM

An Introduction to the Fourier Transform

Before continuing with a general description of enhancement techniques, it seems prudent to review some of the background and techniques of the Fourier transform. This is timely because the next group of processing techniques to be described includes different ways of filtering. Although filtering and other enhancements can be accomplished in the time or space domain, it is usually much easier first to visualize filtering in the frequency domain.

Chapter 3 introduced and used the concepts of the Fourier series, the Fourier integral, and the more familiar Fourier transform with its modern incarnation, the FFT. As a foundation for the following sections in this chapter, these principles will be restated or expanded upon.

1. The Fourier series of a signal can be presented in both its trigonometric and exponential form. Almost any real-life continuous, periodic frequency-domain (analog) signal can be broken down into a series of sine and/or cosine waveforms which constitute the spectrum of the original waveform function. This Fourier series includes coefficients that show the frequency and amplitude of the sinusoidal components.

2. If a waveform is continuous but aperiodic, it can be broken into spectrum components using the Fourier integral. The components and forms of the Fourier integral are somewhat similar to the Fourier series.

3. A discrete (sampled) time-domain signal can be broken into its frequency components using the Discrete Fourier Transform (DFT).

4. The FFT is a more rapid method of calculating the DFT.
5. Just as two or more signals may be mixed or multiplied in the frequency domain, in a similar manner, signals can be convolved in the time domain. Signal convolution in the time domain is the equivalent of signal multiplication in the frequency domain.

It is interesting to note that, using contemporary high-speed computers and the highly efficient FFT algorithms, if two time-domain signals are to be convolved, it is often faster to: (a) use an FFT to transform them to the frequency domain; (b) multiply them in the frequency domain; and (c) use an Inverse Fast Fourier Transform (IFFT) to then send the product back to the original time domain, rather than to solve the convolution integral in the time domain.

Fundamental Image-Processing Techniques—Frequency-Domain Image Filtering

There are a number of image anomalies that can be reduced by filtering. Probably the most obvious is the effect of noise that can often be lessened by using a low-pass filter. To use an FFT to convert an image to the frequency domain, and then filter it and subsequently transform the result back to the time domain with an IFFT is one of the most common processing routines in electronics and communications. Also, it should be remembered that image filtering must be regarded as a three-variable process (u, v, and h) rather than the two-variable process that is common in electrical filters—this is illustrated in Figure 6.7.

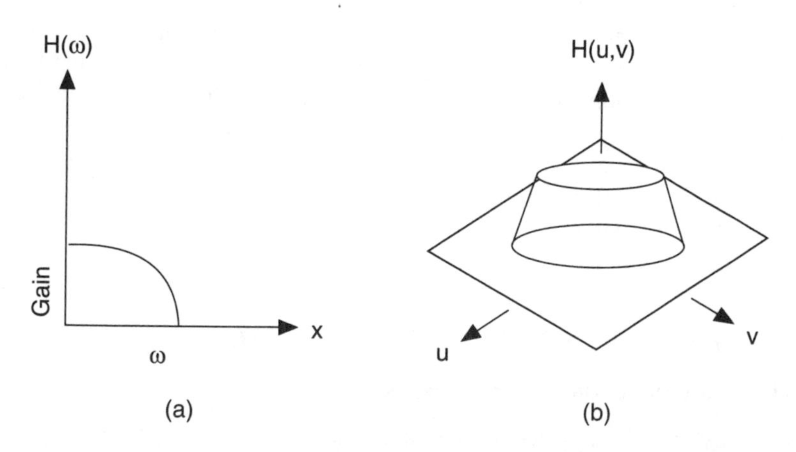

Figure 6.7 A comparison of a two-dimensional frequency response plot and a three-dimensional perspective plot of image low-pass filtering. The roll-off of the electrical filter in (a) or the spatial filtering in (b) can, in theory, be analog or digital. However, digital filtering can produce a much steeper cut-off, without disturbing the phase and transient response, than can analog filtering.

The problems with using filtering as a signal-processing technique, whether it is used to filter communication or image signals, are the unwanted side effects that may occur in the form of in-band ripples or the signal distortions caused by nonlinear phase shifts or improper transitions—poor transient response. The frequency-response plot in Figure 6.7 (a) shows a completely smooth response before the roll-off begins (the in-band portion) and a straight and smooth roll-off in the cut-off or out-of-band region. It can also be assumed that the in-band phase and transitions will not be distorted. However, if a sharper roll-off is desired to get rid of the high-frequency noise, care must be exercised to prevent adding anomalies that are more disturbing than the original noise. Ringing in a processed image appears as added concentric circles around points or additional phantom lines radiating from both sides of lines or edges.

The solutions to the trade-off between sharp cut-off for noise reduction and acceptable ringing, phase, and transient response comes in the form of several different algorithms for the filters. The use of polynomial filter types such as Butterworth, Bessel, Chebyshev, or a general class termed elliptical filters gives the user a wide choice of well-defined filter characteristics. For instance, Butterworth filters are simple and resemble cascaded electronic RC filters. They have excellent transient response but have relatively slow roll-offs. Bessel filters are maximally flat in the bandpass region and guarantee linear phase response while Chebyshev filters have a relatively steep roll-off but are not flat (they ripple) in the passband. The coefficients for these filters are well known and are easy to store in look-up tables so that interactive processing is very practical. Also, producing these filters in the digital domain gives users filters that very closely approach their mathematical ideal, where their analog counterparts were often limited by noise and component tolerances.

The high-pass implementations of these polynomial filters can be used to sharpen the edges in images. This processing can be done on a local or global basis. A little mental reflection on the fundamentals of filtering will reveal that the process of high-pass filtering is close to the process of differentiation. It may be remembered that active, analog high-pass filter circuits very closely resemble electronic differentiating circuits in analog computers. Filters that use the process of differentiation to sharpen transitions are often termed gradient filters. (See the Gonzalez and Wintz reference.)

6.5 MEDICAL IMAGE PROCESSING—X-RAYS

An Introduction to the Characteristics and Use of X-Rays

The discovery of x-rays was purely an accident. In 1895 the German physicist Wilhelm Roentgen was studying the phenomenon of cathode rays (electron beams) in a contemporary vacuum-tube device called the Crookes tube. In the course of his experiments, he accidentally noticed that an unused fluorescent screen would glow brilliantly whenever the tube (cathode ray) beam was on. Further investigation showed that a black paper would not prevent the glow but a metal plate would.

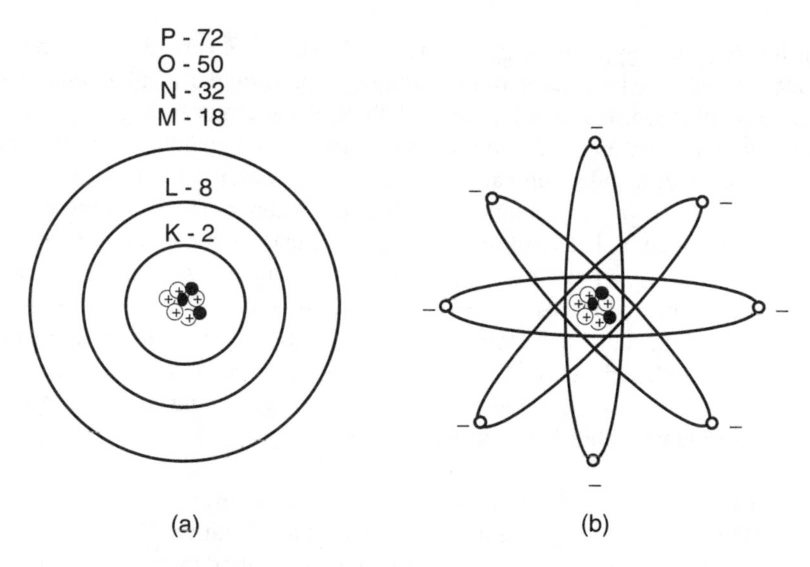

Figure 6.8 The elemental Bohr atom. (a) shows the atomic "solar system" with protons and neutrons in the very heavy nucleus with "orbits" or shells for the outer electrons. Six of the seven possible shells are also indicated by letter with numbers indicating the maximum number of electrons in that shell. (b) is a more contemporary depiction showing the three-dimensional nature of the orbiting electrons. These graphics should be used only as conceptual models. The scales shown in them have no relationship to the relative dimensions of actual atoms.

After more research, he produced the now famous first medical-type x-ray of his wife's hand. Before going on with the use of x-rays and x-ray image processing, it is necessary to detour slightly and review some of the basics of atom structures and elementary atomic physics.

Atomic Models

Even the ancient Greeks postulated that there were natural building blocks or atoms that were used to create matter. One of several modern models of the atom that serves to explain the elementary functions of x-rays was developed by the Danish physicist Niels Bohr and is termed the Bohr atom (see Figure 6.8). The simple Bohr atom postulates the atom as a miniature solar system with its "sun" being a nucleus composed of protons and neutrons and with "planets" of electrons orbiting this nucleus. This nucleus is much, much heavier than all of the orbiting electrons.

The Production of X-Rays

When an electron is "shot" into the atoms of an appropriate material, such as tungsten, there is an interaction that may produce x-rays. An electron can possess certain levels of energy, depending on the accelerating voltage. If the cathode-to-anode electric field (this accelerating voltage) imparts only a small amount of energy to an electron, if it hits the outer P shell of the tungsten it gives up all its energy to heat—infrared radiation. If an electron has enough energy to penetrate into the K shell, the electron can produce the maximum possible x-ray energy. However, the x-rays are produced in an indirect manner. The incoming electron will dislodge a normal electron from the K shell, leaving a temporary hole in the shell. Immediately, an electron will be attracted by the nucleus and will drop in from the higher L shell. This new incoming electron will give up some energy that will be transformed into a photon (a packet) of x-ray energy. If an incoming electron bumps another electron in an outer shell, the same process will take place but the x-rays will be "soft," i.e., they will possess less penetrating energy than x-rays produced from collisions in the K shell.

NOTE: It is often difficult to present introductory information that is understandable and still contains adequate scientific rigor. Both the Bohr atom model and preceding textual explanations are somewhat simplified by contemporary standards. Likewise, just as the study of noise and some of the aforementioned image-processing techniques are based upon probabilities, so too is the production of x-rays. Although this chapter provides an adequate background to understand x-ray image processing, some readers may wish to delve deeper into the pertinent subject of *quantum mechanics.* The Herbert book is recommended as an introductory reference.

The X-Ray Spectrum—Two Different Approaches

Just as a microwave or a beam light is electromagnetic radiation, so are x-rays, and they can be placed in the electromagnetic spectrum as presented in Chapter 4, Figure 4.18. If this figure were extended to the right, x-ray would extend about three to four more decades (10 times frequency or 0.1 wave length) up to about 10^{19} Hz or 10^{-11} meters. This four-decade spectrum would also include gamma rays. After that, cosmic rays occupy about three more decades which take the commonly quoted contemporary spectrum to about 10^{22} to 10^{24} Hz or 10^{-14} to 10^{-16} meters.

The x-ray spectrum, if defined as a range of values, can also be characterized and graphed in a different way in terms of energy. It was noted earlier that the energy in an (anode) attracted electron is a function of the cathode-to-anode voltage. This energy can be indicated in keV (1,000 electron volts) or in the mks unit of energy, the joule (J). (One eV equals 1.6×10^{-19} J.) There are usually two approaches and thus two types of graphs used to describe the energy in any x-ray beam. First, the energy in a universal or average x-ray beam is assumed continuous and takes a distribution that is skewed to the right of the familiar bell-shaped curve. Second, the energy in a beam that is produced from the shells in a tungsten target (shells K through P) is produced at discrete keV energy levels. The plot of

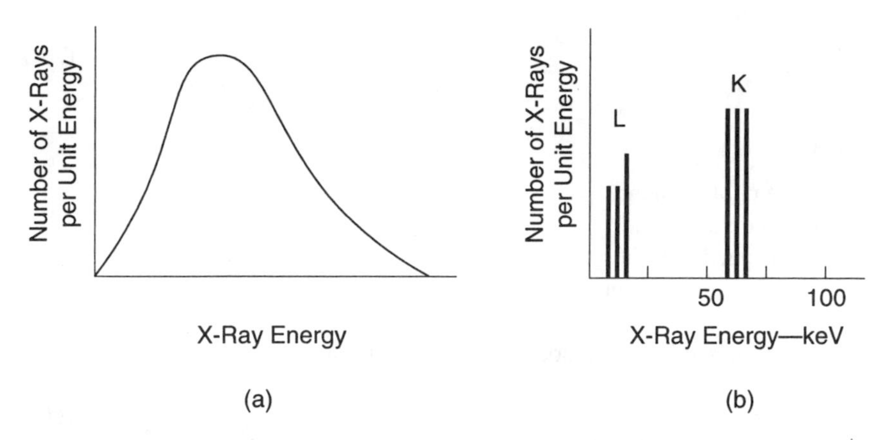

Figure 6.9 Sketches of the energy distribution in an x-ray beam. (a) is the average or continuous distribution and (b) is the discrete distribution that is representative of the beam from a tungsten anode. K and L denote the particular radiating shells.

either of these two spectrums involves a vertical ordinate in number of x-rays per unit energy and a horizontal abscissa in keV x-ray energy—see Figure 6.9.

X-Ray Beam Energy

The energy contained in an x-ray beam, like that of a beam of visible or invisible (infrared or ultraviolet) light, can be described as discrete bundles (photons) of energy or it can be described in terms of waves or waveforms with appropriate wave lengths. (An old joke stated that light was a wave phenomenon on Monday, Wednesday, and Friday and a particle phenomenon on Tuesday and Thursday and science is working on what happens on Saturday and Sunday. See various works on quantum mechanics including the Herbert reference.) To investigate the energy in an x-ray photon or quantum, most modern models start with Einstein's equation:

$$E = mc^2 = hv \qquad \text{(E 6.6)}$$

where:

E, the energy, is in joules
m is the mass in kilograms
c is the speed of light = 300×10^6 meters/second
h is Planck's constant = 4.14×10^{-15} eV-seconds
v is the particle or photon frequency in Hz

From the standpoint of wave phenomenon, x-ray energy is inversely proportional to the photon wave length.

$$E = \frac{hc}{\lambda} \quad \text{(E 6.7)}$$

where:

E is the photon energy
h is Planck's constant
c is the speed of light
λ is the wave length in meters

Since maximum energy is associated with a minimum wave length, from E 6.7:

$$\lambda_{\min} = \frac{hc}{E} \quad \text{(E 6.8)}$$

These somewhat esoteric equations can be used to provide the foundations for finding the practical emissions from x-ray tubes and how the tube parameters affect the intensity and distribution of the emitted x-ray spectrums—see Figure 6.10.

The X-Ray Tube Current

The tube current is governed by the amount of electrons that are boiled off (attracted) from the tube's cathode. The quantity of electrons (some books prefer quantity of charge) is proportional (within limits) to the current in the filament. A tapped transformer or a rheostat is used to control the filament and the beam current. Changing the current changes the energy amplitude of the spectral components in the beam but does not change their relative spectral position.

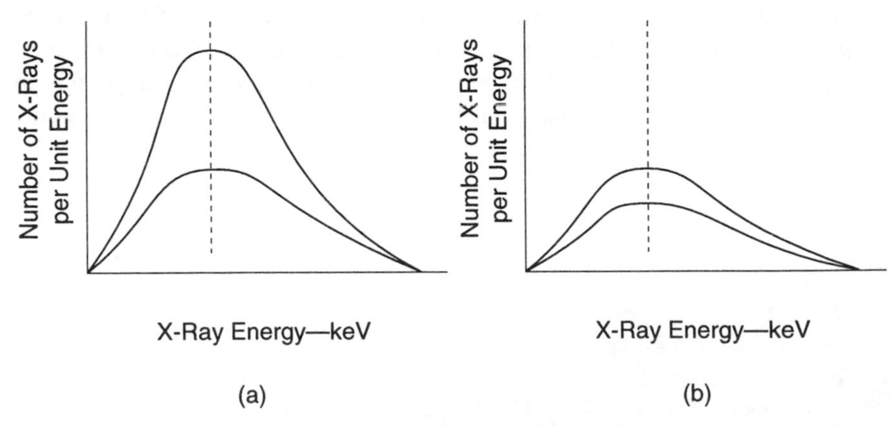

Figure 6.10 Sketches of the effect of changing the beam current and the high voltage for a continuous x-ray spectrum. In (a), the beam current is approximately doubled and the resulting amplitude of the x-ray emission also approximately doubles. In (b), the high voltage was increased about 20 percent. Note that the energy in the lower part of the keV scale has increased and the peak energy has increased and moved to the right This spectrum shift does not occur in a discrete spectrum.

The X-Ray Tube Cathode-Anode Voltage

Changing the tube high voltage will change the energy in the beam and it will also change the shape of the spectrum. By increasing the voltage, the entire spectrum will move to a higher energy level and to the right. The x-ray energy is approximately proportional to the beam current but increases as the *square* of the high voltage.

6.6 X-RAY BEAM AND IMAGE PROCESSING

X-Ray Generation and Scattering

Most contemporary medical x-rays are produced with tubes that have motorized rotating anodes. By having an anode that moves, the effect of the beam current spot-heating is materially reduced. One of the first processing steps is to use some type of aperture or shield to restrict or concentrate the effective area of the beam. It must be remembered that the beam that comes from the tube is usually a cone rather than a flat fan as shown in most pictures, such as Figure 6.11. This cone usually must be altered to restrict the beam to a given patient area and to reduce a phenomenon known as scattering. As the x-rays go through the target (the bone and tissue), the photons interact with the target's atoms and produce secondary x-ray photons that may or may not have different energies than the primary photon. Also, they may travel in different directions. These secondary photons have complex paths, depending on the energy of the primary beam but, because they do often move in random directions, they cause a smearing or fog in the film. (See any reference on Thompson and Compton scattering.)

There are several techniques and devices that are used to counteract the problems of scattering. If the beam is restricted or collimated, scattering can be reduced. Rectangular apertures and camera-like round apertures are used both in the beam and just above the film (see Figure 6.11). Collimators are also used to assure that the primary beam is aimed at the correct area of the object and at the central part of the film. Automatic collimators with motor-driven doors are used to assure that the primary beam is aimed at the central part of the film cassette. (See the Bushong reference.) Other specialized but nonelectronic x-ray image- processing methods include the use of grids and, of paramount importance, the use of specialized film for a particular diagnosis.

Scattering can be reduced and therefore picture contrast enhanced by the judicious use of grids between the object and the film cassette. The grids vary in the area and thickness of their material and are constructed of several materials, such as lead and aluminum with plastic spacers. The grid is designed to pass only the direct rays and inhibit the scattered rays. Grids come in a large variety of options for grid-material (usually lead) height and width. The choice of which grid to use depends on a knowledge of and experience with a particular application. One disadvantage of grids is that along with significantly reducing the scattered x-rays, they also absorb some of the direct rays. That means that the dosage must be increased—see Figure 6.12.

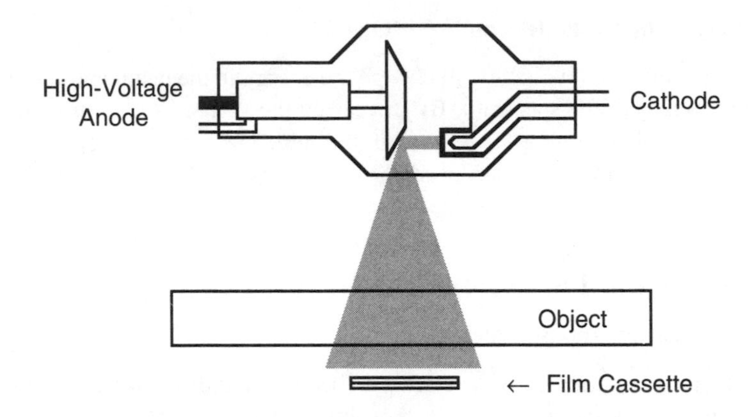

Figure 6.11 A contemporary x-ray tube with a motor-driven rotating anode to keep the beam from continually concentrating on one heat-producing spot. The cathode cup surrounding the filament is used to concentrate the beam on the anode. (Not to scale.)

X-Ray Intensifying Screens

Intensifying screens can be used to improve the intensity and sometimes the contrast of an x-ray picture (a radiograph). The sheet or screen is composed of plastic with a layer of phosphor coating and can be placed on top of the film or sometimes two screens are used to sandwich the film. The phosphor has the ability to convert the x-ray photons into photons of visible light. Since most x-ray film is still inherently more sensitive to visible light, the use of the intensifying screens can improve the effective efficiency of the x-ray film from about 1 percent to about 30 percent. Unfortunately, the price paid for this improved efficiency, which usually means a lower x-ray dose, is a slight film fogging.

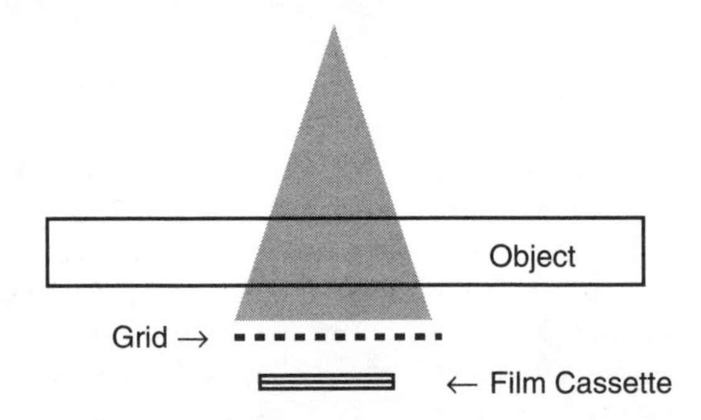

Figure 6.12 An illustration of the use of a grid to reduce x-ray scattering. The proportional scale of the grid is much too large in this graphic since the usual dimensions are in the order of μms (10^{-6} meters).

Electronic Image Intensifiers

The electronic image intensifier (EII) tube uses some of the same principles as the intensifying screens. The EII used a three-step process to convert x-rays into visible light with an overall gain of several thousand. Although the operation of the EII tube is complicated because of the interactions of the anode, the accelerating high voltage, and electrostatic lenses, its principles can be described by showing the three main elements in Figure 6.13. First, an input phosphor receives x-ray photons and converts them to visible-light photons. Second, a photo cathode, which is near or bonded to the input phosphor, converts the light photons to electrons. Third, the electrons hit an output phosphor that converts the electrons to visible light that can be viewed or picked up with a TV camera.

Soft Tissue Radiography—Mammography

The equipment and techniques required for mammography-type x-rays differ considerably from many other types of diagnostic x-rays. Whereas many radiographs rely on the differences between bone, muscle, and fatty tissue to produce contrast, the similarity of the muscle and the fatty tissue in the breast makes the production of a safe x-ray with any meaningful contrast exceedingly difficult. To understand the processing required to produce credible mammograms, a brief review of the interaction between x-ray photons and atoms is required.

When a target atom's electron is bumped by a primary x-ray photon, the electron is excited and, in turn, releases energy in the form of a secondary photon. If there is low incident energy (10 keV or less), the emitted secondary photon has the same energy and wave length λ as the incoming x-ray photon. The process of producing the low-energy secondary photons is termed classical or Thompson scattering. If the incoming x-rays possess moderate energy in the order of 50 to 75

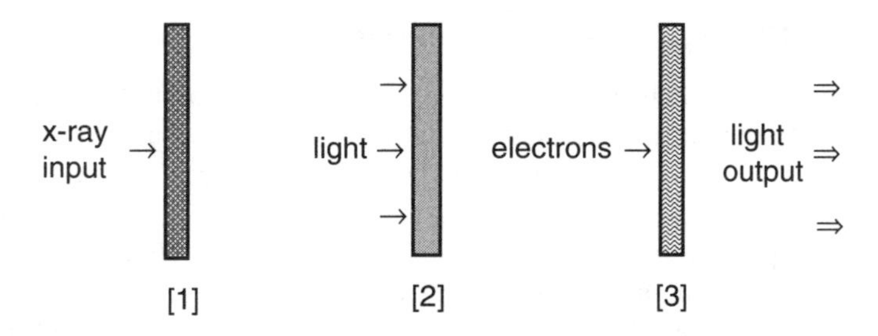

Figure 6.13 The operating scheme of an image intensifier tube. The input phosphor [1] converts the incoming x-rays to visible light. The photo cathode [2] converts this light to electrons and the output phosphor screen [3] converts the electrons back to the visible output light. For simplicity, the tube's glass envelope as well as the other elements required for operation are not shown.

keV, when it bumps an outer shell electron, it knocks it out of the shell—a process termed ionization. This process, which is termed Compton scattering, means that some of the incoming x-ray's energy has been absorbed by the target (tissue's) atom. The secondary x-ray photon will have less energy and a longer wave length. The higher the energy of the incoming x-ray, the more the Compton scattering and the more energy that is absorbed by the tissue.

Since it is well known by both the medical professional and the lay person that x-rays can cause tumors as well as diagnose them, there are conflicting goals in mammography. Because the density of the muscle and fatty tissue is quite similar, and thus it is difficult to produce a mammogram with enough contrast, it would seem logical to use high-energy x-rays. However, to reduce absorption, the x-ray energy should be kept low to minimize ionization. Yet, if low energy (low keV) x-rays are used, then the contrast is low and the tumors are difficult to detect. Again, with these low energy x-rays, the contrast could be increased by increasing the number of x-rays by increasing the beam current—again a potentially dangerous procedure.

There are several techniques that can help strike a compromise between adequate picture quality and patient safety. One of the major differences between mammography and most other form of diagnostic x-ray is the use of tubes with molybdenum anodes. The spectrum of molybdenum has a sharp, narrow peak at about 20 keV that provides an excellent source of safe, low-energy x-rays and, with its low output on either side of this peak, keeps the lower energy scattering to a minimum. An additional sheet or filter of molybdenum or aluminum between the x-ray source and the patient, as well as the judicious use of grids, will further reduce the effects of scattering. Likewise, the distance and angle of the x-ray source, as well as the exact patient's position, can improve the quality of the mammogram.

6.7 COMPUTERIZED TOMOGRAPHY

Computerized Tomography—An Introduction

Computerized Tomography (CT) is an x-ray technique that, to a large degree, overcomes the contrast problems encountered with normal radiographs. Whereas conventional x-rays see through the body and produce a photograph derived from the attenuation caused by the bones, muscles, and tissues, CT produces an image that is essentially a picture of a perpendicular slice of the body. Although the fundamental mathematical algorithm for CT was developed in 1917, it was the British scientists Hounsfield and Perry (and others) at EMI who demonstrated the first successful CT scanner in the early 1970s.

A generic CT scanner produces a highly collimated x-ray beam about the diameter of a pencil. The beam is sent through a small portion of the body and received by a small, sometimes also collimated or filtered x-ray detector. In the first models, the x-ray tube and the detector were directly connected and would move together. To produce a scan, the tube and the detector would successively move in a horizontal direction across the body. Next, the connected tube and de-

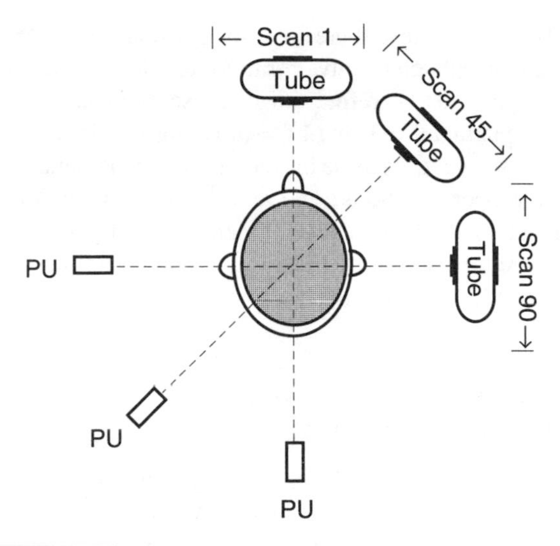

Figure 6.14 The scheme of an elementary CT brain scan. For each scan, the tube and pickup will start at one end and move across parallel to the object. Next, the assembly will rotate about 1° and again move from the start to the finish of that scan and repeat this procedure until it has scanned in 180 positions. This graphic only shows representative positions for the 0°, 45°, and the 90° scans.

tector would rotate typically 1° and produce a second slice. Again the tube and detector assembly would rotate another degree and produce a third slice, etc., until 180°-slices had been traversed and the detector readings had been indexed, quantized, and stored in the computer's memory—see Figure 6.14. Since the original EMI (usually pronounced Emy) design, there have been several generations of CT scan machines produced by companies in Europe, Japan, and the United States. Not only has the single detector been replaced by multiple detector arrays, which materially decrease the required scan time, most contemporary designs have a complete circle of arrays so that only the x-ray tube source is required to rotate.

Elementary CT Matrixes and Their Equations

In computed tomography, the area of the body that is being investigated is mathematically divided into one or more slices and these slices are in turn divided into matrixed rectangles. Often it aids in the understanding of how these matrixed areas are used, and how it is possible to obtain values for elements inside the matrix, by starting with a very simple 3 by 3 matrix. Although the different parts of the body areas under investigation will have different coefficients of attenuation, for this simple exercise, it will be assumed that the attenuation is 0 or 1. Using Figure 6.15, it is possible to write the horizontal binary equations such as, for abc, 0+0+1 = 1. Vertically, it is possible to write three more binary equations such as, for adg, 0+0+0 = 0. Likewise, there can be five diagonal equations such as, for g,

0 = 0 or for aei, 0+1+0 =1. Since there are only nine unknowns, the total of eleven equations is more than enough to solve for these unknowns. Of course, by inspection, adg = 0 and ghi = 0. Therefore, all of these elements are zero. Likewise, using a diagonal equation, c = 1 and bf = 0. Using this intuitive approach, it is possible to see how the value of these buried or hidden elements, such as e, can be derived using the proper equations. The real-world matrices started in the early research instruments with about 80 by 80 elements and have increased with the development of more sophisticated CT instruments and their microcomputers where at present a slice of 512 × 512 elements is common.

The Mathematics of CT Image Processing— The MTF or Modulation Function

It is interesting to compare the analysis of a CT image with that of a TV image. One of the first comparisons is the manner in which most CT scientists analyze spatial picture definition. The checkerboard and its idealized square wave shown in Figure 3.8 can be compared to the idealized image profile that might result from a simplified series of 0 and 1 patterns in a CT slice, as pictured in Figure 6.15. In Chapter 3, Figure 3.9, along with the accompanying text, shows that these idealized up and down transitions are limited by bandwidth. The CT literature usually uses a representation similar to the black rectangles at the top of Figure 3.9 to indicate line pairs. A line pair will include the black rectangle (line) and its

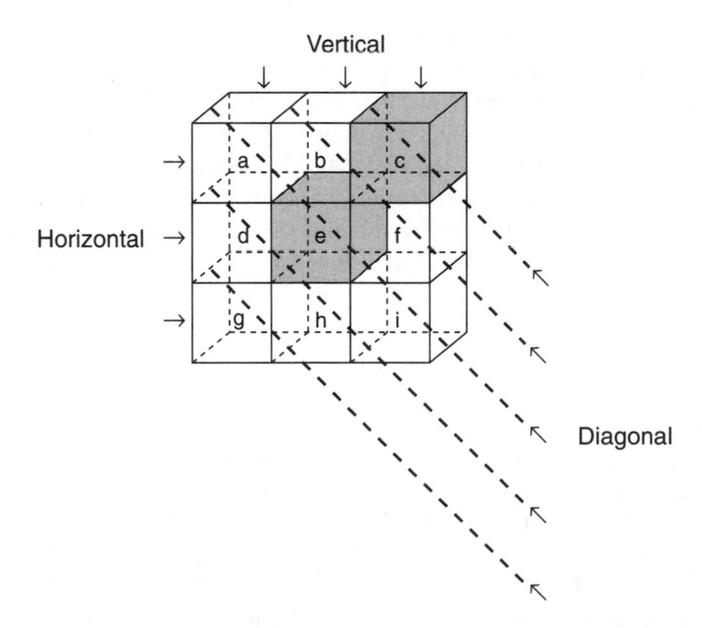

Figure 6.15 An example matrix slice that can be used to develop an understanding of how CT digital images are composed. In binary notation, boxes c and e are 1 and all other boxes are 0.

adjacent white space. A series of rectangle or line pairs can be used to describe the resolution, i.e., the spatial frequency response, of an image—see Figure 6.16. The more line pairs per unit length (usually a centimeter) the better the resolution of the CT system. It is important to remember that CT systems with low MTF values indicate the presence of picture blurring. This blurring can be visualized in two dimensions as the sin x/x "skirts" (see the side ripples of Figure 6.18) that result from band limiting or it can be visualized in three dimensions when it is realized that a three-dimensional rectangle will be transformed into a Mexican sombrero—see also Figure 6.17. (Note: There is a rather interesting but highly involved relationship between the MTF and the three-dimensional Fourier transform. The Krestel reference is an excellent starting point.)

Processing for the CT Image

The final CT image that is displayed on the television-type monitor or saved on photographic film is the result of several mathematical transformations and the multitude of mathematical calculations. The process is conceptually built on the idea illustrated in Figure 6.15, but is much more involved.

The keystone for CT processing is to create a universal contrast scale that can be used as a reference for a majority of images and can be (usually daily) conveniently and accurately calibrated. The attenuation (μ) of the body parts covers a very wide range, from dense bone to some nearly transparent lung tissue, to bubbles or pockets of air. Luckily, air, water, and dense bone can be considered constants and the attenuation scale, which ultimately relates to contrast, can be constructed around these references. Thus, once a generic image (termed an image profile or J) is produced, it is normalized to the density of water (J_0). A common relative scale, but not a universal one, is to assign dense bone a value of 3,000, water a value of 0, and air a value of –3,000.

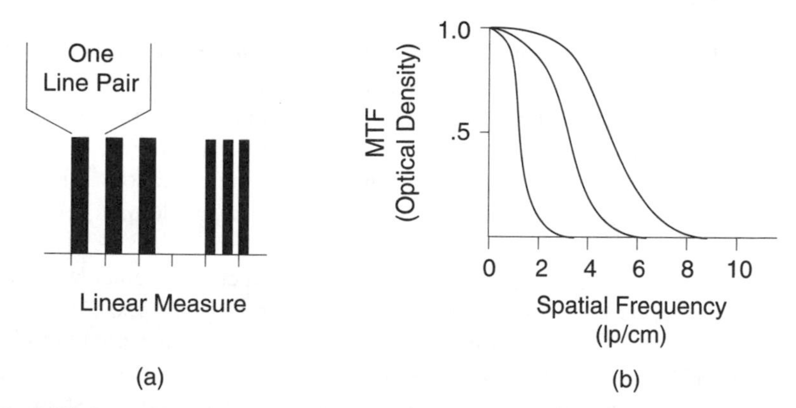

Figure 6.16 A graphic relating the width of one line pair to the description of spatial frequency. A line pair includes both a black line and a white space—(a) shows both one and two line pairs per unit length, (b) shows how the output density (the spatial frequency response) varies as a function of line pairs per centimeter.

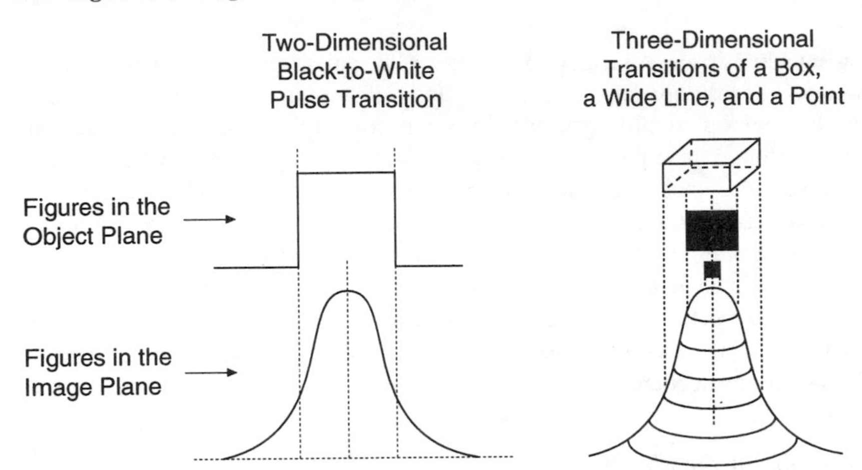

Figure 6.17 A graphic, reminiscent of Figure 3.9, comparing the two-dimensional transient response of a band-limited pulse with those of a three-dimensional box, a line or rod, and a point. Note that in 3-D, any 2-D response is revolved through 360°. The final revolved image of any relatively small object or any edge will be degraded and show blurring by the spatial bandwidth limiting of the CT system. Also, compare with the two-dimensional Figure 6.18.

For relative attenuation:

$$\mu_{rel} = k \frac{\mu_{object} - \mu_{H_2O}}{\mu_{H_2O}} \qquad \text{(E 6.9)}$$

where k is usually 1,000 or 3,000

Next, the μ_{rel} values are converted to a natural logarithmic scale by the algorithmic equation $I_{Ln} = Ln\ \mu_{rel}$.

Once the volume elements of the slices have been normalized to water and these have been converted to a logarithmic attenuation scale, it is possible to mathematically collect the term and produce a picture in the image plane on the display. However, there is the omnipresent problem of blurring. Before the common CT solution to blurring is presented, it seems prudent to review some of the background relating to functions (images) in the time or space domain and the frequency domain.

Again referring to Figure 3.9 in Chapter 3, a perfect rectangular wave in the time domain transforms, courtesy of Monsieur Fourier, to a sin x/x, or Mexican hat, with scrolled edges in the frequency domain. A two-dimensional sin x/x plot is shown in Figure 6.18. If a two- or three-dimensional rectangular figure in the time or space planes can be transformed into a band-limited sin x/x shape in the frequency or image plane, it would seem reasonable that with an additional modifying mathematical function, the characteristic hat shape could be transformed

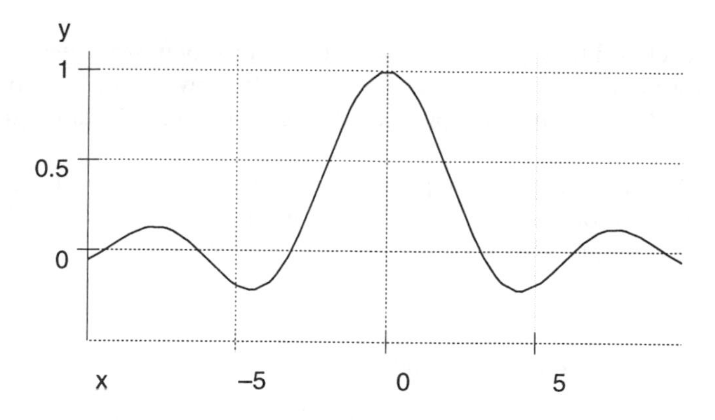

Figure 6.18 A computer-generated graphic of sin x/x. Although not shown, the side "skirts" or ripples go on and on to cause blurring in a CT picture.

back to its (more or less) original rectangular shape. This is true and it can be accomplished in several ways, depending on the wishes and the constraints placed on the designer. One approach would be to first Fourier transform the object-plane image to the frequency domain. There it could be multiplied by the appropriate function to distort it so that when it was inverse transformed back to the object plane, and then used to construct the composite image, that final image would have less sidelobe ringing and thus less blurring.

A similar process uses time- or space-domain convolution. It will be remembered that convolution in the time or space domain is the equivalent of frequency-domain multiplication. Thus, if the Js, the attenuation factors, have been [1] normalized to water's J_0 and [2] converted to logarithmic numbers, they can now [3] be convolved with the appropriate factors, which will remove most or all of the sidelobes or skirts that cause the blurring. As a final step these slice elements that have just gone through the convolution are, with the proper matrixing mathematics, summed together to form the proper image-plane pictures.

6.8 NUCLEAR MAGNETIC RESONANCE AND MAGNETIC RESONANCE IMAGING

An Introduction to Nuclear Magnetic Resonance

Medical magnetic resonance imaging (MRI) uses completely different principles and technologies than x-ray and CT imaging. Whereas both x-ray and CT primarily derive their images from the way x-rays react on the outer-shell electrons in the tissue, fat, or bones in the body, magnetic resonance imaging derives its images from a complicated reaction between the (usually) hydrogen atomic nucleus in the body parts and the specially applied magnetic and radio-frequency fields.

MRI was developed to improve the resolution of low-contrast images. Whereas CT images were a significant improvement over the low contrast of normal x-rays, MRI, in turn, can produce pictures that are much more detailed than CT scans.

MRI is based upon the techniques of the earlier chemical and biochemical nuclear magnetic resonance (NMR) for determining molecular spectrums. Because this represents the fundamental work, the complex technology of NMR will first be described and the imaging part will then be detailed.

More Details of the Bohr-Atom Model

Figure 6.8 was used to introduce the ideas of how a prototype atom seems to be constructed. The (a) part of this figure can be expanded to help explain the processes involved in NMR. It appears that the protons and the neutrons in the nucleus spin on their axes like a toy top. Likewise, again somewhat like a top, the nuclei also process or wobble about their axes as they are spinning. This spinning motion produces a minute magnetic field with a north and south pole somewhat analogous to the magnetic poles of the earth. The north-south magnetic direction or moment of the billions of individual nuclei is randomly directed in most materials—including those in the human body. However, when atoms, such as in the body, are placed in a static DC magnetic field, they align themselves with the applied magnetic field much like the needle of a compass.

In the light of quantum mechanics, the above explanation is overly simple. It seems that some of the magnetic moments align with the northerly direction of the applied magnetic field, some align against it, and some are still left in a random alignment. Likewise, thermal agitation continually causes nuclei to go in and out of alignment. Nevertheless, the external magnetic field always produces a net magnetization of the material's nuclei.

In an applied magnetic field, the wobble or procession of an element's nucleus can be used as a fingerprint to identify that particular element. In fact, this is the fundamental basis for identifying the frequency of procession that in turn leads to the science of magnetic resonance spectroscopy. For an atom of material in a magnetic field, its characteristic resonate frequency can be determined by:

The Larmor equation for the Larmor procession frequency:

$$\omega = \gamma \beta_0 \qquad \text{(E 6.10)}$$

where:

ω = the magnetic resonant frequency in MHz

γ = the gyromagnetic (spin) ratio in MHz/T. Each element will have its characteristic gyromagnetic ratio.

β_o = the static magnetic field in tesla (T) (1 T = 10,000 Gauss)

The process of magnetic resonance spectroscopy involves one more factor—an RF pulse. If [1] molecules containing a pertinent element, such as hydrogen, are [2] placed in a static magnetic field and [3] a very high-frequency RF pulse is injected, the alignment of the atomic nuclei will be changed. The addition of the RF pulse will add energy that will change the orientation of the aligned nuclei. When the pulse is removed, the nuclei will return to their former alignment and, in doing so, give up or emit the energy they just absorbed. This emitted energy is in the form of a decaying sine wave, termed a free induction decay or FID. This emitted energy can be detected using an appropriate pickup coil and amplifier. Once the time-domain energy pulse is captured, it can be converted (using a Fourier transform) to the characteristic frequency-domain spectrum for that element or for compounds containing that element.

The Parameters Influencing Nuclear Magnetic Resonance—Spin Density

There are three pieces of information that affect the production of an NMR spectrum and likewise an MR image. The first of these, spin density or SD, is the relative number of hydrogen (or other elements) nuclei involved in a given measurement of the relaxation or FID signal. For instance, if two similar experiments are attempted, and the FID signal is twice as intense in the second, it would indicate that the hydrogen concentration was twice as intense in the second substance.

The Parameters Influencing Nuclear Magnetic Resonance—T1 Relation Time

T1 is one of the measured parameters related to the aligned nuclei returning to equilibrium after an RF pulse. Figure 6.19 can be used to help understand this process. It will be remembered that, in the absence of the magnetization B_0, all the nuclei are randomly directed. When the static field is applied, although not all the moments align with B_0, there is a resulting net nuclear vector. In fact, some texts indicate one large vector in the y-z plane and/or the x-y plane to indicate the cumulative or net value of all the aligned nuclei. (See the Bushong reference.)

For clarity and convenience, Figure 6.19 shows the aligned nuclei in the y-z plane and the flipped or spiraled nuclei in the x-y plane, 90° away. The RF pulse can be shaped to move the net vector to any desired angle including 90°, 180°, or even into the negative x-y plane. It is important to remember that these flips involve energy. When the Larmor frequency RF pulse is applied, energy is given to the nuclei to move them. When the pulse is shut off, energy is released or returned. This released energy is in the form of an exponentially decaying sine wave at the Larmor frequency. It is thus similar to the discharging of a capacitor in a shorted RC circuit. For the electronic circuit, T = RC for the decay of energy; for the FID, when nuclei reorient to the y-z plane, T = T1.

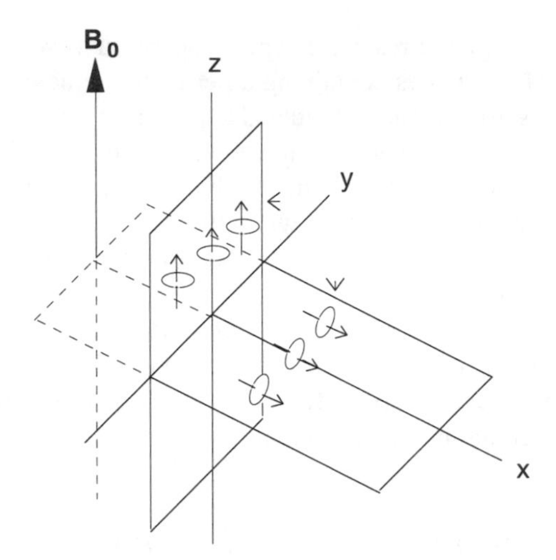

Figure 6.19 A three-dimensional graphic showing nuclear magnetic moments, aligned by B_0, in the vertical y-z plane and also those that have flipped into the horizontal x-y plane because of an RF pulse. Note that in the real quantum mechanical world, all the nuclei are spinning and processing, and when they go to or from the x-y plane, they actually spiral rather than simply flip.The Parameters Influencing Nuclear Magnetic Resonance—T2 Relation Time

The Parameters Influencing Nuclear Magnetic Resonance—T2 Relation Tine

T2 is a second measured parameter related to the aligned nuclei returning to equilibrium after an RF pulse. When the RF pulse is injected and the net nuclear moments align in the 90° (and any other predetermined angular) plane, they are also processing in synchronism—they are phase coherent. However, the magnetism (the magnetic moments) of each individual nuclei tends to affect each other and eventually they will cause each other to drop out of what is known as this coherent phase. This dephasing or spin-spin interaction at the Larmor frequency causes the field to weaken. Indeed, after a time the net field would decrease to zero. One of the methods to counteract this T2 dephasing is to apply a second RF pulse that is itself at a different phase.

6.9 PRODUCING THE MAGNETIC RESONANCE IMAGE

Comparing Nuclear Magnetic Resonance Spectroscopy and Medical Magnetic Resonance Imaging

The previous section detailed NMR as a foundation for MRI. With that background, the focus will be turned to magnetic resonance as a vehicle for producing detailed images of human body sections. However, before we completely change the thrust, it is interesting to note that, just as the physical chemist will use NMR to analyze molecules in a test tube, termed in vitro, so can the physician investigate molecules, or the change of molecules, in the human body, termed in vivo.

On first inspection, it is presumed that the Larmor frequency of an element is an absolute, fixed quantity. However, theoretical reasoning and experiments have shown that the Larmor frequency of an element as part of a molecule or compound will change slightly from the measured frequency on the element alone. Likewise, an element's Larmor frequency will vary slightly from compound A to compound B. For instance, the processional frequency of the hydrogen in water is about 5 ppm (parts per million) different than it is in fatty-acid tissue. This is termed chemical shift and is defined as:

The chemical shift equation:

$$ppm = \frac{\nu_1 - \nu_0}{\nu_0} 10^6 \qquad \text{(E 6.11)}$$

where:

ν_1 = processional frequency of a given element in this compound
ν_0 = the reference processional frequency

likewise:

$$\nu_0 = \omega_0 / 2\pi$$

The chemical shift is caused by the magnetic interactions between the orbiting electrons in the excited (by thc RF) atom or molecule. Recall that any electric current, which produces a magnetic field, can be defined as a flow of electrons. The electrons do not have to move (flow) in a wire. By using this principle of chemical shift, physicians can use MRI apparatus, but in an NMR mode, to check, for example, certain body molecules to determine if the metabolism of a specific area in the body appears normal or abnormal.

MRI versus NMR

First, the apparatus for human MRI is big. The circular body inspection cavity and its movable patient's couch is surrounded by a large supercooled magnet that, in turn, is enclosed in an insulated and shielded housing. Inside this special outer housing are also vacuum compartments and special cylindrical compartments that contain liquid nitrogen and liquid helium or (with contemporary designs) just liquid helium.

Besides the giant coil for the primary B_0 magnet, there are shim coils to linearize the primary magnetic field. In addition, there are the all-important gradient coils that provide the MRI with the ability to produce the image slices and the transmit and receive coils (sometimes one coil can both transmit and receive) for the RF energy. External to the main MRI unit is the processing computer, the image archiving equipment, and the control and display consoles.

The Primary, Superconductor Magnet

The magnet field for diagnostic MRI must be intense and extremely uniform. There are three possible methods to create the B_0 magnetic field. The first is, of course, a permanent magnet. Although developments in ceramic magnets have greatly increased their flux density per unit volume and weight, any practical permanent magnet for MRI would produce far less than one tesla, would be excessively heavy, and would not produce the required uniform field.

Wire-wound or magnetic solenoids, although possible, have two major disadvantages. First, like permanent magnets, the maximum practical field strength is well under 1 T. The second drawback is the amount of power, with its inherent heat, that is required. A practical wire-wound (termed resistive) MRI magnet requires almost 100 kilowatts. Even with intense water cooling, resistive magnets, although used in very early designs, would produce only marginal results.

Starting in the 1960s, materials were developed that lost almost all of their resistance when cooled down to approximately absolute zero or 0° Kelvin (K). When these superconductors are used, practically no heat is generated and once the current flow is stated, because there is no IR drop, the current continues to flow almost indefinitely. The field strength is between 1 and 2 T. The price that must be paid, literally as well as figuratively, for the breakthrough superconductor technology is the highly insulated vacuum vessel (termed a dewar) required for the magnet housing and the expense and work required to maintain the supercold or cryogenic material. (The liquid nitrogen will eventually return to a gaseous state and must be drained, recompressed, and the dewar again loaded.)

The Problems of Magnetic Field Uniformity—Shielding and Shimming

Equation 6.11 shows a difference in the processional frequency measured in parts per million. If these or other MRI measurements are to be accurate to the low parts per million (1 ppm or less is common), the uniformity and stability of the magnetic field must be kept in this range. So far, it is impossible to produce these giant magnets that have fields that are this uniform over the range of the entire body. The solution is to add uniformly spaced auxiliary or shim coils that can be used to adjust the trim of the primary magnetic field. The individual currents in these coils can be computer controlled and adjusted to produce the necessary uniformity along the axis of the field.

The ability to produce a magnetic field that is precise and stable to 1 ppm, which also represents an accuracy of 20 bits in the digital domain, is a tribute to the outstanding work of electrical, mechanical, and thermal engineering design teams from all over the world. For historical perspective, at the end of the all-analog period, circa 1970, the lens current and the high voltage stability of production electron microscopes was about 5 ppm, with the ability to tweak some laboratory instruments to 2 to 3 ppm. It is interesting to note that, besides using such wonderful techniques as the FFT, successful signal and image processing also depends upon choosing materials with the proper mechanical and thermal coefficients.

The internal diagnostic fields, both magnetic and RF, must be confined to the MRI instrument and shielded from externally generated fields. The room housing the MRI, as well as the instrument itself, must be extensively shielded from external radio and TV signals (this special shielding is referred to as a Faraday cage) and the magnetic disturbances that large, moving metallic objects, such as trucks and buses, could produce.

The Gradient Coils—These Make It Work

A review of the Larmor equation $\omega = \gamma\beta_0$ shows that if the magnetic field B_0 changes, the processing frequency ω will shift. For NMR spectroscopy, this would cause a shift in the spectrum (peak)—usually a very unwanted effect. However, for MRI, this same result can be used to distinct advantage. By judiciously using the computer to change the magnetic field very slightly, a different peak or perspective of the object can be obtained. These peaks can be used to create an image slice, much like those created using CT. These slight shifts can be accomplished by separate gradient coils which, like the shimming coils, have their individual DC currents carefully computer controlled. The formation of the slices is much more complicated in theory and application than in CT, since there are coils to produce gradients in the x, y, and z planes. Theory and experience give the physician the setting for a given malady. Likewise, the vast majority of user-controled consoles and displays are completely interactive.

Processing to Produce the MR Image

Since CT and MRI use such entirely different technologies, it is often misleading to compare the results of their processing techniques. However, one comparison is usually considered valid. In both x-ray and CT radiography, the major determining factor for picture contrast is the attenuation of the input beam. (Scattering is also a degrading factor.) Thus, to a first approximation, the ability to produce picture contrast is dependent upon any ability to overlay or over-project areas that may differ only about 1 percent in their attenuation coefficients. In comparison, using the entirely different technology of MRI results in a contrast, using like tissue samples, of over 30 percent.

MRI has three major pieces of processing information that can be used alone or in concert to optimize the resolution of low-contrast structures. Referring back to the section on NMR, the controllable variables include the aforementioned spin density, T1, and T2. Since the spin density is an indication of the hydrogen concentration, it can, at times, be used as a processing gain factor. However, the ability to control the degree and the angle of the spinning nuclear magnetic moments, plus the use of controllable gradient field, is the basis of the outstanding definition and contrast of MRI. The algorithms used to control the RF pulses are very complex and depend upon the type and combination of bone, muscle, and tissue that is being investigated. For an example of how the processing is accomplished, there are two or three classic wave shapes that are used in the literature that show some of the processing strategies. (Also see practically any reference text on MRI. For instance, the Stark and Bradley reference is very thorough.)

Figure 6.20 illustrates some of the basic techniques used to enhance the contrast of MR images and to counteract interactions of the north-south magnetic moments such as the aforementioned spin-spin. Part (a) of Figure 6.20 shows the simplest method of producing the FID RF signals. Although somewhat exaggerated, it indicates the second decaying burst may be smaller than the first. In (b), a special extra 180° pulse is added to counteract the spin-spin cancellation effect. Part (c) illustrates another one of many techniques to counteract the effect of dephasing. By repeatedly bumping the net magnetic moment back into phase coherence with 180° pulses, the effective coherent time is expanded and repeatedly extends T2 from the first 90° pulse to the second 90° pulse.

Composing the Final MR Image

When first using the very simplified analogy, an MR image can be compared to the prototype CT image grid as show in Figures 6.14 and especially 6.15. However, the actual compilation and the calculations are much more involved. If the

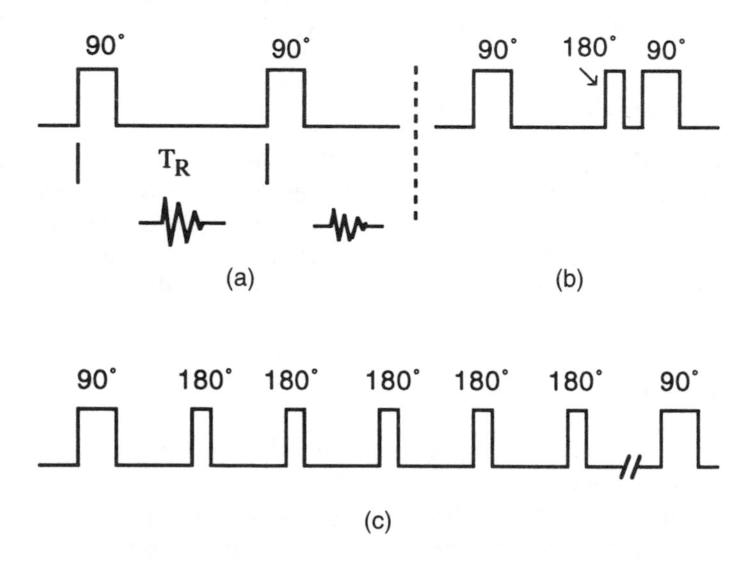

Figure 6.20 Methods of controlling MRI picture contrast. (a) shows a simple pulse repeated every T_R. Note that the second FID wave shape is smaller than the first. (b) illustrates a method to counteract spin-spin interactions. (c) is detailed in the text.

concept of the slice with individual pixels (or three dimensional voxel) is used, each of the MR voxels can be defined by rotating the magnetic gradient rather than rotating the x-ray tube and/or pickup. Producing and keeping track of the intensities of these B_1 magnetic gradient values is, of course, in addition to maintaining a running record of the B_0 value, the RF pulse amplitude and phase, and their resulting effects on T1 and T2. These values are usually assembled and used in the much more complex two- or three-dimensional versions of the Fourier transform.

6.10 SIGNALS FROM SPACE—PROCESSING SIGNALS FROM EARLY SATELLITES

The Space Program—The Search for Perspective

This still seems to be the greatest exploration, the most magnificent adventure, in the history of humankind. At times, we may get bogged down thinking about Sputnik I, the so-called space race, and all of the other peripherals and forget how and why it all started. From the standpoint of science and technology, both the Sputnik series and the U.S. Vanguard series, which was initiated as a contribution to the 1957 International Geophysical Year, started an elegant quest for knowledge from the perspective of space. (Vanguard I, the 6.4-inch "grapefruit" satellite, was the first to use solar cells and distinguished itself by discovering the Van Allen radiation belts.) With the rocketry and other technology developed during and after World War II, we now had the resources to look beyond.

The Space Program—Information Pickup and Retrieval

The initial investigations included such subjects as solar and other radiations, magnetic fields, cosmic rays, and biological and medical experiments. Each of these quests required both devices to gather the respective information and a means to get this information back to earth for analysis. In general, the pickup devices fall into three categories including a wide range of scientific instruments such as magnetometers and Geiger counters, still photography and motion-picture cameras, and TV cameras. The information retrieved by these devices was either brought back to earth or sent back by radio, i.e., radiotelemetry. The photographing techniques as well as the cameras have varied widely over the years. The Hasselblad camera, using 70-mm film, has been the utility still camera on most manned flights along with numerous 35-mm cameras. (It is reliably reported that John Glenn took one of his most important pictures with a 35-mm still camera that he purchased at a local drugstore.) Likewise, throughout the years both the unmanned and manned flights have included 16-, 35-, and 70-mm motion-picture cameras.

The collected scientific information is usually radioed (micowaved) back to earth on a telemetry link. However, like many tasks in space, this is not easy. Many times it is impossible to relay information directly back because of the position of the space vehicle. For instance, data cannot be transmitted from the dark side of the moon. Also, especially when using deep-space probes, it is often necessary to have the earth station facing in signal source. Such problems as these prompted the early inclusion of ultra-compact and ultra-reliable magnetic tape machines that would both record and play back on signals from earth. Another complication in this process is the time delays involved in both sending the commands and receiving the data. The delay from most synchronous communication satellites, which are typically in orbits about 22,500 miles from earth, is in the order of 100 ms, whereas the delay from Mars is about twelve minutes. Deep-space signals can take hours to send or receive. Also, Doppler shift can be a problem for some types of telemetric or communication signals.

Radiotelemetry from Space

The signals involved in telemetry are somewhat reminiscent of the signals involved in telecommunications (see Chapter 4). Although practically all of the scientific signals are slow, and thus require narrow bandwidths, many of these are recorded and later sent to earth in a much more rapid non-real time. Likewise, many of the command and control signals for the spacecraft need to be executed or received as soon as possible. Therefore, it was necessary, especially as the space program expanded, to provide some multiplexing schemes to accommodate both narrow- and wide-band signal transmission.

The preferred final microwave modulation was FM. However, the multiplexed subchannels feeding into the final carrier modulation included analog single sideband, frequency shift keying (FSK), time or frequency division multiplex, and, especially later, PCM. Some communications satellites also use a form of two-signal quadrature modulation that enables one antenna to be used in both its horizontal and vertical polarization for separate signal transmission. It is important to remember that space communications, like (and including) telecommunications, use both synchronizing and error-checking signals.

6.11 SIGNALS FROM SPACE—TIROS AND MARINER

The Tiros Meteorological Satellites

Tiros introduced TV imaging for information gathering in space. (The 1959 Vanguard II used photocells to produce cloud-cover pictures.) Tiros I (Television Infra-Red Observation Satellite), launched April 1, 1960, began a family of meteorological satellites that spanned almost three decades. It had two TV cameras, one for high resolution (by those early-day standards) and one to produce wider angle shots. Both cameras were slow scan and produced 500 lines per frame with bandwidths of only 62.5 kHz. The cameras always pointed toward the earth since the unit was spin-stabilized. The transmissions could be controlled from the ground and could be received from stations in Kaena Point, Hawaii, or Fort Monmouth,

New Jersey. This was truly a pioneering satellite since it introduced storm tracking to the scientific community. Later Tiros II, launched in November 1960, provided additional infrared information about ground and water-surface temperatures.

An Introduction to the Planets—The Mariner Spacecraft

The space ships were growing up. The Mariners weighed about 500 pounds and carried much more varied and sophisticated scientific instruments than the near-earth satellites. Mariner I had to be destroyed seconds after liftoff, on July 22, 1962, because of a booster-rocket malfunction. Mariner II was launched as a Venus probe, August 27, 1962, and Mariner IV was launched to investigate the surface of Mars. A brief review of the way scientific and TV data was processed and transmitted on these two spacecraft, plus the later sections on the Apollo missions, will provide a perspective into the way that signals were processed in the early part of the U.S. space program—see Table 6.2.

Mariner II was truly a scientific probe. There were eight to ten major scientific instruments that gathered data and had it faithfully transmitted, some 30 million miles back to earth, with a single three-watt, 960 MHz transmitter. The literature on the Mariner II mission also lists the principle U.S. scientists connected with the following experiments—including Dr. J. Van Allen of the State University of Iowa.

On to Mars—Capturing the Images with Television

Starting with Mariner IV, TV cameras were added to see and better understand the planet Mars. By following the progression of these probes, it is possible to see the development of space telecommunications as a prelude to the Apollo project and its culmination with humans walking on the moon.

The following six paragraphs are taken from "The Surface of Mars" by Robert B. Leighton. © May, 1970 by Scientific American, Inc. All rights reserved. Used by permission.

> The single camera on Mariner 4 used 200 lines per frame, noninterlaced, with 200 pixels per line for a total of 40,000 pixels per picture. The video output was encoded with 6 bits or 64 brightness levels. This produced a picture with an area of about 300 kilometers square and could resolve areas about five kilometers in diameter. Mariner 6 and 7 had a wide-angle, 53-millimeter focal-length camera and a narrow-angle, close-up camera with a focal length of 508 millimeters. Since these spacecraft went closer to the surface of Mars, the close-up camera covered an area about 80 kilometers square and could resolve areas down to about 300 meters across. The cameras in these later probes used 704 lines per frame with 935 (active) pixels per line. Likewise, these camera signals were digitized with 8 bits that produced about *five million* bits per frame. Thus, the definition was somewhat better than commercial B&W television.

Table 6.2 The Principal Scientific Experiments of Mariner II

Microwave Radiometer	Used to determine the temperature of the planet surface.
Infrared Radiometer	Used to determine the temperature distributions in the clouds.
Magnetometer	Used to measure magnetic fields.
Ion Chamber and Geiger-Mueller tubes	Used to measure high-energy cosmic radiation.
Anton Special-Purpose Tubes	Used to measure Van Allen-type radiation.
Cosmic Dust Detector	Used to measure the flux of cosmic dust.
Solar Plasma Spectrometer	Used in solar wind experiments.

The processing in Mariners 6 and 7 was much more complex than in Mariner 4. The tape recorders in Mariner 4 could store about 5 million bits or 22 pictures. These were later transmitted back to earth at a rate of 8.33 bits per second. The transmission of a set of 22 pictures took more than *eight hours.*

Mariners 6 and 7 chose a system involving both a 4-track analog video recorder and a similar 4-track digital recorder. Because of the limited amount of record time on both these units, especially the digital machine, a rather convoluted process was devised to accommodate both the limitations of the recorders and the radio telemetering capacity of the spacecraft.

On Mariners 6 and 7 the analog signals from the Vidicon cameras were treated in two ways. One sample was first translated into an 8-bit code. The first two bits were averaged over several scanning lines and were transmitted directly to the earth in real time as part of the telemetry stream of engineering data. The remaining six bits from *every seventh* pixel were recorded by the digital tape recorder and later played back to the earth, producing 1,777 pictures, each one-seventh complete.

Simultaneously another sample of the Vidicon analog signal was passed along to the analog tape recorder, but first it was modified in two ways. In order to enhance the visibility of small-scale features, the analog signal was automatically controlled so that its average value was approximately constant by using automatic gain control (AGC). In addition the signal was put through a circuit with a cube law response, which enhanced the local contrast by a factor of about three. The ultimate effect of this procedure was to provide a more finely graduated signal when the analog recording was subsequently converted into a 6-bit digital signal for transmission to the earth.

Later the first two bits, which had been averaged over several scan lines and returned earlier to earth, were recovered by computer processing and merged with the analog picture data to yield a resultant picture with the full 8-bit (256 levels) encoding range. If the brightness levels of successive pixels had varied from 0 to 255 in entirely random fashion, this scheme would not have worked. We knew, however, from the Mariner 4 results that the surface of Mars would not resemble a random-number table, and that it would be very low contrast. It can also be noted that later Mariner probes had the luxury of increased transmitter power and better transmitting and receiving antennas, along with better error-checking that allowed a transmission rate of 16,200 bits per second.

6.12 THE APOLLO MOON PROJECT

The Apollo Mission—A Prelude

It took 400 years of trial and failure, from da Vinci to the Wrights, to bring about the first flying machine, and each increment of progress thereafter became progressively more difficult.

But nature allowed one advantage: air. The air provides lift for the airplane, oxygen for the engine combustion, heating, and cooling, and the pressurized atmosphere to sustain life at high altitude. Take away the air and the problems of building the man-carrying flying machine mount several orders of magnitude. The craft that ventures beyond the atmosphere demands new methods of controlling flight, new types of propulsion and guidance, a new way of descending to a landing, and large supplies of air substitutes.

Now add another requirement: distance. All of the design and construction problems are recompounded. The myriad tasks of long-distance flight call for a larger crew, hence a greater supply of expendables. The functions of navigation, guidance, and control become far more complex. Advanced systems of communications are needed. A superior structure is required. The environment of deep space imposes new considerations of protection for the crew and the all-important array of electronic systems. The much higher speed of entry dictates an entirely new approach to descent and landing. Everything adds up to weight and mass, increasing the need for propulsive energy. There is one constantly recurring, insistent theme: everything must be more reliable than any previous aerospace equipment, because the vehicle becomes in effect a world in miniature, operating with minimal assistance from earth. Such is the scope of Apollo.

(From a NASA/North American Rockwell statement to the public, circa 1969.)

The Apollo Mission—The Equipment for Signaling and Signal Processing

During the later flights, the following includes a general equipment complement. There were ten antennas on the command and service modules for RF communication using the X-band, S-band, and VHF. Some of them were omnidirectional, some directional, and some used vary narrow beams. The modulations were primarily PCM; FM was used for TV. The telemetric data streams included biomedical data on the astronauts, environmental information about the Command Module (CM) or the Lunar Module (LM), and audio communication plus data from the investigations using scientific instruments. Radar was used for navigation and ranging.

Starting with Apollo 7, October 1968, monochrome TV systems were included as part of the information-gathering equipment The Apollo 11 mission in July 1969, which included the famous first moon walk, included the first Apollo color camera. The sections to follow will outline the technical and environmental problems that had to be solved to provide operating systems for both the monochrome and color systems.

The Apollo Television Cameras—System Requirements

The cameras can be divided into three groups: [1] the monochrome camera that was used for surveillance and information gathering on the initial, unmanned missions, [2] the monochrome camera that was used on early manned missions, and [3] the color camera that was later used to include the activities on the lunar surface.

It should again be emphasized that along with the difficult problems of camera design and production were the solutions required to integrate the video signals into the communication systems. The earlier Apollo spacecraft had not included storage or mounting space for TV cameras. The bandwidths of the communication channels, including the bandwidths of some of the spacecraft and earth receiving antennas, were certainly not designed for TV. For a time, the low light levels in space seemed to make TV impractical. Some of the routing and switching systems had to be modified as well as the techniques for recording the video information. The later cameras had to be designed so that suited astronauts could manipulate them, including changing lenses.

Once the decision had been made to include a TV system, several intense and quite scholarly research projects were initiated to find the best solutions (or best compromises) to both the system and specific technical problems. Early investigations included (circa 1962) an all-digital system. Although it was abandoned because of bandwidth limitations, auxiliary studies investigated video compression techniques—again, in the early 1960s.

The Signals and Processing Systems for the First Monochrome Surveillance Camera

The first monochrome camera was designed to operate with a video bandwidth of 500 kHz. To accommodate this narrow bandwidth, the scan rate was only 10 frames per second. There were 320 noninterlaced lines per frame with a 4:3 aspect ratio. (This same general slow-scan technique has been used in the past and is currently revived again to send video [Videophone®] over the narrow-band telephone lines.) Because of the uncommonly narrow bandwidth, this slow scan signal could be used by the Apollo in-flight tape recorders. Special slow scan-rate monitors had to be used. This first RCA color camera was the one that showed the famous "and one giant step for mankind" shot.

On earth, if the slow-scan signals were used with commercial broadcast-type monitors or tape recorders, or supplied to the commercial networks, the scan rate had to be converted to 30 frames per second. The principle of frame-rate conversion was first based upon the old mechanical intermittent pull-down technique used to play 24-frame per second motion pictures. Likewise, there was a method to electronically convert back and forth between U.S. and European TV scan stanardards. The basic component of any scan conversion scheme is a read-write memory—be it a delayed mechanical pull-down, the screen persistence of a cathode-ray tube, or the more contemporary semiconductor memory—see Figures 6.21 and 6.22.

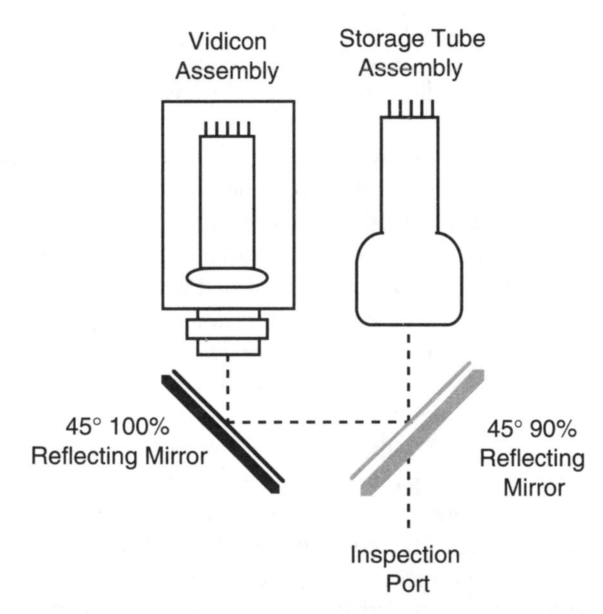

Figure 6.21 A scheme for scan conversion similar to the one used for the Apollo TV signals. The slow-scan video is written into the long persistence storage tube that is read by a Vidicon pickup that is scanned at a much faster, usually commercial, TV rate. The 10 percent feed-through of the storage-tube mirror makes possible direct visual monitoring.

The Second Monochrome Apollo Camera

When the decision was made to use a video camera for observation and some limited surveillance on the lunar walks, a program was started to upgrade an earlier Apollo-program camera for the rigors of climate on the moon's surface. Some of the requirements included satisfactory operation in the temperature extremes of ± 250° F in a vacuum of 10^{-14} torr (1 torr = 1 mm of mercury). It also had to withstand a 100-percent salt-contaminated humidity and a 100-percent oxygen atmosphere plus the accelerating, decelerating, and landing shocks of the CM and LM.

The camera modifications included more massive heat shield and heat sink—within the confines of the stringent weight limitations—and improved shock mounts. A high-resolution mode was added that included 1,280 lines per frame at a rate of 0.625 frames per second. This high-resolution mode, plus the existing 320-line/10-frame mode, could still be accommodated in the 0.5-MHz bandwidth. Difficulties were experienced in obtaining a stable synchronization that could operate over the required environmental extremes. The technique of using sync pulses was exchanged for one of using a series of very stable sine-wave bursts.

The Apollo Color Camera

Around 1968, investigations started to find a suitable color camera that would work on the CM, the LM, and out on the lunar surface. Again, the technical, environmental, and weight limitations had to be considered. Likewise, it was realized that for color, the ground-station facilities would have to be modified and upgraded.

The Westinghouse camera that was selected was modified from a monochrome camera that had been used on an earlier space program and had been adapted for the requirements of space. The method of producing and transmitting color was similar to the original CBS color-wheel, field-sequential system (see Chapter 3). A known monochrome space camera was modified with a color-filter wheel and its clocks and synchronizing system were converted to the NTSC standards. The color bandwidth was decreased to 2 MHz. Although this did not give commercial quality color, it did provide both acceptable motion rendition and definition of the terrain.

Because of the wider bandwidth, plus the problems of phase and transient response (see Chapter 3), the preparations for the color transmissions required significant changes in the ground station's receiver and signal-routing systems. Likewise, the transmission systems had to have special equalization to compensate for the Doppler delays. As with most space systems, this color system was exhaustively "breadboarded" and simulated, both in the laboratory and at the actual sites.

Just as the original CBS system was not compatible with either the older EIA black and white standards or the NTSC color standards, so the Apollo field-sequential color (FSC) was not compatible with NTSC. In order to use commercial monitors and videotape machines at the ground station, as well as for feeding the

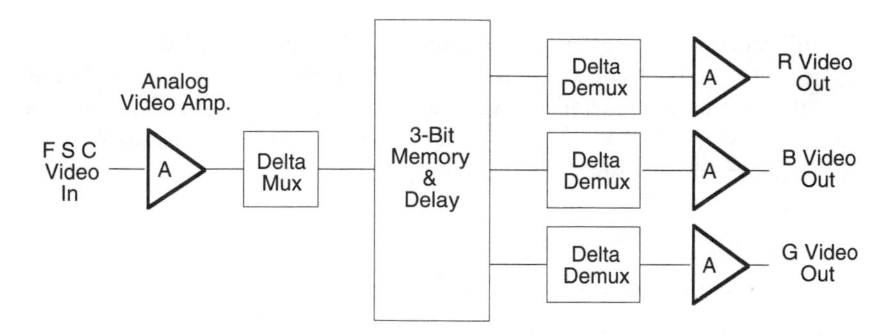

Figure 6.22 An early semiconductor-memory device for converting the Apollo field-sequential color video to NTSC RBG. Note that the analog FSC signal is digitized for storage using delta modulation (see Chapter 4), since the technology had not advanced to permit faster, flash ADCs. For simplicity, the clocking and synchronizing circuits are not shown.

networks, it was necessary to convert from the Apollo FSC to NTSC. However, this was made easier because the basic scan rates of the FSC matched NTSC. It is interesting to observe that this complex scan could be accomplished by using the then very new, and somewhat unproven, technology of integrated circuits.

A Perspective on Processing Signals from Space

After the Apollo project, the spacecraft and space probes continued to produce information and images that staggered the imagination of even the most sophisticated scientists, let alone the general public. It is difficult to decide which voyage or which image processing was the most important. Certainly this one, done by totally involved and dedicated scientists at JPL, should be considered.

The following paragraphs are taken from "The Photographs from Mariner IV" by Robert B. Leighton. © April 1966 by Scientific American, Inc. All rights reserved. Used by permission.

> As the Mariner 4 signals arrived at the ground station they were recorded on magnetic tape to provide a permanent record, and they were also typed out simultaneously as a sequence of 0s and 1s on paper tape that resembled adding-machine tape. Many people were clustered around the machines producing these tapes. It was an exciting experience to realize that we were actually receiving knowledge from a manmade machine almost 150 million miles away. Of course we were seeing only a sequence of bare numbers. What would the picture look like? Eight hours (of computer processing) seemed an eternity to wait.
>
> Then someone conceived the idea of cutting the tape from the printers into short lengths, each containing a series of 200 numbers representing the light intensity of one line of the picture. These sections of tape could be stapled together, one next to the other, to build up a two-dimensional picture of the numbers. To make the picture readable, each element was filled in with one of

five different colors of crayon, depending on the light level indicated by its numerical code. Each color of crayon was applied by a different person. In this way the first closeup picture of Mars emerged line by line in the form of a hand-colored mosaic. Within the day it was framed and presented to William H. Pickering, director of the Jet Propulsion Laboratory.

6.13 IMAGE PROCESSING—THE CHAPTER IN RETROSPECT

An Introduction to Image Processing

This chapter first presents fundamental models for an image and for image processing and shows some general models for image restoration and enhancement. These models are followed by examples of a number of techniques that can be used to provide restoration of enhancement. These techniques include using histograms and the statistics-related probability density functions. Also illustrated is the use of both two- and three-dimensional Fourier transforms. To illustrate the principles and techniques of image processing, examples are presented from the technologies of medical image processing and the processing of images from space.

Medical Image Processing—X-Rays, CT Scans, and MRI

In each case, a brief and very simplified description of how the particular radiation is produced and an overview of the processing required to produce a satisfactory image is presented. Next, the spectrum of each signal is, in its appropriate section, described and the required processing is presented.

For x-rays, the energy and wave length relationships to accelerating voltage and beam current are given. Enough fundamental information is presented so that the reader can relate the processes to Einstein's fundamental equation and to the ideas of photons of energy. The problems of beam scattering are presented and remedial processing techniques are discussed. The very interesting and difficult problems and processing in mammography are also presented.

Computerized tomography is introduce with its techniques of producing a small pencil-like beam of x-rays and, by careful manipulation, producing "slices" of parts of the body that are mathematically combined to produce an image of the area of interest. One of the important ideas presented in the sections on CT is the technique of measuring spatial resolution termed MTF or the modulation transfer function. This is described in the context of CT and relationships are shown to the theories and techniques of measuring resolution in television and a reference is given that shows the relationship between the MTF and the three-dimensional Fourier transform.

Nuclear magnetic resonance imaging is also presented with its reliance on RF energy bursts to produce "spins" in some of the body's hydrogen atoms that, in turn, can be detected and used to produce very high-resolution images of body areas. Among the details of the processing required for an MR image are the computer-generated RF pulses, with characteristics predetermined by the area of the body and the type of image desired; the pickup of the resulting hydrogen-atom resonance RF spins; and the very involved computer processing of these resulting RF bursts from the resonance spins to compose an MR image.

Processing Signals from Space

This chapter concludes by describing the processing of some signals that are transmitted back from space including temperatures, data on magnetic fields and radiation, the measurement of cosmic dust, information on solar winds, the terrain and the composition of planets, and TV pictures of the solar bodies, the space vehicles, and the involved personnel. These signals are returned to earth on photographic film, microwave telemetry links, and analog or digital TV modulation on microwave carriers.

The section on space signals and their processing starts with the signals from the first U.S. Vanguard series and the Russian Sputnik. Next, the signals and their processing from Tiros and Mariner are detailed and some of the methods of signal transmission and the signal bandwidths are presented.

Soon after the Mariner series started the investigation of Mars, rudimentary television was added. The TV in the Mariner and the Apollo spacecraft present a different set of problems, goals, and necessary processing than was presented in Chapter 3. For one example, it is shown that even the very early B&W TV camera signals were digitally encoded—six bits at first. Likewise, these digital data streams were usually recorded on tape for later transmission to earth. The rather convoluted process of how the early digital codes were broken up to send some of the bits back by telemetry, to record other bits on analog tape machines (for later transmission to earth), and likewise to record still others on digital tape machines is an exciting tale of the innovative compromises that often had to be made in the early space missions and serves as a precursor to the complex digital streams that are common in contemporary telecommunications.

The part television played in the odyssey of the Apollo moon missions is covered in some detail. Besides the aforementioned constraints on encoding and bandwidth, the problems of how the high vacuum and temperature extremes encountered in space are related to the design and operation of the various Apollo cameras are described. Both the Apollo B&W and color cameras are described as well as their processing and transmission of signals back to earth for both the scientific community and the general public.

Archival and Cardinal References

Cortright, Edgar M. *Exploring Space with a Camera.* Washington, DC: NASA, 1968.

James, J. N. "The Voyage of Mariner II." *The Scientific American,* July 1963.

Le Galley, Donald P. (ed.) *Space Science.* New York: John Wiley & Sons, 1963.

Leighton, Robert B. "The Photographs from Mariner IV." *The Scientific American,* April

1966.

———. "The Surface of Mars." *The Scientific American*, May 1970.

Staff of the Jet Propulsion Laboratory. *Mariner Mission to Venus.* New York: McGraw-Hill, 1963.

Contemporary References

Bushong, Stewart C. *Radiologic Science for Technologists.* St. Louis, MO: C. V. Mosby Co., 1988a.

———. *Magnetic Resonance Imaging: Physical and Biologiccal Principles.* St. Louis, MO: C. V. Mosby Co., 1988b.

Gonzalez, Rafael C., and Wintz, Paul. *Digital Image Processing.* Reading, MA: Addison-Wesley Publishing Company, 1978.

Herbert, Nick. *Quantum Reality Beyond the New Physics.* Garden City, NY: Anchor Press/Doubleday, 1985.

Inglis, Andrew F. *Satellite Technology: An Introduction.* Boston, MA: Focal Press, 1991.

Kak, A., and Slaney, M. *Principles of Computerized Tomographic Imaging.* New York: IEEE Press, 1988.

Krestel, Erich (ed.) *Imaging Systems for Medical Diagnostics.* Berlin and Munich: Siemens AG, 1990.

Macovski, A. *Medical Imaging Systems.* Englewood Cliffs, NJ: Prentice Hall, 1983.

Marshall, Christopher (ed.) *The Physical Basis of Computed Tomography.* St. Louis, MO: Warren H. Green, 1982.

Pennebacker, William B., and Mitchell, Joan L. *JPEG Still Image Data Compression Standard.* New York: Van Nostrand Reinhold, 1993.

Pratt, W. K. *Digital Image Processing.* New York: John Wiley & Sons, 1991.

Stark, David, and Bradley, William G. *Magnetic Resonance Imaging.* St. Louis, MO: C. V. Mosby Co., 1988.

Strum, Robert D., and Kirk, Donald E. *First Principles of Discrete Systems and Digital Signal Processing.* Reading, MA: Addison-Wesley, 1988.

Terrell, Trevor J. *Introduction to Digital Filters* (2d ed). New York: John Wiley & Sons, 1988.

7

Microprocessors and Microprocessor Systems

7.1 THE DEFINITION OF A MICROPROCESSOR—AN EVER-MOVING TARGET

What we now term a microprocessor was an entirely new and almost indefinable electronic component in the early 1970s. The first one, which became the 4-bit Intel 4004, was primarily developed using the relatively new technique of small-circuit integration to replace a larger circuit board. Although this new integrated circuit was, in fact, unique and clever for its time, it is doubtful if the designers anticipated the industry it helped to create. The 4004 was soon replaced with an 8-bit 8008 that, again relatively soon, was superseded by the very famous 8080A. In very abbreviated terms, this is how it all began.

We now have 32- and 64-bit microprocessors running with clock speeds in the order of 100 MHz, which process instructions in the few tens of nanoseconds. Contrast this with the original 8080A running with a clock speed of about .5 MHz and an average instruction time of two to nine microseconds. Besides the ability to process instructions much faster, the more contemporary microprocessors possess circuits and processing techniques that make it difficult to show a completely connected rational progression of ideas and technologies. The development of these little electronic machines represents some of the most creative work in all technology. The future looks bright for more innovation and, of course, unprecedented and radical change.

So, what is a microprocessor? The following are a few elementary conceptual models that will help in the never-ending quest for the definition of a microprocessor.

What Is a Microprocessor? Some Elementary Microprocessor Definitions

- A microprocessor is a complex integrated circuit that substitutes software instructions for hardware logic whenever possible. It can be programmed, by using its unique instructions (its instruction set), to perform arithmetic functions, produce control pulses, and change or create data based upon either programmed or input information. This newly created and/or processed data can be output in such forms as new pulses, coded or Arabic numbers, words, and graphics.

- A microprocessor is a complex integrated circuit composed of logic elements (gates), flip-flops, counters, storage registers (some called accumulators), etc., which are configured to use programmed instructions and input data to provide outputs that control external devices or output computations, text, and graphics.
- A microprocessor uses a list of carefully chosen instructions (a program) and sometimes input signals or other information to input instructions and data to its central processing unit, usually termed an *ALU*—an arithmetic/logic unit. Usually the instruction or input information is first routed to a temporary register termed an *accumulator.*

What Is a Microprocessor? What Are Its Rudimentary Operations?

The fundamental operation of a microprocessor unit (MPU or micro) is to retrieve and carry out the request of a list of stored instructions. Although any given micro may use several hundred available instructions, most of these instructions can be placed into two categories: [1] instructions that move data and [2] instructions that process data. The instructions that move data include those that write or read to and from registers, memory, or peripheral devices (I/O instructions.) Those that process data include arithmetic and logical operations.

In its rudimentary form, each instruction is divided into two operations: fetch and execute. One after one, the MPU retrieves the stored instructions and, in turn, carries out their mandate. In it simplest form, the time to fetch and the time to execute would be equal. In fact, some real-world MPUs operate in such a mode. However, in most MPUs, the instruction time of the fetch and execute cycles will be different from the time of the instruction cycle. Simple fetch-execute operations contain two groups of clock pulses termed machine cycles. The more complex fetch-execute instructions will contain one fetch machine cycle group and usually a longer execute machine cycle group. The clock pulses in these machine cycles are often called T_1, T_2, etc.

What Is a Microprocessor? A Model of Some Very Simple Microprocessor Instructions

The following illustrates the use of some simple instructions. Each instruction is fetched to a temporary instruction register where it is decoded and then the execution cycle is carried out. The use of the MPU's accumulator is illustrated. Each instruction is given in assembly code with the explanation presented in the comments that follow.

```
LDA  2050H    ;Load the accumulator with the contents of the memory at its
              (hexadecimal or Hex) address of 2050. (Some systems use the
              $ to denote hexadecimals.) In many micros the accumulator is
              also called register A.

STA  2051H    ;Store the contents of the accumulator at address 2051H.

ACI  22H      ;Add the Hex number 22 to the present contents of the accumu-
              lator and store the sum back in the accumulator. (The "C" in
              the ACI code has to do with any arithmetic carry.)

MOV B,A       ;Move the contents of register A to register B. It should be noted
              that most microprocessor MOV (move) instructions are really
              copy instructions—the contents of the source register, memory,
              port, etc., are not altered.

JMP  2070H    ;The program sequence is transferred to the (program counter)
              address 2070H.

ORA  C        ;The contents of the accumulator are logically ORed with the
              contents of register C and the result is stored back in the accu-
              mulator.

OUT  2        ;This instruction assumes data XX in the accumulator is to be
              moved (copied) into I/O port number 2. This port number would
              be the next entry after the OUT instruction.
```

NOTE: While these instructions come from the general Intel set, most micros have very similar assembly-language codes.

What Is a Microprocessor? A Model for Simple System Timing

One of the fundamental premises of a microprocessor is that it, like the larger mainframe computers, operates by reading (fetching) and implementing (executing) a list of stored instructions. In general, these continuing fetch-execute cycles constitute the fundamental operation of the microprocessor system. An elementary fetch-execute cycle operates as follows:

1. A continuous signal—the clock—is used as a reference for all other timing pulses. Some MPUs require two clock waveforms. See Figure 7.1 for the clocking and other control waveforms.

2. A control pulse, or a combination of pulses, may be used to synchronize or start a fetch action. There may be a similar or second pulse termed the READY pulse.

3. There may be one or two control pulses called MEMORY READ and/or MEMORY WRITE. These signals are often combined on one line (one circuit) where a Hi on the line would indicate a READ instruction and a Lo on the line would indicate a WRITE instruction. The symbol would be RD/$\overline{\text{WR}}$.

4. If the low-order address bus and the data bus are multiplexed, a latch is required for the low-order address. Once the data is valid on the bus, it is latched with a control signal such as address latch enable (ALE).

5. One or two pulse lines will indicate whether the instruction involves memory or input/output. Again, often these are selected with one high or low signal such as IO/$\overline{\text{M}}$.

6. Some real-life systems cannot use the control pulses as they come from the MPU. Many systems use special controller or common logic chips to produce more usable control signals.

The clocking control address and data pulses normally proceed along a given path or bus in most elementary microprocessors, as shown in Figure 7.2. Most micros send and receive data from memory using control signals as shown in Figure 7.3. The early units sometimes required external memory and I/O logic as shown in Figure 7.4.

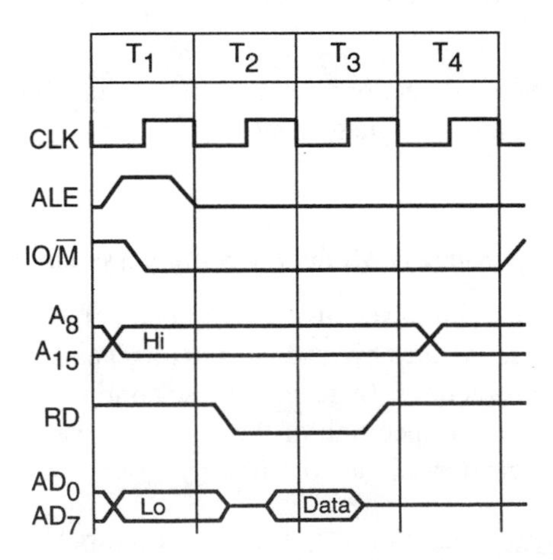

Figure 7.1 The control timing model for an elementary microprocessor. These are examples of signals that are used in the fetch part of the instruction.

NOTE: The CAD rule for microprocessor busses: As a memory aid, it is useful to remember that the fundamental busses for a micro are the Control, Address, and Data busses. (The control bus is not a group of similar lines like the Data and Address busses but rather a group of control functions.) There are often several Status outputs (pins) that can be included in the acronym CADS.

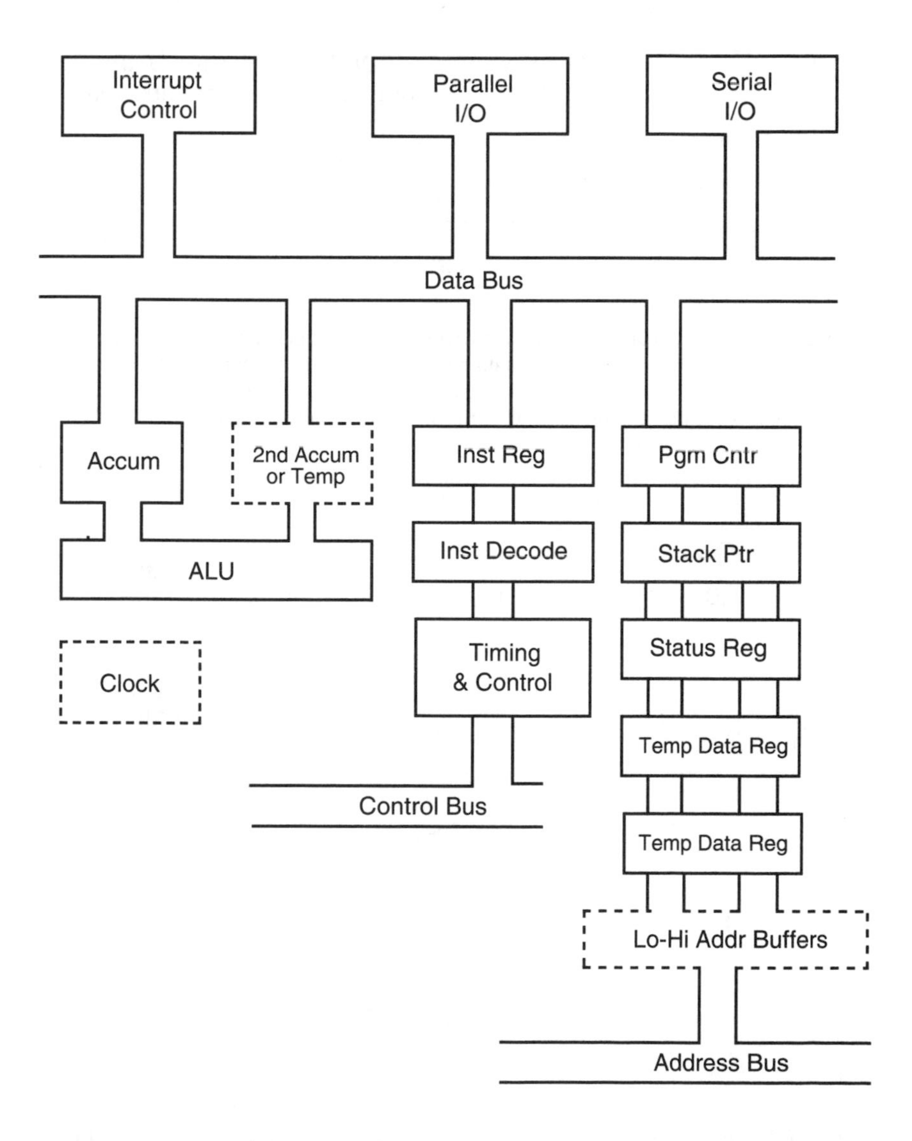

Figure 7.2 A model of an elementary microprocessor. This model is not necessarily intended to represent a working configuration or any idealized design, but rather to indicate the elements that can be used for comparing commercial designs.

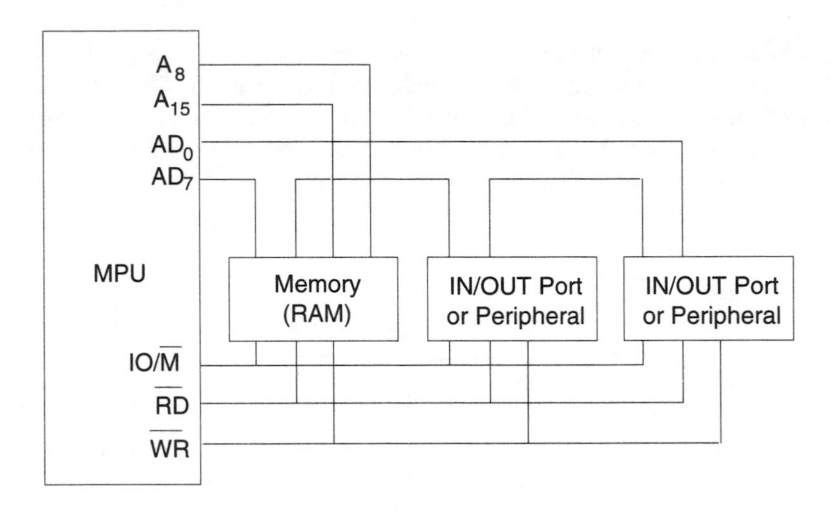

Figure 7.3 The model of a very simple microprocessor system. This configuration assumes a multiplexed low-order address and data bus. Note that the memory is addressed by both the low- and high-order parts of the bus, but the I/O ports only use the low-order portion. This scheme is similar to those used in some working systems.

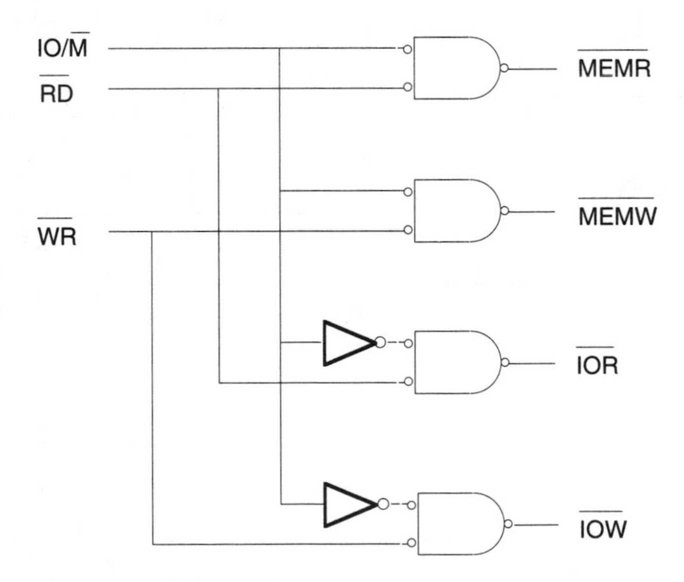

Figure 7.4 An illustrative model of how logic circuits can be used to change the format of MPU control signals. In many designs, this or similar logic is built into a dedicated controller chip or even may be part of the MPU.

What Is a Microprocessor? Elementary Memory Addressing Techniques

The simplest memory and memory addressing model is the old-fashioned pigeon-hole mailbox numbered 1, 2, 3, ... N. Thus, a computer system could place or retrieve its data simply by going to any numbered box. A more sophisticated and workable scheme is to form the boxes in a matrix. Such a configuration might have a numeric horizontal index and an alphabetic vertical index that would give the upper left-hand box an address of 1A, etc. In fact, most simple addressing uses some form of this row and column approach. A more complex addressing system will also include some grouping scheme. Quite often in large memory capacity systems, these groups are called pages, with each page representing 2^{16} or 64KB of memory.

Immediate Addressing

An addressing method where the instruction itself includes the data (the operand) as: MVI B, 55 means load 55 (the data) into register B.

Direct or Absolute Addressing

An addressing method where the address required by the instruction is part of the instruction, as: STA 1234 means store the contents of the accumulator in memory location 1234H. (H stands for Hex or hexadecimal.)

Indirect Addressing

An addressing method where the location required by the instruction is part of the instruction as: MOV B, M means move the contents of the memory location M to the B register.

Indexed Addressing

An addressing method usually used to move related data in increasing or decreasing steps, such as in a matrix. Each address located in the instruction was or can be modified by the contents of an index register to find the next actual address. In the instruction MOV A, M, it is assumed that location M is actually pointed to by a register M. Thus, the instruction really says move the contents of the memory location pointed to by register M (for the 8080-85 series, registers H and L) to the accumulator. In indexed addressing, there are usually other instructions that are used repeatedly to increment or decrement this indexing, pointer register.

What Is a Microprocessor? Status Registers

The results of an MPU operation are often indicated as bits in a status or flag register. These bits indicate if the result of an operation was zero, or there was an arithmetic carry and/or sign, or there is even or odd parity. One or more of these bits may be brought to the accumulator and tested for use in the next instruction. For example, the program would inspect the status register for the presence or absence of a carry bit before the instruction "Jump on Carry."

What Is an Microprocessor? The Progressively Changing Role of Microprocessor Software Including RISC Processors

Even though this chapter is primarily concerned with the hardware configurations of microprocessors, it is now prudent to review briefly some of the attendant chronology of their software developments. In the early days of microprocessors most, if not all, of the designers used software instructions to support the devices as controllers. The original instructions were two- or three-letter machine-language codes (the very simple instructions that are unique to a particular computer) or assembly-language mnemonics (slightly higher-level codes in forms that are usually easier to remember). If more clever instruction codes were added to a given micro, in most cases this seemed to provide added versatility and thus made the device more useful. In fact, the first few years of microprocessors included a horsepower race to see which manufacturer could produce an 8-bit micro with the most instructions.

As micros expanded from their primary use as miniature controllers, which were usually programmed with machine or assembly languages, the design goals changed. Higher-level languages such as BASIC and FORTRAN, and later COBAL, Pascal, and C, along with many others, placed serious limitations on the supposed versatility of the original lengthy machine and assembly codes. Such functions as word processing and especially graphics forced traditional hardware microprocessor designers to collaborate with software experts so as to provide later-generation micros that produced much more efficient and faster software-driven true microprocessors.

The ever-growing desire to increase the instruction speed brought several design innovations in addition to the obvious increase in clock speeds. One of the radical changes in microcomputer thinking was the RISC or reduced instruction set computer. The fundamental idea is based on a statistical analysis of just how often instructions are or are not used. If only the most commonly used instruction in a conventional set—termed primitives—can each be designed to operate in one clock cycle, the overall processing time can be improved by as much as fivefold. Although RISC processing also must include other hardware techniques such as pipelining, many of the contemporary micro designs incorporate this technique—see the references on RISC processing.

7.2 COMMERCIAL MICROPROCESSORS

Representative Microprocessors—The 8080A and Its Family of Chips

The preceding text and graphic models of a prototype microprocessor are purposely simplified. They are designed to be both building blocks and to help define and explain the more complex units to follow. They do not contain such information as cost, speed, and addressing techniques, nor do they give any hint of what other chips are required to configure a workable microprocessing system. As an example, the 8080A required at least an 8224 clock and an 8228 controller to allow it to operate in even the simplest system.

The 8080A/85 N-MOS Technology

The semiconductor material and the techniques of using that material largely determine the micro's maximum operating speed. Although the exact processing and fabrication can vary from manufacturer to manufacturer, it is common in the semiconductor industry to simply use the term technology to indicate a given semiconductor mix and the arts used in its fabrication. Thus, the 8080A used N-MOS technology.

The Power Supply and Clock

The 8080A, shown in Figure 7.5, required an external two-phase clock and three power sources supplying +12, +5, and –5 VDC. Not only did the two-phase clock have to be supplied by a second chip, the complex clocking sometimes produced additional interfacing problems. The three power supply requirements increased the complexity and expense of a complete system. The 8085 was updated to contain its own internal clock circuitry and to require only one +5- VDC power supply.

The 8080A MPU possessed a true 8-bit data bus (8 pins) and a 16-bit address bus (16 pins giving 2^{16} addresses). However, the 8085 time multiplexed the eight low-order address bits (A_0–A_7) with the eight data bits. When the ALE control signal was high, it indicated a low-order address on the bus and also was used as a latching signal. When the ALE line was low, it indicated data on the multiplexed lines. The memory addressing capabilities were quite simple compared to later micros—see Figures 7.6 and 7.7.

The 8080/85 Serial Data

The 8080A did not have any direct method to input or output serial data.

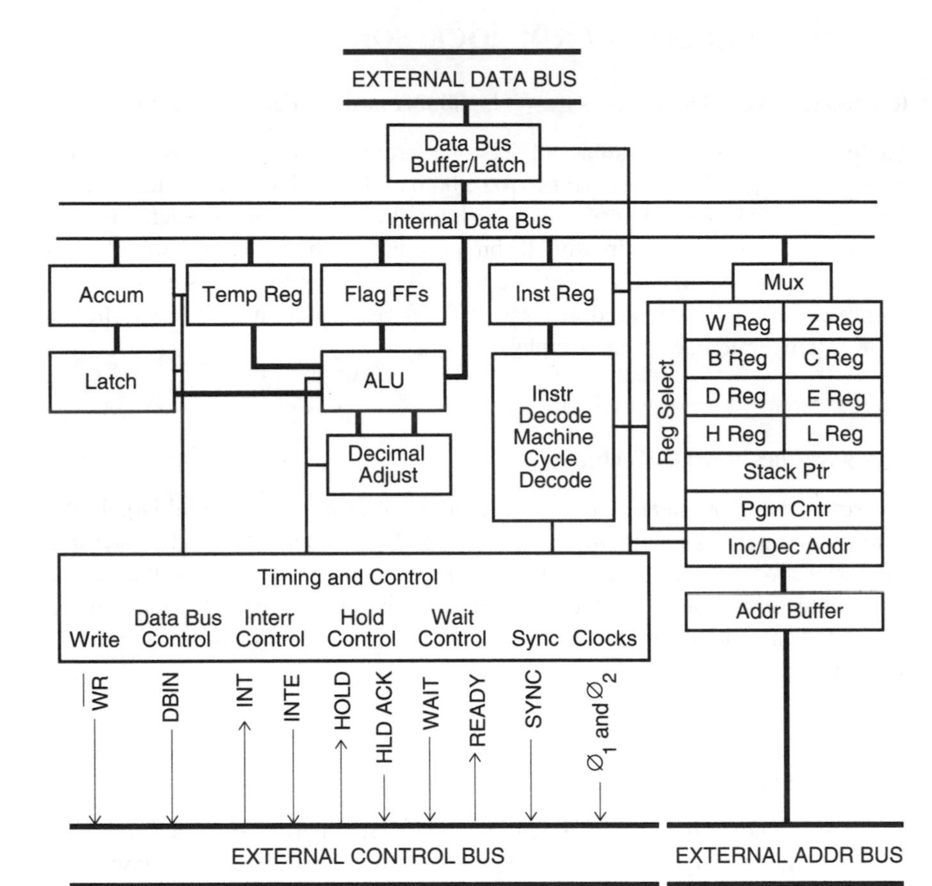

Figure 7.5 The block diagram of the 8080A microprocessor showing its connections to the external system's Control, Address, and Data (CAD) busses.

Table 7.1 The 8080A signal pins.

+15, +5, –5, V_{SS}	Pwr In	SYNC	Machine Cycle Sync
A_0–A_{15}	Addr Bus—*Tristate*	DBIN	Data Input Strobe
D_0–D_7	Data Bus—*Tristate*	$\overline{\text{WAIT}}$	Make MPU Wait
$\varnothing_1$	Clk Phase 1 In	WR	Write Data
$\varnothing_2$	Clk Phase 2 In	HOLD	Enter Hold State
INT	Interrupt Request	HOLDA	Hold Acknowledge
READY	Data Input Stable	INT	Interrupt Request
RESET	Reset MPU	INTE	Interrupt Enable

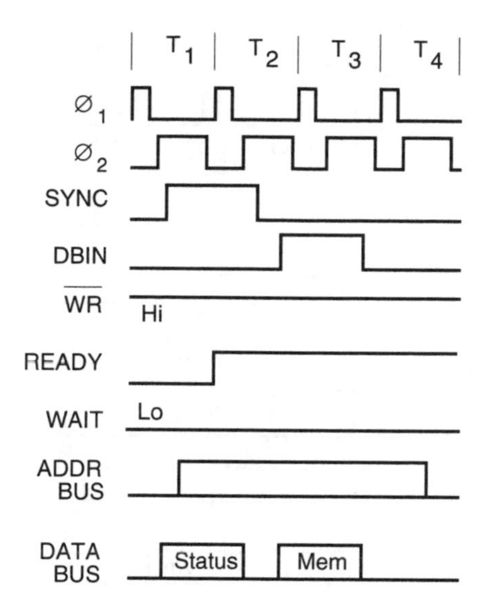

Figure 7.6 The 8080A timing and control signals for the FETCH portion of an instruction.

The 8080A FETCH—memory READ timing and control signals:

$\varnothing_1$ and $\varnothing_2$	The two clock signals furnished by an external chip.
SYNC	A pulse to signal the beginning of a machine cycle.
DBIN	The Data Bus is ready for data.
$\overline{\text{WR}}$	Data is stable and ready for writing.
READY	Indicates the MPU is ready for read or write.
WAIT	Frees external memory or logic from MPU timing. WAITing was necessary for (the older) slower dynamic memories that often needed refresh time.
ADDR BUS	The address is stable during this period.
DATA BUS	Status word indicates an Interrupt, Halt, Out, Stack, etc. Mem indicates the data that was fetched.

NOTE: For the external logic, there is no difference between this FETCH timing and a READ (memory read) instruction. However, the status word will indicate a difference.

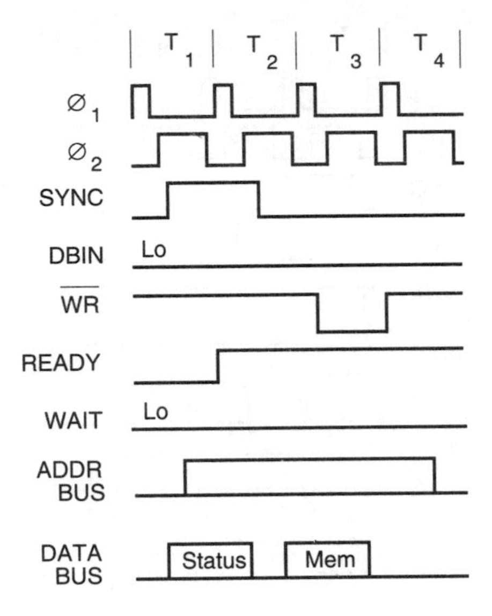

Figure 7.7 The 8080A timing and control signals for the EXECUTE (memory write) portion of an instruction.

The 8080A Programmable Registers

A register is a temporary storage for a computer word. Often registers are configured to store an 8-bit word—a byte. The 8080A has both 8- and 16-bit registers. The A or accumulator is 8 bits as are the B, C, D, E, H, and L registers. However, the B-C, D-E, and H-L register pairs can be combined to store a 16-bit word. Registers A through L can be accessed and controlled by software. Registers W and Z as well as some other temporary registers are integral to the routine operation of the MPU and are not accessible with software instructions. There is also a flag-bit register that indicates whether the result of an arithmetic operation was zero, had a carry, even or odd parity, etc.

The 8080A External Interrupt Processing

The interrupt is one of several pivotal inventions that make computers and microcomputers really work. By using some prearranged hardware and software routines, it is possible to stop the normal flow of the fetching and executing of program instructions and to administer one or more special, more urgent, sequences.

If the 8080A sees a high for three or more clock periods on its INT pin, the MPU will, at the end of its present instruction cycle, usually enter into an interrupt cycle. (The internal interrupt flip-flop must have been set/enabled by software.) If the INT signal is recognized, the following routine will take place:

1. The next instruction cycle is special—it is an interrupt fetch. During T_1, the status byte is changed to reflect the interrupt condition.
2. The INTE (interrupt acknowledge) pin is set low to disable any further interrupts. This pin and the interrupt flip-flop also can be set high or low by software.
3. The program counter logic is suppressed and its contents (the next instruction address) are saved on the stack.
4. Using external logic to provide a special address on the data bus, the MPU system executes a CALL instruction by incorporating an 8228 System Controller. This CALL will execute the preprogrammed interrupt service routine.
5. The end of the service mini-program must include the resetting of the interrupt flip-flop. Also, it must bring the next instruction from the stack to the program counter, where it is used to start executing the normal program again.

The 8080A Ready and Wait Signals—Slow Memories

When the 8080A sends out a memory address, it should receive a ready signal—typically from the 8224 clock generator. If it does not see a high on the ready line, it assumes a slow system condition, such as with a dynamic memory. It then ceases normal operation and goes to a Wait state and sends out a Wait signal. This Wait state and its signal continue until it sees a high READY line again. When the MPU is in a Wait state it places a high signal on the Wait pin.

The 8080A Hold and Hold Acknowledge Signals—DMA

A high HOLD signal will tristate the MPU's address and data bus drivers and allow external devices to access these two busses. Once the HOLD state is in effect, the system is notified with a HOLD ACK signal. One common use of the HOLD signal is for Direct Memory Access (DMA). When DMA is used, the memory read-write functions do not go through the usual MPU fetch-execute cycles and thus the reading and writing is much faster. Floppy and hard disk reading and writing are usually done as DMA operations.

Representative Microprocessors—The 8085 and Its Family of Chips

The 8085 combined the three basic IC chips of the 8080A into one single microprocessor. It included an on-chip clock, internal system control, and it required only one power source. The 8085 also included serial I/O and a much more sophisticated interrupt scheme. Because these new features required additional IC pins, the 8-bit data bus and the low-order portion of the 16-bit address bus were combined (multiplexed).

The 80885 N-MOS Technology

The 8085, like the 8080A, used N-MOS (N-Channel) technology. However, since the 8085 was developed about two years after the 8080A, when the alloying and fabricating techniques had improved, the 8085 ran slightly faster than the 8080A.

The Power Supply and Clock

The 8085 required only a +5 VDC power supply and its clock ran at about 3 MHz rather than the 2-MHz clock for the 8080A. Likewise, later versions of the 8085 used a 4-MHz clock that then gave an average instruction cycle of slightly under a microsecond. (See this topic in the 8080A section.)

The 8085 Serial Data

The 8085 included the two pins SID (serial input data) and SOD (serial output data). The use of serial input or output circuits requires several lines of programming that involve the shifting and masking of special data in the accumulator. The data bit on the SID line is loaded into the accumulator (bit 7) with the RIM instruction. The output SOD is set or reset by the SIM instruction.

The 8085 Address and Data Busses

The 8085 time multiplexes the eight low-order address bits (A_0–A_7) with the eight data bits. When the ALE control signal is high, it indicates a low-order address on the bus and is also used as an external latching signal. When the ALE line is low, it indicates data on the multiplexed lines. The memory addressing capabilities were quite simple compared to later micros.

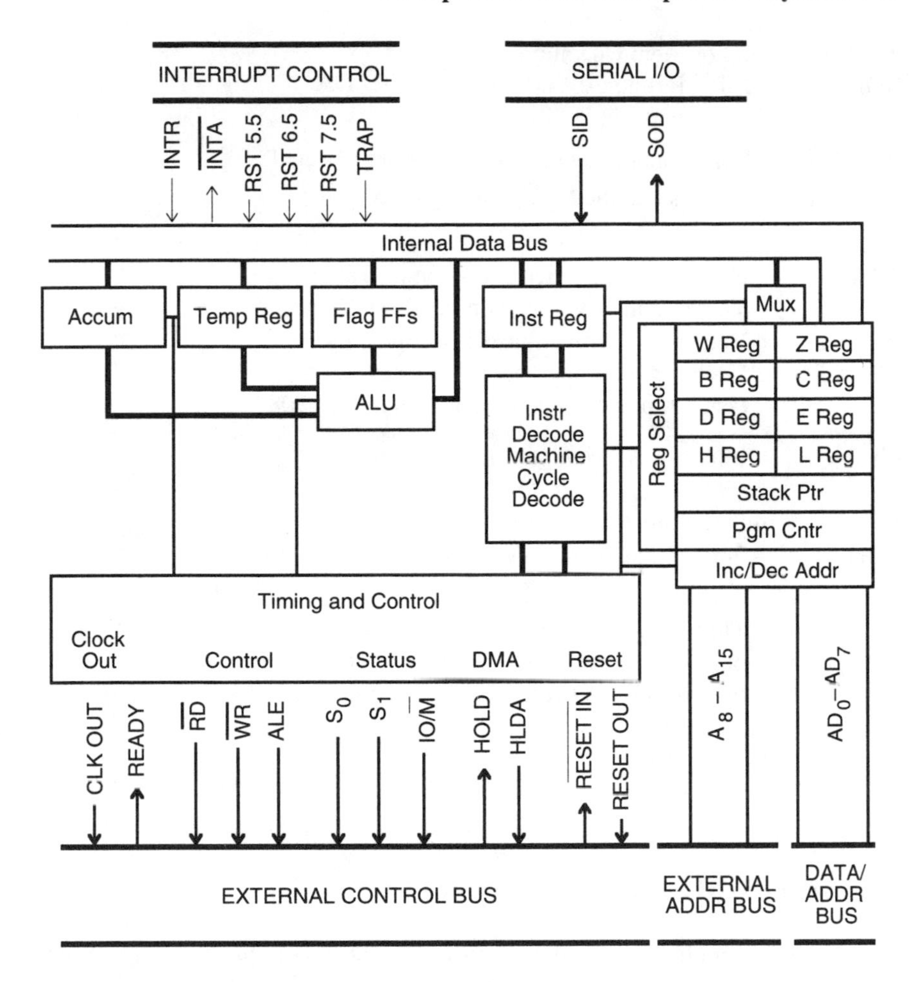

Figure 7.8 The block diagram of the 8085 showing its connections to the external system busses.

Table 7.2 The 8085 signal pins.

+5, V_{SS}	(Not shown in figure)	READY	A wait-state request for DMA, etc.
A_g–A_{15}	Addr Bus–*Tristate*		
AD_8–AD_{15}	Mux Addr & Data–*Tristate*		
ALE	Addr Latch Enable		
$\overline{RD}$	Memory Read	$\overline{WR}$	Memory Write

IO/$\overline{M}$ IO or Mem Operation
HOLD MPU Relinquishes Buses HLDA Signals a Hold Opr

$\overline{\text{RESET IN}}$ Pgm Count is set to zero, busses are tristated
RESET OUT Signals that the MPU is being reset

TRAP Highest priority interrupt—nonmaskable
RST 7.5 Vectored interrupt—second highest priority
RST 6.5 Vectored interrupt—next lower priority
RST 5.5 Vectored interrupt—next lower priority
INTR General-purpose interrupt as was INT in 8080A
$\overline{\text{INTA}}$ Interrupt Acknowledge output signal

SID Serial input data—use with RIM instruction
SOD Serial output data—use with SIM instruction

S_0 & S_1 These Status Timing Signals can identify the timing sequence of various operations. They should not be confused with the status bit indicators in the interrupt mask registers.

The 8085 FETCH and EXECUTE Timing

The use of the multiplexed address and data (AD) bus requires special timing techniques. At the start of FETCH, during the first clock T_1, the low-order address byte is placed on the multiplexed bus so that it can be externally latched and then time aligned with the high-order address byte. Slightly after a high ALE pulse is used to latch this address byte, the data byte becomes valid on the bus during T_2 and T_3. Almost concurrent with the ALE signal during T_1, the IO/$\overline{M}$ line goes low to signal a memory-related operation. During T_2 and T_3, either the $\overline{RD}$ or $\overline{WR}$ line will go low indicating a read or write operation.

For a READ instruction, the EXECUTE cycle is almost a duplication of the FETCH cycle. During fetch, the timing is designed to read memory for the instruction. During execute, the timing is designed to read memory for the data. However, the two cycles are different for a WRITE instruction. Although the FETCH cycle must again read an instruction and thus needs the identical $\overline{RD}$ pulse, the EXECUTE cycle must use the $\overline{WR}$ pulse.

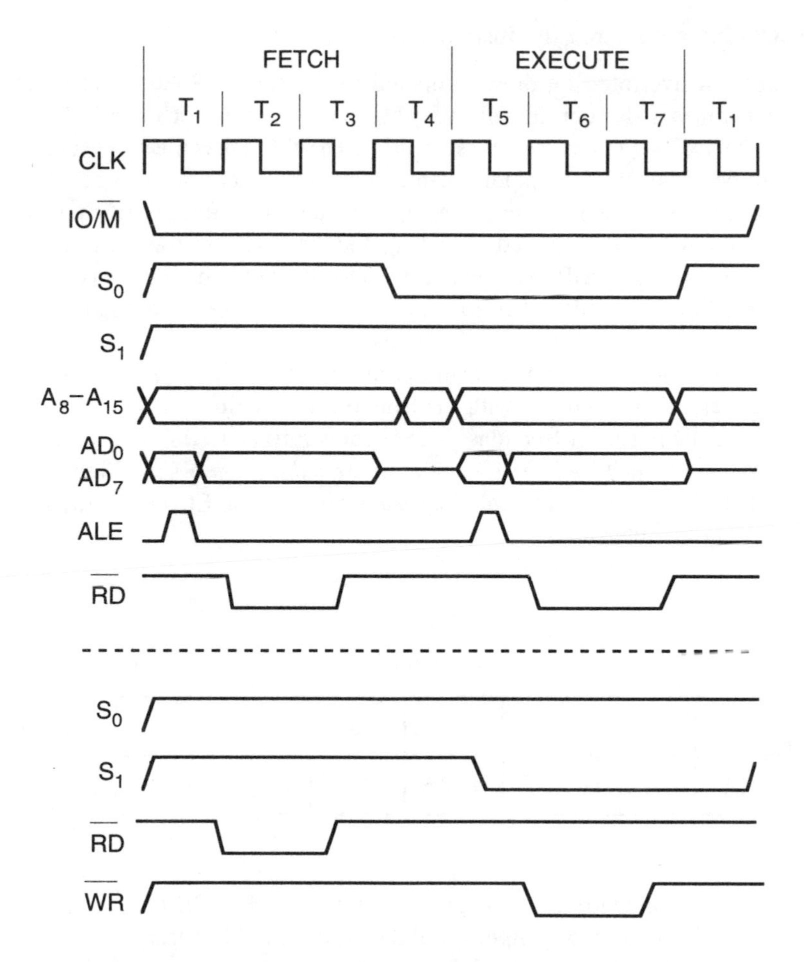

Figure 7.9 The 8085 FETCH and EXECUTE timing signals. The top part of the figure shows the six waveforms required for a memory READ. The lower portion displays only the four signals that change for a memory WRITE.

The 8085 Programmable Registers

The registers in the 8085 duplicate those in the 8080A, including a flag register. The flag register indicates the condition of five flip-flops that are set if there is an arithmetic Zero, Carry, or Auxiliary Carry along with the resulting Sign or if there is a Parity indicated. Some books also include the nonprogrammable registers W and Z. (See the Gaonkar reference.)

The 8085 Interrupts and the Instruction Masks

The 8085 has five interrupt pins, compared to one for the 8080A. (The HOLD input sometimes is also considered a very high priority interrupt.) The INTR is the same as the 8080A and has the lowest priority. The TRAP has the highest priority. It cannot be disabled—it is nonmaskable—and it does not need to be enabled. There are three other RST (reset) interrupts, with descending priorities, that can be enabled or disabled—masked. The Trap and the RST interrupts are automatically vectored (transferred) to a specific memory location where there is usually a JUMP or CALL instruction that goes to a special subprogram termed a service routine.

The control for most of the interrupts comes from the bits set in the instruction mask that can be manipulated with the SIM (set instruction mask) and the RIM (read instruction mask) instructions—see Figures 7.10 and 7.11. The logic for all the interrupts, except TRAP, will include *some combination* of [1] an input (interrupt) signal, [2] an Interrupt Enable (or disable) instruction EI and DI, [3] a Reset signal, and [4] an interrupt mask.

The SIM Interrupt Mask

7	6	5	4	3	2	1	0
SOD	SDE	XXX	R 7.5	MSE	M 7.5	M 6.5	M 5.5

SOD	Serial Output Data—ignored if bit 6 = 0. The RIM instruction will input serial data and the SIM instruction will output serial data.
SDE	If = 1, bit 7 can be output data latch
XXX	Ignored with the SIM instruction
R 7.5	RESET RST 7.5: If = 1, RST 7.5 flip-flop is reset OFF
MSE	Mask Set Enable—If = 0, bits 0–2 ignored. If = 1, their mask is set.
M 7.5	REST 7.5 MASK 0 = available, 1 = masked
M 6.5	REST 6.5 MASK 0 = available, 1 = masked
M 5.5	REST 5.5 MASK 0 = available, 1 = masked

Figure 7.10 The bit pattern for the 8085 in the accumulator after using the Set Interrupt Mask instruction.

The Accumulator Following a RIM Instruction

7	6	5	4	3	2	1	0
SID	I 7.5	I 6.5	I 5.5	IE	M 7.5	M 6.5	M 5.5

SID Serial input Data, if any
I 7.5 Pending interrupts by priority where a 1 = pending
I 6.5 Pending interrupts by priority where a 1 = pending
I 5.5 Pending interrupts by priority where a 1 = pending
IE Interrupt Enable flag where 1 = enabled
M 7.5 Interrupt Mask where 1 = masked
M 6.5 Interrupt Mask where 1 = masked
M 6.5 Interrupt Mask where 1 = masked

Figure 7.11 The bit pattern in the accumulator after using the Read Instruction Mask instruction.

The interrupts, in descending priority, are:

TRAP A nonmaskable interrupt that is vectored to memory location 0024 (Hex). It is not affected by the status of the Interrupt Enable flip-flop or any reset. This is usually a panic interrupt initiated by a power or program failure.

RST 7.5 The highest priority RST instruction. To be recognized, it requires a high input pulse (signal) and the interrupt mask bit 3 must be 1. Also, bit 4 and bit 2 of the interrupt mask must be 0 (not reset), and the Enable Interrupt (EI) must be set in software. It is vectored to memory location 003C.

RST 6.5 To be recognized, it requires a high input pulse, bit 3 = 1 and bit 1 = 0, in the interrupt mask, and IE must be set. It is vectored to 0034.

RST 5.5 To be recognized, it requires a high input pulse, bit 3 = 1 and bit 0 = 0, in the interrupt mask, and IE must be set. It is vectored to 002C.

INTR A high pulse on this pin will allow external hardware to place one of eight different codes on the data bus. Each one of these codes, zero through seven Hex, will, in turn, produce a vector to one of eight memory locations. The interrupts associated with the INTE input also require the EI instruction. The INTE input is the same as the INT in the 8080A.

Representative Microprocessors— The 6800/6802 and Their Family of Chips

The 6800 series introduced several new approaches to the design of an 8-bit microprocessor. Intel, with their 8080, and Motorola, with their 6800, both produced unique designs that showed much imagination. Both of these introductory microprocessors contained many elegant, fundamental principles that are still used in their contemporary products.

Although the 6800 contained several new and different approaches than the 8080, when it was introduced, the first comment one might read or hear was "No input or output instructions." The 6800 communicated with I/O devices in the same way that it did memory. Thus, each I/O device had its own address just like any memory cell (location). At first, this concept somewhat mystified the novice. Since there were fewer control signals than the 8080, the 6800 did not need to multiplex the data and address busses. All of the 6800 machine cycles were the same length. It only needed one 5-volt power supply. The 6800 did not have an on-chip clock but the 6802 did. The instruction set was somewhat simpler than the 8080 because it did not contain many one-of-a-kind instructions.

The N-MOS Technology

The 6800 used a technology similar to the 8080. However, as time progressed, N-MOS became faster and faster. In theory, the 6800 was slightly faster than the 8080A. However, the different instruction techniques, plus the fact that each company made constant technical upgrades, almost canceled any advantage one or the other had because of the semiconductor technology they used.

The Power Supply and Clock

The 6800 and the 6802 had a single +5-V power supply. The 6800 required two clock input signals ϕ_1 and ϕ_2, from a clock generator chip such as a 6870 or 6875. The 6802, shown as a block diagram in Figure 7.12, had two pins for a crystal or an RC network. The ϕ_2 signal, in TTL format, was also used for other system chips. In 6800 systems, this was obtained from the clock generator and in 6802 systems it was obtained from a TTL output pin on the MPU.

The 6800 Serial Data

The 6800/02 did not have any dedicated serial data pins. They used external chips in the 6800 family.

The 6800 Address and Data Busses

Both the 6800 and the 6802 had eight data pins and sixteen address pins. There was a wide variety of addressing methods, including the direct addressing of all

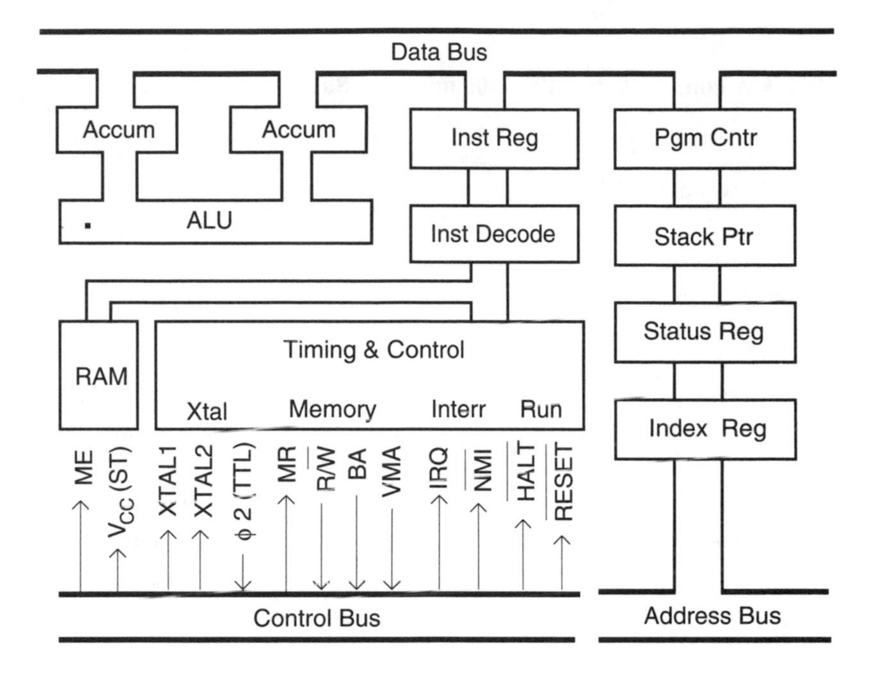

Figure 7.12 The block diagram of the 6802 MPU.

64K memory locations with 3-byte instructions, page addressing with 2-byte in structions, and the ability to address the first 256-byte page with 1-byte instructions. The 6800/02 could use indexed and relative addressing. It could also use the stack pointer as an address register.

The 6800 used the Data Bus Enable (DBE) to tristate the data bus for DMA operations—the 6802 did not have a provision for DMA. The 6802 could accommodate slow memories by a process called clock stretching using the Memory Ready (MR) signal. Although a normally low signal, when MR was forced high, the clock would stop to provide extra time for external memory. Table 7.3 shows the pins of the 6800 and 6802. Figures 7.13 and 7.14 show their READ/WRITE timing.

NOTE: In this chapter, all the microprocessor block diagrams are presented in a form that is derived from the original Sample MPU Graphic Model. This provides a uniform method to show the evolution of these units. Although every effort is made to assure overall accuracy, it is possible that some minor details may be very slightly altered. Always refer to the manufacturer's literature for specific design particulars.

Table 7.3 A comparison of the 6800 and the 6802

6800		6802	
+5, V_{SS} (Power and GND)		+5, V_{SS}	
V_{SS}(ST) Internal RAM Power*			
ME Internal RAM Enable*			
MR Memory Ready*			
D_0–D_7	Data Bus—Tristate	D_0–D_7	Data Bus—*Tristate*
A_0–A_{15}	Addr Bus—Tristate	A_0–A_{15}	Addr Bus (*not Tristate*)
ϕ_1	Clock Input	XTAL 1	Crystal Input
ϕ_2	Clock Input	XTAL 2	Crystal Input
TSC	Three-State Control		
DBE	Data Bus Enable		
R/$\overline{\text{W}}$	Read/Write—*Tristate*	R/$\overline{\text{W}}$	Read/Write—*Tristate*
VMA	Valid Memory Addr	VMA	Valid Memory Addr
BA	Bus Available		
$\overline{\text{IRQ}}$	Interrupt request	IRQ	Interrupt request
$\overline{\text{RESET}}$	Restarts MPU	$\overline{\text{RESET}}$	Restarts MPU
$\overline{\text{NMI}}$	Nonmaskable Interr	$\overline{\text{NMI}}$	Nonmaskable Interr
$\overline{\text{HALT}}$	Ceases MPU Oppr	$\overline{\text{HALT}}$	Ceases MPU Oppr

* Used in 6802 only. The V_{SS}(ST) input supplies standby power to the first 32 bytes of a 128 byte on-chip memory in the 6802. The ME signal is used to enable/disable this RAM.

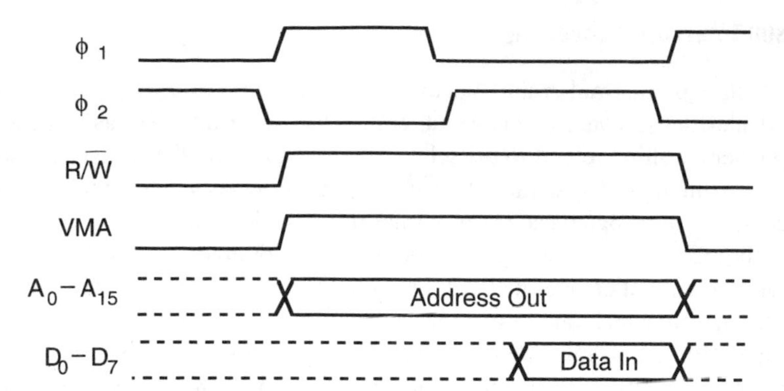

Figure 7.13 The 6800/02 READ machine cycle.

The 6800 Programmable Registers and Status Flags

The 6800 and the 6802 contain six programmable registers, including two 8-bit accumulators, a 16-bit index register, a 16-bit program counter, a 16-bit stack pointer, and an 8-bit status register. All appropriate instructions apply to both accumulators except those that move information between accumulator A and the statusregister. The flags in the status register include Carry, Overflow, Sign, Zero, and Auxiliary Carry.

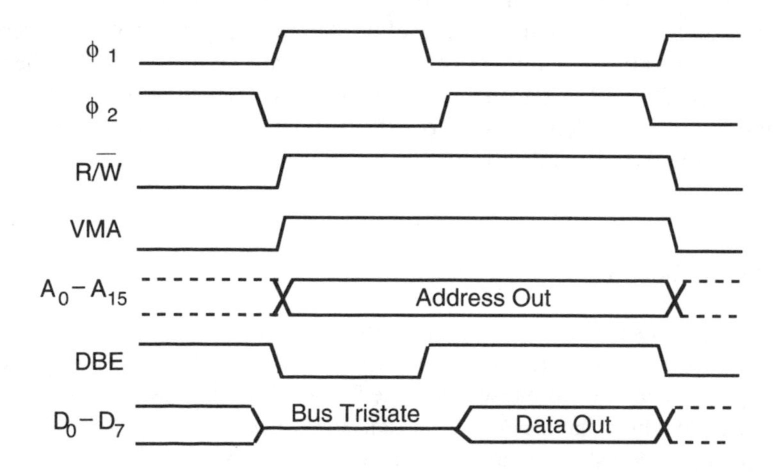

Figure 7.14 The 6800 WRITE machine cycle. The $\Phi 2$ and the DBE signals shown are identical. If external slow memories require more time, the DBE pin can be forced Hi to temporarily stop (stretch) the system clock. The 6802 does not contain the DBE.

The 6800 Interrupt Processing

The 6800 uses polling rather than vectoring for interrupt processing. This makes the 6800 interrupt software-intensive rather than hardware-intensive as in the 8080 series. Generic polling relates to the scheme of going around the table to question each participant about its position. For interrupt processing, the technique is used to successively interrogate each appropriate input device. However, the processing usually incorporates a form of double polling or double questioning because any or all of the input devices may send an interrupt request. At first, the software must find, in turn, which one(s) sent the interrupt signal. When there is a yes indication, the information must be input. The input priority must be set in the software and not in hardware as is the usual case in vectored interrupts. The NMI is not maskable. The interrupt requests are acknowledged at the end of the current instruction.

The 6800 DMA Processing

A low DBE will tristate the address and data busses affording an opportunity for DMA. The 6802 cannot accommodate DMA.

Other 8-Bit Microprocessors

Very quickly after the pioneering microprocessors were developed, other producers created units that were unique and useful. The following descriptions describe only a sampling of the 8-bit microprocessors that were marketed in the middle and late seventies. The following micros are included to show both some other approaches that were applied and also to indicate how the marketplace rapidly became a factor in the design thrust.

The Z80 Microprocessor

The Zilog Z80 was software-compatible with the 8080A but also included eighty more instructions. However, it was not hardware-compatible. Even though parts of its single-chip architecture resembled the 8080A three-chip system, it is overly simplistic to conclude that the Z80 was just a single repackage of the 8080A. Even a casual analysis will show major differences and additions.

The Z80, like the 8085, operated from a single 5-V power supply and used a single externally supplied clock. At a time when slow, dynamic memories were in vogue, it included memory refresh logic and provided a signal to automatically refresh external dynamic memory without stealing cycles.

Table 7.4 The Z80 signal pins

+5, V_{SS}	Power Input	ϕ	Clock Input
D_0–D_7	Data Bus	A_0–A_{15}	Address Bus
$\overline{RD}$	Memory Read	$\overline{WR}$	Memory Write
INT	A Maskable Interrupt	$\overline{NMI}$	Nonmaskable Interrupt
$\overline{MREQ}$	Memory Access in Progress	$\overline{IORQ}$	Valid Port Address
$\overline{M_1}$	In Fetch Cycle		
$\overline{IORQ}$	When both are low,		
& $\overline{M_1}$	there is an Interrupt		
RFSH	Refresh Dynamic Memories		Acknowledge
BUSRQ	Tristate Bus for DMA	BUSAK	Bus Ready for DMA
WAIT	Like the 8080A READY	HALT	Executes NOPs

The duplicate registers, including the two accumulators, permitted simple and rapid register-to-register transfer instructions. The IX and IY registers allowed sophisticated indexed addressing. The interrupt vector register provided for the use of vectored call routines similar to those in the 8085 and the memory refresh counter register kept track of an automatic refresh sequence that continually renewed the contents of any slow but inexpensive dynamic RAM. The status words (flags) were similar to the 8080A—see Figure 7.15.

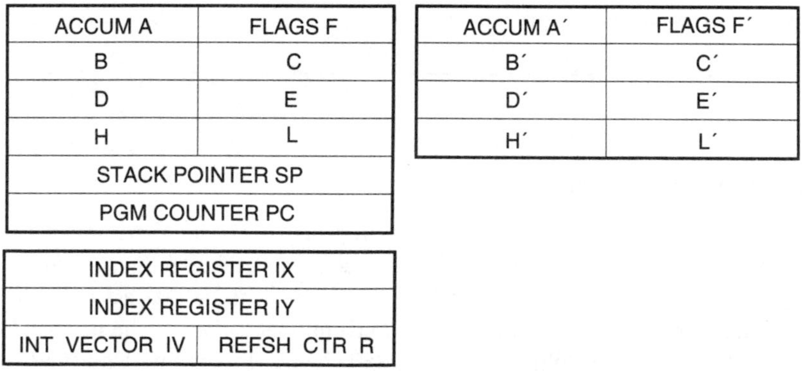

ACCUM A	FLAGS F
B	C
D	E
H	L
STACK POINTER SP	
PGM COUNTER PC	

ACCUM A′	FLAGS F′
B′	C′
D′	E′
H′	L′

INDEX REGISTER IX	
INDEX REGISTER IY	
INT VECTOR IV	REFSH CTR R

Figure 7.15 The Z80 programmable registers and status flags. Note that there are two accumulators as well as two B through L registers. The registers in the dark outline are duplicates of the 8080A that made the Z80 software compatible.

The 6502 Microprocessor

Just as the Z80 was in some ways similar to the 8080A, the 6502 was in some ways similar to the 6800/6802. However, the 6502 is primarily of interest because of its selling price. In a period when the single unit price of a general 8-bit microprocessor was in the order of $150 to $200, MOS Technology offered the 6502 for $20. This meant that the rest of us could afford to purchase a micro and its associated chips for experiment and learning, which, in many cases, resulted in a salable product. Whether it was the KIM-1, which was the first completely assembled, reliable, working single-board microcomputer, or the later Apple products, many people had a 6502 in their first micro system.

The 6502 was neither hardware nor software compatible with the 6800. For example, it had an on-chip clock and only one accumulator. The 6502 was superior in its addressing functions (modes) but lacked some of the 6800's valuable processing instructions. Even so, from the perspective of almost twenty years later, many mass-marketed products, as well as smaller quantity control and research products, were very successfully created using the MOS Technology's 6502. See Table 7.5 and Figure 7.16.

Table 7.5 The 6502 signal pins.

+5, V_{SS}	Power In	$\overline{NMI}$	Nonmaskable Interrupt
ϕ	Clock in	$\overline{IRQ}$	Maskable Interrupt
ϕ_1, ϕ_2	System Clock Outputs	$\overline{RESET}$	Reset as in 6800
DB_0–DB_7	Data Bus—*Tristate*	RDY	Wait for slow memories
AD_0–AD_{15}	Addr Bus—*Not-Tristate*	SO	Set overflow flag
$R/\overline{W}$	Mem Read/WRITE	SYNC	Identify fetch cycle

The F8 Microprocessor

Besides the market pressure for lower microprocessor costs, there was a similar demand for lower overall system costs that was translated into reduced chip counts. One of the interesting entries was the Fairchild F8. The F8 was a family of four chips that were configured in a very unconventional manner. The normal functions of the MPU were split among the 3850 Central Processing Unit (CPU) and the 3851 Program Storage Unit (PSU), which contained the addressing functions. The PSU also contained 1,024 bytes of Read Only Memory (ROM) and an interrupt controller. The ROM was mask programmable at the factory. The PSU contained an internal program counter and stack register to keep track of its own 1,024-

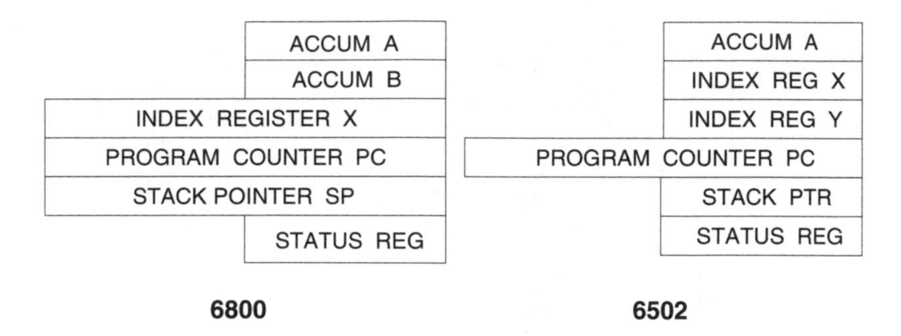

Figure 7.16 The programmable registers of the 6800 and the 6502 microprocessors. The longer boxes indicate 16-bit registers and the shorter boxes indicate 8-bit registers.

byte internal memory. Thus, several PSUs might be used in a given system. The CPU and the PSU both contained two 8-bit I/O ports.

This combined 3850 and 3851 MPU could be augmented with several combinations of three other chips to form a complete F8 system. These three chips included the 3852 Dynamic Memory Interface (DMI), the 3853 Static Memory Interface (SMI), and the 3854 DMA controller. Like the 3850 and the 3851, the function of these other three chips was not at all conventional. The 3852 DMI could address and control up to 64K of ROM or RAM. It had its own chip program counter and dynamic memory refresh capability. The 3853 SMI could address and control up to 64K of static memory. The 3854 DMA could provide direct memory access for up to 64K of memory. The F8 was very unconventional and that, along with the factory programmable ROM, made it unattractive for hobbyist and low-volume producers. However, the F8 proved to be very versatile and popular with high-volume producers for many applications. A block diagram of the F8 system is shown in Figure 7.17.

The 8048 Microprocessor

The introduction of the Intel 8048 was another result of the market pressure to reduce the chip count. The 8048 was actually a family of one-chip microcomputers that varied according to the choice of on-chip memory, which included no ROM, 1,024 bytes of ROM, or 1,024 bytes of EPROM (Erasable, Programmable, ROM), There were also 64 bytes of on-chip RAM. Even though the introduction of the 8040 family was quite significant in the history of microprocessors, since the MPU was derived from the 8080A architecture, only its general block diagram and major signals will be presented. The model 8048 system is shown in Figure 7.18.

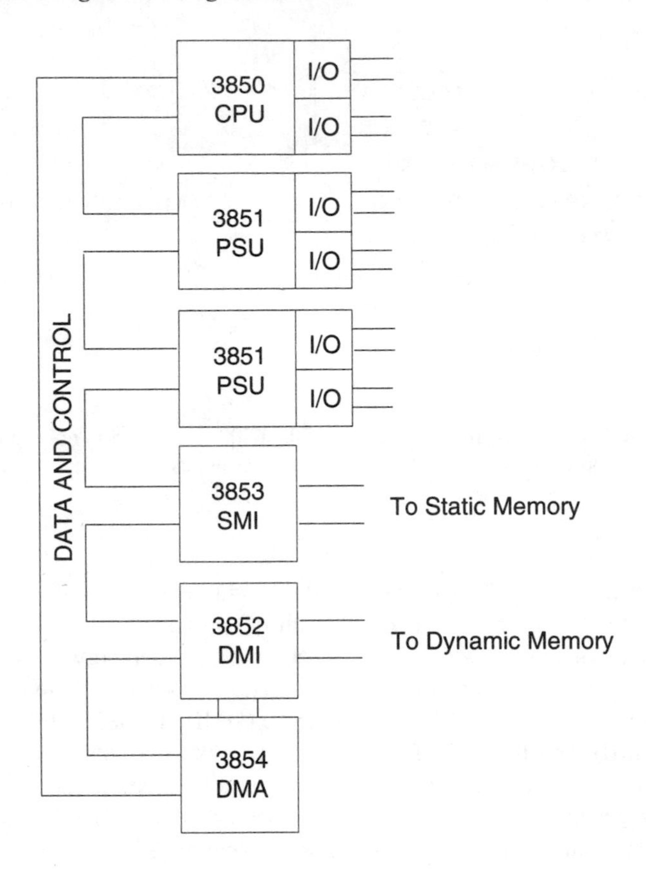

Figure 7.17 An elementary model of an F8 system.

Table 7.6 The 8048 signal pins.

$\overline{\text{INT}}$	Interr Request	$\overline{\text{RESET}}$	System Reset
PROG	Strobe for external chip	EA	External Pgm Mem Access-Debug
X1, X2	Crystal		
ALE	External Mem Latch and Clock for External Logic		
PSEN	Strobe for external mem data	SS	Single step debug
$\overline{\text{RD}}$	Read external memory	$\overline{\text{WR}}$	Write external mem
T0	Test In or Clock Out	T1	Test or Event Cntr In
V_{DD}	+24 volts used on 8748 to program the EPROM		

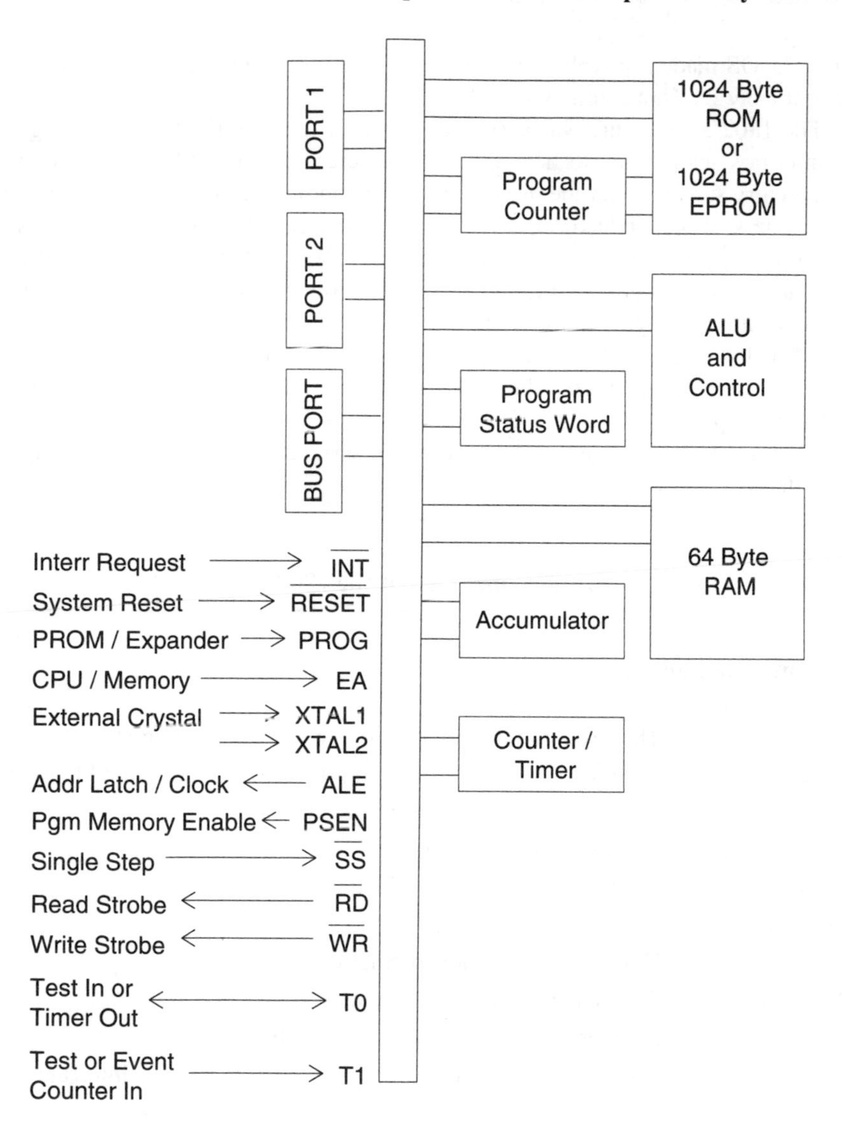

Figure 7.18 An elementary 8048 model. The microcomputer will contain ROM, EPROM, or nothing depending upon which family member it is.

The 1802 Microprocessor

The RCA 1802 (sometimes called COSMAC) micro found a market because of its extremely low power requirements. At a time when most microprocessors used N-MOS Technology, the 1802 used Complementary Metal-Oxide Silicon (CMOS), which meant it could operate satisfactorily at one-fifth to one-tenth the input power.

This CMOS made it possible to operate the 1802 and other members of the chip set with any DC input from 3 to 12 V.

The 1802 architecture was very unconventional. It had no dedicated program counter or stack pointer. Its accumulator was referred to as the "D" register. As a satisfactory substitute for the more commonly committed storage or control areas or registers, it had sixteen 16-bit general-purpose registers that could be designated as program counters, stack pointers, etc. While this was a bit confusing, especially to the programmer familiar with the more common micros, its visibility soon became an ally. It could access up to 255 I/O devices and 64K of ROM or RAM. It had a convenient interrupt structure and could support both maskable and nonmaskable interrupts. The SC0 and SC1 state signals exactly indicated, for external chips, either a Fetch, Execute, DMA, or Interrupt sequence. The 1802 programmable "Q" output was convenient for serial outputs. Table 7.7 and Figure 7.19 describe the 1802.

Table 7.7 The registers and signal pins of the 1802.

The internal registers:

R(0) through R(F)	The 1802 had sixteen 16-bit general-purpose registers that could be divided into an 8-bit Hi register, termed R(n) .1, and an 8-bit Lo register, termed R(n) .0.
D or Accum.	The MPU accumulator
ALU	The MPU arithmetic-logic unit
DF (1-bit)	ALU (Carry) Data Flag
I (4-bits)	Holds the high-order instruction digit
P (4-bits)	Designates which R(n) is the Program Counter
X (4-bits)	Designates which R(n) is the Data Pointer
T (8-bits)	Normally holds old Hi byte or the P byte after an interrupt

The signal pins:

DATA BUS	A conventional 8-bit in/out data bus
ADDR BUS	A conventional 16-bit address bus
CLEAR & WAIT	Provide the functions of Run, Load, Pause, and Reset
$\overline{\text{XTAL}}$	One input for an external clock crystal
$\overline{\text{CLOCK}}$	The other crystal input or an input for an external clock.

	The clock is internally divided by eight to give eight clock pulses per machine cycle. Most instructions are two machine cycles.
EF1, 2, 3, 4	Inputs (flags), sampled each execute cycle, that can be used for vectored interrupts.
Q	A single bit, software controlled serial-data output.
$\overline{\text{MRD}}$	Used for memory read and I/O.
$\overline{\text{MRW}}$	A conventional memory write pulse.
INTR	A nonmaskable interrupt
DMA In, DMA Out	DMA controls
TPA, TPB	Timing control pulses for system chips
SC0, SC1	Pins that can be combined to show the MPU states
N0, N1, N2	Signals used to control output devices

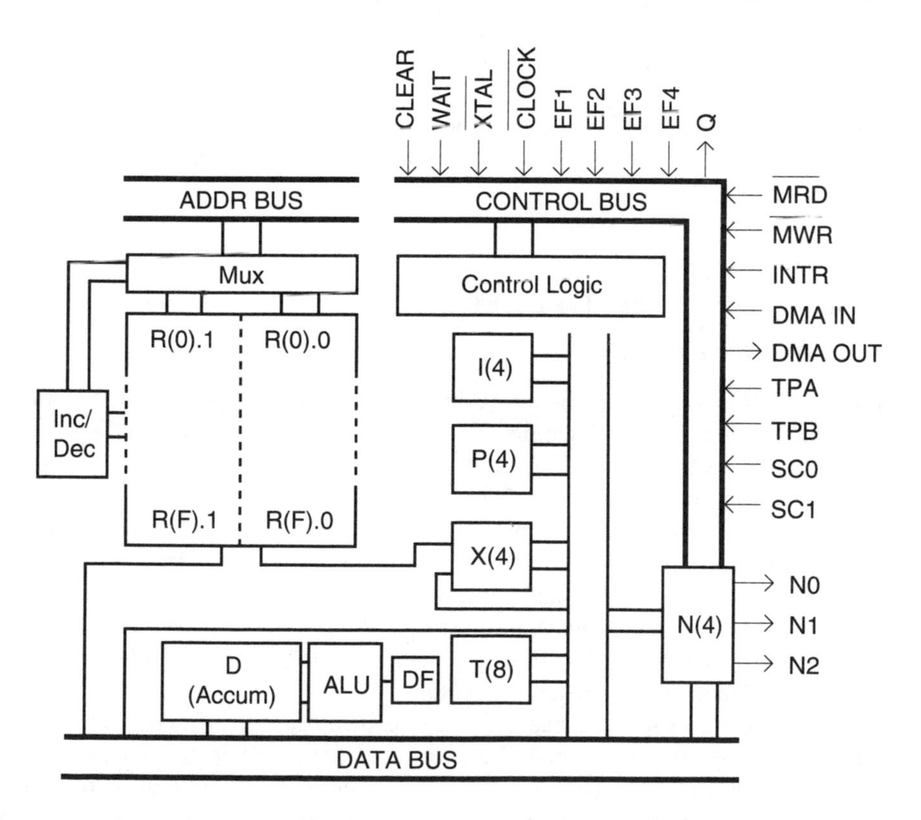

Figure 7.19 The block diagram of the unconventional, but quite elegant, RCA 1802 microprocessor. As in other MPU diagrams, certain simplifications are incorporated.

7.3 16-BIT MICROPROCESSORS

Microprocessors—A Second Look

The definition of a microprocessor was covered in detail at the start of this chapter and several models were proposed in order to craft a foundation for its understanding. When the focus is now switched to the more advanced 16-bit units, it may be necessary to rethink both the fundamental meaning and uses of microprocessors. The 16-bit MPUs contain much more than simply larger registers and faster execution time. They can do much more than expanded 8-bit units. Before various 16-bit units are detailed, some of the ideas behind the microprocessor modes will be revisited and upgraded. Although the sections to follow will concentrate on hardware, a discussion of the need and uses of software will first be presented.

The very early micros were fundamentally miniaturized logic circuits. The instructions were at first switches used to conveniently reconfigure this logic or to have it do relatively simple logic or arithmetic operations. These micros were controllers and their relationship to languages had little influence on their early designs. Later, the 8-bit micros, and certainly the 16-bit units, included the concept that, for efficient program execution, several features of the architecture (such as the registers, the fetch and execute procedures, and the memory management) must be tailored to the requirements of high-level programming languages. This led to not only expanded registers and memory, but new approaches to their use. Some of the new 16-bit designs were thus begun with the premise "Let's start from square one."

16-Bit Microprocessors—The Z-8000

The Z-8000 General Overview

The Z-8000, like most other 16- and 32-bit micros somewhat resembles the architecture and operation of minicomputers. The registers were opened up and became more general purpose. The registers include sixteen 8/16-bit, eight 32-bit, and four 64-bit units—see Figure 7.20. The Z-8000 and other 16-bit micros to follow include registers that are primarily designed for user programming and registers that are primarily designed for system housekeeping. The Z-8000 will recognize seven types of data including BCD (binary coded decimal), what they call digits (4 bits), bytes (8 bits), words (16 bits), long words (32 bits), byte strings (8n + 8 bits), and word strings (16n + 16 bits). This dual string capability is especially helpful in high-level matrix processing.

The Z-8000 can support virtual memory. Its memory management techniques permits dividing all of the system memory into what are often called chunks or segments. The concept of virtual memory presupposes that these chunks or segments include both physical and perceived (virtual) program and data storage. It permits both the programmer and the system to postulate storage units (segments) in "space." (64K segments are common but not necessarily typical.) Each segment of virtual memory is given a starting address by the memory management. In somewhat oversimplified terms, if an application program selects a virtual memory location out of the bounds of physical RAM, the memory management system first will use the appropriate stacks, etc., to save the location of the present sector. Next, it will temporarily store this present sector to disk or to another part of RAM. Finally, it will load the virtual sector into physical RAM and do the required processing. When the need for the code and date in the virtual area is over, it is written back to its specified segment and the original segment is returned.

The Z-8000 Registers

Different operations can utilize register combinations from 8- to 64-bits. The 8- and 16-bit registers are available as general purpose accumulators and most are also available as index registers_see Figures 7.20, 7.21, and Table 7.8.

The Z-8000 can have up to twenty-four address lines, which will address up to 16,777,216 (2^{24}) bytes—see Table 7.8. When an additional memory-management chip is added, the addresses are divided into 64K segments and, likewise, the total virtual capacity is increased to over 96 megabytes. Including several methods of indexing, there are eight addressing modes as follows.

Table 7.8 The Z-8000 memory addressing techniques.

1. IMMEDIATE: The operand (number), i.e., the address of the data, is in the instruction.
2. DIRECT: The operand is in the location addressed by the instruction.
3. DIRECT REGISTER: Operand is the contents of the register.
4. INDIRECT REGISTER: The operand is the contents of the location addressed by the instruction.
5. INDEXED: The address is in the location addressed by the instruction plus or minus the index (offset) number.
6. BASE ADDRESS: Similar to indexed.
7. BASE INDEX: The offset is the sum of an index and a base number.
8. RELATIVE ADDRESS: The operand is the contents of the location whose address is the contents of the program counter offset by the displacement in the instruction.

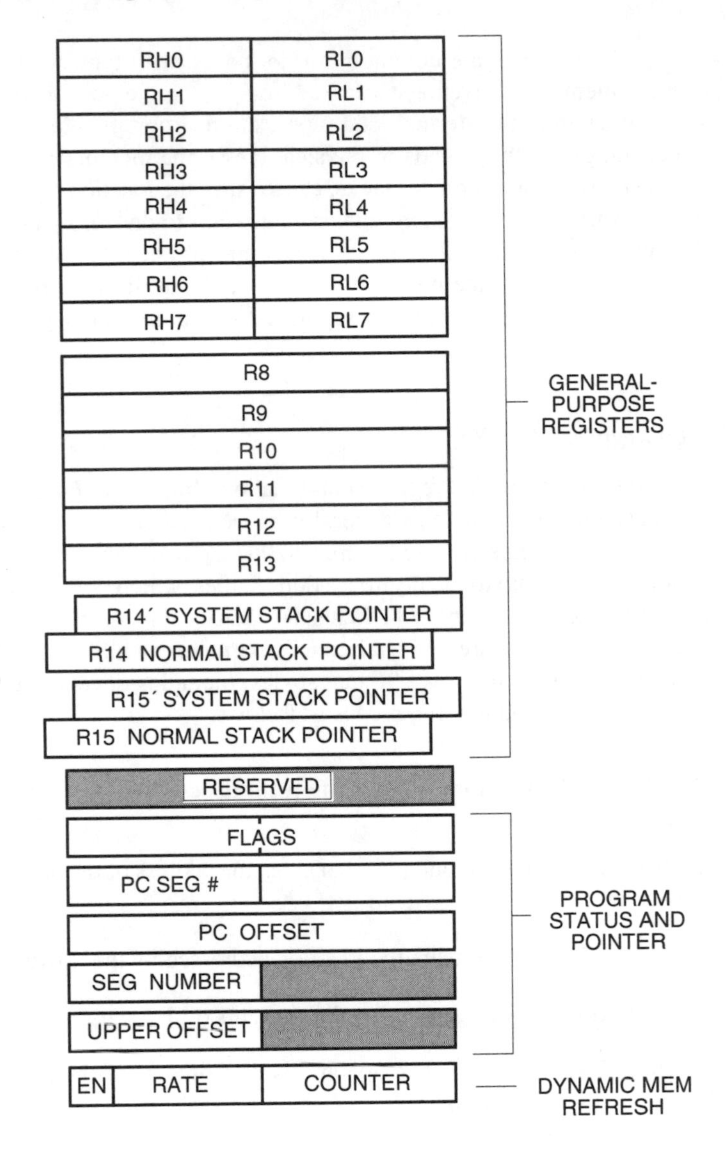

Figure 7.20 The Z-8000 registers.

Table 7.9 The Z-8000 address and interrupt control pins.

AD0–AD15	Multiplexed Address and Data bus
SN_0–SN_6	Segment number
$\overline{\text{SEGT}}$	Segment Trap

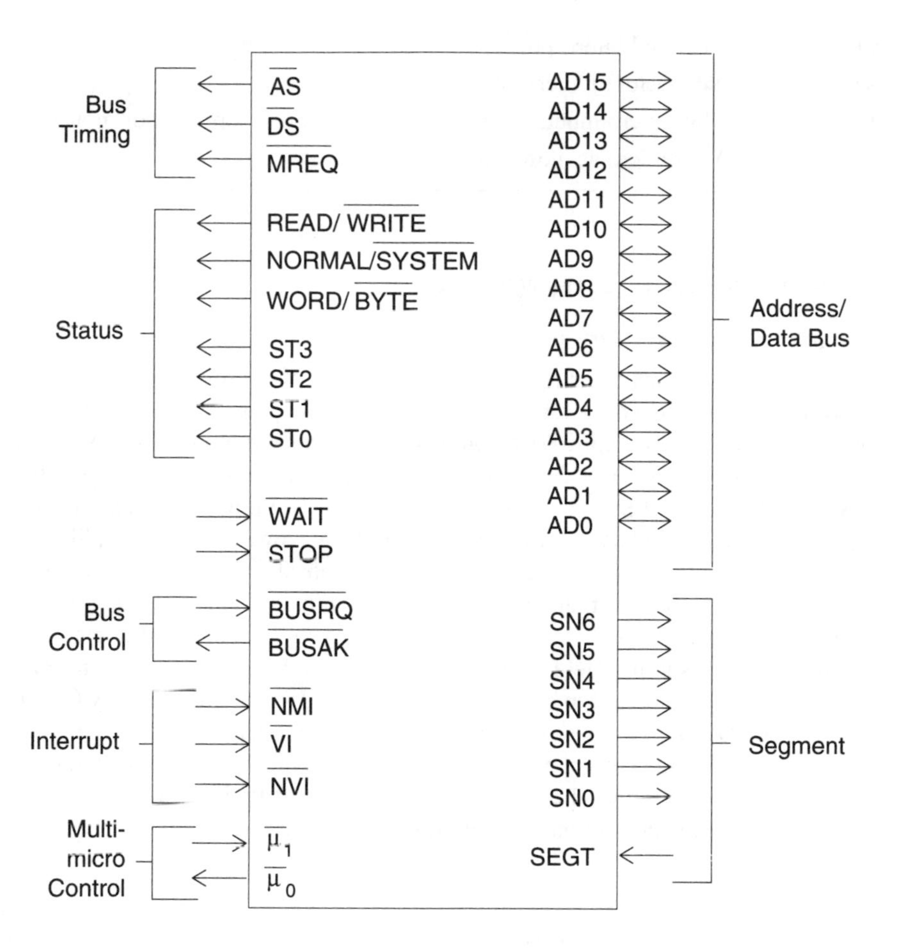

Figure 7.21 The major pins of the Z-8000.

$\overline{\text{AS}}$ Address strobe to indicate addresses are valid
$\overline{\text{DS}}$ Strobe to time the data in and out of the MPU
$\overline{\text{MREQ}}$ Memory request signal to interface dynamic memories
ST0-ST3 Outputs sixteen status codes
$\overline{\text{WAIT}}$ Indicates that memory or I/O is not ready
$\overline{\text{STOP}}$ Stops normal execution—goes to single step
$\overline{\text{BUSRQ}}$ Requests the bus from the processor
$\overline{\text{BUSAK}}$ Bus has been relinquished
$\overline{\text{NMI}}$ Nonmaskable interrupt

$\overline{VI}$	Vectored interrupt
$\overline{NVI}$	Nonvectored interrupt
$\overline{\mu 1}$	Multimicro input. Tests for state of multimicroprocessor inputs
$\overline{\mu 0}$	Multimicro acknowledge

16-Bit Microprocessors—The 8086/8088

The 8086/88 General Overview

At first inspection, the introduction of this Intel series of microprocessors seemed to present only two major advantages. First, they could address one megabyte or more of memory. Second, the 8088 could run programs written for its new 16-bit architecture as well as the older 8-bit programs designed for the 8080A and the 8085. The 8086 and 8088 micros both have identical 16-bit internal data processing paths and the 8086 has a 16-bit memory interface. However, the 8088, like the 8080A-8085, has an 8-bit memory interface. The introduction of this new series of MPUs gave the designer, and ultimately the user, many more advantages. The 8086/88 series use the technique termed pipelining to do some of the fetch and execute operations in parallel rather than all in series, thus giving faster program execution. The arithmetic can be written and executed in ASCII or Binary Coded Decimal (BCD). Both of these microprocessors can be operated either in a minimum mode, where the control pins function in a conventional MPU manner, or in the maximum mode where the MPU operates in concert with external controller chips to manage the performance of much larger systems.

The Configuration of the 8086/88

The overall MPU structure is divided into two general sections, the Bus Interface Unit (BIU) and the Execution Unit (EU), as shown in Figure 7.22. The BIU fetches instructions, reads the operands, and writes the processed result. The EU executes the instructions. As stated before, this division of responsibility, among other advantages, significantly speeds up the overall operation by pipelining the instructions in a group of instruction-queue registers.

The segment registers are used to help direct the storage and retrieval of different information to and from the segmented memory. Each segment register points to one of the four currently addressable segments. Programs obtain access to code and data in other segments by changing the segment registers to point to the desired segments. The CS register points to the current code segment, DS to the data segment, SS to the stack segment, and ES to an extra or auxiliary segment. The IP is the instruction pointer. Each address segment can contain up to 64K, 16-bit bytes of memory. The 8086/88 signal pins are presented in Table 7.10 and Figure 7.23.

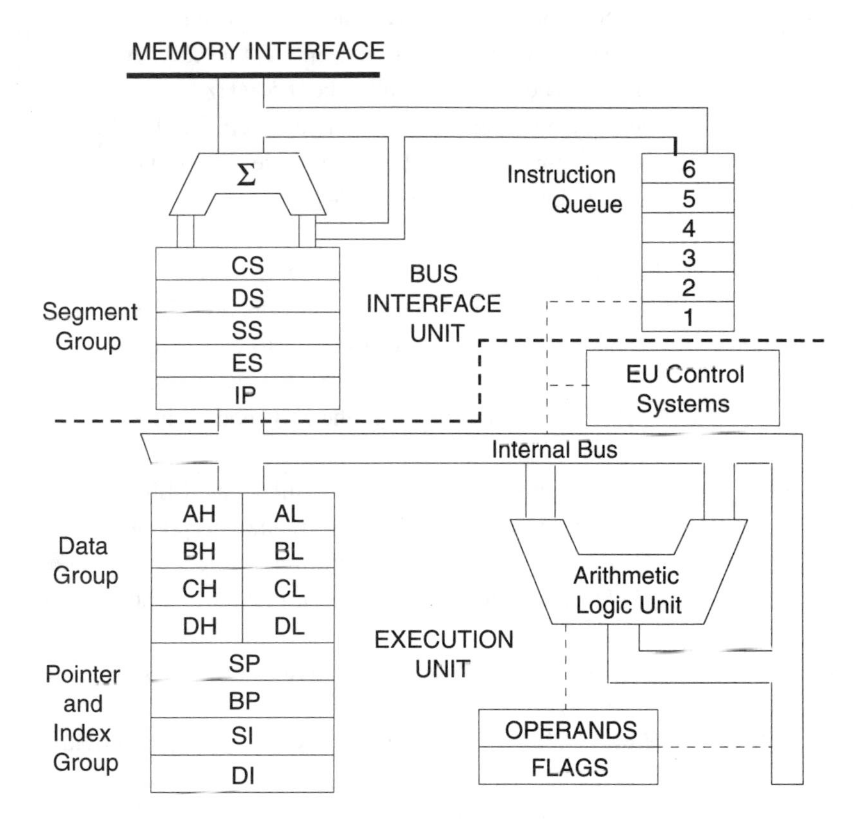

Figure 7.22 The data paths in the 8086/88 microprocessor. The 16-bit EU is identical in both micros. Likewise, the BIU in the 8086 is also 16 bits wide while the 8088, for downward compatibility, is only 8 bits wide.

Table 7.10 The 8086 / 8088 signal pins.

V_{CC}, GND	+5 V Power In
AD_0–AD_{15}	Multiplexed Address and I/O Bus in the 8086
A_8–A_{15}	Memory and I/O high-order bits in the 8088
AD_0–AD_7	Multiplexed Address and I/O low-order Bus in the 8088
NMI	Nonmaskable Interrupt
INTR	The input is usually from the 8529A Programmable Interrupt Controller. When the INTR is activated, the MPU takes different actions depending on the state of the Interrupt Enable Flag.

Signal	Description
CLK	The System Clock is usually provided by an 8284 Clock Generator, which also provides inputs for READY and RESET. The clock is usually about 5 MHz.
A16/S3 A17/S4 A18/S5 A19/S6	During the first part of the bus cycle, A16–A19 address high memory. Later in the cycle, these pins provide status information, given as S3–S6.
$\overline{BHE}$/S7	Bus High Enable is used to select external memory. S7 provides the upper coded status information signal.
SS0 (High)	An 8088 pin providing one of three special status bits for the 8288 Bus Controller. Pin 34 is always a status (HIGH) when the MPU is in the MX mode (pin 33 is strapped low).
MIN/MAX	Strapping this pin to a high (+5) will set the MPU in its minimum-system (small system) mode. Strapping it low (GND) will place it in the maximum of large-system mode. (GND) will place it in the maximum of large-system mode.
$\overline{RD}$	Memory read
HOLD ($\overline{RQ}$/$\overline{GT0}$)	HOLD is an MN-mode external request for the system bus. ($\overline{RQ}$/$\overline{GT0}$) is an MX-mode request for the system bus. However, in the max mode, the ($\overline{RQ}$/$\overline{GT0}$) signal provides both the bus request and the bus grant information.
HLDA ($\overline{RQ}$/$\overline{GT1}$)	A hold acknowledge signal in the MN mode and a combined request/grant signal in the MX mode. This is a lower priority signal than the one above.
$\overline{WR}$ ($\overline{LOCK}$)	An MN write signal and an MX bus arbitration signal.
M/$\overline{IO}$ IO/$\overline{M}$ (S2)	Memory or I/O (MN) or S2 (MX) used for the bus controller. Note the reversed polarity between the 8086 and the 8088.
DT/$\overline{R}$ (S1)	Data Transmit/Receive is used with external bus transceivers such as the 8286/87. S1 is another bus-controller signal.
$\overline{DEN}$ ($\overline{S0}$)	$\overline{DEN}$ is also used with transceivers. S0 is another control.
ALE (QS0)	Address Latch Enable and a (MX) bus-controller signal.
$\overline{INTA}$ (QS1)	Interrupt Acknowledge (MN) and controller signal (MX).
$\overline{TEST}$	An input signal, used with the WAIT instruction to synchronize the MPU with other chips such as an 8087 coprocessor.
RESET	Reset the MPU.
READY	A Wait State control signal.

8086

Signal	Pin	Pin	Signal	Max mode
GND	1	40	V_{CC}	
AD14	2	39	A15	
AD13	3	38	A16/S3	
AD12	4	37	A17/S4	
AD11	5	36	A18/S5	
AD10	6	35	A19/S6	
AD9	7	34	$\overline{BHE}$/S7	
AD8	8	33	MN/$\overline{MX}$	
AD7	9	32	$\overline{RD}$	
AD6	10	31	HOLD	$\overline{RQ}/\overline{GT0}$
AD5	11	30	HLDA	$\overline{RQ}/\overline{GT1}$
AD4	12	29	$\overline{WR}$	($\overline{LOCK}$)
AD3	13	28	M/$\overline{IO}$	($\overline{S2}$)
AD2	14	27	DT/$\overline{R}$	($\overline{S1}$)
AD1	15	26	$\overline{DEN}$	($\overline{S0}$)
AD0	16	25	ALE	(QS0)
NMI	17	24	$\overline{INTA}$	(QS1)
INTR	18	23	$\overline{TEST}$	
CLK	19	22	RESET	
GND	20	21	READY	

8088

Signal	Pin	Pin	Signal	Max mode
GND	1	40	V_{CC}	
A14	2	39	A15	
A13	3	38	A16/S3	
A12	4	37	A17/S4	
A11	5	36	A18/S5	
A10	6	35	A19/S6	
A9	7	34	SS0	(HIGH)
A8	8	33	MN/$\overline{MX}$	
AD7	9	32	$\overline{RD}$	
AD6	10	31	HOLD	$\overline{RQ}/\overline{GT0}$
AD5	11	30	HLDA	$\overline{RQ}/\overline{GT1}$
AD4	12	29	$\overline{WR}$	($\overline{LOCK}$)
AD3	13	28	IO/$\overline{M}$	($\overline{S2}$)
AD2	14	27	DT/$\overline{R}$	($\overline{S1}$)
AD1	15	26	$\overline{DEN}$	($\overline{S0}$)
AD0	16	25	ALE	(QS0)
NMI	17	24	$\overline{INTA}$	(QS1)
INTR	18	23	$\overline{TEST}$	
CLK	19	22	RESET	
GND	20	21	READY	

Figure 7.23 The pins of the 8086 and the 8088.

The 8086/8088 Memory Addressing Techniques

Background for the addressing options of the 8086-8088 micros includes the following:

1. These units have the equivalent of twenty address lines and therefore can address 2^{20} or 1,048,576 bytes (1 megabyte) of memory.

2. The 1-Mb space can be divided into 64K address segments. These segments may contain up to 64K of memory and the segments may be contiguous or they may overlap. Any overlapping is reconciled by memory management. Thus, any physical memory may contain many segments of virtual memory.

3. For large microcomputer systems, the MPU can be strapped into a special MX (maximum) mode where the external chips, including I/O and memory, are controlled by special pin signals.

16-Bit Microprocessors—The 68000/08/10

The 68000 General Overview

To go from an 8-bit micro to a 16-bit version requires much more than twice as many (sometimes multiplexed) address and date pins and, likewise, a linear expansion of registers. As all manufacturers, Motorola did much soul searching to find the best compromises between developing the ideal next-generation product and a unit that would be downward compatible with much of its older hardware and software. In fact, they instituted the Advanced Computer System on Silicon (MACSS) project to develop what later became the 68000 family. The MACSS development produced, among many features, a microprocessor that had an extremely versatile register structure. It produced a concept that could address 2^{24} (16 Mb) or, in later versions, 2^{32} (4,096 Mb = 4 gigabytes) of physical memory. The MACSS concept also provided very sophisticated virtual memory and multiprocessing techniques. The use of high-level languages significantly influenced the concept and development of the processor family.

The hardware was exceptionally well configured for rapid instruction processing. Not only did it include an instruction queue, this queue was intelligent—it would inspect the incoming instruction and choose the fastest processing method, depending on the instruction's type and size. The register and register instructions permitted direct register-to-register transfers as well as direct memory-to memory transfers. The nongeneral registers that were used were also somewhat intelligent. For instance, the data registers affect the status and condition flags differently than the address registers.

The 68000 User and Supervisor Programming Models

The 68000 has two levels of control when it executes instructions. The second, or lower, user level is where most application programs execute. However, there is also a higher, protected supervisor level, isolated from the user, that is used to protect the system from faults such as improper code. All resets, interrupts, and other non-programmed or improper operations will initiate this more secure system-management level.

The 68000 Series—Structured Programming and Program Testing

The 68000 series is designed to use the advantages of higher-level structured languages such as Pascal. The instructions make it easy to transfer parameters between and within software modules that operate on a reentrant and recursive basis. (To be reentrant, a routine must be usable by an interrupted or a normally driven program without loss of data. A recursive routine may call or use itself.)

Several instructions, including LINK and TRAP, assure that these parameters can be transferred without the loss of data or memory location clashes. There are also several hardware TRAPS that detect illegal addresses and instructions, divided by zero, overflow conditions, register out of bounds, etc.

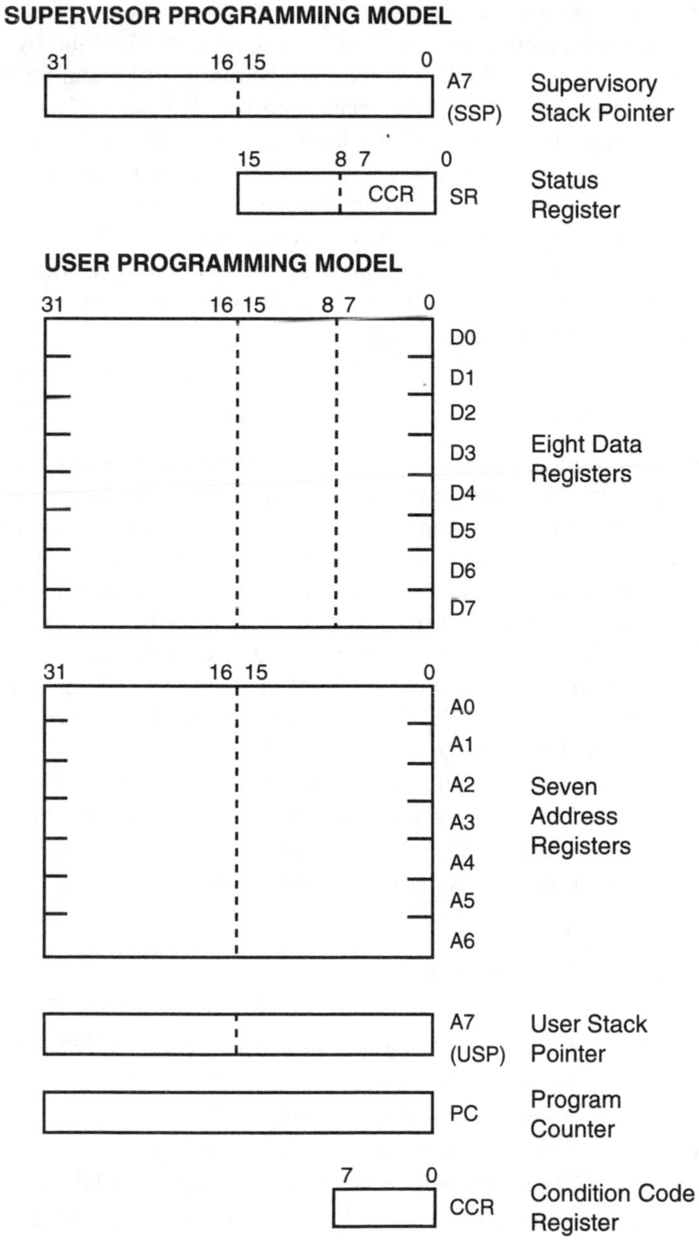

SUPERVISOR PROGRAMMING MODEL

31 16 15 0
A7 (SSP) Supervisory Stack Pointer

15 8 7 0
CCR SR Status Register

USER PROGRAMMING MODEL

31 16 15 8 7 0
D0
D1
D2
D3
D4
D5
D6
D7
Eight Data Registers

31 16 15 0
A0
A1
A2
A3
A4
A5
A6
Seven Address Registers

A7 (USP) User Stack Pointer

PC Program Counter

7 0
CCR Condition Code Register

Figure 7.24 The 68000 two-register model.

The Register Configuration of the 68000 Series

Most of the conventional, dedicated registers were replaced by general-purpose registers. These general-purpose registers could be used with 8-bit bytes, 16-bit words, or 32-bit long words. To speed up the 16-bit by 16-bit multiplication and other processing functions, the appropriate registers were made 32 bits wide. The major allocation of these registers was divided between data and addresses. This division guarantees that the housekeeping, such as the entries in the status register, is kept accurate when either data or addresses are processed.

The 68000 has three separate 16-bit arithmetic logic units (ALUs) that work in parallel. One does the data calculations and the other two perform the address calculations. This makes it possible to do a 16-bit data calculation and a 32-bit address calculation at the same time.

The 68000 Microcoding

The logic in the 68000 series micro is implemented with a technique termed microcoding rather than the more conventional technique of random logic. Random logic is the method of using gates, buffers, and transistors to implement the processing required in circuits such as controllers and microprocessors. Although this is a very compact way to accomplish the logic, the expanded microprocessor layout required the use of the very complex VLSI (very large scale integration). Not only is VLSI very expensive to produce, there is another high cost if the layout ever needs to be corrected or changed. While microcoding is less efficient and may take more real estate on the chip, any corrections, changes, or upgrades are much easier to implement and track.

Microcoding substitutes registers, shifters, flags, busses, etc., for logic gates. It then uses a special microprogram (microcode) to call and control the various internal operations of the processor. Thus, microcode permits the exchange of less costly software for the very expensive VLSI hardware.

Table 7.11 The 68000 signal pins.

V_{CC} and GND	The 5-V power inputs. Some packages have duplicate pins that should be used to reduce glitches.
CLK	An input for an 8- to 16-MHz TTL clock
FC_0–FC_2	State (user or supervisor) and cycle-type codes
E	The 68000 series enable signal for peripheral devices
$\overline{\text{VMA}}$	Valid Memory address signal for peripheral devices
$\overline{\text{VPA}}$	Valid Peripheral Address must be in sync with the E signal
$\overline{\text{BERR}}$	Bus Error in peripherals, memory management, interrupt, etc.
$\overline{\text{RESET}}$	This may be an externally generated input or a software generated output for peripherals.

$\overline{HALT}$	Another bidirectional signal that may be externally generated or it may originate in the processor to signal that the processor has stopped.
$D_0 - D_{15}$	The 16-bit, bidirectional, three-state data bus
$A_1 - A_{23}$	The 23-bit, unidirectional three-state address bus
$\overline{AS}$	The address is stable on the bus.
$R/\overline{W}$	Read/Write
$\overline{UDS}$, $\overline{LDS}$	These signals, along with $R/\overline{W}$, control the data on the data bus. They are decoded to show valid 0–7 or 8–15 data bits.
$\overline{DTACK}$	Data Acknowledge indicates that data transfer is complete.
$\overline{BR}$	Thc input signal indicates that an other device wishes to become the bus master (on large systems).
$\overline{BG}$	The processor will release the bus at the end of the current cycle.
$\overline{BGACK}$	Some other device has become the bus master.
IP_0–IPL_2	Indicate interrupt priority level

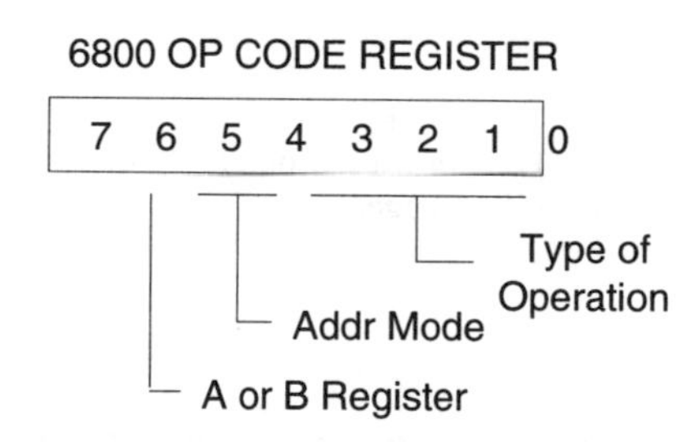

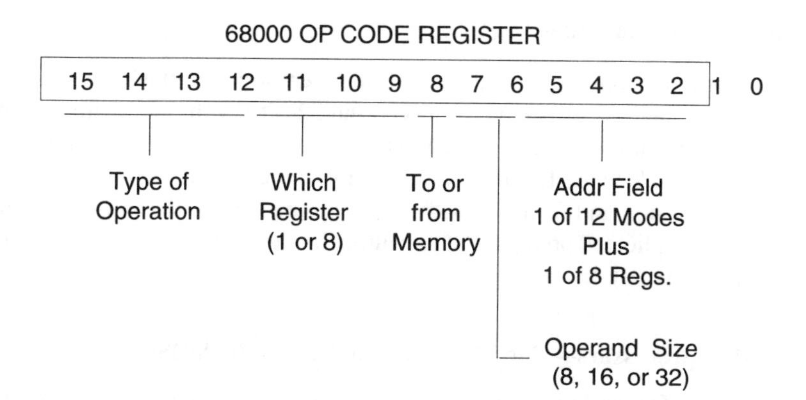

Figure 7.25 A comparison of the 8-bit op code used in the 6800 and the 16-bit op code used in the 68000.

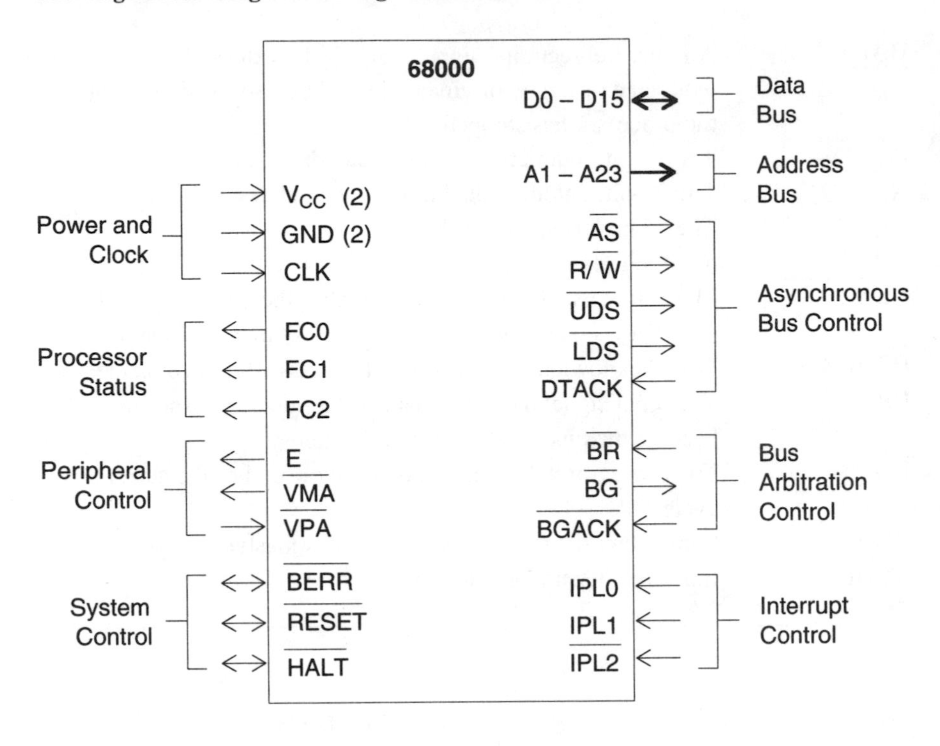

Figure 7.26 The pins of the 68000. Pin numbers are not included since the 68000 series comes in several packages.

7.4 32-BIT MICROPROCESSORS

Microprocessors—A Third Look

The quest for more addressable memory, faster operation, multitasking, and more efficient use of high-level languages forced the 32-bit revolution. With the new microprocessors came new terms that had to be added to or exchanged in the fundamental models. The ALU sometimes became the computation unit (CU) the addressing, queuing, and fetching instructions were given to the instruction unit (IU) and the complicated processes of executing instructions were concentrated in the execution unit (EU).

32-Bit Microprocessors—The 80386, 80386SX, and the 80286

The Intel 80386DX includes the registers, data and addressing busses, and processing and instructions to make it a true 32-bit micro. The 80286—which is not detailed in this chapter—followed the compatible 16-bit register scheme of the 8086 and had a 24-bit data bus. The Intel 80386SX is nearly the same as the DX but, for downward compatibility, has a 16-bit data bus. The 386 can process 8-, 16-, and 32-bit data. One of its most unique features is its advanced instruction pipelining. This fast instruction fetch, plus its on-chip address translation and large bus speed, contribute to its very rapid instruction execution time. The 80386 also has increased its clock speed—approaching 40 MHz—by using the more advanced CHMOS III and CHMOS IV technologies. See the 80386 model in Figure 7.27, its pins in Table 7.12, and its registers in Figure 7.28.

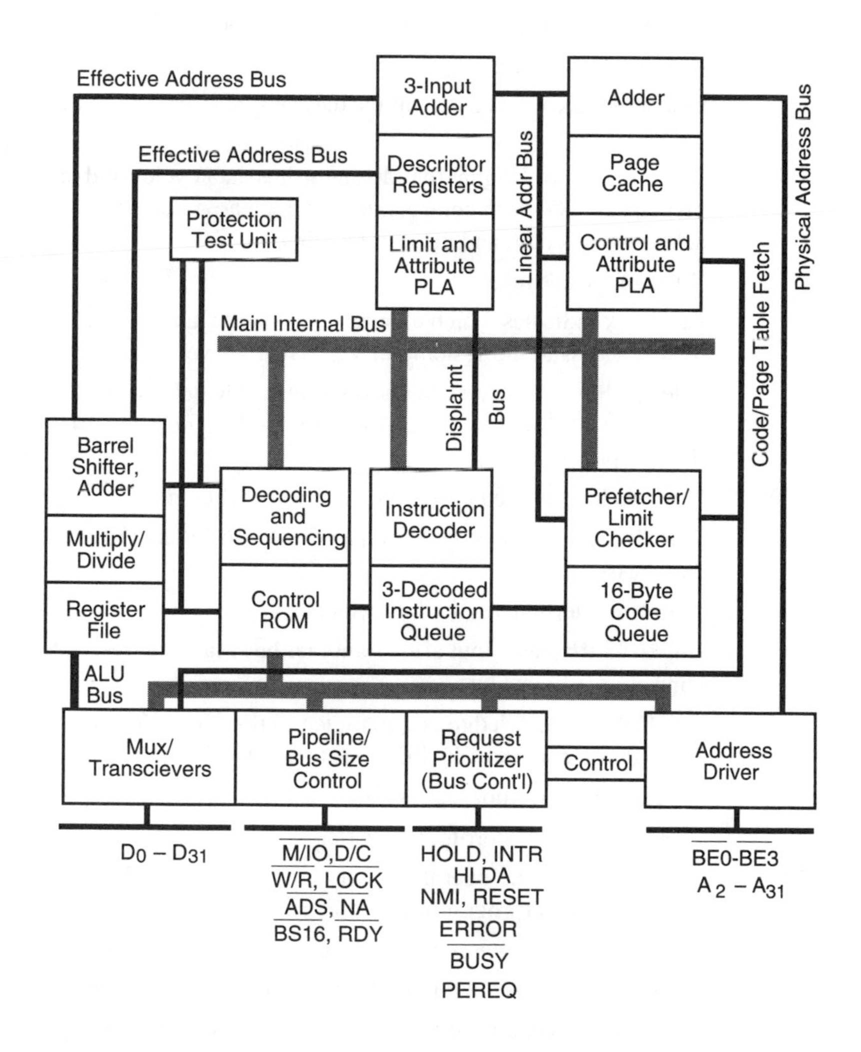

Figure 7.27 A simplified model of the operation paths in the 80386.

Table 7.12 The pin description of the 80386.

NOTE: The 386 series comes in several packages so only the pin descriptions, rather than pin numbers, are given.

Pin	Description
D_0–D_{31}	Data Bus inputs data during memory, I/O, and interrupt acknowledge read cycles and outputs data during memory and I/O write cycles.
A_2–A_{31}	Address Bus outputs physical memory or ports I/O addresses.
$\overline{BE_0} - \overline{BE_3}$	Byte Enable indicates which data bytes on the data bus take part in the bus cycle.
$\overline{M/IO}$	Indicates which data bytes on the data bus take part in the bus cycle
$\overline{D/C}$	The Data Control signal that distinguishes data cycles, either memory or I/O, from control cycles, which are interrupt acknowledge, halt, and instruct fetching.
$\overline{W/R}$	Write or Read cycles
$\overline{LOCK}$	Bus Lock indicates which external system bus masters are denied access to the system bus while it is active.
$\overline{ADS}$	Address Status indicates that a valid bus cycle definition and address ($\overline{W/R}$, $\overline{D/C}$, $\overline{M/IO}$, BE_0–BE_3 and A_2–A_{31}) are going to the 386 pins.
$\overline{NA}$	Next Address is used to request address pipelining.
$\overline{BS16}$	Bus Size input allows direct connection of 16-bit and 32-bit data busses.
$\overline{READY}$	Bus Ready terminates the bus cycle.
HOLD	Bus Hold Request input allows another bus master to request control of the local bus.
HLDA	Bus Hold Acknowledge output indicates that the 386 has surrendered control of its local bus to another bus master.
INTR	A Maskable Interrupt request
$\overline{NMI}$	A Nonmaskable Interrupt request
RESET	Reset suspends any operation in progress and places the 386 in a known (programmed) reset state.
$\overline{ERROR}$	Error signals an error condition from a processor extension.
$\overline{BUSY}$	Busy signals a busy condition from a processor extension.
$\overline{PEREQ}$	Processor Extension Request indicates that the processor extension has data to be transferred by the 386.

The 80386 register groups:

General Registers	The functions of the general data and address registers parallel those of the 80286 and the 8086. However, the eight 386 registers are all 32 bits.
Segment Registers	The 386 has added two registers to the four status registers (Code Segment Selector, Data Segment Selector, Stack Segment Selector, and Extra Segment Selector) in the 8086 and the 80286. The new FS and GS are also data segment selector registers.
Flag Registers	The 32-bit flag register includes, in bits 0–15, all the flags of the 8086 and the 80286 for downward compatibility.
IP	The Instruction Pointer (Program Counter) is similar to the ones in the 8085, 86, and 286 but is now 32 bits.
Control Registers	The 32-bit Control Registers contain machine state information including the Machine Status Word.
System Address Registers 286 or 386.	These four registers (Global Descriptor Table, Interrupt Table, Local Descriptor Table, Task State Segment) are defined to reference the tables or segments supported by the
Debug and Test Registers	These registers aid in the on-chip testing such as specifying breakpoints and debug status. The Test Registers are used in testing the on-chip memories.

32-Bit Microprocessors—The 80486 and 80486SX

The 80486 is similar to the 80386 with an 80487 FPU (Floating Point or Math Coprocessor Unit) and an MMU (Memory Management Unit) on the 486 chip. The register set for the processor is the same for as for the 386 but also includes the appropriate support registers for the FPU.

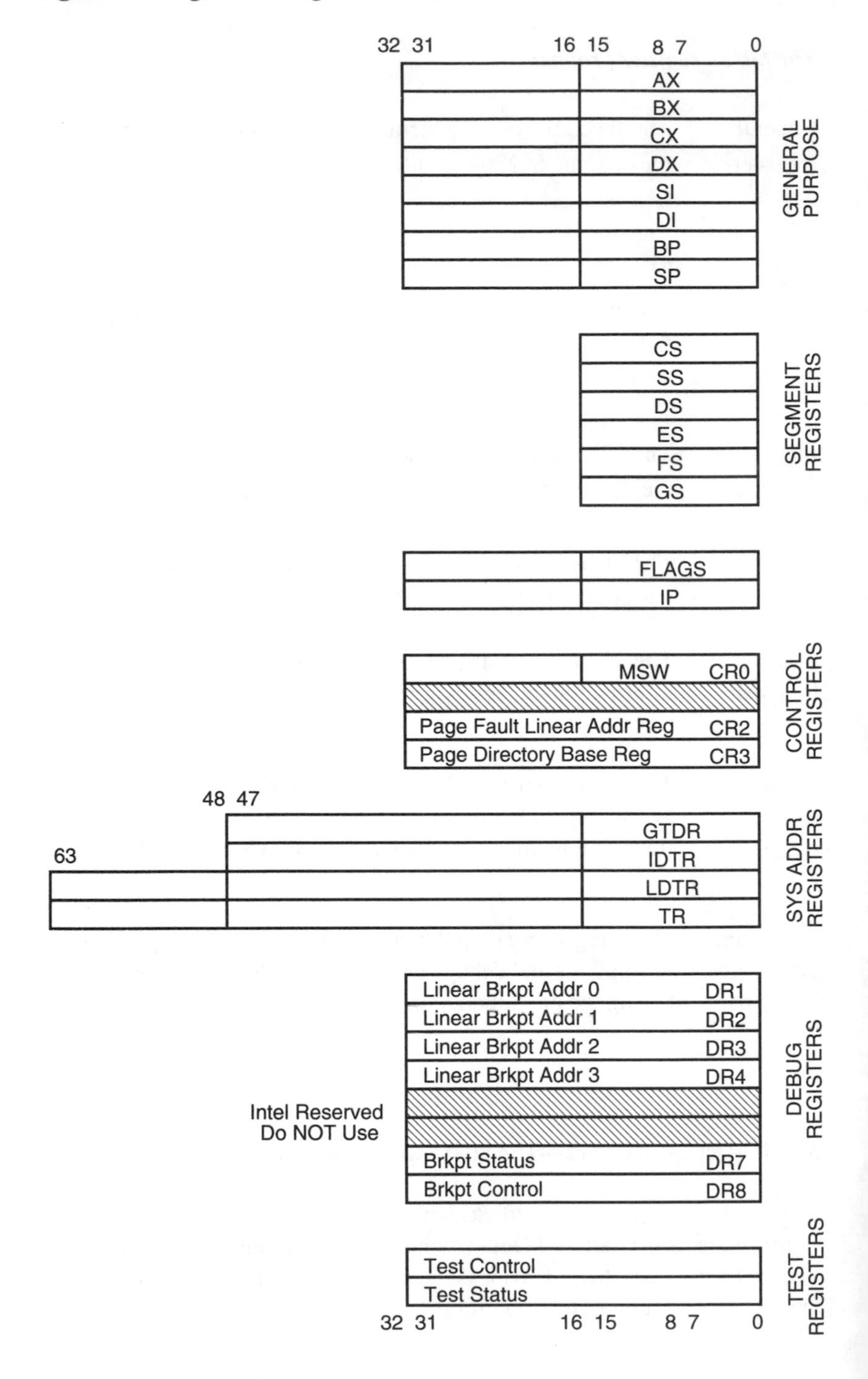

Figure 7.28 The Intel 80386NN 32-bit microprocessor

32-Bit Microprocessors—The 68020

The 680XX General Overview

The Motorola 68020 is a true 32-bit micro with full support of virtual memory and a virtual machine—it can address at least 2^{32} cells of memory. The addressing modes were also enhanced for improved support of high-level languages. This micro includes bit data fields that accelerate bit-oriented (mapped) graphics.

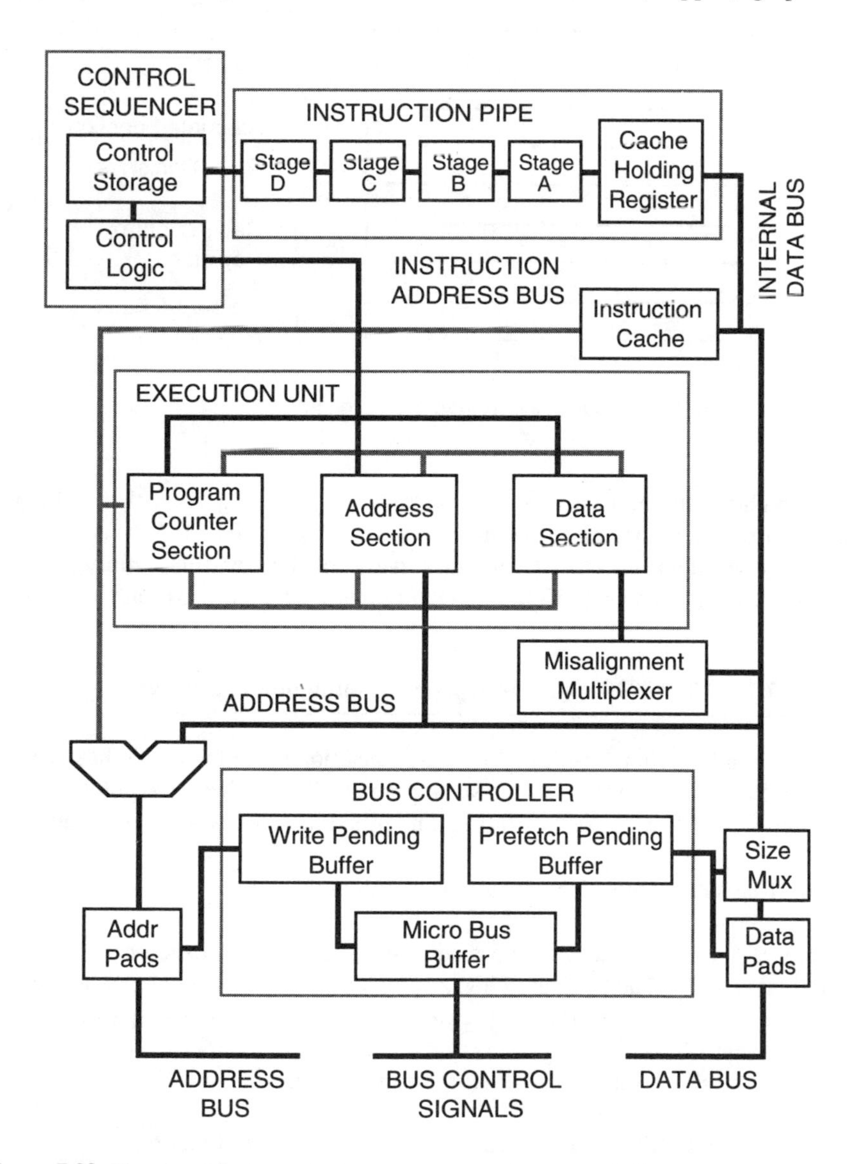

Figure 7.29 The simplified model of the operation paths in the 68020.

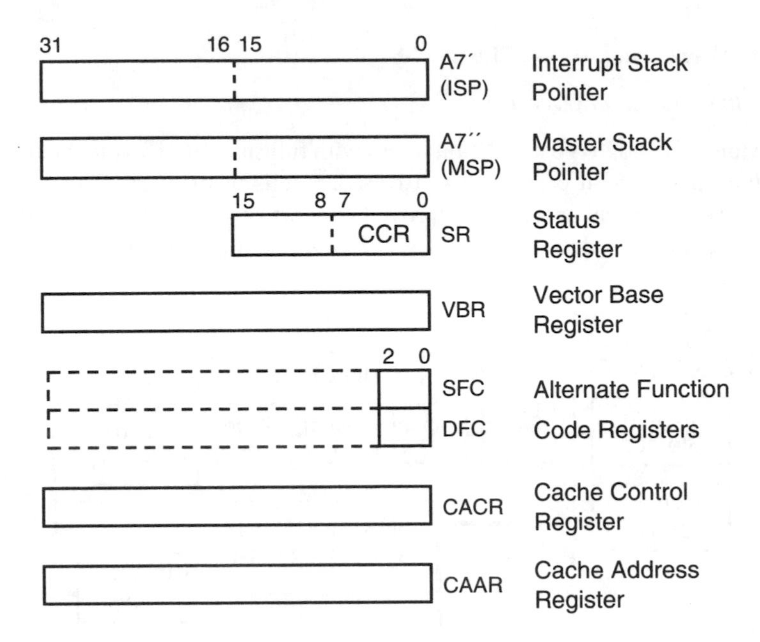

Figure 7.30 The 68020 supervisor programming model.

The on-chip instruction cache speeds instruction processing. Also, this 32-bit processor is designed to work with other true 32-bit chips such as an FPU and a Paged Memory Management Unit. Different versions of the 68020 can run with clock speeds ranging from 16 to 33 MHz. See Figures 7.29, 7.30, 7.31, and Table 7.13.

Table 7.13 The 68020 supervisor programming model registers.

Interrupt and Master Stack Pointer	In the 68020, the single stack pointer of the 68000 has been replaced by two stack pointers. The *active* stack pointer (be it ISP or MSP) is again termed the Supervisor Stack Pointer.
Status Register	Duplicates the use and form in the 68000.
Vector Base	Contains the base address of the exception vector table.
Alt Function	The SFC and DFC are 3-bit registers that act to extend the 32-bit linear addresses to eight 4-gigabyte address spaces.
Cache Control	Controls the on-chip instruction cache.
Cache Address	Stores the address of the Cache Control function.

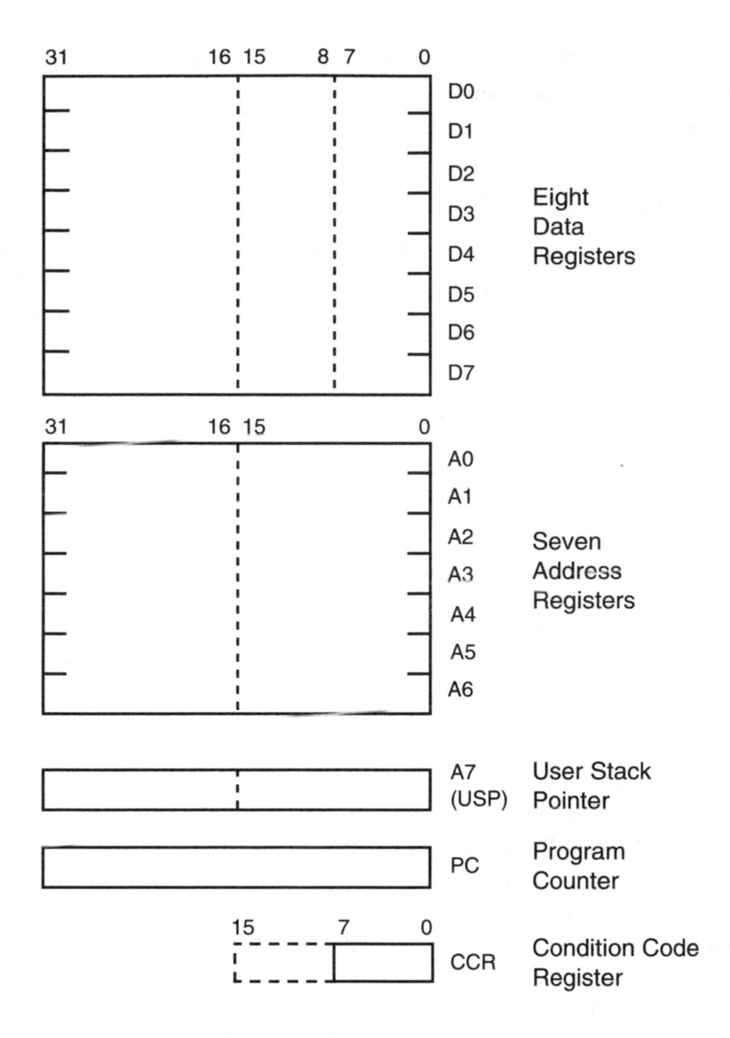

Figure 7.31 The 68020 user programming mode.

The 68020 user programming model registers:

Duplicates the use and form in the 68000.

The 68020 status registers:

Duplicates the use and form in the 68000 except for the addition of an M-bit for toggling between the Master and Interrupt state and a second T or trace bit to implement the extended trace (debugging) mode. The 68020 supports seven data types (64-bit Quad Words and 32-bit consecutive Fields are added) compared to five for the 68000. It also supports eighteen address modes compared to fourteen in the 68000.

32-Bit Microprocessors—The 68040

The 68040 follows the 66000 family pattern but again has several upgrades and innovations. The most significant is the incorporation of an on-chip 68881/68882-compatible FPU. Besides this there is a 68030-compatible Integer Execution Unit, independent instruction and data Memory Management Units, and expanded instruction and data caches.

7.5 MICROPROCESSORS AND MICROPROCESSOR SYSTEMS—THE CHAPTER IN RETROSPECT

An Introduction to Microprocessors

This rather long and detailed chapter provides the reader with an insight into the "why" as well as the "how" of the general design, uses, and especially the progression of the hardware, a little of the software, and the changing applications for microprocessors. Several pages and a number of different perspectives are used to attempt to answer the conceptual question, "What is a microprocessor?"

Using a somewhat genealogical approach, the micro is shown to advance from the Intel 4004, and its immediate offspring the 8008, through the Intel 80486 and the Motorola 68040. To provide continuity, and to always give the reader a source to common and readily available hardware and literature, the products of Intel and Motorola form the spinal column of the chapter. However, along the way, innovative units from a number of different venders (some of them no longer in business) are presented for insight and comparison. In most cases, the micros are divided into classes—such as, but not necessarily, 8-bit, 16-bit, etc.,—and a primary model for that class is first presented with its general design goals or performance, its architecture, and its input and output signals. As the micros, and thus their descriptions, became more complicated, such things as interrupts, masks, traps, etc., are described. In most cases, read and write timing diagrams are included.

Archival and Contemporary References

Gaonkar, R. S. *Microprocessor Architecture, Programming and Applications with the 8085/8085A*, (2d ed.) Columbus, OH: Merrill Publishing Co., 1984.

Hunter, Colin and Banning, John. "DOS at RISC." *Byte*, November 1989.

Intel, *Microprocessors,* Bellevue and Mt. Prospect: Intel, 1990.

———. *386SX Microprocessor Programmer's Reference Manual,* Bellevue and Mt. Prospect: Intel, 1990b.

———. *486SX Microprocessor, 487 Math Coprocessor.* Bellevue and Mt. Prospect: Intel, 1991.

Johnson, Thomas L. "The RISC/CISC Melting Pot." *Byte*, April 1987.

Larson, David G.; Rony, Peter R; and Titus, Jonathan A. *The Bugbook III— Micro Computer Interfacing.* Derby, CT: E & L Instruments, 1975.

Markowitz, M. C. "EDN's 18th Annual mP/mC Chip Directory." *EDN*, November. 21, 1991. See *EDN* Nov. 26, 1992.

Marshall, Trevor. "A Calculating RISC." *Byte*, May 1990. (This article introduces the idea of a RISC coprocessor.)

Motorola, *MC68000 Programmer's Reference Manual,* Phoenix, AZ: Motorola, 1989a.

———. *MC68020 32-Bit Virtual Memory Microprocessor.* Phoenix, AZ: Motorola, 1989b.

———. *MC68040 Third-Generation Microprocessor.* Phoenix, AZ: Motorola, 1989c.

———. *MC68000 Family Reference Manual.* Phoenix, AZ: Motorola, 1990a.

———. *MC68020 32-Bit Microprossor User's Manual,* Englewood Cliffs, NJ: Prentice Hall, 1990b.

Osborne, Adam. *An Introduction to Microcomputers, Volume II.* Berkeley,CA: Adam Osborne and Associates, 1976.

Patterson, D., and Hennessy, J. *Computer Architecture, A Quantitative Approach.* San Mateo, CA: Morgan Kaufman, 1989.

Robinson, Phillip. "How Much of a RISC." *Byte,* April 1987.

Titus, John. "Build the Mark 8 Minicomputer." *Radio-Electronics*, July 1974. (This was the article that started the personal computer.)

8

Networking—Transmitting and Processing Data between Computers

Networking Models

8.1 AN INTRODUCTION TO NETWORKING

In this chapter, some liberties will be taken with the definition of the word networking. Using a very broad perspective, networking can be modeled as [1], communicating and processing data inside a single microcomputer or mainframe (see Figure 8.1), [2] exchanging data between two like computers, [3] exchanging data between two unlike computers, [4] exchanging data and using the peripherals, such as printers or fax machines, from the other computer, and [5] exchanging data and the use of facilities of a number of both like and unlike computers. This last model can include both Local Area Networks (LANs) and Wide Area Networks (WANs).

Mechanical Data Processing—A Little Historical Background

The wide use of mechanical data and information communication may be visualized as nominally starting with Gutenberg's printing press (circa 1425). There was a multitude of discoveries and inventions that influenced and expanded this art, including the overall development in the wider fields of radio and TV. However, the three developments that were direct precursors to contemporary data communication and processing were Morse's telegraph (circa 1845), the perfection of the typewriter for mass production (circa 1875), and Baudot's teleprinter (circa 1875), which evolved into what we commonly know as a teletype.

Mechanical Data/Word Processing

Morse incorporated the technique of *coding* into electromechanical data transmission and processing. The processing, including the coding and decoding, was done by the human operator using a quasi 5-bit code for the alphanumerics. At first, only upper-case letters were available. The typewriter gave a very readable, uniform copy but did not contain many of the features we now find necessary in word processing. For many years, the only fonts were pica and elite with both upper- or lower-case letters. The exchangeable type-balls or cylinders, which permitted a wider choice of type styles, came later. The printed page was the only storage medium. The teleprinter used several clever mechanical devices to decode the dots and dashes and then either punched holes in a paper tape or actuated the appropriate letter arms, similar to those in a typewriter, to print on role-feed paper.

8.2 INTERNAL NETWORK PROCESSING—PASSING DATA BETWEEN COMPONENT PARTS

Internal Electrical Data/Word Processing—Keyboard Data

The information that a particular key has been actuated is usually recognized by the computer's CPU in one of two ways. One method is to regularly scan the keyboard input (port) looking for a circuit closure. When one is detected, a predetermined software process will establish which key it is. Further processing will then decode that particular key's position and, usually with a look-up table, send its unique alphanumeric code (usually ASCII) to the accumulator for further processing. A second, more contemporary method uses a key closure to produce an interrupt. The resulting interrupt's software routine will first determine which key this is. It will then decode the key's intended information and send the appropriate code to a register or the accumulator.

The PC (here meant to include the IBM PC and clones) recognizes and interprets a key closure with its Basic Input Output System (BIOS). The BIOS is the collection of routines (somewhat akin to the ToolBox in the Macintosh) that handles the housekeeping in the PC. There are about ten service routines (depending on the particular model) that read, interpret, and store the keyboard information for further processing. There are several character tables that service 83-, 84-, and 101/102-key keyboards as well as recognizing U.S. or foreign configurations.

The original Macintosh and the Mac Plus scanned the keyboard four times a second looking for a key closure. If there was a closure, the keyboard's internal chip would send a signal to the Mac's Versatile Interface Adapter (VIA). The VIA would in turn generate an interrupt and the CPU would retrieve the key code from the VIA. Also, all Macs and PCs contain some form of rollover. Consequently, if you press two keys at nearly the same time, both will be recognized in turn.

The keyboards for the later Macs are placed in series (daisy chained) with the mouse and other input devices on the Apple Desktop Bus (ADB). The ADB is a

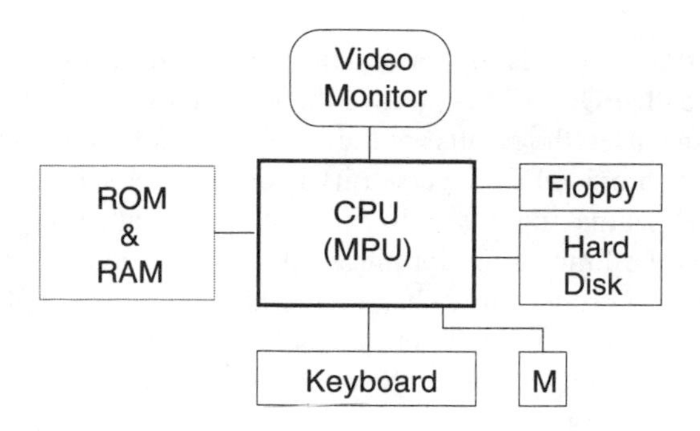

Figure 8.1 The CPU and peripherals using the internal network. The disk drives may be internal or external to the physical computer and may be connected with several types of networks including an SCSI port. The mouse may be connected to a communication port or may be daisy chained in serial with the keyboard (as in the Macintosh).

two-way, single-master, multiple-slave bus. When the devices on the bus wish to input, they send a coded request down the bus and the CPU recognizes the device and returns a code that inputs the data (the key code, mouse pointer's relative position, graphics tablet pointer's position, etc.).

Internal Electrical Data/Word Processing—Video Monitor Data

The PC again uses the omnipresent BIOS, together with a Video Adapter, to send alphanumeric and graphic information to the monitor. The acronym and definition for some of these video adapters are:

MDA	Monochrome Display Adapter—A mono monitor for text-mode displays of 80 characters per line with 25 lines.
HERC	A third party Hercules monochrome adapter card that would support 720 × 350 graphics but was not supported by ROM BIOS.
CGA	Color Graphics Adapter—a low-resolution color monitor with a resolution of 640 columns × 200 lines of pixels per screen. Simple graphics are produced by "connecting the dots." As a comparison, laser printer usually has a resolution of 300 × 300 dots per inch (dpi).
EGA	Enhanced Graphic Adapter—a screen resolution of 640 × 350.
VGA	Video Graphics Array—a screen resolution of 640 × 480.
Super VGA	Video Graphics Array—a screen resolution of 640 × 600.
Paper White	Mono monitors and display adapters with resolutions of 1024 × 960 to 2048 × 1536.

Conceptually, the original IBM PC printed graphics to the video screen much as a typewriter prints a character on paper. The screen resolution was a function of the monitor and the video adapter card. The Macintosh, however, by using a ToolBox routine called QuickDraw, draws the symbol of each character on the screen. To the casual user, the results are the same. However, the Mac's symbolic technique originally provided more versatility in the number and size of fonts that can be used as well as making it easy to produce a wide range of graphics. One major drawback is the limit of 72 dpi placed on most Macintosh monitors. For comparison, most laser printers print characters with a resolution of 300 to 600 dpi and textbook characters are printed with a resolution of up to 2,400 dpi. (Note: Chapter 3 gives more of the details on the operation and problems of both monochrome and color monitors.)

Internal Electrical Data/Word Processing—Disk Storage

The technology and applications of both the so-called floppy and hard disk drives have had a wonderful metamorphosis since the introduction of the PC and the Macintosh a little over a decade ago. The floppy disks have evolved from single sided to double sided and from low density to double density and then to high density. The maximum capacity of a 5-1/4" diskette has gone from the original 160K to 1,200K (1.2 Mb) and the capacity of the 3-1/2" version has expanded from 360K to 1.44 Mb.

Like other peripherals, reading and writing to and from these disks depends on the proper combination of hardware and software. The disk grooves or tracks do not spiral as do those on a phonograph record, but are concentric. The tracks on the floppy disks are divided into pie-wedge pieces called sectors. Each track section (sector arc) will contain a like amount of data ranging from 128 to 512 (8-bit) bytes depending upon the vintage of the drive. Different models of the PC have contained 9, 15, and 18 sectors per one concentric track. Most of the older PC data is on 5-1/4" diskettes—3-1/2" PC disks came later.

The drives for the floppy disks closely resemble those of the audio Compact Disk presented in Chapter 1. The floppy disk is motor driven and rotates at 360 RPM. The double-sided disk is accessed by a read/write head on each side. Each head is driven by a stepping motor that moves it from track to track to access the data sectors. Erase/write is accomplished by the head somewhat like that of a magnetic tape recorder. In the read mode, the head senses the charges in magnetic flux that are then amplified and converted to coded 1s or 0s. In the write mode, the head magnetizes the track in a manner corresponding to the source digital code. Each disk drive requires a disk controller. The controller must decode the disk address information and move the head to the proper track and sector. Each disk has a certain amount (a few percentage) of space set aside for index/address information. When a file is stored on a disk, its name, time and date of creation, and the size and the location of its first sector are all stored in the disk directory. In addition, a File Allocation Table (FAT) is created that shows the sectors belonging to each file and the unused sectors. When a new file is written or a file is erased, the disk controller interrogates the disk directory and the FAT to determine if space is available, where it is, or what name should be deleted. (When a file is supposedly completely erased, actually only its name is erased. This is why, sometimes, an erased file that has not yet been overwritten can be reclaimed.)

Internal Electrical Data/Word Processing—Hard (Fixed) Disks

The fundamental principles used in the read/write of hard disks are similar to those of floppies but the implementation differs somewhat. A given hard disk will, in fact, contain several platters or disks. Likewise, each separate disk will be accessed by several read/write heads on each side. These movable heads do not contact the disk surface as they do with floppies. Whereas floppy disks spin precisely at 360 RPM, hard disks usually spin at 3,600 RPM. The combined tracks on each disk of heads in the same position are termed a cylinder. Hard disks and their disk controllers are designed to transfer data between RAM and the platters as fast as possible. Data is usually transmitted using DMA. The BIOS in the PC and the ToolBox in the Mac have special service routines for hard disks.

8.3 EXTERNAL NETWORK PROCESSING—PASSING SIMPLE DATA AND INSTRUCTIONS TO THE OUTSIDE WORLD

External Electrical Data/Word Processing—Using a Printer

A printer is usually the first purchase after the basic PC. Many of the printers for both the early mainframe and, later, PCs were teletypes. The introduction of the wire pin (dot matrix) printers usually gave a higher print speed and permitted the use of low- to medium-resolution graphics. The original ASCII character set supported only very limited graphics so with the introduction of the matrix printers came the introduction of new graphics printing techniques. With a teletype-style printer, the code of the alphanumeric character was usually sent from the microprocessor's accumulator through an output port to actuate the printing mechanism (usually called hammers). However, the additional complexity of the multiwire print head required additional processing. Teletype-style printers most often used serial data input. Dot-matrix printers increased the data flow speed by going to multiwire parallel inputs. This parallel input scheme is commonly termed the Centronix input. However, different wire-head printers with a different number of wire pins and different printing schemes soon led to the use of software print drivers.

The introduction and rapid acceptance of laser printers produced more changes. Most often they use parallel inputs for PCs and serial (the AppleTalk network) for the Macintosh. The introduction of page-description languages (PDLs) required additional hardware and software "smarts" in both the computer and the printer. A PDL such as PostScript must calculate the precise style and size for the printed bit pattern from a catalog of user selected stored font outlines. Not only is the printer's computer required to calculate these patterns for the particular characters involved (including graphics), it must continually keep track of the details of the layout for the entire page. In addition to the style and size of the font, the layout housekeeping includes continually calculating the correct line width, the tracking (the space between letters), the line leading (the space between lines), the paragraph parameters including alignment, indents, and tabs and, if needed, a multitude of color settings.

NOTE: In each of the following external networking topics, a model commercial program is used to show a practical example of the principles and techniques involved. These programs, as true models, are meant to show one example of the application and no special recommendation of these programs, either directly or indirectly, is implied. The model programs are presented so the reader will have a workable starting point for study and comparison.

External Electrical Data/Word Processing—Exchanging Data between Two Like Computers

The most elementary modeling approach to computer networking is the paradigm of two very similar computers communicating over a simple two-wire data path (link). There are a number of software programs, such as LAP-LINK, that can be used to effectively exchange data. LAP-LINK makes it simple and convenient to exchange the contents of data files and directories between a desktop PC and a laptop (or another desktop unit). Both computers must have the LAP-LINK program open and there must be a special null-modem connection between the two computers. This particular program model presents the data-file listings as a two-column display (two side-by-side windows). On the left is the file listing for the

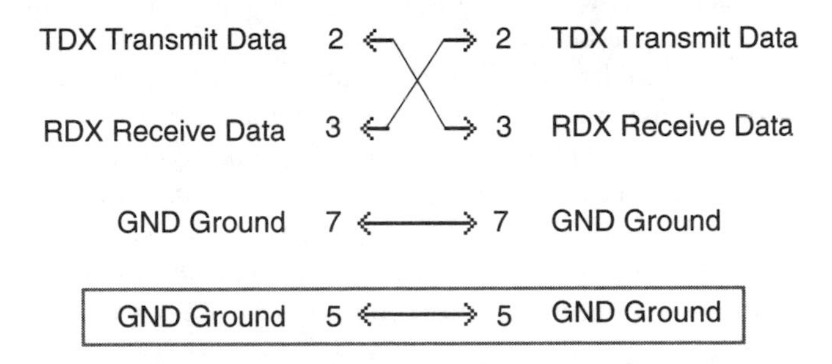

Figure 8.2 The wiring diagram for the usual null modem. Note that the two signal wires cross over so that the output of one computer is always the input of the other computer. Pins 2, 3, and 7 are correct if the connector is a 25-pin DB25. If one or both of the connectors are a 9-pin DB9, the signal pins are still 2 and 3 but the GND must be connected to pin 5 (shown dotted) instead of pin 7.

local computer and on the right is the file listing for the remote computer. (Since this program is truly bidirectional, the local computer, and thus the local listing window, is always the one you are using.)

The options of LAP-LINK make it possible to select the proper signal port (COM1, COM2, etc.), the exchange baud rate (up to 115,200), the disk drives, and various directories—you can choose and manipulate your directory and the remote directory. There are several file management features included such as erase, name, copy, wild-card copy, and the facility to temporarily return to DOS. It is possible to obtain a tree diagram with something like DOS 5.

Linking programs such as LAP-LINK are designed to link two like computers with similar operating systems. They are file processing programs and cannot utilize any of the peripheral printers, faxes, or modems of the other computer.

External Electrical Data/Word Processing—Exchanging Data between Two Unlike Computers

There are also linking programs that make it possible to link unlike computers, such as a PC and a Mac, and to transfer their unlike files to the other computer. MacLinkPlus/PC (ML) is used for the program model to illustrate the software and techniques required to transfer and use files for differing computers. ML can be configured [1] for disk swapping between the unlike computers, [2] to use a directly connected null modem cable as shown in Figure 8.2, or [3] to use modems (see next section) to transfer the data using telephone circuits.

The process of disk swapping requires special software and a compatible disk drive in both computers. Since the formats of the PC and Mac disks are different, normally a foreign disk will not be recognized if it is inserted in the other computer. However, programs like Apple File Exchange or Mac-LinkPlus/PC, provide special software recognition routines. After the disk is recognized and the files are available to the host computer's system, a second part of the special software must be used to translate these files into a suitable format in the host computer. The software translators provide different degrees of translated compatibility. For instance, if the source documents (files) are produced with an uncommon program, the translation will probably produce the proper alphanumeric characters but the formatting may be incorrect. However, if the source file was written in a common program such as WORD for the PC, it will translate quite well to WORD for the Mac or vice versa. DATAVIZ, the publishers of Mac-LinkPlus/PC, also include several file translators for PC to Mac and Mac to PC. After one has experimented with several types of files and their possible translators, the process becomes more graceful and accurate. This disk swapping is the simplest and fastest way to learn the translators and their idiosyncrasies. Most programs like ML include versatile cables that usually make linking the unlike machines quite easy. For using a modem, ML includes its own communications subprogram that gives the user the ability to set the usual items such as the desired port, the baud rate, the password, etc.

Networking with Modems

8.4 EXTERNAL NETWORK PROCESSING—COMPUTER-TO-COMPUTER COMMUNICATION WITH MODEMS

This section starts a series of topics dealing with communications between computers that are truly remote and generally use transmission mediums other than a direct computer-to-computer cable. The modem (a contraction of modulator-demodulator) is one of the most convenient devices for utilizing external paths such as telephone circuits.

Modem Communication—Tone Modulation

The early data-transmission paths usually were continuous wire circuits. Except for normal line losses, the same voltage or current amplitude, including a zero value, could be sensed anywhere along the network. A logic one (1) could be some plus (or minus) value and a logic zero (0) could truly be a zero value. (The terms one and zero are used in this work although in the past a high and a low, or a mark and a space, roughly indicated the same things.) Even if a true zero value was not part of the code, it was very important to be able to sense plus or minus DC values. However, most telephone circuits do not carry the DC values. Telephone circuits use amplifiers having coupling capacitors and transformers, fiber optics, and satellites, all of which remove the original DC or zero component. A technique was needed to ensure that, regardless of the original level values of the coded 1s and 0s, these codes could be transformed and transmitted accurately and, a moment later, detected and properly decoded—across town or later across space. Enter the modem.

The modem operates by modulating, and later demodulating, a high-frequency carrier in a special way. FSK or Frequency Shift Keying is one common type of modulation used in modems. The carrier is switched to one frequency for a 1 and to another frequency for a 0. These modulated signals are tones, in the audible frequency range, and are reminiscent of the familiar audible telephone dialing tones. These data tones are produced in sets—one tone for a 1 and a second for a 0. When your local modem is set in the originate mode, the most common setting, the modulator will usually produce tones of 1,070 and 1,270 Hz. If the remote modem, such as a bulletin board, is set in the answer mode, it will usually send a tone set of 2,025 and 2,225 Hz. For proper communication, each modem must be able to detect and demodulate the input signals of the other modem.

Modern modems usually come with the rear panel equipped with a 25-pin D connector for the RS-232 serial-data computer interface, a J11 modular jack for the telephone line, and a second J11 (a series loop-through) that can be used to reconnect a telephone on that line. The modem should contain a speaker to monitor the connecting and, if desired, the communication signals. There is often a volume control for the speaker. The front panel of most contemporary modems contains

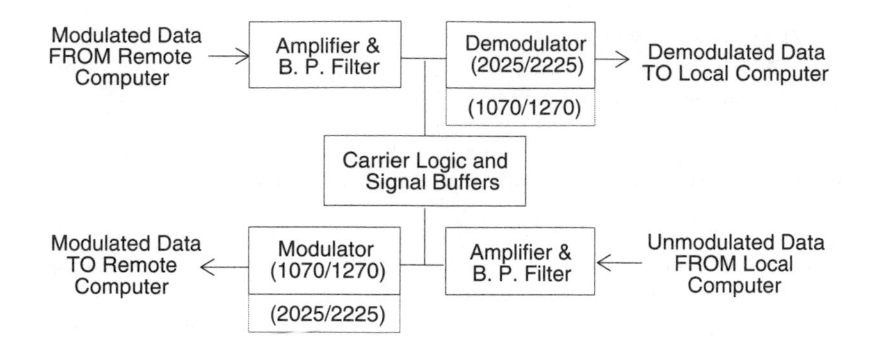

Figure 8.3 A block diagram of an elementary modem set for originate only.

several LED status indicators. At first these are just more flickering lights but they soon become a much-used aid in monitoring and troubleshooting the communication exchange.

In the Originate Only mode, the modem can receive and demodulate the incoming 2,025 and 2,225 Hz Answer Only remote signals to produce the 1s and 0s for the computer. Likewise, the local or originate 1s and 0s would be sent from the computer to the modem to produce modulated output frequencies of 1,070 and 1,270 Hz. If the modem had been set for Answer Only, the appropriate send and receive signal frequencies would be exchanged and are shown in the dotted boxes in Figure 8.3. The carrier-sensing logic and the signal buffers prevent collisions between the transmitted and received signals.

The nomenclature for the modem status lights is shown in Table 8.1:

Table 8.1 The Abbreviations and Explanation for the Modem Status Indicating LEDs

HS (High Speed)	Lights when in the 1,200 or 2,400 baud mode. (not on all units)
AA (Auto Answer)	Lights when modem is set to Auto Answer mode for incoming calls. Part of the optional setup—see any particular modem's instruction book.
CD (Carrier Detect)	Lights when the modem receives a carrier tone from a remote modem.
OH (Off Hook)	Indicates that the modem is drawing DC current from the telephone line as an off-the-hook telephone would.
RD (Receive Data)	Light flashes as data bits are received.
SD (Send Data)	Light flashes as data bits are sent. It will flash each time you press a key.
TR (Terminal Ready)	The terminal is ready to dial out or receive data.
MR (Modem Ready)	The modem is ON and ready to send or receive data.

NOTE: When a working modem is first switched on, MR and HS (if available) will light. In addition, when there is an operating communications program loaded in your computer, TR will also light.

Modem Communication—Dumb and Smart Modems

The early modems (circa 1970) did not connect their terminals directly to the telephone line but contained a small loudspeaker and a microphone to send or pick up signals to and from a telephone handset. The handset was placed in special rubber cups that formed two semi-noise-free acoustical couplers. After the breakup of the Bell System in 1984 and the attendant changes in the FCC rules, it was permissible to connect directly to the lines. The internal circuit configurations of the early modems resembled those shown in Figure. 8.3. To establish a connection, one manually dialed the appropriate number in a normal manner and, when a dial tone was heard, quickly installed the handset in the acoustical couplers, and hit RETURN on the terminal. In response, the terminal would print out or display a short message such as CONNECTED.

The advent of mini and personal computers, rather than dedicated dumb terminals (with little or no processing or error-checking capability), brought the communications driver programs and, later, modems with internal microprocessors that were termed intelligent or smart. The first driver programs would allow one to use the computer's keyboard to set the baud rate and the parity and stop bits. It also allowed you to enter the desired telephone number and have it automatically dialed. A typical entry would be:

ATDn 123-4567	Where n would be the letter P for Pulse dialing or T for Tone dialing.

Soon after the introduction of the communication driver program came the facility to store several numbers and other data for automatic dialing, which later became known as the phone book.

In the foregoing dialing example, the first two letters, AT, stood for the command attention. They were (and still are) included in a group of commands that form the AT Command Set. These AT commands (having nothing to do with the IBM Model AT), which are given in most modem instruction books, permit unique dialing instructions, allow special modem configurations, and initiate modem checkout procedures. The use of this AT command set, in a very special way, is synonymous with the concept of the smart modem.

For a simplified definition, a smart modem is a modem that can send or decode the AT command set with software commands rather than needing hard-wired controlling connections.

Since the AT software-generated commands are in the data stream, several of the hard-wired connections are not necessarily needed with smart modems. However, most computer-to-modem cables include all of the normal (old) RS-232C or RS-232D wires. (The D denotes the 25-pin D connector.)

To be entirely accurate, RS-232 is an older form of the EIA (Electronics Industry Association) standard for serial communication which we know as RS-232C. RS stands for Recommended Standard and the symbols 232, 232A, 232B, and the contemporary 232C and 232D, are different historical versions of this EIA standard.

Referring to Figure 8.4, the top dumb-modem cable connection requires eight wires plus the common to exchange the data and control signals. However, with software encoding for the AT control signals, the bottom smart-modem cable theoretically requires only two wires plus a common (if used) although it may also physically contain eight wires. The "to DCE" and "to DTE" notations denote the signal directions.

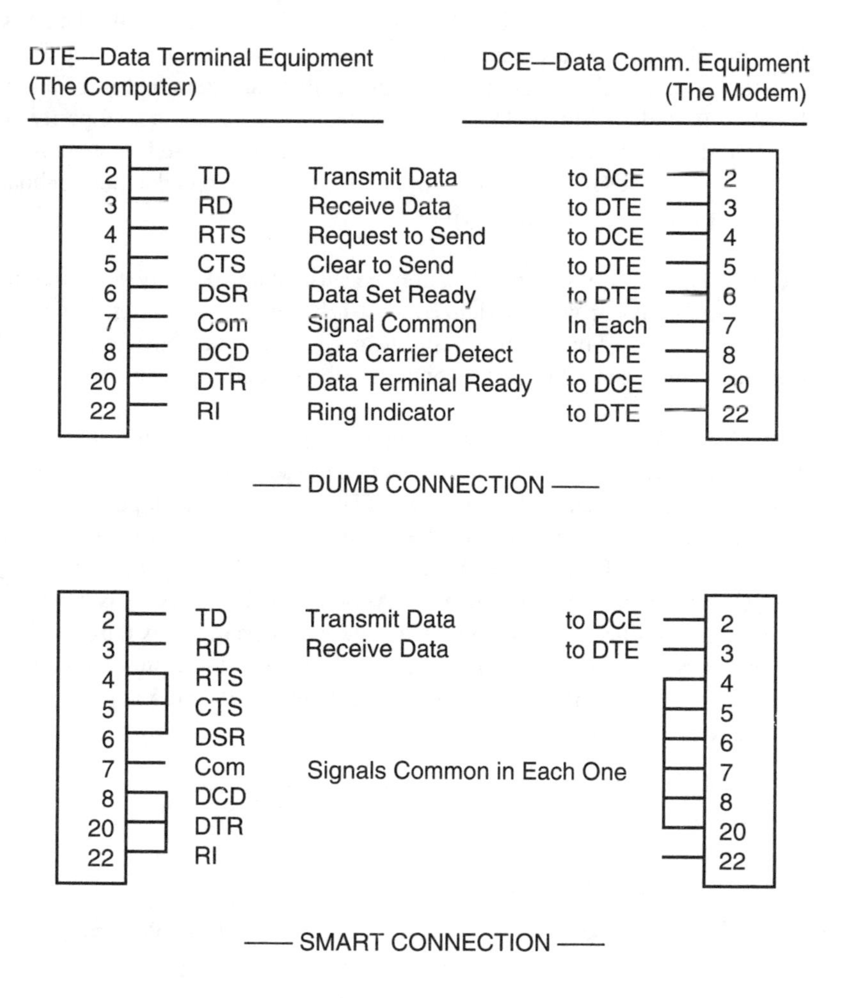

Figure 8.4 The RS-232 connections between a computer and a dumb modem and a smart modem.

Modem Communication—Error Checking

The movement of data from one location or medium to another is not completely error free. When the transmission medium is telephone lines, the error rate may increase dramatically because of the electrostatic and electromagnetic noise pollution in our industrial world. The recognition that there is a definite possibility of transmission errors leads to these two questions: first, how can these errors be detected and, second, how can some or all of these errors be corrected?

There are many data communications error-checking algorithms in use—several of them are for special computer situations and are proprietary. The two most common are character parity and checksum. Both of these have several variations.

Referring to Figure 8.5, each transmission will begin with a start or "wake-up" bit followed by seven data bits if parity (correction) is chosen or eight data bits if parity is not chosen. The user can also choose to provide one or two stop bits. In the figure, the diagonal lines are used to indicate that each data bit can be a 1 or a 0. A common modem setup is 1,200 or 2,400 baud, eight data bits, no parity, and one stop bit. If you cannot see what you are typing (sending), use HALF duplex. If you see double (DDOOUUBBLLEE), use FULL duplex. To call a bulletin board, you would most often use FULL duplex.

NOTE: This form of Figure 8.5 is used because that is the way converted parallel data would appear when serialized in a shift register. However, this is conceptually backwards. For actual serial transmission, the start bit is sent first, the first or least significant bit (LSB) second, then the second LSB, etc.

Parity is usually considered a system that checks each character as it is transmitted. In essence, a special code may be added to each character as it is transmitted and this code is checked at the receiver. (There are also block parity checking schemes where one or more bits are added to a block [128 or more bytes] of data for parity checking.) If the received code is found to be in error, the receiver can ask the sender to retransmit that character. Although parity checking is quite accurate, it requires additional processing time that can materially slow the overall rate of transmission. Likewise, the parity bit cannot be used if it is desired to send 8-bit characters such as graphics. The parity bit is derived to fulfill the requirements of one of the following parity algorithm rules:

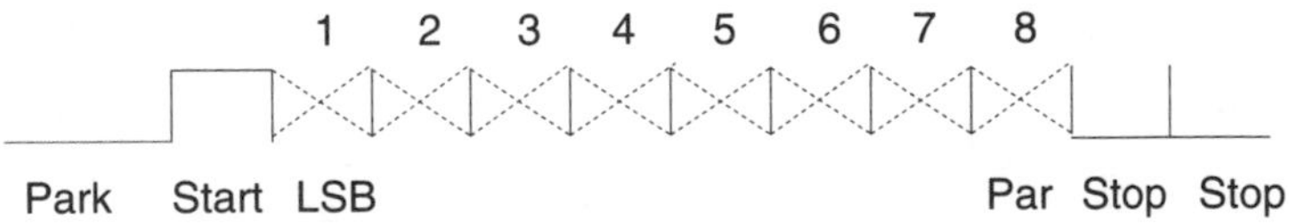

Figure 8.5 The prototype configuration for RS-232-type serial data transmission.

Even Parity — *The sum of the number of ones (1s), including any parity bit, shall be even.* Thus, the sum of the 1s in the code 1101001 is even and no 1 in the parity position is required. However, for the code 1101000, the character sum is odd and a 1 is required in the parity position to make the total sum of all the 1s, including the parity bit, even.

Odd Parity — *The sum of the number of ones (1s) shall be odd.* The sum of the 1s in the character code 1100100 is odd and no parity bit is required. However, for the code 1110100, a 1 in the parity position is required to make the sum of all 1s, including the parity bit, odd.

When parity is used, the sending computer calculates the requirement for a parity bit and, if needed, adds it to the data stream in the parity position. The receiving computer, which has been preset for the same type of parity as the sender, analyzes the data stream, including the parity bit, and determines if there is an error. If it finds an error, it requests that the sending computer retransmit the character. There are usually constraints upon the number of retransmit requests.

A second general error-checking technique known as checksum is as accurate (sometimes more accurate) as character parity and steals much less transmission time. There are many checksum algorithms, but in their simplest form the total of the 1s in a block of data (128 or more bytes) is obtained, complemented, and added to the original data block for transmission. The receiver again counts the 1s in the original data block and adds it to the complemented data byte. If the sum of the calculated value of the 1s and the complemented byte is zero, the receiver knows that block is error free and thus signals the sender to transmit another block of data. It should be noted that when the sender originally counted the 1s, the total may have accumulated to over eight bits. If this was so, the overflow was discarded and only the low 8-bit (byte) was transmitted. This checksum scheme is termed cyclic redundancy checking or CRC.

Modem Communication—Transfer Protocols

Modem communication, like human communication, goes more smoothly if there is some agreed-upon etiquette involved. Thus, the exchange of data between computers is governed by a series of rule and procedure structures that are designed to permit rapid, low error rate data communication. In fact, the two major questions that should be asked are: first, what is the relative speed efficiency of a given protocol and, second, what is the error rate? The speed efficiency has to do with the overhead (nonproductive transmission time) built into the particular data transmission system. For a simple example, refer to Figure 8.5. Assuming a setup for eight data bits (no parity) with one start bit and one stop bit, ten time spaces must be used to transmit eight data bits—20 percent of the transmission time is overhead. If there are errors, additional overhead time is required.

Kermit Protocol

Kermit (yes, legend says it was named after the frog) is perhaps the oldest and best known data-transfer protocol. It was created at Columbia University in 1981 as a versatile system that could be used by all their various computers, both large and small, including their time-sharing facilities. It is very complex, both in concept and in implementation, and contains such niceties as being able to work with IBM's Extended Binary-Coded-Decimal Interchange Code (EBCDIC) and many operating systems including the (then popular) CP/M operating system.

Transfers using Kermit contain data packets that are conceptually similar to data blocks. The different fields in the data packet not only transmit the chunks of data, but aid in the error-detection process. It is this technique of error correcting the packets, rather than each individual character, that improves the speed efficiency of most protocols. It must again be emphasized that Kermit is a very complex protocol and is used primarily where several different mainframe systems, as well as smaller computer systems and time-sharing terminals, are involved. Figure 8.6 gives a brief outline of Kermit's make-up and many functions.

XMODEM Protocol

XMODEM, like Columbia University's Kermit, was one of the pioneering data-transfer protocols. Kermit and XMODEM (written in 1977 by Ward Christensen) are in the public domain for noncommercial users. XMODEM has been changed throughout the years in many ways by many people. One of the significant upgrades

MARK	char LEN	char SEQ	TYPE	DATA	CHECK

MARK	This first character is the only nonprintable ASCII character (ASCII 32-126) in any of the fields. This MARK or SOH (start of header) control character (ASCII 1) is permitted because it is recognized in most operating systems.
LEN	The total number (length) of ASCII characters in the rest of the packet—not to exceed 94.
SEQ	The packet sequence number that helps the receiver do its bookkeeping. This number goes from 0 to 63 (decimal) and starts again.
TYPE	A single, literal unencoded ASCII character that defines the type of the following DATA, such as D for data, A for acknowledge, etc.
DATA	The main reason for this packet. Its type must conform to the TYPE nomenclature.
CHECK	The check (checksum) value. (See the Campbell and the de Cruz and Catchings references.)

Figure 8.6 The six data fields that make up the Kermit packet.

SOH	SEQ No.	1s Comp Seq No.	DATA 128 bytes	Check-sum

SOH	Start of Header byte
SEQ No.	One of 256 possible numbers starting with 1.
Comp.	The 1s complement, for ease in a receiver calculation, of the SEQ No.
DATA	This field must be 128 bytes but the bytes can be text or binary.
Checksum	A 1-byte arithmetic sum of the contents of the DATA field.

Figure 8.7 The five fields of the XMODEM packet. All fields except DATA are one byte

changed it to permit multiple file transfers—as Kermit incorporates. It is really the basic generic microcomputer file-transfer protocol. From this solid substructure have come several interesting and very useful extensions.

XMODEM-CRC Protocol

Similar to the generic XMODEM but incorporating the CRC error-checking scheme.

XMODEM-1K Protocol

The DATA byte is expanded to 1,024 bits.

YMODEM and YMODEM-1K Protocol

YMODEM made it possible to send multiple files and YMODEM-1K expanded the DATA block to 1,024 bytes. It is wise to review carefully the instruction book for any particular version of YMODEM to understand the way this protocol appends names to the transmitted files.

YMODEM-G Protocol

This version does not do any error correction; therefore it is faster than the other version. When using YMODEM-G, the transmission lines should be *very* clean or the modem should incorporate its own error-correcting mechanism such as MNP, LAP-B, or X.PC. Again, know what any particular modem does and does not do. (Note: Some modem software, such as Flash by White Knight, have special, very fast transfer schemes that rely upon the error correction in the modem.)

ZMODEM—The (Nearly) Ideal File-Transfer Protocol

This public-domain protocol has most of the features anyone would want. It is fast, almost as fast as YMODEM-G, and also offers error correction. ZMODEM sends data in 512 or 1,024 (8-bit) byte blocks without waiting for acknowledgments (ACKs) but it does look for not acknowledgment (NAK) after each block transmission. It also has an excellent interrupt (crash) recovery feature.

XON/XOFF—A Data Flow Protocol

Computers use a special section of memory as a buffer to aid in file transfers. Often, especially in mini and mainframe computers, the buffers of the transmit or receive computer will tend to become overloaded if the data flow is not stopped until further processing can take place. When this happens, if the XON/XOFF protocol is selected, the appropriate computer will sent the XOFF signal (CTRL-S), which will stop the data flow until the processing catches up and the buffer is emptied. At that time, the computer will send an XON (CTRL-Q), which is a request for the other computer to resume data transmission. Although most personal computers do not routinely use XON/XOFF, it is handy if one wishes to interrupt transmission to temporarily read the incoming text on the screen.

MacBinary

Macintosh disk files are configured differently than other microcomputers because of the way the Mac uses such graphics as icons, dialog boxes, and other types of pictures. A Macintosh disk file contains a data fork for text data, a resource fork for the graphic information, and finder information for indexing. By agreement among Macintosh communication software developers, these three types of data are formatted into a format called MacBinary. Not only is this a very acceptable way to exchange data files between Macintosh computers, but these files can also be received and, if need be, used or retransmitted by PCs without any loss of formatting or data.

Elementary Local Area Networks

8.5 EXTERNAL NETWORK PROCESSING—COMPUTER-TO-COMPUTER COMMUNICATION WITH LOCAL AREA NETWORKS

An Introduction to Computer Networks

This section starts the description of data transmission and processing using real networks. It will start with networks of like computers and then progress to the more difficult task of configuring systems that will allow unlike computers to talk to each other. Most networks can be divided into two general categories—those that are used primarily for file exchanges and those that are used for file exchanges and for sharing peripherals such as a printer, fax, or modem. In addition, many networks include some form of electronic mail—e-mail. Most networks require one dedicated computer to be an exclusive, dedicated file server. This usually means that the server computer and its hard disk are the major repository for the data files used by all the rest of the users. If the network is to contain ten or more computers, this may be a wise utilization of resources—if the network would only have five computers, it may not make sense.

Choosing a network for your computers is quite simple in theory but usually quite complex in practice. In theory, all you must do is [1] choose the type of hardware cablings you require (seemingly only a speed versus cost trade-off) and [2] choose the software network operating system to complement your cabling choice. When a practical network is considered, the problems of cost, speed, physical installation, as well as network management and maintenance, must be considered. Each computer may require an interface card, costing in the order of $250; a network connecting box, costing $50–$100; and the interconnecting cabling. The cost of the cable will vary from a few cents per foot for a twisted pair to over a dollar a foot for coax or fiber-optic cable—the higher the data transfer speed, the higher the cost. For example, the Macintosh AppleTalk network is inexpensive and convenient but relatively slow—230 Kbps (230,400 bits per second)—whereas Ethernet transfers data at a speed of 10 Mbps. A simple (up to twenty-five computers) LAN usually will not require cabling to be run in the walls or need special junction/utility rooms. However, the costs of planning, installation, and maintenance increase rapidly as the size and complexity of the network increase.

Protocols, Bridges, Routers, and Gateways

Each type of network software will incorporate a protocol—the specific rules for data transmission and reception. This protocol will usually include procedures for error correction and for the prevention of message interference (sometimes termed crashing) as indicated in Figure 8.8. There may be extra software for the management and maintenance of the network. If an area needs more than one network, but still requires that most or all of the computers (including the server) in network "A" be able to talk to the computers in network "B," you may need a network-to-network interface device termed a bridge. A bridge interfaces hardware to dissimilar hardware. For instance, a PC can be used as a bridge (with the appropriate software) to connect a network using coaxial cable to another network using twisted wires. If there are two networks, each with differing protocols, a router that knows both protocols must be used to interface the data between the networks. A router interfaces between dissimilar protocols. Sophisticated computer networks use interfacing devices called gateways. A gateway is a protocol converter.

An Introduction to Computer Networks—Cabling and Frames

Cables and other parts of the network hardware can be classified by performance, i.e., by data speed, by cost, and by the convenience of installation and maintenance.

Sender ID	Dest. ID	Frame Type	Data (Message)	Error (CRC)

Sender ID	The address of the sender.
Dest. ID	The address of the message destination.
Frame Type	An identification for the contents of the frame.
Data	The actual message being sent.
Error	The bits used for error correction. The correction scheme is usually CRC.

Figure 8.8 An elementary prototype for the message frames used in networking protocols.

The three most common cable types are unshielded twisted pair, coax, and fiber optics. Most of the recommended networking cables are defined and specified by IBM or other computer manufacturers. For instance, there is the recommended AppleTalk/LocalTalk cable, the AT&T Premises Distribution System recommendations, and Digital's DECconnect cabling. Likewise, the software, and its protocol for each network system, is usually defined by its rudimentary communication packet or frame.

Networking Models: The Macintosh AppleTalk—A Built-In LAN

Most recent Macintosh microcomputers have the AppleTalk LAN hardware built in. Likewise, some of the peripherals, especially the Apple (and several other manufacturers) laser printers also include AppleTalk hardware. Thus, to use a Macintosh with a laser printer, the only other hardware that is required is a special AppleTalk cable. If this simple network is expanded to include more than the one Macintosh, a special AppleTalk connector box with its appropriate cabling must be added for each new Mac, as in Figure 8.9. Each task that the AppleTalk network is to perform requires some appropriate software. If the network uses a laser printer, each computer must contain the correct print-driver software. If the network is to also provide file sharing, there must be appropriate software to facilitate the file transfers.

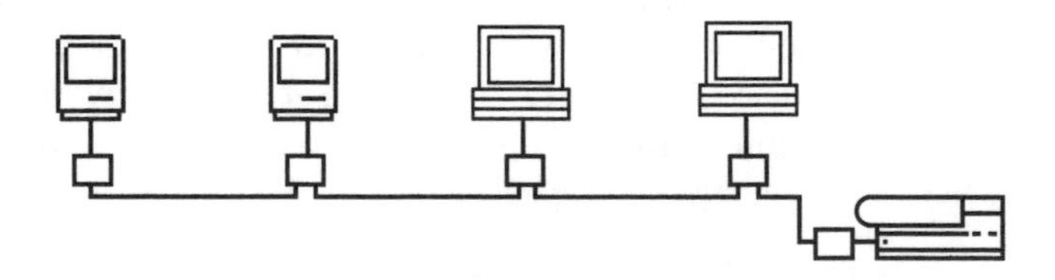

Figure 8.9 A simple AppleTalk network showing four computers and a laser printer with the junction boxes and the connecting cables. Such a network, be it this simple or much more complex, must include the appropriate file transfer and driver software as well as the hardware shown.

Cabling for a network such as AppleTalk (the hardware part is also called Apple LocalTalk) must conform to some standard. Such standards specify the resistance or, more commonly, the characteristic impedance. For instance, even a simple twisted pair should conform to the telephone or Bell standard for copper quality. Likewise, the coaxial cabling will usually have a characteristic impedance of about 50 or 75 ohms. All network lines should be properly terminated to permit the lowest error rates for pulse transmission. Most junction boxes have built-in single line terminators. However, they are usually automatically disconnected if the box is used to connect two devices in a daisy-chain configuration.

Networking Models: TOPS For the Macintosh—Simple but Very Workable Networking Software

The TOPS software has the advantage of simplicity combined with a reasonable amount of versatility. It is inexpensive, easy to install, and easy to learn. One major advantage for the small network user is that it does not need a dedicated computer as a file server. Each computer can publish (make available) any or all of its files and, likewise, each computer can mount the published files of any other computer. If a file is mounted on a given computer, that file (or folder) is displayed on the desktop and can be opened in the normal manner. Of course, the other serving computer must be on to make the file available. TOPS has its own file access protocol and, because some application programs need the AppleTalk APF file transfer protocol, there are a few file types that cannot be transferred with TOPS.

Networking Model—The AppleShare File-Transfer Software

To complement its AppleTalk hardware, Apple has also provided a software package called AppleShare. Although this software is more expensive than TOPS, it may provide better control and management for a larger network. AppleShare (and most other networking software) requires a dedicated file-serving computer and is much more structured and regimented than TOPS. Access can be divided and restricted into groups and users. The security is well done and access may be limited by files or folders. Likewise, the permission to make changes may be restricted to the owner, to certain groups, or, in some cases, available to anyone.

Networking Models—Network PC Configurations

Figure 8.9 shows one of the simplest possible network configurations. Different network programs and different computers or combinations of computers require different connection structures (topologies). For instance, if an LAN uses a central, dedicated server, quite often each user computer will have a line going directly to that server in a so-called star connection. Another very popular IBM network is in the form of a loop or ring.

A network with one given topology can be connected to another network with a different topology using a connecting bridge. The bridge is used to sort out the addresses—only data transmissions that are targeted for the other network are recognized and allowed to pass. Bridges allow internetwork data transfers but still allow each of the individual nets to maintain its own reliability and security. A gateway is used to connect two or more LANs with different protocols. Gateways, of course, are complex devices and have the built-in smarts to interpret and translate the data from one protocol to another.

Networking Models—The General PC Token-Ring Network Configuration

At this writing, the IBM PC line and the compatibles do not have networking software or hardware built in per se. Its software, NetBIOS, was originally part of the PC BIOS but is now a part of several third-party PC networking software packages. (See the later section on the OSI networking model.) The de facto topology for networking PCs is the Token-Ring, although other configurations, such as Ethernet, are also widely used.

The Token-Ring topology shown in Figure 8.10 is a closed-series system that connects each computer (sometimes called a station, a workstation, or a client) to the next one down the line and then to the next and so on until the circle is closed on the originating computer. Each station is given an address number, although these numbers do not have to be in the same sequence as the computers. A special message packet —the token—containing a recipient address, as well as a message and possibly a second destination address, is sent down the line. Each station in turn will intercept and interrogate this message token. If it does not contain that station's address, the token is immediately retransmitted to the next station. If the token does contain that station's addresses, the message is read and any appropriate action is taken. The appropriate action may be to store or process the incoming message data or it may be asked to send a file to some other station, which it will do. Once the required action is completed, the token is now free and is sent down the ring for consecutive interception until a station adds another address and data to the token.

The housekeeping for the Token-Ring in carried out by one of the stations that is designated as the active monitor. This monitor will assign address numbers, control timing, and do some routine diagnostics. Each station on the ring may go off line without disrupting the overall operation of the system. The major claim to fame for the Token Ring system is its inherent ability to prevent collisions on the network. Two or more Token Rings can be interconnected using a bridge. A bridge will connect two different networks operating with the same protocol. If a computer in network A desires to communicate through a bridge to a computer in network B, all that is required is that the housekeeping in both the A and B networks be preset so that the addresses for the bridge, and thus for the other network, are correct.

Networking Models—The ARCnet Token-Ring Network Configuration

ARCnet is a Cheapernet-type product (see below). It is less expensive than products like Novell's NetWare and, while it does not exactly adhere to the IEEE net-

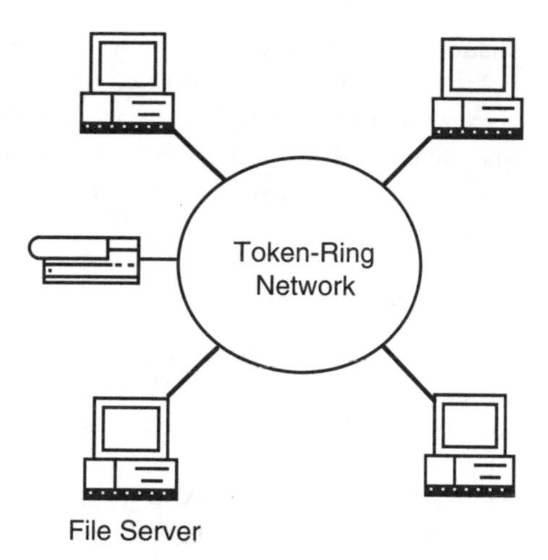

Figure 8.10 The Token-Ring topology that is common in PC LANs. Note that one of the computers has been assigned as the network's file server while the other units can be termed clients or workstations.

work standards or the OSI Model (see later), it performs quite well and is widely accepted for limited, small LANs. There are versions of this product for both the PC and the Macintosh.

Networking Models—The PC and the Ethernet (also the Cheapernet) Configuration

Ethernet is probably the most widely used LAN for the IBM and its clones. There are several versions of it that have been derived from the original developments of Xerox and were later adapted as the IEEE 803.3 network standard. The components for the baseband LAN (the network was also developed for RF transmission—thus the name "ether") are now supplied by a number of vendors. In fact, many of these vendors refer to the LAN as *Ethernet DIX,* indicating the three early supporters including Digital Equipment Corporation, Intel, and Xerox. Ethernet generally uses very high quality coaxial cable and has a data rate of 10 Mbps. This will permit a segment length of up to 500 meters without a repeating amplifier. It will also permit up to 100 nodes (computer T connections) per section. There are several variations from this generic system. There is the aforementioned Cheapernet that uses less expensive coax and nodal connection. The Cheapernet does not necessarily reduce the data rate but it does, because of attenuation, reduce the maximum segment length to about 200 meters. There are also twisted-pair versions, although this is not common.

The transmitted data frame in Ethernet usually contains six fields. In each frame, an 8-byte preamble is first sent for recognition, synchronization, and framing. Next comes the unique destination address that is followed by (for reference) the source address. These three fields are followed by a short two-byte type field. The type field is usually ignored except in large, multiple-network installations where several protocols are incorporated. The fifth or data field can contain from 46 to 1,500 bytes of information. The last field contains CRC error correction.

Networking Models—Novell's NetWare

If it is possible to reduce all of the excellent networking software to two model programs, for study and comparison, it would be TOPS by Sitka Software, including DosTOPS and MacTOPS, and NetWare by Novell. TOPS is an inexpensive, small system LAN controlling program that includes many convenient, rudimentary networking features. For comparison, NetWare is super deluxe with all the features, including network management and diagnostics. (All the features come with a higher monetary price, as well as a higher price required for learning, operation, and maintenance.) As has been stated several times, this book does not endorse any product but it does try to present representative models for investigation and comparison.

1. NOS—NETWORK OPERATING SYSTEMS:

   Along with the certainty of death and taxes, it is assured that even the best planned network will later expand. With this in mind, it is wise to carefully compare several network operating systems before making the final financial and sweat-equity commitment. The main item is compatibility with your mode of operation. Will the management activities add to the productivity of your enterprise, as they should, or will they overly complicate the operations? NetWare has gone through several generations that have tended to make it more easy to learn and use. It provides a high degree of redundancy that adds to the reliability. It also has a number of user classifications, which speed up the overall network and can be combined with its extensive security protection to protect both the user and the files.

2. LAN CONNECTIVITY AND INTERCONNECTIVITY:

   NetWare, and other large networking programs, provide access to most, if not all, of the so-called platforms including the IBM and clone PC, the Macintosh, the NeXT, and most higher-level workstations including those of Sun, Hewlett-Packard, DEC, etc., as well as most mainframe computers. NetWare is well supported by third parties that provide compatible copper and fiber-optic cabling and fittings, as well as repeaters, bridges, routers, and gateways. These last three items may do quite sophisticated processing and their use (and potential problems) should be considered in detail when planning an LAN or a WAN.

3. LAN APPLICATION SOFTWARE—GROUPWARE:

   If a large, sophisticated network is planned, the matter of compatible, network software should be considered. If applications are written so several different operating systems (platforms) can be used, they will require less translation and thus speed up the networking processes. The choice of Data Base Management Systems (DBMS) is especially critical. (For example, see the Corrigan and Guy reference.)

4. NETWARE CACHING:

   NetWare has both file and directory caching. The file server sets aside cache memory to retain files that are requested from disk so that they can be re-accessed more rapidly. Likewise, directories are kept in memory for more rapid access. NetWare has very elaborate printer management facilities. There are several queuing and spooling options that include some ability to use the printers of other workstations as well the those of the server.

5. GROUPS AND USERS—CREATING AND LIMITING ACCESS RIGHTS:

   Many times it is more efficient to configure users with similar requirements into a group rather than simply dictating the rights and/or limits to each individual. Thus, it often makes sense to combine all the users in manufacturing in one group, accounting in another group, engineering in another, etc. This may speed up operations and it certainly makes it easier to modify the access, passwords, user cost information, etc., for a group rather than each individual user. NetWare provides a multitude of access right options.

6. USING NETWARE'S FILE-SERVER MANAGEMENT RESOURCES:

   Besides being the omnipotent network traffic controller, the file server (console) can provide housekeeping information, assemble statistics, monitor the overall operations and status of the network, and provide diagnostics. In small networks, some of these activities might be overkill. However, in a complex LAN or WAN, some form of these management activities is a must. For instance, NetWare provides a technique to include printer queue sequences in the AUTOEXEC file. It can send time and date messages, as well as special information messages, to any or all stations. By using a special command, some of the diagnostic routines can be conducted from a given workstation rather than from the server. Some of the major server management services are to gather and display network statistics, including the routes used, the time of day and the length of time used, details on such things as the packets used and their routing and the amount of memory used. There are usually provisions to print various reports.

8.6 EXTERNAL NETWORK PROCESSING—COMMUNICATING BETWEEN DESKTOP AND MAINFRAME COMPUTERS

An Introduction

The early mainframe computers used punched cards and printers as I/O devices—they were, at the time, the fastest, most efficient way to go. Later, improved systems allowed interactive, multiusers with one or more teletypes and/or video terminals. The introduction of the personal desktop computer established the potential for, first, dumb terminal use and, later, a more versatile smart terminal use. However, there were several levels of development required before the desktop could become a terminal for either a minicomputer or a mainframe. Also, as personal computers became much more powerful and mainframe-like, some limited peer-to-peer applications have developed. Thus, desktops can be used to link with mainframes (including minicomputers) as: [1] (dumb) terminals, where most of the processing is done by the larger computer, [2] smarter terminals, where some of the data and processing resides in the desktop, and [3] smart or very smart terminals, where a one-to-one processing and data storage can virtually exist to the point where the mainframe may originate the link and use the application and data in the desktop.

The magic word for elementary personal computer utilization is emulation. The first, and still a commonly used, desktop computer emulator is for the teletype or the "TTY." This is common because the software and operating system in most larger computers is designed to converse with an emulated TTY using the basic ASCII character set. What is not so apparent is that this (relatively slow but quite wonderful) TTY emulation is not bothered by the fact that the desktop might (in the past) be an 8-bit machine successfully communicating with a 32-bit mainframe. If it is desired to increase the usefulness of the emulation process, there are several major problems that must be solved. The problem of an n-bit desktop talking to an m-bit mainframe has been introduced and will be expanded upon later. A more sophisticated terminal emulator must be capable of producing most, if not all, of the unique character set of the mainframe. Thus, a VTXXX emulator must be able to use the special characters of a Digital's VAX and a 3270 or a 5250 emulator must understand the character set of the appropriate IBM. Also, many of the larger IBM computers do not use ASCII but communicate in the 8-bit EBCDIC.

Desktop to Mainframes—The Mac to VAX And Back

As a starting point, both the late Macintoshes (the Mac-II series) and Digital's VAX family of computers are 32-bit machines. Although it is certainly not necessary to have equal n-bit computers for successful emulation, with equal 32-bit word lengths one less translating process is required, and it is especially helpful when emulating complex terminals. This is not meant to imply that there are not protocol and bridging problems that must be solved in going from the Macintosh AppleTalk network to the VAX DECnet. However, there have been some very imaginative solutions developed for these problems.

Programs such as Makeasy and MacNix add Mac-like icons to VAX files, whether they are using the VAX VMS operating systems or Digital's version of UNIX called Ultrix. Several vendors have used Macintosh's HyperCard to create smooth links to database programs and their databases on the VAX. There are also programs that will allow the Macintosh to access the VAX when it is running programs in its DECwindows version of X Windows. X Windows, developed at MIT, has become widely accepted as a universal windowing and graphics system.

Both Apple and Digital have made changes to aid the other company's products in the more sophisticated operations of network interconnections and file and application processing. Applc offers toolbox items that make it possible to interface with DEC's Ethernet Local Area Transport (LAT) and DEC offers converters to use MacWrite, MacPaint, and the PICT format on the VAX. (See the Meng reference.)

Desktop to Mainframes—An Introduction to Some of the Networking Protocols

Figure 8.8 shows an elementary prototype of a message frame for terminal-to-computer or computer-to-computer data transmission. Some of the very early protocols did not completcly isolate the main data from the control parts of the frame. Thus, in many cases, some of the special control portions would print as text. Later, improved protocols added special control characters as flags that would turn the control portions of the frame ON or OFF and thus make them isolated or transparent from the printed text. The following transmission protocols present the basic models for the development and improvement of general transmission protocols. (See the Jordan reference.)

1. The DoD TCP/IP:

   The Transmission Control Protocol/ Internet Protocol is an early (circa 1960) Department of Defense protocol for interconnecting dissimilar computer systems. Although it has not been embraced by many vendors, it has gained wide acceptance with education, engineering, and manufacturing, as well federal government users.

2. THE IBM BSC:

   In the early 1960s, IBM initiated its BSC or Bisync (Binary Synchronous Communication) protocol. This has proven to be their flagship protocol and was used in many products. It contained data transparency and would accept 7-bit ASCII, 6-bit Transcode, and 8-bit EBCDIC. It was synchronous and contained block-check error detection. Its major limitations were its complexity. Because it supported the three codes, it did not support full-duplex transmission, and it did not work well in ring configurations.

3. THE IBM SDLC:

   Because of the growth of cable-loop (ring) configurations, IBM redesigned the previous protocol and, about a decade later, introduced the Synchronous Data Link Control (SDLC) protocol. This new standard would support both half and full duplex and reduced the control characters to accommodate ring configurations. It also was so configured that the code, be it ASCII or EBCDIC, was completely isolated and had nothing to do with the transmission. Thus, even in a ring, any code could be used.

4. THE IBM SAA:

   In March 1987, IBM announced its Systems Application Architecture as a new unified systems design and implementation ideology to expand the compatibility and upward mobility of its various computer systems. It was specifically designed to reduce the system and architectural disparity between the PCs, the System/400, and the System/370. The SAA would provide a common base for the systems, their application software (whether it be IBM or from a third party), and the communication between the various IBM computers. It was to provide commonalty of interfaces, conventions, and protocols for data and user access. Because of the (then) memory limitations of the PC and its single tasking operating system, SAA features were incorporated in the new OS/2 SE operating system (with Microsoft Corporation) and later IBM's exclusive, extended OS/2 EE.

5. THE ISO HDLC:

   The International Standards Organization (see later) adopted the High-level Data Link Control (HDLC) for public networks in 1971.

6. THE CCITT LAP-B:

   The International Telephone and Telegraph Consultative Committee (Comité Consultatif Internationale de Téléphonique et Télégraphique) made minor modifications in the design of HDLC and published the results as Link Access Protocol-Balanced (LAP-B). Both the IBM SDLC and LAP-B are now considered subsets of the ISO HDLC. Most of the differences in these subsets relate to the link diagnostics and the hierarchy of control in a communication network. LAP-B, X.25, X.400, and Microcom Network Protocol (MNP) are hardware protocols that are used for diagnostics, control, and error checking. (Remember that XMODEM, ZMODEM, etc., are software error-checking protocols. See the Banks and Jordan references.)

Desktop to Mainframes—Networking the PC and the PS/2 to Interface with the Big Blue Mainframes

The original IBM PC was primarily designed for the individual as a personal computer. Its use as an emulated terminal or as one of the participants in a network came later. The appropriate communication programs gave it a terminal emulation capability but, with its original memory limitations, it could do little processing of host programs, let alone peer-to-peer processing. As the PC evolved from its original 8088 version and thus could support more memory (and processing), it became more realistic, under some conditions, to use these more contemporary PCs as emulating terminals. There can, however, be speed limitations. A very common IBM terminal to emulate is the 3270. This terminal is a synchronous terminal. If it is required to emulate the speed and precision of a 3270, it is best to purchase a synchronous modem, a terminal emulation board, and a direct or leased telephone line. However, if speed is not the dominant factor or if the PC is only occasionally used as a 3270 (a 3101, 5250, etc.), it is appropriate to use the more common asynchronous modem and a communications program such as PROCOMM Plus. However, if an asynchronous modem is used, a protocol converter is required to change the asynchronous to a synchronous signal for the mainframe. There may also be a problem of keyboard mapping with some terminals. Keyboard mapping refers to the relationship between the key that is hit and the actual symbol for that key that is shown on the monitor screen or sent out the communication line. For some emulated terminals, an extra mapping program is required. (See the Bruce reference.)

Advanced Networking Models

8.7 NETWORK PROCESSING STANDARDS—A COMPARISON OF MODELS FOR NETWORKING ARCHITECTURE

The explanation and models in the previous sections of this chapter were somewhat pragmatic—they were presented to best illustrate the physical aspects of the appropriate processing. This section will deviate from this primarily practical approach and illustrate how some of the protocols can be referenced to the widely accepted OSI theoretical networking model. The major problem with any universal model is that it is just that, universal, and, in so being, purposely does not define the details. These models, especially the OSI model, were often produced by a committee of international or academic authorities as a reference for the development of transmission techniques and protocols by many diversified corporations and institutions in many countries. It is up to the specific developer to fill in the blanks to produce a workable networking system. However, if developers realistically follow the framework guidelines of the theoretical model, the problems of connecting from one type of network to another can be minimized. The following paragraphs will define some of these theoretical models and later show how the structure of some functioning networking systems are based upon the theoretical foundations.

The OSI (Open System Interconnection) Model

The OSI model is usually presented as a layered block structure. To follow the sequence of intended or inferred processing, one can start at the top application layer (the document or file being produced) and progress down through six more layers of possible processing until this application is sent out on the communication channel. Once the transmitted document is received, the message is sent up the layer chain for reverse processing to again retrieve the document on the top OSI layer.

Table 8.2 The Elementary Definitions of the Layers in the OSI Model

Application Layer	Provides access for the data from application programs, files, e-Mail, etc., to access the OSI environment. The frame might now have the form of [AH][DATA] where the AH stands for Application Header. For a general reference, see Figure 8.8.
Presentation Layer	Defines special service codes for the application such as compression or encryption. This layer will initiate translations and byte reordering between different type of computers. If it is used, this layer would add a [PH] header to the frame.
Session Layer	Coordinates the communication between two workstations exchanging messages during one particular session. If it is used, this layer would add an [SH] header to the frame.
Transport Layer	Regulates the sequencing and flow of traffic. For instance, it can detect and discard a duplicate message packet. If it is used, this layer would add a [TH] header to the frame.
Network Layer	Specifies the addressing, switching, and routing of a message. If it is used, this layer would add an [NH] header to the frame.
Data Link Layer	Defines the protocol that detects and corrects transmission errors. This layer is divided into MAC (Medium Access Control) for either token passing or collision sensing and LLC (Logical Link Control) which sends and receives the actual data messages. This layer will append a final Link Header [LH] and, after the message, a final Link Trailer [LT].
Physical Layer	Defines the physical and electrical characteristics of the medium such as RS-232C, fiber optics, repeaters, etc. This is the hardware specification layer but it does not include the actual channel wire itself.

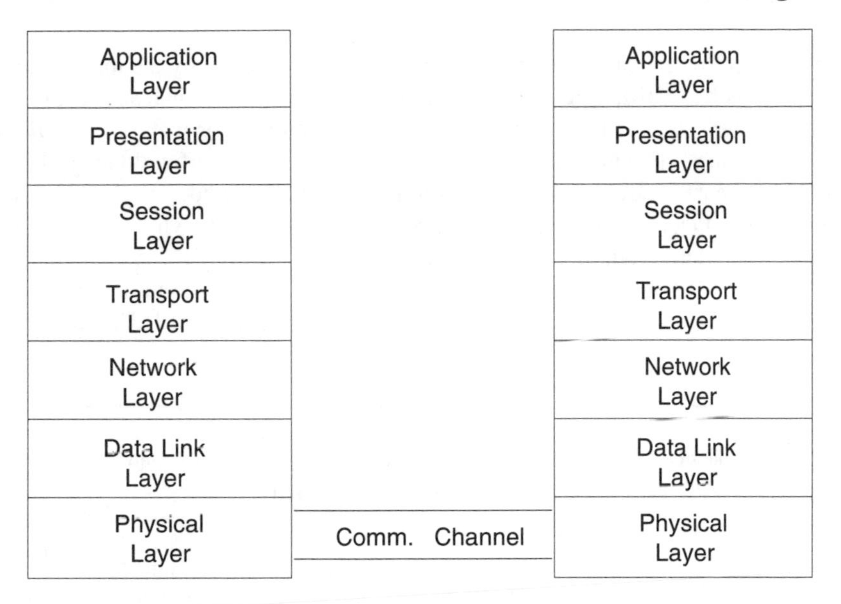

Figure 8.11 The OSI reference model(s) for two connected systems. The Communication Channel is not part of the OSI specification—it might be a twisted pair, a coaxial cable, a fiber-optic link, etc. Table 8.2 defines the layers in the theoretical model.

It must be remembered that the implementation of each of these OSI layers may require many lines of code to completely describe the protocol, translation, or control signals used to successfully transmit or receive the file or document message.

For perspective, this model is sometimes considered overkill to the point that it is ridiculously complex. Some simple transmission processes simply do not need all the steps (layers) the OSI model proposes.

For comparison, Figure 8.11 shows the complete OSI model and Figure 8.12 presents the minimum number of layers that are required for message transmission and reception. There are numerous network architectures that fall somewhere between the complete OSI model and the minimum model in. These architectures

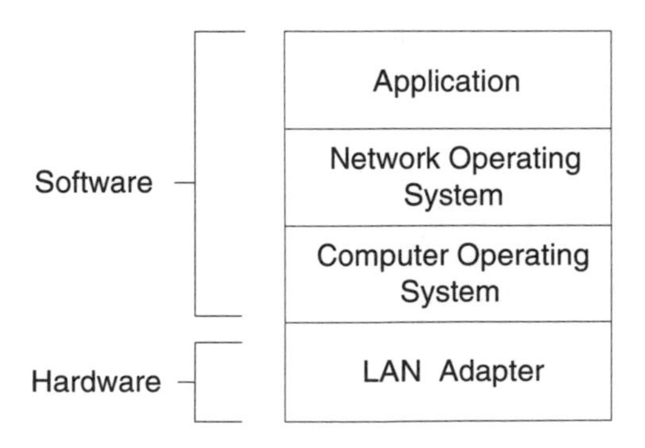

Figure 8.12 The minimum layers required for an elementary LAN.

are designed by standards committees such as OSI or CCITT, by computer manufacturers such as IBM, DEC, Apple, etc., by common carriers such as AT&T, and by large users such as universities or large corporations. For instance, the IEEE has created several standards known as 802.x. Boeing Aircraft and General Motors have created standards that are used in industry. The Federal Government's Department of Defense has produced the combined protocol known of TCP/IP. (See the Jordan and the Stallings references.)

In Figure 8.13, note that there is some overlap in the corresponding layer functions. (See the Jordan reference for a further comparison with the IEEE 802.x protocols.)

OSI Model (Reference)	SNA Model (Comparison)	
Application Layer	Transaction Services	Provides application services such as distributed data base access and document interchange.
Presentation Layer	Presentation Services	Formats data for different presentation media and coordinated the sharing of network resources.
Session Layer	Data Flow Control	Sychronizes data flow and groups related data into units
Transport Layer	Transmission Control	Paces data rate to match processing capacity. Will encipher if security is needed
Network Layer	Path Control	Routes data between source and destination and controls data trafic in the network
Data Link Layer	Data Link Control	Transmit SDLC transmission frames between adjacent network nodes
Physical Layer	Physical Layer	Provides physical connectivity between adjacent network nodes

Figure 8.13 A comparison between the OSI model and the IBM system architecture model. Note the overlap with the elements in the OSI model.

8.8 ALL DIGITAL NETWORK PROCESSING—THE ISDN

ISDN—An Integrated Services Digital Network

Since the vast majority of the information, data, and graphics that are now created and used are digital, it seems rather odd that a significant portion of these are still transmitted using analog techniques. Of course, the primary reason for these remaining analog transmissions is the historical existence of the local, national, and international analog telephone circuits. However, with the proliferation of wideband fiber optic and satellite links, the analog circuits and their switching interconnects are rapidly dwindling. Even so, the dreams of all-digital communication need one or more guiding models, such as ISDN, just as the protocols and application of local and wide area networks needed overall models like the OSI.

The Goals of ISDN

In the phrase Integrated Services Digital Network, the word services is especially significant. The all-digital network should provide more than just the conduits for various forms of digital data. The services, to be expanded upon later, should include such things as fax, telephone caller identification, and video Teletex. Like OSI, it should provide transmission standards without dictating all the details. Some of the overall objectives are:

1. The standardization of interfaces.
2. The transparency to network users. Transparency is defined as the ability to transmit and control different types and rates of data without placing any additional design or programming burden on the user. The internal switching, routing, and protocols of the network should not concern the sender or the receiver. ISDN should allow all users, regardless of their particular requirements, to (within certain limits) simply plug in.
3. It should support all existing communication systems as well as accommodate new systems as they are developed. This support for existing systems should include analog telephones, modems, etc. This ability to progress from the old to the present and then to the future technology is often called migration or a migration path.
4. It should support many types of communication systems, such as private or local telephone exchanges (PBXs), public telephone systems, computer networks, video services, etc.
5. Just as important as the technical models are the modeled guidelines for establishing the line-rate structure. This is an international standard that can be used in countries that have state-operated systems, completely privately operated systems, and countries like the United States, with its FCC, that have private systems operating with some government rules. One of the goals is to have the user price based on the cost of providing the service rather than on the type of service.

The ISDN System Architecture

In general, ISDN can be defined as a complete communication system that contains a set of services, interfaces, and controls for, and including, digital communication channels. Although it is completely different, both in concept and implementation than OSI, there is a slight resemblance to its lower Network, Data Link, and Physical layers.

ISDN is composed of a multiplicity of three types of channels. The A-channel is, for reasons of historical compatibility, an analog channel with a 4-kHz bandwidth. There are also some quasi or hybrid analog/digital channels that have been proposed but have not yet been adopted as standards.

The two fundamental channels for ISDN are the B-channel and the D-channel. The main communication or B-channel can operate at either a low rate of 32 Kbps or a high rate of 64 Kbps. (The B-channel can also be operated down to 16 Kbps.) There is also a D-channel that usually operates up to 16 Kbps for signaling and control although it too can go to 64 Kbps. The very elegant concept of these channels is that they can be combined to increase the overall throughput. Thus, two 64- Kbps B-channels and one 16-Kbps D-channel can produce a combined throughput of 144 Kbps if needed. A user requiring a very large throughput can combine channels, mostly Bs, to increase the aggregate rates to several megabytes per second.

The types of service are usually divided into a basic access service and a primary access service. The basic service, for minimum requirement users, usually consists of two B-channels and one D-channel. It is interesting to note that with one set of wires, the user can simultaneously place a voice telephone call, communicate between a PC and a remote computer or bulletin board, send or receive a fax, and allow the electric utility to manage energy use in the home. Compare this with the functions that can now be simultaneously performed through a single telephone line. The primary service is designed for large throughput users. A 24-channel primary service, which will be typical for North America and Japan, will provide data at 1.544 Mbps. Also, a 31-channel primary service, typical for the other parts of the world, will provide a data rate of 2.048 Mbps. (See the Jordan reference.)

The building-block components of ISDN are divided between the type of (input) terminal equipment, the terminal adapters, and the network terminations—see Figure 8.14, as well as Tables 8.3 and 8.4.

The Potential Uses of ISDN

There is every reason to believe that an all-digital network such as ISDN will become the world standard. It is being embraced in Europe probably faster than in the United States. One of the first differences the users of personal computers will see will be the disappearance of the old analog modem. Likewise, with this much speed and convenience, user data networks should expand at an exponential rate. There are also some subtleties associated with the growth and use of these wideband, all-digital networks. One of the realities of their acceptance, along with the

Table 8.3 ISDN Terminals—Their Description and Use

ISDN Terminals	Description and Use
Analog Telephones	Older-style equipment that must be serviced by ISDN until, by normal attrition, they are replaced with all-digital units.
TE1 Terminal Equipment Type 1	Any device that complies with ISDN interface specifications without assistance from external conversion devices. Digital telephones and all-digital input terminals and fax machines are examples.
TE2 Terminal Equipment Type 2	Terminals that comply with CCITT recommendations but do not completely comply with ISDN recommendations Some public data networks as well as some modems and PCs with asynchronous, serial ports qualify as TE2.
TA Terminal Adapter	A device that will convert the physical, electrical and protocol characteristics of TE2 devices to operate as TEI devices.

Table 8.4 ISDN Network Terminations—Their Description and Use

ISDN Network Terminations	Description and Use
NT1 Network Termination 1	The NT1 function group provides the electrical and physical termination of the ISDN network and includes the features and functions that fall within the Physical Layer of the OSI Model. NT1 provides the interface between customer equipment and the ISDN transmission system in the same sense that a telephone company's central office switch provides an interface between your telephone and the transmission system that connects to the central offices.
NT2 Network Termination 2	The NT2 group provides a single connection for ISDN-compatible devices. NT2 provides the features and functions of the lower three layers of the OSI model.
NT1,2 Network Termination 1,2	A group that combines the functions of both NT1 and NT2. Because of FCC requirements, NT1,2 will probably not be used in the United States. (See the Jordan reference.)

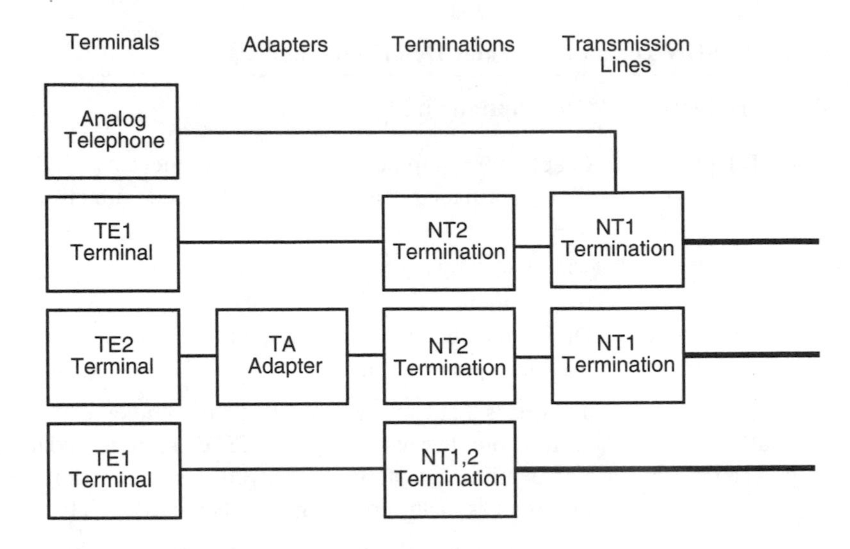

Figure 8.14 The elementary ISDN reference model. Note that the NT1 provides an interface between a user terminal and the transparent ISDN in a manner that is similar to the one function of the Physical Layer in the OSI model. The NT2 is more complex and provides features and functions similar to the three Physical, Data Link, and Network layers of the OSI reference model. (See the Jordan reference.)

digital circuitry and controls, is the channel bandwidth. And, of course, greater bandwidth directly equates to, in most cases, fiber optics. These fast networks will permit the exchange and use of information that will amaze even the most starry-eyed among us. Another interesting result will be the competition and/or overlapping of services provided by the telephone companies and the present television cable network. Who, in the future, will provide the potential services, many of them interactive, that the ISDN-type of communication conduits can provide?

8.9 NETWORKING—THE CHAPTER IN RETROSPECT

What Is a Network—The Personal Computer

This chapter takes a very broad perspective by defining a network in several different ways including the way a network is configured and the way it is used. Once a given network is presented, it can then serve as a foundation on which to build and expand the next higher level configuration.

The first broadly defined network is the computer itself. A graphic model is presented that describes the normal desktop or personal computer as a "network" that requires communication between the primary CPU and its ROM and RAM, its keyboard and mouse, its floppies and hard disks, and its video monitor. Examples are given of the ways the ROM, RAM, and disk data flows and the way the keyboard and the mouse may be interrogated. The collection of routines in the

Macintosh ToolBox and the PC BIOS are mentioned. Also, the various PC monitors are defined and described. There is also a discussion of the operation of the disk drives. This includes the capacity of the various disks along with the processing needed to successfully read and write data to these disks.

Networks External to the Personal Computer—Simple Exchanges of Data with the Outside World

The history, current status, and use of modems is presented along with examples of communications programs such as LAP-LINK. Several details about the working of modems are presented so that the reader should understand their setup and use in different situations. Error checking is covered including the use of parity, checksum, and their variations in such protocols as Kermit, XMODEM, etc.

Local Area Networks

These sections present an introduction to "real" computer networks, with their data transfer protocols, and the problems of communicating with both unlike personal computers, such as a PC and a Mac, and the additional problems of these computers communicating with mainframes. Such hardware and software devices as bridges, routers, and gateways are introduced and examples are given. The Macintosh AppleTalk and the PC Token Ring networking configurations are detailed and the application of various programs for networking is illustrated.

High-Level Network Processing Models

Two of the high-level standards that have been developed to hopefully ease the transfer of data (or telecommunications) at very high rates and with minimum inconvenience in going from brand, system, or country A to brand, system, or country B include the OSI (Open Systems Interconnection) model and ISDN (Integrated Services Digital Network). Both of these models are detailed and examples of both their complete and abbreviated use is given.

Archival and Cardinal References

Banks, Michael, A. *The Modem Reference: The Complete Guide to Selection, Installation, and Applications.* New York: Brady/Simon & Schuster, 1991.

Bell, C. Gordon, *Computer Structures—Readings and Examples.* New York: McGraw-Hill, 1971.

Bruce III Walter R. *Using PROCOMM PLUS.* Carmel, IN: Que Corporation, 1991.

Campbell, Joe. *C Programmer's Guide to Serial Communications.* Carmel, IN: Howard W. Sams & Co., 1989. (This book contains an excellent bibliography.)

Chorafas, Dimitris N. *Local Area Network Reference.* New York: McGraw-Hill, 1989.

Corrigan, Patrick H. and Guy, Aisling. *Building Local Area Networks with Novells NetWare.* Redwood City, CA: M&T Books, 1989.

da Cruz, Frank, and Catchings, Bill, "Kermit: A File-Transfer Protocol for Universities." Parts 1 and 2. *Byte,* June and July 1984.

Dortch, Michael. *The ABC's of Local Area Networks.* San Francisco: Sybex, 1990.

Fortier, Paul J. *Handbook of LAN Technology.* New York: McGraw-Hill, 1989. (This book contains an excellent communications bibliography.)

Jordan, Larry E. *System Integration for the IBM PS/2 and PC.* New York: Brady/Simon & Schuster, 1990a.

———. *System Integration for the IBM PS/2 and PC.* New York: Brady Books, 1990b. (One of the better reference books. It includes a nice section on protocols.)

———. *Qmodem.* Cedar Falls: The Forbin Project, Inc., 1990c. (THE instruction book for the PC communication software.)

———. "Apple Sharing—Here's What You Need to Know to Work with the AppleTalk Filing Protocol." *Byte*, October 1991, p. 247.

Kosiur, David R. "Expanding the Conversation—An Overview of AppleTalk Network Products." *MACWORLD,* May 1988.

———. "Network Connections." *MACWORLD*, November 1989. (AppleTalk and Networking)

———. "File Service—How to Move and Share Files over an AppleTalk Network." *MACWORLD*, August 1990.

———. "Going the Ethernet Route." *MACWORLD*, April 1991. (Excellent information on routers and router tests as well as Ethernet.)

———. "Building a Better Network—How to Avoid Growing Pains as Your Network Needs Increase." *MACWORLD*, November 1991. (Read this before you design, let alone install, any network!)

Meng, Brita. "Mac to VAX and Back." *MACWORLD*, May 1990.

Nance, Barry. *Network Programming in C.* Carmel, IN: Que Corporation, 1990.

Putman, Byron. *RS-232 Simplified.* Englewood Cliffs, NJ: Prentice Hall, 1987.

Rodgers, Mike, and Bare, Virginia. *Hands-On AppleTalk.* New York: Brady Books, 1989.

Segal, Mark L. "Toward Standardized Video Terminals." *Byte*, April 1984.

Stallings, William. *The Business Guide to Local Area Networks.* Carmel, IN: Howard W. Sams & Co., 1990.

Veljkov, Mark D. *MacLANs: Local Area Networking with the Macintosh.* Glenview, IL: Scott, Foresman and Co., 1988.

Watson, Scott. "An EitherPort in the Storm." *MACWORLD*, May 1988.

———. *White Knight.* Beaver Falls, PA: The FreeSoft Co., 1989. (THE instruction book for the Mac communication software—the successor to *Red Ryder.)*

9

Integrated Digital Signal Processors

9.1 PROCESS CONTROL—A HISTORICAL PERSPECTIVE FOR THE CIRCUITS THAT MODIFY AND CONTROL SPECIALIZED DYNAMIC PROCESSES

An Introduction

The circuits used to capture, control, and display the signals from industrial and medical processes have usually been given a unique place in electronic science and technology. These process control configurations are not just more computer circuits, although they may use very complex computational algorithms. One reason for this specialized category is that most, if not all, of these industrial and medical signals originate as analog voltages or currents and, for the first half of the twentieth century, they were extracted, processed, and displayed by discrete analog circuitry. The filters, mathematical processors, and displays were analog. However, because of the diligence and creativity of the involved engineers and scientists—they worked. They worked in spite of the propensity of analog circuits to be affected by hum and other electrical disturbances, to drift with temperature and age, to often have linearity problems, and to be somewhat unreliable because their vacuum tubes would age or burn out. Besides originating as analog inputs, these signals very often need to be processed in real time. Although many of the early process-control signals were DC or quite low frequencies, where the rates of both analog and digital processing are not taxed, more contemporary inputs are in the video-frequency region and require very advanced circuitry.

Examples of the industrial signals include speed, temperature, pressure, magnetic field strength, liquid level, viscosity, and the mass spectrum, as well as natural and synthesized speech and music. The medical signals include the direct body functions such as heart beats, pulse rate, blood pressure, etc., as well as a variety of analysis signals such as pH and blood count. The design of the pick-up devices for the signals, usually classed as transducers, includes a highly specialized technology unto itself.

Signal Conditioning—Linearization

Chapters 1 and 2 give several examples of the problems that are encountered when circuits or devices are not linear. Process control systems can exhibit unwanted nonlinearities at the component level and at the system level. For instance, a thermocouple (a temperature-sensing device) is usually nonlinear over a wide temperature range and, if it is to be used to provide accurate, wide-range temperature readings, a special circuit or, in a digital system, the appropriate software, must be added to linearize the device. Likewise, in a complex process control system, the nonlinearities of the individual components can compound to produce an unreliable or even unworkable overall system. In the analog domain, system linearization has historically been done by feedback. While this is an excellent and well-documented technique, it is usually difficult or very expensive to reduce an unwanted distortion, including linearization, to less than one part in 10,000, that is, a resolution of 0.01 percent or about 13 bits.

Signal Conditioning—Processing Functions

Analog signals can be processed in a wide variety of ways. The simplest processing is probably linear amplification, as typified in Figure 9.1. More complex changes usually can be specified by some mathematical function such as subtract, multi-

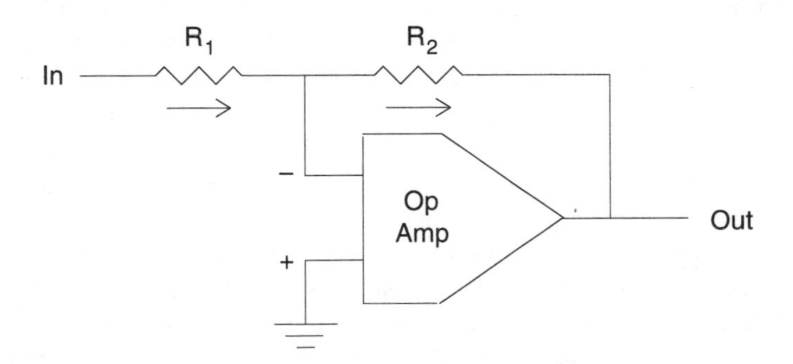

Figure 9.1 A prototype operational amplifier circuit using a high-gain DC amplifier and various parts in the feedback circuits. For instance, if several of the same resistors are placed in parallel with R_1, the circuit becomes an analog adder and if a capacitor C is used to replace R_2, the circuit becomes an analog integrator, etc. (See the Libbey reference.)

ply, or divide; taking its root or raising its to a power; finding its integral or derivative; or using more complicated mathematical techniques such as deriving a signal's spectrum by using a FFT. Many of these functions can be accomplished or approximated with simple circuits using basic resistors, capacitors, or inductors. More accurate function processes can be obtained by using the basic electrical components as elements in the feedback circuits of high-gain DC amplifiers called operational amplifiers. A number of these operational function circuits can be combined to form what is known as an analog computer. Before the perfection of the digital computer, circa 1945, any complex mathematical computations or functions that were done with electric circuits used an analog computer such as is shown in Figure 9.2. Unfortunately, even the best analog computers suffered from the shortcomings of their analog circuits.

Signal Conditioning—Filtering

Most analog signal sources need some type of filtering before they can be amplified, displayed, modulated, converted, etc. Perhaps a conditioning low-pass filter is most common, followed by a bandpass filter. High-pass filters are less common because, since most noise occurs at higher frequencies, a high-pass filter will often decrease the signal-to-noise ratio.

Filters can be categorized in many different ways. Table 9.1 summarizes the general types of filters and notes their pertinent characteristics.

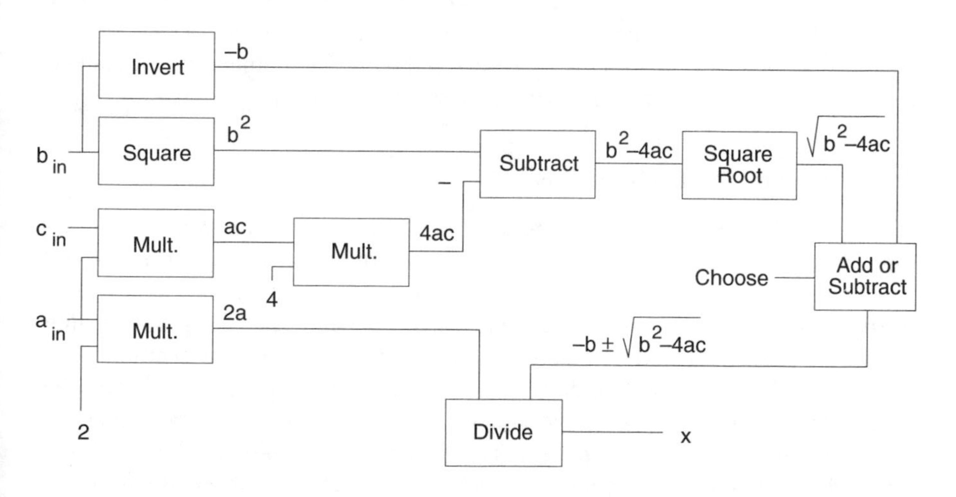

Figure 9.2 An elementary analog computer that is being used to solve the equation

$$x = \frac{-b \pm \sqrt{b^2 - 4ac}}{2a}$$

Table 9.1 Electrical Filters and Their Characteristics

Analog	Digital
Component Filters: RC, RL, and RLC type circuits	Component filters can be simulated by digital delays.
Transmission-Line Derived Filters: Examples include constant-K and M-derived	Using software programs (or hardware), transmission-line filters can be transferred from analog designs or originated in the digital domain.
Polynomial Filters: Used to produce a sharper cut-off and/or better phase response than transmission-line derived filters	Using software programs, polynomial filters can be transferred from analog designs or originated in the digital domain.

Simple RC, RL, or RLC Filters

These combinations can be configured as low-pass, high-pass, or bandpass filters and are commonly used as analog interstage coupling and tuned circuits. They normally produce excellent phase response but are limited (by noise) in the amount of cut-off that can be obtained—20 to 30 dB per octave (double the frequency) is usually the upper limit. These prototype component filters are often used as conversion examples in texts on digital filters, although their simple digital equivalents are rarely implemented in practical designs.

Transmission-Line Derived Filters

Transmission-line derived filters are a legacy from the early research and development done for telephone transmissions. These circuits, including constant-k and m-derived filters, are really a conceptual extension of transmission lines. They were usually implemented with discrete parts, although later designs in analog music synthesizers included operational amplifiers in their circuits to produce active filters. Transmission-line derived filters are designed to match or equal the characteristic impedance of the transmission line and to be used as both a design parameter and a working component. These filters also assure that the maximum in-band power is transferred down the line. It is interesting to note that most early books and papers on the design and implementation of transmission-line derived filters do not discuss their phase response. This was not due to a lack of scholarship but rather due to the fact that phase response was not a dominant factor in the transmission and reproduction of early telephone-type audio.

Polynomial Filters

This class of filters was developed to give designers more control over the frequency cut-off response and the phase and transient overshoot response of their circuits. Butterworth, Chebyshev, Bessel, and Elliptical (Cauer) are the most common members of this family of filters. All these filters are some variation on the idea of a general input-output transfer function that can be represented by this polynomial equation:

$$\frac{e_{out}}{e_{in}} = H(s) = K \frac{(s-z_1)(s-z_2)(s-z_3)\cdots}{(s-p_1)(s-p_2)(s-p_3)\cdots} \qquad \text{(E 9.1)}$$

where K (sometimes termed *G*) is the gain value at DC, "*z*" is a zero, and "*p*" is a pole in the *s*-plane.

Likewise, for polynomial filters only:

$$\frac{e_{out}}{e_{in}} = H(\omega) = K \frac{1}{S^n + a\,S^{n-1} + b\,S^{n-2} + \cdots + d\,S + 1} \qquad \text{(E 9.2)}$$

Analog polynomial filters played an important part in the development of communications systems such as TV, stereo, and radar. The digital implementations of these filters are even more powerful than the original analog designs. For instance, while the frequency cut-off of practical analog polynomial filters is limited to about 40 dB per octave, digital polynomial filters that drop off at over 100 dB per octave are quite common. For an example, sharp cut-off digital filters are being used as signal conditioners to eliminate hum interference from sensitive medical pickups. In addition, many polynomial implementations of digital filters allow the tailoring of the phase response as well as the frequency response. Thus, with these digital circuits, it is possible to have very sharp cut-off and linear phase response.

9.2 EXAMPLES OF INTEGRATED SIGNAL PROCESSORS—THE TEXAS INSTRUMENTS TSS400

An Introduction to Integrated Signal Processors

Any signal processing system requires some lower limits of its short- and long-term precision coupled with the highest possible conversion speed. However, the specific speed or accuracy is usually a function of the input signal and the required type and the intended use for the digitized output. For instance, an analog video signal may only require 8 bits of precision but may require a complete conversion in under 100 nanoseconds. A temperature or pressure measurement may require a precision of 12 bits but the conversion and display or transmission of the data may only require a conversion time in the millisecond range. In addition to the precision and speed, such parameters as circuit size, cost, and power drain may also influence the choice of an integrated processor.

Integrated processors also have a hidden advantage that makes them superior to other systems—the use of programmable software to help accomplish, adjust, or customize the functions of the processor. It is the use of this versatile software that makes most integrated signal processors adaptable to so many different applications.

The TI TSS400 Sensor Signal Processor

The TSS400 family is used as an introduction to integrated signal processors (ISPs). The generic TSS400 is a small, very low power 4-bit ISP that can be used for a variety of measurement and recording or display tasks in industrial, commercial, and medical applications. Some of the features of this small processor include:

- Different members of the TSS400 family will operate with either 3 or 5 VDC. Because of its extremely low current drain and its power-saving software routines, one version of the TSS400 can operate for ten years with the recommended 3-volt lithium battery as its power supply.
- An internal R-C system clock of about 500 kHz is included. An external clock crystal can also be used.
- A separate, precise crystal-controlled timer can be used to schedule measurements and outputs in real time.
- Four multiplexed analog inputs.
- A 12-bit successive-approximation, double conversion ADC. Software routines can be used to match and linearize temperature, pressure, and movement sensors for the ADC inputs.
- The sensors can be powered by a precise, adjustable constant-current source.
- Functions such as test and display can be initiated from software or a keyboard using the internal key-scan routines. The output R lines can be used as switch matrix drivers or to drive relays, external logic gates, LEDs, etc. The K lines can be used for general I/O functions including inputting switch status information. There is also a dedicated serial I/O line that can be used to communicate with a host computer.

The embedded standard system software includes routines such as start-up and shut-down; read/write and I/O; keyboard scan; A/D conversion; integer arithmetic including add, subtract, multiply, divide, and square root; and battery monitor. In addition, there are routines to correct for the offset and nonlinearities of the recommended sensors. If the user has special sensors, there are small programs that will develop and store the appropriate linearizing and offset correction factors for these special sensors in nonvolatile RAM. A special interpreter is included that converts the easy-to-use micro instructions of any user-written routine into executable machine code.

One of the joys of this little processor is the use of the development tools that are available. These tools include a free (at the time of this writing) PC disk that provides a simulation of a working TSS400 configured as a thermometer. The PC screen displays the registers, the ports, the counter, the status flags, I/O, the four A/D inputs, and the simulated output digital display. In addition, the measurement program is presented. This program can be executed using "free run" or "single step." Up to ten program break points can be set. The contents of the registers,

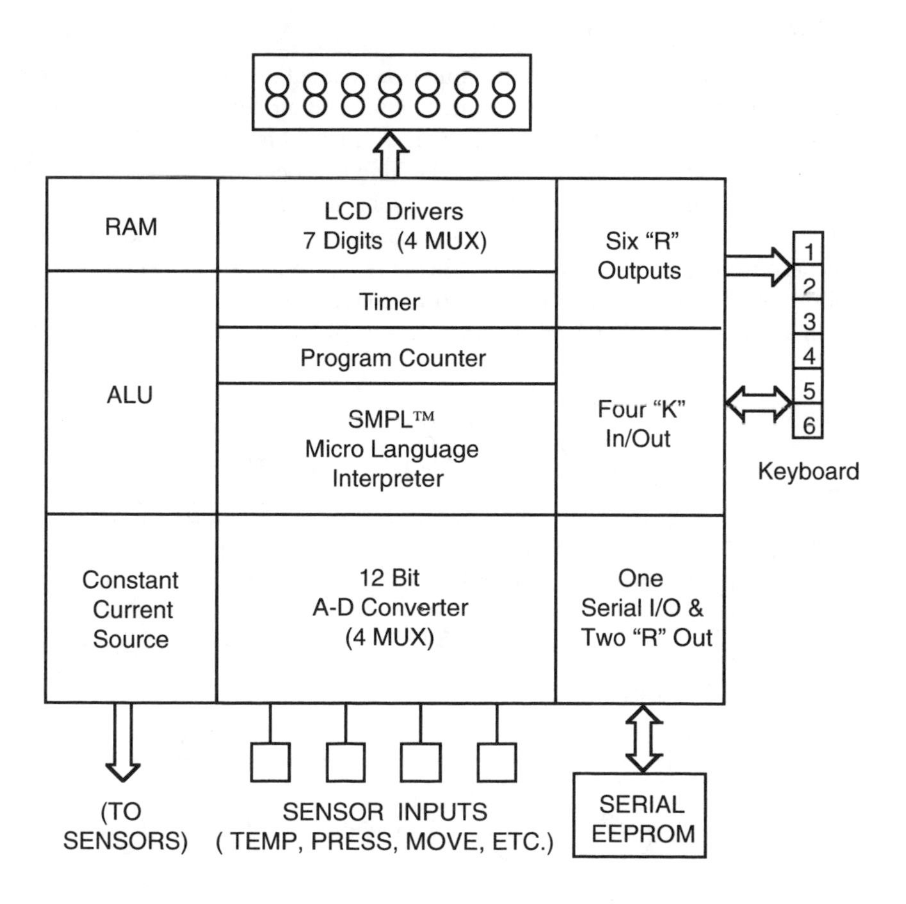

Figure 9.3 A two-chip sensor processing computer. The EEPROM contains the setup, measurement, display, and, if used, communication programs tailored for a given sensor measurement system. Up to four 512 × 8-bit EEPROMs can be accommodated.

flags, etc., can be changed to observe their effect on the program and its results. Any one who has paid $10,000 to $20,000 for a microprocessor development system appreciates the use of the PC as a contemporary development tool.

A second-level development package (about $50) includes another, more powerful simulation disk and reams of information and sample simulations. A third-level developer includes the software, a PC interface plug and cable, and a personality (prototyping) board to facilitate the development and execution of a complete TSS400 system in real time. The personality board includes a keypad, an LCD, and a convenient area for additional parts. This third, more complete, TSS400 development system includes the facility to program EEPROMs.

A TI TSS400 Application Example

Figure 9.4 shows the diagram of a simple two-channel digital temperature meter that might find an application in a home heating and air conditioning system, an industrial processing plant, or in a scientific or medical laboratory. The only major required additional parts, besides the signal processor, are the LCD, the EEPROM (containing the program), the sensors, and the battery.

The program development could be started with only a PC using the 400 development software package for about $50. In fact, Figure 9.4 shows one of the sample applications that TI includes. The macro language SMPL™ is easy to learn and use. In addition, the video display shows the position of the program counter, the

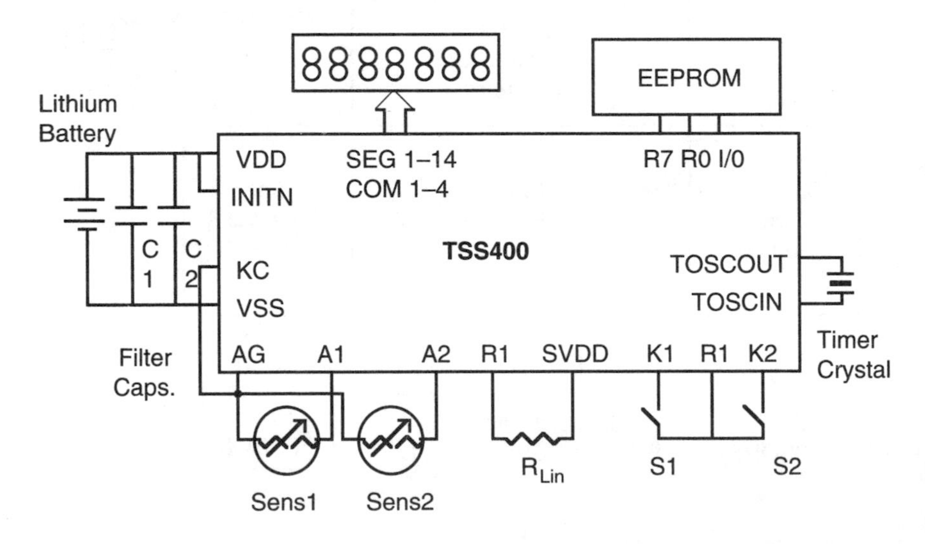

Figure 9.4 The prototype of a two-channel digital temperature gauge using the TI TSS400 sensor signal processor. R_{Lin} is used to select the proper current for the sensors and the switches can choose the type of operation. AG is the analog ground. The chip designers were very careful to provide the proper power and analog returns (grounds).

instruction or data in the program memory's cells, and the op code and mnemonicfor each instruction. The program is divided into seven sections including:

- Thermometer Adjustment—set the 0°C and the 80°C points to be certain that the sensor is linearized and there is a correct compensation for the offset.
- Thermometer—this section will test to see that the sensor, whether simulated or real, is reading the correct temperature and displaying it properly.
- Battery Adjustment and Battery Check—this routine checks to see if the battery check program is operating properly. There is a battery adjust feature that allows for fine-tuning of the software reference.
- Input Any 7-Digit Number—Test system by entering and displaying any 7-digit number.
- 24-Hour Watch—set the system's real-time clock.
- Display Hex Numbers—System will display all numbers from 0 through F in the 7-digit display.
- Current Measurements— displays the word "curr" and alternatively turns the output R1 and SVDD (the switched supply voltage) on and off.

After a few hours with the PC and the demo disks, one is ready to do constructive work with the Advanced Development Tool (ADT) TSS400 circuit emulator and the prototyping board.

9.3 EXAMPLES OF INTEGRATED SIGNAL PROCESSORS—THE TEXAS INSTRUMENTS TMS320 SERIES

The 320 Compared to the 400

The study of the TMS320 series produces an interesting contrast when compared with the TSS400. The 400 is highly specialized and very efficient in one area of signal processing. It is conceptually easy to learn and work with. One of its major features is its built-in ADC. However, the 400 might be described as a very clever, specialized 4-bit microprocessor with a built-in ADC rather than a general digital signal processor. The 320 series are real digital signal processors. Incorporated in the hardware architecture of these units (the detailed architecture will vary among members of the family) are DSP-related features such as μ-law and A-law companders (see Chapter 4), baud-rate generators, a 32-bit ALU and a 32-bit accumulator, and a 16 × 16-bit multiplier. The accompanying software routines will include implementation of FIR and IIR digital filters (see later), FFT and convolution algorithms, matrix multiplication, and floating-point arithmetic. Besides these all-important bread and butter DSP routines, there are arithmetic and bit-processing routines similar to, but greatly expanded compared with the TSS400. Likewise, the TMS320 series can be programmed in the C high-level language as well as a much more versatile and complex assembly language.

From its introduction in the early 1980s, the 320 series has been continually expanded and updated—see an example in Figure 9.5. Some models (at this writing) now have instruction cycles in the order of 35 ns. The later implementations include parallel processing and pipelining. As with general-purpose microprocessors, it is impossible to list all the applications for these programmable signal processors—the range of uses is limited only by the imagination of circuit designers. Some of the most common areas of application include telecommunications, image processing, voice processing, industrial processing, and automotive and consumer electronics.

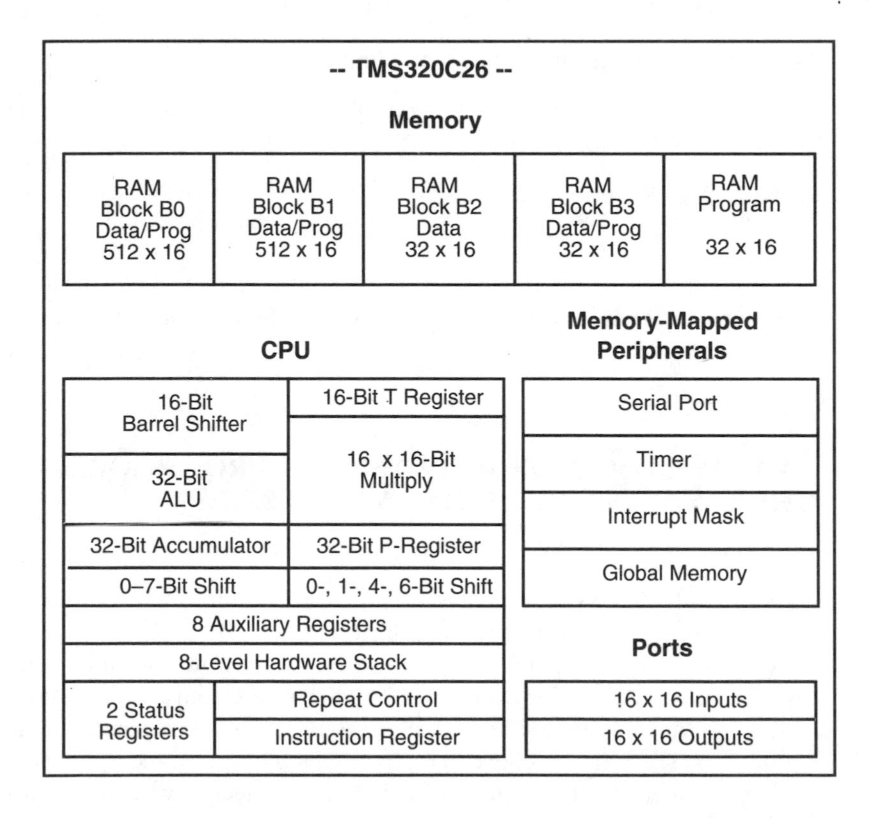

Figure 9.5 The architecture of one of the second-generation programmable digital signal processors in the TI TMS320 series. This series has now developed into a wide, versatile product line.

9.4 TYPICAL PROCEDURES FOR DIGITAL SIGNAL PROCESSING

Digital signal processing, like analog signal processing, encompasses a vast variety of applications that are limited only by the imagination of the designer. There is, however, a small collection of algorithms that continue to be the foundation of most DSP. The following section will outline three of the routines that are important in integrated, as well as general, signal processing.

DSP by Finding the Spectrum of a Signal in the Digital Domain—The FFT or Fast Fourier Transform

One of the guiding principles of analog frequency-domain signal theory is that almost any real, physical, periodic signal can be represented by a combination of sine and cosine wave forms using a Fourier series. There are well-established mathematical techniques that will either analyze (decompose) a given signal into its sine and cosine components or synthesize (compose) a desired waveform from a predetermined set of sine and cosine components. Analog music synthesizers relied upon this latter technique. If the signal is not periodic, it can still be reduced to its component frequencies—its spectrum—by using an allied method using the higher mathematics of the Fourier integral to obtain a Fourier series.

The analysis and synthesis of periodic, sampled signals, termed discrete time-domain signals, can be decomposed using a similar technique called the discrete-time Fourier series. If the discrete-time signal is aperiodic, the discrete-time Fourier transform is used.

NOTE: In the literature of electronic communication and processing, the terms Fourier integral and Fourier transform are often used interchangeably. While this is not precisely correct, they can be used interchangeably, in relation to aperiodic signals, without a significant loss of rigor.

While the Discrete-Time Fourier Transform (DFT) is a very workable method for decomposing—finding the spectrum of—a sampled signal, even with the much faster contemporary computers and processors, it takes an inordinate amount of computer time. When digital computers were in their infancy and were quite slow, this time factor made it almost impossible to do any real-time spectral analysis. Luckily, in 1965, James Cooley and John Tukey developed an algorithm for materially reducing the time required to obtain the DFT of a signal. Their research showed that the conventional methods of doing a DFT required many calculation steps that were redundant and thus unnecessary. Their fast method, which eliminated these calculations, and other approaches that followed, truly revolutionized both the philosophy and techniques of digital signal processing. Their FFT is a very special, fast form of the DFT.

For an example of the utility of this fast method, remember that the convolution of two signals in the digital domain is the same as multiplying their equivalents in the frequency domain. However, the frequency-domain multiplications are much faster than the time-domain convolution. Thus, to convolve two time signals, it is usually much faster (sometimes by a factor of 1,000) to follow this procedure:

1. Take the FFTs of the two original time signals—this puts them in the frequency domain.
2. Multiply the two results in the frequency domain. Note that in most of the 320 family there is a fast 16×16-bit hardware multiplier.
3. Take the Inverse Fast Fourier Transform (IFFT) of this subsequent frequency-domain multiplication to again obtain a time-domain signal. This complete, faster process is the equivalent to the convolution of the original signals.

Thus, although digital convolution and other similar digital processing techniques are still used, the FFT and IFFT are the dominant techniques used for the rapid equivalent of convolving or for just finding the frequency-domain spectrum of time signals.

DSP by Using Floating-Point Arithmetic—Floating Point-Convenience and Precision at the Speed of Fixed-Point Processing

The algorithms that a given digital computer system or computer programs use to calculate, distribute, and display data depend on a variety of factors. For instance, most computers, including hand calculators, do their arithmetic in binary—yet, both the input and output information is most often displayed in decimal. This means that the system software or a particular routine must convert the digital keyed inputs to binary, perform the calculation, and reconvert the outputs to digital. Going one step deeper into the process, binary subtraction is most often done in two's complement arithmetic that precludes a facility for using negative numbers. Thus, the computations are usually done in an n-bit number range of -2^{n-1} to $(2^{n-1}) - 1$ (for example -128 to $+127$). There are many other number-crunching choices that must be made by any serious system or program designer.

For maximum accuracy, most signal processing calculations are done using floating-point arithmetic. However, most computers normally use fixed-point. This means that the DSP program designer must carefully examine various fixed-point algorithms in order to get the optimum precision. The problem with floating-point arithmetic (sometimes referred to as exponential scientific notation) is overflow. For instance, using a simplistic nonbinary example with only three significant figures, the fixed-point number 456 would be $.456 \times 10^3$ in floating-point notation. If two numbers are added, there is a chance of an overflow error, as:

$$\begin{array}{r} .456 \times 10^{3} \\ +\ .678 \times 10^{3} \\ \hline 1.134 \times 10^{3} \end{array}$$

There is an overflow if only three places are used. If the program tries to take care of the overflow by going to $.113 \times 10^4$, then, by still using only three significant places, the answer is off by 0.0004. This is an example of the types of errors that the program designer and user must consider.

DSP Using Digital Filters

A preceding section introduced the subject of digital filters and outlined some of their characteristics and advantages over their analog counterparts. Digital filters fall into two general classes, termed Finite Impulse Response (FIR) and Infinite Impulse Response (IIR). FIR filters are feed-forward filters that do not use feedback and thus are unconditionally stable and have a guaranteed linear phase response. However, they are more complex and require more hardware or software than IIR filters. IIR filters do use feedback and, like some analog operational circuits, there is a distinct possibility of oscillation. Also, they may or may not be linear phase. Many times, if a filter requirement does not require linear phase, an IIR design will usually exhibit a cost or execution-time saving over its FIR counterpart. Likewise, in a few cases, an IIR filter can be specially configured for linear-phase response.

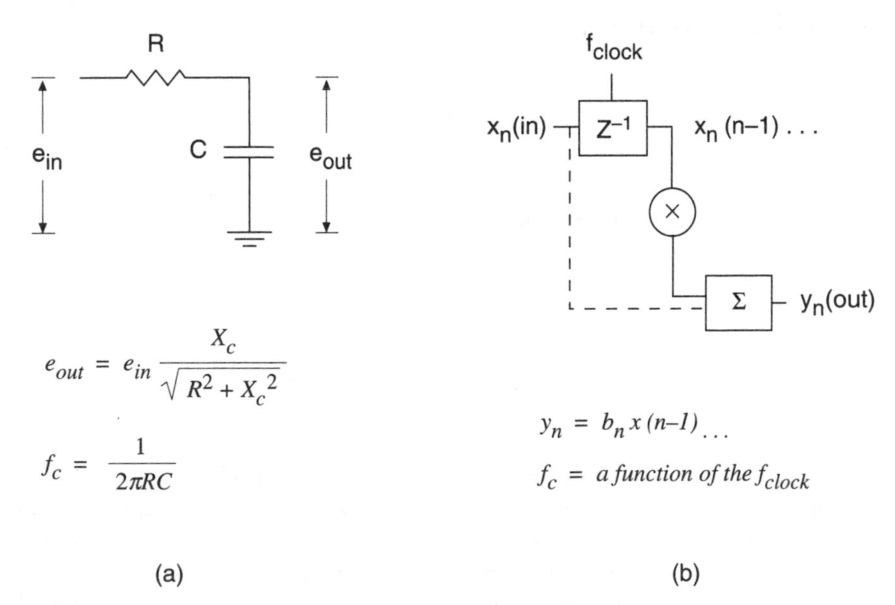

Figure 9.6 A comparison of the elementary building blocks for (a) a simple RC analog filter and (b) the component parts of an FIR digital filter.

The elemental building block of an FIR filter contains a digital delay (shown as Z^{-1}) device or instruction and a multiplying (shown as x) device or instruction. It somewhat resembles, but certainly does not duplicate, the RC building block in an analog low-pass filter—see Figure 9.6.

Conceptually, digital filters change a string of binary numbers that represent the incoming digital-domain waveform just as the circuit elements change the voltage or current amplitude and phase of the incoming analog-domain waveform. In very simple analog filters, the action of the filter is broadly dependent on the relationship between the time constants of the components and the input frequency. The action of digital filters is broadly dependent upon the frequency of the clock, the amount of delay in each block, and the a_n or b_n coefficients that are used in the multipliers, as shown in Figure 9.7. These coefficients somewhat resemble the a, b, c, . . ., coefficients for the more complex analog polynomial filters given in E 9.2. In fact, using the proper mathematical translation, such as with the bilinear transformation (see the Lynn and Fuerst reference) or the calculated look-up tables, the time-domain a_n, b_n ... equations for digital polynomial filters can be derived from the a, b, c, ... coefficients in the analog domain. It is these coefficients that give each filter, as the such as Butterworth, Chebyshev, Bessel, etc., its particular shape. Also, the number of the components that are used, and thus the number of blocks that are used, dictates the shape of the response. The following specific equations for a Butterworth and a Chebyshev are in the form of the general polynomial equation E 9.2. The b coefficients are derived by repeatedly changing the value of n.

For a normalized (K=1) Butterworth:

$$H(\omega) = \frac{1}{\sqrt{1 + (\omega/\omega_c)^{2n}}} \qquad \text{(E 9.3)}$$

For a normalized (K=1) Chebyshev:

$$H(\omega) = \frac{1}{\sqrt{1 + \varepsilon^2 C_n^2 (\omega/\omega_c)}} \qquad \text{(E 9.4)}$$

Note that the additional constants ε and C will make the calculations of the coefficients more complex.

Once the coefficients are known, it is possible to go through the rather laborious process of manually doing an n-taps calculation. However, there are many computer filter design routines, including the one in the TI development systems, that will look up or calculate the appropriate coefficients and do an n-tap design for most common digital filters. A digital filter has x_n samples of the input waveform. The filter output is the sum $y(n)$ of all the converted x_n samples. The equation that the digital filter solves is part or all of the following:

$$y(n) = a_1y(n\text{-}1)+a_2y(n\text{-}2)+\cdots+b_0x(n)+b_1x(n)+b_1x(n\text{-}1)+b_2x(n\text{-}2)+\cdots \quad \text{(E 9.5)}$$

Recursive Terms (for an IIR filter) — Nonrecursive Terms (for an FIR filter)

If the filter is FIR, $y(n)$ = the right-hand or b_n part of the equation. If the filter is IIR, $y(n)$ = includes the a_n part of the equation. It is possible to combine both FIR and IIR sections into one common filter although this is not common.

DSP to Improve the Shape of Digital Filters—The Gibbs Phenomenon and Windowing

Returning to fundamentals for a few moments, the basic idea incorporated in producing a digital filter is to construct a series of digital numbers that represent a Fourier series—this will represent the filter transfer function. In theory, almost any filter response can be obtained if there is the luxury of an infinite number of Fourier coefficients. However, in real-world filter designs, it is necessary to use a limited or truncated number of coefficients.

For the desired (theoretical) transfer function Hn_d:

$$H_d = \sum_{n=-\infty}^{\infty} C_n z^n \quad \text{(E 9.6)}$$

For the approximate, real-world transfer function H_a:

$$H_a = \sum_{n=-\infty}^{\infty} C_n' z^n \quad \text{(E 9.7)}$$

where C_n' = only truncated coefficients

Thus, if the input waveform is represented as $x(n\text{–}k)$ and the transfer function H_a is now defined in lowercase as $h(k)$, the input-output action (the convolution) of a digital filter can be represented, in its simplest form, as:

$$y(n) = \sum_{k=0}^{N} h(k)\, x(n-k) \qquad \text{(E 9.8)}$$

Many parts of the resulting truncated filter closely approximate the desired response. However, there are other portions of the response that deviate from the desired response, especially at or near points of discontinuity (the impulse response points) such as the starting (zero frequency) point and the zero output points in the stop band. At or near these points in the response curve, the filter response will oscillate up and down, i.e., it will ripple. The mathematics of this occurrence was originally described by J. W. Gibbs in 1900. Several researchers have devised methods to alter the filter's response in these discontinuous or impulse regions by again modifying the transfer function coefficients. These methods are termed windowing. With windowing, the truncated coefficients are again changed such that:

$$C_n'' = C_n' \, w(n) \qquad \text{(E 9.9)}$$

where $w(n)$ is the windowing factor

As examples, several of the common windowing sequences are included as:

The Hanning window $w(n) = 0.5 + 0.5 \cos(2n\pi/N)$ (E 9.10)

The Hamming window $w(n) = 0.54 + 0.46 \cos(2n\pi/N)$ (E 9.11)

The Lanczos window $w(n) = \dfrac{\sin(k\pi/n)}{(k\pi/n)}$ (E 9.12)

The development system for the TMS320 series includes software for producing both FIR and IIR filters. Most of the application literature from companies such as TI assume that the user has a rudimentary background on the theory and use of digital filters. The Chassaing and Horning reference includes an excellent section on the theory and implementation of digital filters.

9.5 INTEGRATED DIGITAL SIGNAL PROCESSORS—THE CHAPTER IN RETROSPECT

General-Purpose versus Integrated Processors

Chapter 7 presents great detail about the operations and versatility of general-purpose microprocessors. However, at times the use of one of these large micros

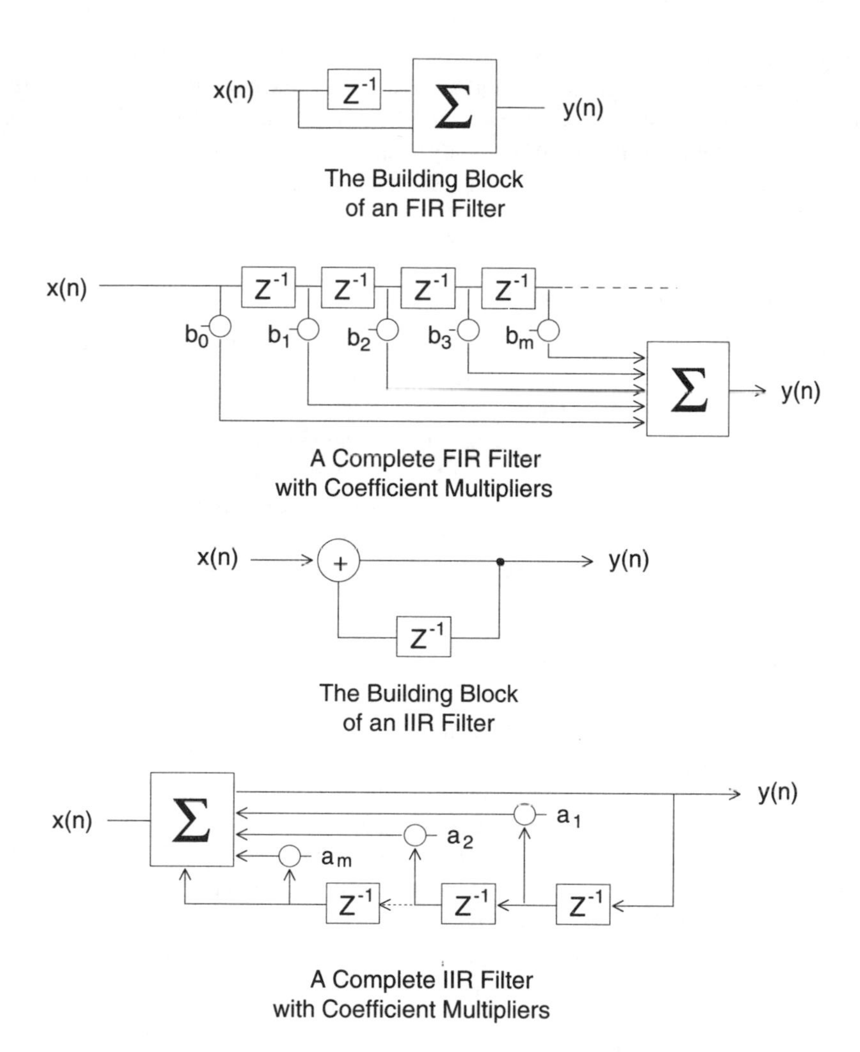

Figure 9.7 The block diagrams of rudimentary FIR and IIR digital filters. Although these figures imply hardware components, most of the digital filters used in integrated signal processors are created (using the same general schemes) as software routines.

for a specific (small) amount of processing can represent gross overkill in the programming, the system considerations, and cost. Likewise, if the input signal is a physical parameter such as temperature, pressure, motion, etc., additional circuits are often needed for preconditioning or preprocessing before the main signal processor can do its job. Also, many times a physical signal, which is analog, may require an input ADC and fast, efficient built-in processing functions such as an FFT. In the past, microprocessors have needed such auxiliary analog conditioning circuits as component or operational-amplifier configured filters. Various contemporary integrated processors already contain some of these conditioning circuits, the ADCs, and many pertinent, very efficient processing functions such as the FFT.

The Contents of This Chapter

The chapter begins by reviewing the circuits required for preprocessing from simple RL and RC filters through transmission-derived filters, polynomial filters, and analog computing circuits. The chapter next describes the application of two common integrated processors and it shows that, although these devices are relatively simple to use, they represent the best of contemporary high technology.

The Texas Instruments TSS400 is presented as an example of a very low cost, low power measurement processor. Depending on the desired operations and the sensors that will be used, the system designer can choose from such software routines as A/D conversion, keyboard scanning, sensor poling timed by the normal computer or a real-time clock, and special linearization routines tailored to the specific sensors used by the system. The easy-to-use and relatively inexpensive development system is described and an application example is given.

For comparison, and increased insight, the TI TMS320 series of integrated processors is presented. These processors represent much more sophistication and speed than the TSS400. The TMS320 processors do many operations in hardware than the slower 400 did in software. Its overall speed and efficiency make it an excellent processor for speech and medical signals. Whereas the TSS400 was a 4-bit processor, the TMS320 is a "real" 16-bit processor that includes the ability to do a 16 × 16-bit multiply. The block diagram and the development and operating systems are described as well as the use of the various built-in processing functions.

Contemporary References

Bahar, H. *Analog and Digital Signal Processing*. New York: John Wiley & Sons, 1990.

Chassaing, Ralph, and Horning, Darrell W. *Digital Signal Processing with the TMS320C25*. New York: John Wiley & Sons, 1990.

Johnson, David E., and Hilburn, John L. *Rapid Practical Designs of Active Filters*. New York: John Wiley & Sons, 1975.

Kloker, K. "The Architecture and Applications of the Motorola DSP56000 Digital Signal Processing Family." Proceedings of the IEEE International Conference ASSP, paper 13.13, April 1987, pp. 523–526.

Libbey, Robert L., *Handbook of Circuit Mathematics for Technical Engineers.* Boca Raton FL: CRC Press, 1991.

Lynn, Paul A., and Fuerst, Wolfgang. *Introductory Signal Processing with Computer Applications.* Chichester and New York: John Wiley & Sons, 1989.

Motorola. DSP56000 Technical Summary—*A 56-Bit General-Purpose Digital Signal Processor.* Phoenix, AZ: Motorola Semiconductor Products, 1986.

———. *DSP56000/DSP56001 Digital Signal Processor User's Manual.* Phoenix, AZ: Motorola Semiconductor Products, 1990.

Strum, Robert D., and Kirk. Donald E. *First Principles of Discrete Systems and Digital - Signal Processing.* Reading, MA: Addison-Wesley, 1988.

Texas Instruments, *TMS32010 User's Guide.* Dallas, TX: Texas Instruments, 1983.

Texas Instruments. *TSS400 Family—Sensor Signal Processors.* Dallas, TX, and Berlin Germany: Texas Instruments, 1991.

Texas Instruments. *TMS320 Family Development Support Reference Guide.* Dallas, TX: Instruments, 1992.

Texas Instruments. Digital Signal Processing Group. *Digital Signal Processing Applications with the TMS320 Family.* Dallas, TX: Texas Instruments, 1986.

Weiss, Daniel P. "Experiences with the AT&T DSP32 Digital Signal Processor in Digital Audio Applications." *Proceedings of the Audio Engineering Society Seventh International Convention.* New York: The Audio Engineering Society, 1990.

10

Artificial Intelligence and Neural Networks

Processing Information Based upon Human Thought Models

An Introduction to Intelligent Machine Processing

10.1 AN INTRODUCTION TO HUMAN AND MACHINE THINKING

Artificial Intelligence (AI) is not a mainstream science or technology. It has been developed by a unique band of investigators with exceptional curiosity, tenacity, and intuition. Its history has included many exhilarating victories as well as many more agonizing and humiliating defeats. Artificial Intelligence is derived from some models of how human beings create and process information and ideas—things that we still know very little about. Even so, the challenge of postulating these rudiments and then attempting to transfer their projected principles into the operation of computers or computer-controlled machines is one of the most intriguing odysseys of twentieth-century science.

AI is about using artificial (nonhuman) devices to solve problems in a manner similar to the way the human brain would. The interesting twist is that no one is very certain about exactly how the brain really solves problems. Thus, AI has drawn upon the research and insights of psychology as well as mathematics and computer science. In fact, one of the major advantages and challenges of AI is the breadth of knowledge it requires. Fundamental artificial intelligence spans an extremely wide area of science and technology—from the learning and cognitive (knowing, being aware) sciences to the mathematics, computer science, and electronics required in the field of robotics. Some understanding of the physiology and medicine of the human senses, especially sight, sound, and touch, is as necessary as a knowledge of optics and mechanics. The researcher and the implementer both need to continually ask, "How is this task, i.e., this process, accomplished by a human and can we, in turn, develop a satisfactory (or improved) 'artificial' software or electromechanical model?"

One of the major uses and advantages of AI has been its use in reflecting back information about the possible mechanisms of the natural brain by building artificial brains. Likewise, AI contributes to the designs of artificial brains by the continuing and diligent study of the natural brain functions. This results in what might be called double mirroring. When an artificial application is contemplated, the investigator will usually first look into the way the anticipated process is accomplished in the human being to obtain that reflected knowledge. Likewise, once the artificial application is completed, the new knowledge gained from that development may, in turn, provide additional insight into the fundamental methods first considered as being used by the human being.

10.2 A HISTORICAL PRELUDE TO ARTIFICIAL INTELLIGENCE AND NEURAL NETWORKS

Serious investigation on how the brain works, i.e., how do people think, certainly goes back to the writings of the ancient Greeks and Chinese. Not only did they discuss and write about this subject, there are records of brain operations and conclusions about the functions of the brain going back at least 2,500 years. Certainly Western thought was extremely influenced by the logic, philosophy, and ethics of Aristotle (circa 350 B.C.). The general subject of reasoning, including both deductive and inductive, continued to be investigated by philosophers, theologians, and scientists for the next 2,000 years. Likewise, the physiology of the brain, as well as its human interpretations, using psychology and psychiatry, continued to grow or was invented. In more recent times, such mathematicians and philosophers as Kurt Gödel, Alfred North Whitehead, and Bertrand Russell have made significant contributions to the ideas of how human beings think and calculate. By the early part of the twentieth century, all the thoughts, discoveries, and projections of these historical thinkers had produced a passionate desire and initiative for a nonhuman "thinking" device. By the 1920s and 1930s, the concept of a mechanical giant brain had permeated both the technology and the literature of most of the Western world.

The two individuals who most directly influenced the beginning of what was to be known as AI were John Von Neumann and Alan Turing. It was Von Neumann who produced the general architecture for the first stored-program digital computer. The Von Neumann design was based on sequential programming and primarily used the four functions of add, subtract, multiply, and divide. By contrast, the British mathematician Alan Turing proposed with equal conviction a computer architecture based on the principle of symbolic processing that would include such functions as "if," "and," and "or." Turing also was probably one of the first to give serious, detailed thought to using the brain as a model for computer architecture. Although the majority of digital computers have been designed using the Von Neumann approach, much of the early AI theoretical studies related to the concept of symbolic processing as proposed by Alan Turing and his conceptual symbolic computer, the Turing Machine.

Although the exact genealogy of the digital computer is always subject to some question, the following items are of interest in the context of AI. Charles Babbage (circa 1830) is credited with creating the first general-purpose (mechanical) digital computer which was termed an analytical engine. The first electronic digital computer, which (in general) followed the Von Neumann stored-program architecture, was the ENIAC (Electronic Numerical Integrator and Calculator). It was developed about 1945 at the University of Pennsylvania by John W. Mauchly and J. Presper Eckert, Jr. The first experimental electronic symbolic processing computer was developed by Xerox (circa 1970) while the first commercial symbolic computer was the Apple Macintosh (circa 1982).

In addition to the work of Von Neumann and Turing on computer architecture, H. S. Black, Hendrik Bode, and later Norbert Wiener made major contributions to the field of feedback and cybernetics, and Claude Shannon and H. Nyquist made major contributions in the field of information theory. There were also many contributors in the fields of medicine, physiology, and psychology that helped to build a starting foundation for AI.

Artificial Intelligence

10.3 THE BEGINNINGS OF ARTIFICIAL INTELLIGENCE

Although there had been much AI-type theoretical and applied research before the 1950s, the scientific discipline of AI formally started as a result of the development of the electronic digital computer. Historically, the birth of AI is usually associated with a 1956 Dartmouth College Summer Research Project sponsored by the Rockefeller Foundation. This distinguished gathering of many of the brightest researchers in the field, along with the investigations and papers that soon followed, formed the thrust that launched AI. It is interesting to note that the name "artificial intelligence" was, and always has been, used by default rather than by acclamation. Few people in the field really liked (or now especially like) the name, but no individual or group invented a more attractive substitute.

Like the choice of the name, there is also no unanimous agreement on exactly what AI really is. Since the field is so diverse and extends into so many established disciplines, it is difficult to meld all the many perspectives into one coherent statement defining its field of investigation and/or its use. The following definitions include samplings from several very respected authors in the field and conclude with the one that will be used as the primary model for this book.

- Artificial Intelligence is the study of how to make computers do things that, at the moment, people do bettter. Also, AI can be termed information-processing psychology. (See the Rich reference.)
- Artificial Intelligence is the study of ideas that enable computers to be intelligent. (See the Winston reference.)
- AI is a research field concerned primarily with studying problem solving in the abstract. (See the Harmon and King reference.)
- Artificial Intelligence is the theoretical and applied science of teaching computers and computer-controlled machines to process information, solve problems, and perform tasks in a manner similar to the way(s) a human being would accomplish these functions.

10.4 SOME SIGNIFICANT EARLY AI PARADIGMS

Significant Paradigms—An Introduction

The following section presents a number of fundamental paradigms (very important or archetypal models) in the literature of artificial intelligence. These techniques and their illustrative programs represent both some very early and some later, more advanced approaches to AI information processing and problem solving.

Usually the first questions that are asked about any AI program (especially in the early days) are, "Is it intelligent?" and "How do you measure its intelligence?" Both the definition and the measurement of natural and artificial intelligence is controversial. One of the more common methods of measuring human intelligence is the IQ test. The classical method of determining if a machine is intelligent was proposed by Alan Turing in 1950. Turing suggested that the test should include both an alleged intelligent machine and a human placed behind a curtain. A human interrogator, in front of the curtain, would ask a series of questions, and if he or she cannot tell whether it is the computer or the human that is answering the questions, the machine is truly intelligent. One can readily see that such an experiment poses a number of difficult housekeeping problems including:

- Both the computer and the human must possess a large amount of knowledge—a large knowledge base.
- If the responses are spoken, the computer must possess a human-like voice, not the computer-synthesized voice that was typical in the late 1970s or early 1980s.

- The responses of the computer or the human must be properly timed. For instance, even though the computer would be able to do most calculations infinitely faster than the human, it must possess the proper human-like response timing. Likewise, if the test is more complex and uses some visual pick-up scheme, the human may have to delay its response. Many problems, such as recognizing faces, may require much more computer time than human response time.

Significant Paradigms—Pattern Recognition Using Analogy or Describe and Match

Pattern recognition problems are ideal tests for the comparison of machine responses with those of humans. For example, pattern recognition, with wooden blocks or printed geometric figures, is well accepted in human intelligence tests. The typical problem will be, "Find how Figure A is similar to Figure B and then, with this knowledge, find how Figure C is similar to D"—where there are several patterns or figures to choose from to select the correct Figure D. Pattern recognition usually uses the principle of analogy which is one of the fundamental techniques used in human thought.

A program that will select by analogy the D figure which has the *closest* corresponding relationship to C, must have at least these three types of rules:

- A description of how any single figure relates to any additional figures inside and thus are a part of the larger figure. See Figure 10.1 (C), (D2), and (D3).
- A description of how the second major object relates to the first major object as the small circle in Figure 10.1 (A). Is this second object larger or smaller, above or below, to the right or the left, or some combination of these relationships?
- A description of how any figure, including any inside object, changes from one figure to the next. For instance, is it larger, smaller, inverted, or has it changed its relative position? Are there any additions or deletions?

The general outline for a describe and match analogy program will include [1] produce a set of rules that will describe the objects, [2] find the rules that describe each of the objects, and [3] from the selected rules, choose the rules that most closely describe the analogs between the objects. Each program designer may take a different approach to the methods of describing the shape and position of the different major and minor inside objects. Some type of implied grid structure can be used. Likewise, each figure can be divided into sections such as quadrants to include descriptions such as x above or below y, x to the right of y, etc. One of the major problems in pattern recognition by analogy is ambiguity. For instance, B is rotated 45° from A. C is a larger object than either A or B. When choosing a D object, should the increase in size or the rotation be the dominant factor in choosing a comparison rule? (See the Winston reference.)

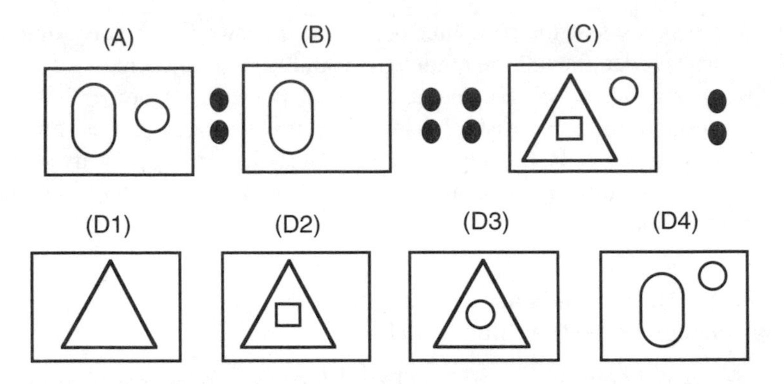

Figure 10. 1 A graphical analogy problem in the form A : B : : C : D. Given a problem like this, the average human can rapidly solve it without thinking about or, in fact, knowing exactly how he or she did it. However, if this type of problem is to be solved by a computer, it must know and use specific rules for determining the relationships between shapes, lines, and positions. In this problem, the B figure is similar to the A figure but has the right object (the circle) missing. Likewise, Figure D2 is similar to Figure C and also has the right (upper) object missing.

Significant Paradigms—Pattern Recognition Using Goal-Oriented Procedures and Trees

Goal-oriented thinking is quite common in our contemporary society. Whether it be a college education, a job, or a career, using the technique of finding a goal is a general thought process. A similar process can be used in programming a computer to solve problems. Elementary goal seeking can be broken down into a two-state process—the present state or situation and the desired or future state. Often there are many intermediate states before the final goal is reached. Goal-oriented procedures are common in both human and AI game playing. For instance, when a checkerboard is set up, ready to play, it is in the first state of the game. After the first move, the board is in its second state, etc. Of course, a winning final or goal state is desired by both players. The overall program of each player is to go from the initial or starting state to the final or goal state as cleverly as possible. The overall program can be broken down into a number of goal states that occur after each move.

Analogy problems can also be solved by a goal-oriented paradigm called goal reduction. In 1977, Terry Winograd of MIT wrote a program to show that a computer could understand commands and questions expressed in English. One of its software routines, called MOVER, controlled a simple one-armed robot while it moved children's building blocks. MOVER worked by using a number of simple symbolic procedures that each specialized in achieving just one kind of goal. Conveniently, the names of the procedures included good mnemonics for what they did:

- PUT-ON arranges to place one block on top of another block. All action starts with a request to PUT-ON. It works by finding a specific place on the target block and by moving the traveling block.
- PUT-AT places one block at a specific place. The place must be specified by a set of coordinates. PUT-AT works by grasping the block, moving it to the specified place, and ungrasping it.
- GRASP grasps the block. If the robot is holding a block when GRASP is involved, GRASP must arrange for the robot to get rid of that block. Also, GRASP must arrange to clear off the top of the object to be grasped.
- CLEAR-TOP does the top clearing. It works by getting rid of everything on top of the specified object.
- GET-RID-OF does the getting rid of. It works by putting objects at locations on the table.
- UNGRASP lets go of whatever the hand is holding.
- GET-SPACE finds space on the top of target blocks for traveling blocks to go. To do its job, it uses either FIND-SPACE or MAKE-SPACE.
- FIND-SPACE finds space. If there is no room on the target block, FIND-SPACE gives up.
- MAKE-SPACE helps out when FIND-SPACE gives up. MAKE-SPACE can do better because it can get rid of blocks that are in the way.
- MOVE-OBJECT moves objects, once they are held, by the moving hand.
- MOVE-HAND moves the hand.

The sequence, as shown in the Figure 10.3 goal tree for moving the blocks shown in Figure 10.2, is as follows.

First, PUT-ON asks GET-SPACE for some coordinates for A on top of B. GET-SPACE tries FIND-SPACE, but FIND-SPACE gives up because D is in the way. GET-SPACE then appeals to MAKE-SPACE, hoping for a better answer. Next, MAKE-SPACE asks GET-RID-OF to help by getting rid of D. GET-RID-OF obliges by finding a place for D on the table, using FIND-SPACE, and by moving D to that place using PUT-AT.

When D is gone, MAKE-SPACE, using FIND-SPACE, now finds a place for A to go on top of B. Recall that MAKE-SPACE was asked to do this by GET-SPACE because PUT-ON has the duty of putting A on B. PUT-ON can proceed now by asking PUT-AT to put A at the place just now found on top of B.

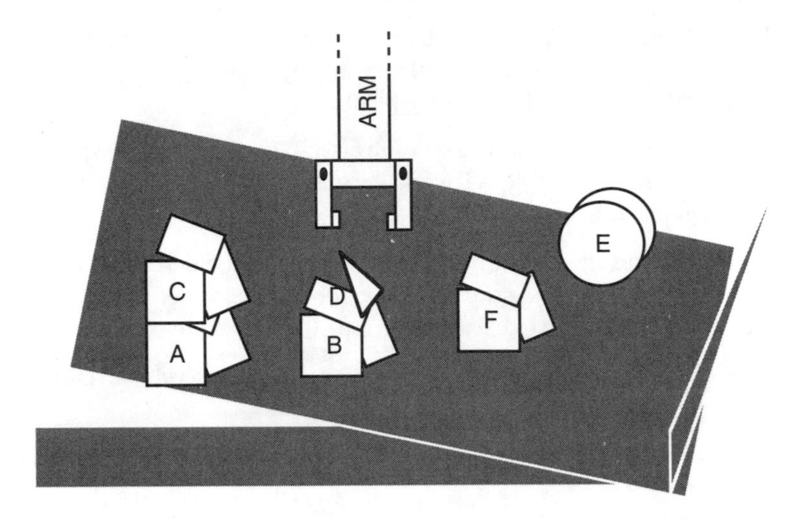

Figure 10. 2 Blocks on a table and the robot arm and hand grasper used with MOVER.

PUT-AT, sensibly enough, asks GRASP to grasp A. GRASP wants to use MOVE-HAND immediately to get to A, but GRASP realizes that it cannot grasp A because C is in the way. GRASP asks CLEAR-TOP for help. CLEAR-TOP, in turn, asks GET-RID-OF for help, whereupon GET-RID-OF arranges for C to go on the table using PUT-AT.

In Figure 10.3, the branches joined by straight lines without arcs are under OR functions. If a function cannot proceed down a given path to the next goal, it will backtrack and either use that same AND or OR function or it will backtrack further until it can find a completed goal.

Note that PUT-AT, at work placing A on B, eventually produces a new job for PUT-AT itself, this time to put C on the table. When a procedure uses itself, the procedure is said to recurse. Systems in which procedures use themselves are said to be recursive.

With A cleared, CLEAR-TOP is finished. If there were many blocks on top of A, not just one, CLEAR-TOP would appeal to GET-RID-OF many times.

Now GRASP can do its job. Once the hand is in the correct position and A is grasped, PUT-AT asks MOVE-OBJECT to move A to the place previously found on top of B. MOVE-OBJECT does this uneventfully with a request to MOVE-HAND. Finally, PUT-AT asks UNGRASP to let A go, which again happens, uneventfully. (The previous text and figures on MOVER are from Winston, Patrick Henry, *Artificial Intelligence,* (2d ed. Reading, MA: Addison-Wesley, 1984. Used by permission.

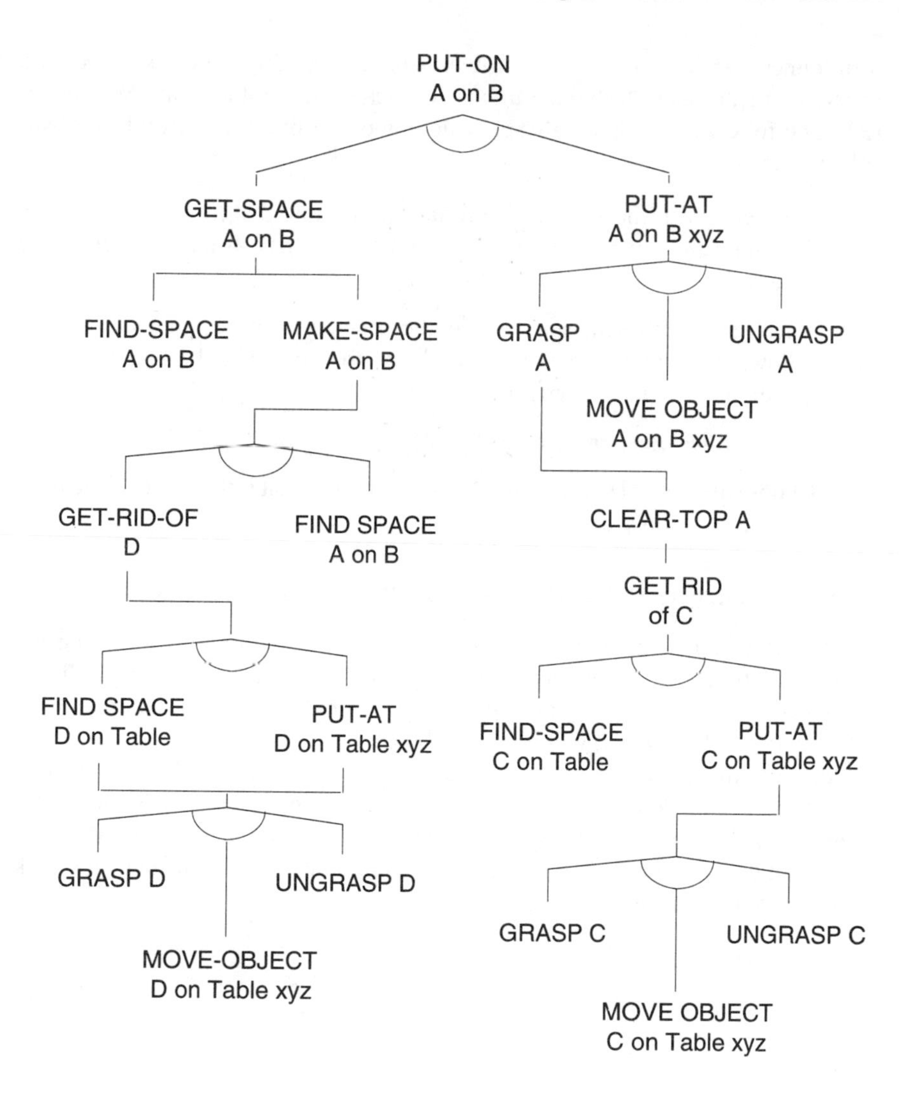

Figure 10. 3 A goal tree for MOVER. Branches joined by diagonal lines and arcs are under AND (node) functions.

10.5 FUNDAMENTAL PROBLEM-SOLVING METHODS—USING METHODS AND KNOWLEDGE

An Introduction

Throughout the life of artificial intelligence, there has been an ongoing debate concerning the relative merits of the methods or procedures versus the merits of knowledge in AI problem solving. The earlier programs focused on methods or form rather than knowledge. The problem-solving paradigms in this section are primarily method- and technique-intense or method-rich rather than knowledge-rich. The following steps usually provide a good outline for a useful problem-solving technique.

- Precisely define the problem. Be certain that the initial or starting state and the final state are completely described. Also, categorize as many intermediate states as possible.
- Analyze the problem and choose the best possible search method. The following methods are typical of those used in AI and can be used as paradigms for many problems.
- List the rules that pertain to each state.
- Choose the control strategy, the rules about rules, that is the most efficient for this type of problem.

Problem-Solving Methods—Forward and Backward Reasoning

If the goal trees in Figure 10.3 were horizontal, rather than vertical, the starting state would be on the left and the final or goal states would be on the right. Thus, the reasoning would normally progress forward from left to right rather than from top to bottom. However, it will be remembered that, at times, it was necessary to backtrack to find the correct path. This principle of going forward or going backward can be extended to the analysis of complete tree diagrams. Some search programs will begin at the initial state and proceed to the final or goal state while others can be configured to start with the assumptions of the goal state and work backward to the starting state.

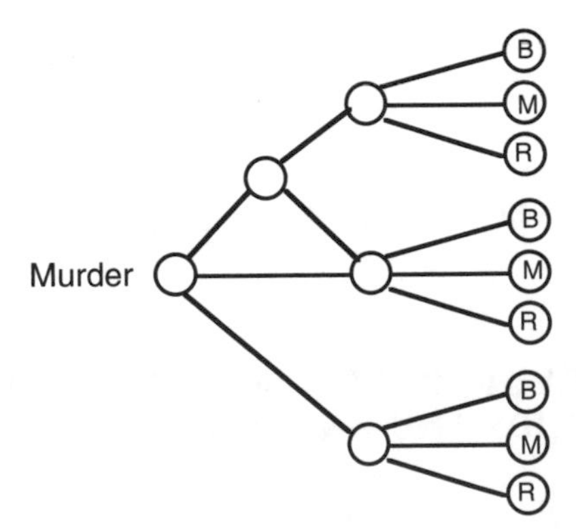

Figure 10. 4 A simplistic goal tree for a theoretical who-done-it. The starting state is the knowledge and rules about the murder and the ending states are the different paths to identifying the butler-B, the maid-M, or cousin Rupert-R as the killer.

A simple-minded "who-done-it" can be used to illustrate these principles. Who killed Aunt Agatha—the butler, the maid, or cousin Rupert? A left-to-right tree, as shown in Figure 10.4, would start with description and rules about the murder, progress through the maze of subgoal nodes and, hopefully, with the clever program you just wrote, end up at one of the final nodes to identify the killer. This process of methodically selecting subgoals or nodes, using the knowledge and rules of the problem from the start state to the end or goal state is termed forward chaining. It should be possible to find the killer by starting at each B, M, or R output node and progressively working backward until the correct path to the murder description is found. In this way, one person would finally be successfully traced back to the starting state and thus identified as the murderer. This reverse, right-to-left process is termed reverse chaining. Doubling back one or more nodes to select a different path is allowed in both forward and reverse chaining. As with the other search techniques described in this section, reverse chaining may not be applicable to all problems. For instance, although it should work to find the who-done-it killer, it probably would not work in trying to follow the clues on a treasure map where the possible final goal locations of the gold are not well defined.

Problem-Solving Methods—Heuristics

The methods that will be outlined in the following sections are all common to a class of AI programs using heuristic search. (Heuristic is derived from the Greek word *heuriskein* meaning "to discover," which is also the root for *eureka.*) Heuristics are procedures or rules that have usually been derived by experiment or trial and error—they are rules of thumb. Heuristic rules will usually work but do not necessarily produce an optimum answer. It will be shown later that it is often difficult to transfer a heuristic rule into an if-then form. Since these heuristic methods are quite often better adapted to a particular type of problem or knowledge domain, not all of these methods will work on every search problem. In fact, one major task is to find the proper method for a particular type of problem. The examples will indicate the general type of problem that can be solved with each described paradigm.

Problem-Solving Methods—Generate and Test

The generate-and-test paradigm can be used when the final goal is well defined. It starts by first generating an answer and then progressively testing this assumption with all the possible solutions. When one uses a dictionary to check the spelling of a word, he or he are using the generate-and-test paradigm. Although this technique is somewhat limited in the types of problems it can solve, it has been used successfully in several very famous AI programs. One of them, DENDRAL, created by the Nobel Prize–winning chemist Joshua Lederberg with Bruce Buchanan and Edward Feigenbaum, is used to predict the molecular structure of an unknown molecule. The program uses the data from a mass spectrogram of the unknown to make a methodic comparison with known molecular structures. ACRONYM is a generate-and-test program used in machine vision. (See the Winston, Rich, and Harman and King references.)

Problem-Solving Methods—Hill Climbing

Hill climbing is an interesting algorithm that, at least in its simplest form, relies on heuristic rules that fall into the category of common sense. It is used in search problems where the final goal may be considered at the top of the hill. However, in all complex search problems, there are impediments to either seeing the top of the hill or knowing what other direction to follow. The key to using hill climbing is to be able to estimate, somehow, the distance to the final goal (the top of the hill) or to estimate the distance of two or more alternate routes. For a practical illustration of the principle of hill climbing, assume that you walk into a strange hotel room and find it too warm. Upon inspecting the thermostat, you find all the markings are worn off. Using your knowledge of the system, you turn the dial a few degrees one way. If both the thermostat and heating system are properly working, within a few minutes you will know whether you turned it the wrong way (it is even warmer) of if you turned it the correct way (it is getting cooler). The problems with the hill-climbing algorithm include the plateau problem and the ridge problem. If there is a plateau problem, you may move in any direction without an indication that you are closer to or farther from your goal. Likewise, a so-called ridge may arise between you and your goal and thus give you a false reading of the distance between your present state and your goal state. Sometimes these problems can be overcome if you use the "big jump" or backtrack. These are special methods that must be included in the program's rule base that let you escape from a position (state) when all normal rules fail. (See the Winston reference.)

Problem-Solving Methods—Depth-First, Breadth-First, and Best-First Searches

Generate and test and hill climbing are termed depth-first searches and indicate moving up or down the branches of a vertical tree, as shown in Figure 10.3. This technique is easy to implement and, many times, can produce a rapid solution. However, if it does not "hit" very quickly, it may waste a lot of time, and, as has been indicated, not find a solution at all.

A second technique, the breadth-first search, is guaranteed to produce a solution—if one exists. Using the breadth-first technique, the search will progressively backtrack and go to another node on the same level. (See Figure 10.3 or mentally rotate Figure 10.4 ninety degrees to visualize the horizontal node levels.) Even though a solution is usually guaranteed, this type of search is likewise very time and memory consuming.

Best-first search is a special combination of the depth-first and the breadth-first techniques. Starting at the beginning node, best-first will drop to the next lower

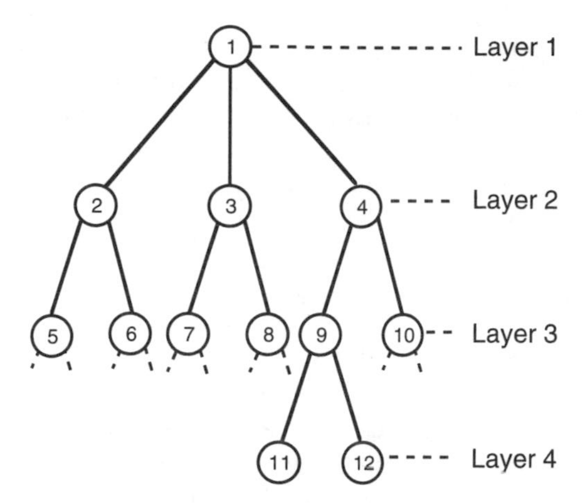

Figure 10. 5 A best-search tree. Starting at node 1, the search will use depth-first to drop to Level 2. Here it will use breadth-first, and a given heuristic algorithm, to estimate the best node in Layer 2 to obtain the goal. This example assumes that of the nodes 2, 3, and 4, node 4 is the most promising. Returning to depth-first searching, the search drops to Level 3 where breadth-first tests nodes 5 through 10. Node 9 seems most promising, so depth-first is again used to go to layer 4. It must be remembered that for any search, a dead end may be encountered and the program will have to backtrack and use a node that was less promising.

level using depth-first. Then, using some heuristic value-estimating algorithm, it will change to breadth-first and test all the nodes on this horizontal level. When the most promising node is found, the depth-search algorithm is again used for another set of tests. If the goal is found, then the search will cease. If the goal is not found, the search will revert to breadth-first and the process will continue(see Figure 10.5).

The literature is full of variations on the above techniques. The number of scholarly papers, magazine articles, and book sections on this general subject is almost endless. For a general discussion see the Rich and Winston references.

Problem Solving Based upon Control

10.6 FUNDAMENTAL PROBLEM-SOLVING METHODS—GPS

An Introduction

GPS, the General Problem Solver, has been given its own section because it was so intellectually intriguing, almost to the point of being seductive. Its implied promise was, "We have done our homework. We now know some of the ways humans and machines operate and think. Now, using this information, we can create a very advanced program, with very special search and control techniques, that will solve a broad spectrum of problems." Although GPS, like several other early AI attempts, did not at all live up to its expectations, its techniques did advance the art of AI and they have since become important references for even contemporary programs. GPS is a control technique. It used a new and, at that time, more sophisticated, strategy to select the path for the next move.

GPS, like several other problem-solving programs, was goal and subgoal oriented. Progress to subgoals or goals may be visualized in several ways. The most obvious is distance—moving up the ladder or down the field. Likewise, in describing an object, the next subgoal should be the next level of overall description. (The leaf is yellow, not green; it is about three inches long, not two inches; it has smooth edges, not jagged edges; etc.) In GPS, the fundamental scheme for finding the next move, and thus progressing toward the goal state, was new and unique. The search algorithm became known as means-ends analysis. To use means-ends analysis, the move may at times go to the right or left or even fall back rather than always proceed in a general forward direction. Because of this, the graphical metaphor of a football gridiron may be more appropriate than the more common family tree configuration. The major thrust in GPS is to reduce differences in a manner similar to the idea of moving up in short increments for a first down in football.

GPS uses symbolic locations to define the states of the problem including:

- The IS or Initial State—the starting point.
- The CS or Current State—you are here now but wish to proceed toward the goal state.
- The AS or nearby Adjacent State—this may be the most convenient stepping stone for reaching the next or final goal state. In GPS, there are usually several nearby or adjacent states. Any particular AS may or may not help. Likewise, even if it does help, it may temporarily set you back. (Using the football metaphor, you may be thrown for a loss but if you gain some positional advantage, it may be worth it.)
- The GS or Goal State—the object of the state search.

The Difference-Procedures Table

A simple but very lucid example of using GPS was given by Winston in his book and also in an excellent MIT taped lecture course. (See the references—used by permission.) Assume that Robbie-the-Robot wishes to travel from his home in Boston to visit Aunt Agatha in Los Angeles. The goal is to go to Aunt Agatha's home—a total distance of about 3,000 miles. However, this major distance can be broken into a number of minor lengths. A second parameter of the problem, the *thus* of the decision-control process, can be the mode of transportation. Alternatives can include the procedures of walking, taking his own car, a taxi, a train, or a plane. All of these procedures for using transportation require the prerequisite or precondition "be there." While the Difference-Procedures Table, as illustrated in Figure 10.6, is not always necessary for composing a GPS program, it is recommended because it greatly clarifies the relationships in the problem.

- Starting at the initial state I (Robbie's home in Boston) he has the five choices of walking, driving, airplane, taxi, and railroad. Since none of the preconditions, such as "be at car" can be met, the first procedure must be walking. However, because the distance limitation on walking is less than a mile, it is assumcd that he cannot walk to the airport, let alone to Aunt Agatha's. Robbie can only choose a taxi or the car.
- Robbie chooses to drive and since he is now "at car," the precondition is met.
- Once at the airport, he must again examine the difference-procedure tradeoffs to get from his parked car to the airplane. Again, the "be at" limitations dictate that he must walk to the airplane.

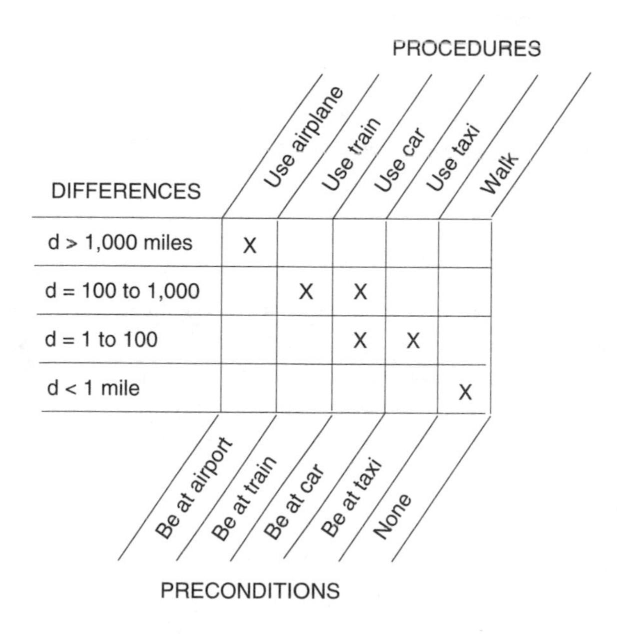

Figure 10. 6 The Difference-Procedure Table is one way to visualize the way decisions relate the differences (distances in this example) to the preconditions and the possible moves.

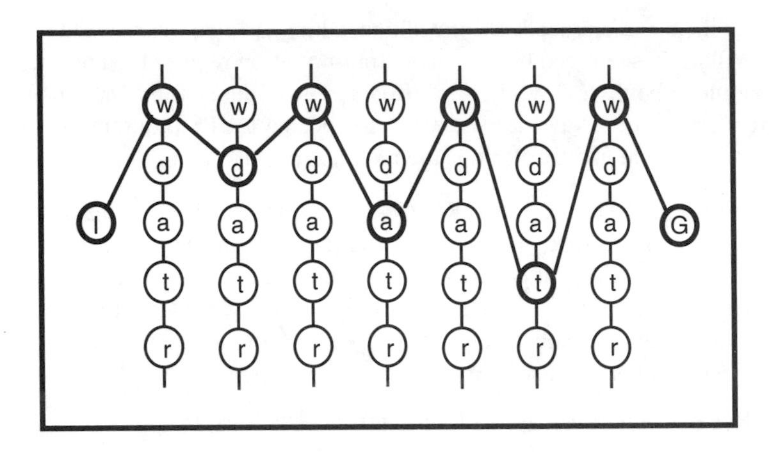

Figure 10.7 A gridiron representation of the Robbie-the-Robot trip from Boston to Los Angeles to see Aunt Agatha. The I in the circle stands for the initial state, G the goal state, and w, d, a, t, and r stand for walk, drive, airplane, taxi, and railroad. The interconnecting lines indicate only the final decision routes the program selected and do not show the recursive try and fallback attempts for the various adjacent states.

A second graphical technique for visualizing the relationships in GPS problems can be the gridiron representation. Figure 10.7 presents the relationships using this technique.

- The "at airplane," and "d > 1,000 miles" not only allow but dictate the airplane as the mode of travel.
- At the Los Angeles airport, the goal is still Aunt Agatha's house but first he must find a mode of transportation from the airport to the house. Since this distance is between one and ten miles, the first choice would be to use the car. However, because of where the car is, this would require an airplane trip back to Boston. The next choice would be to use a taxi but Robbie is not "at taxi." The only possible choice is to walk to the taxi since walk requires no preconditions.
- Now "at taxi," Robbie can take a taxi to his aunt's house and reach the final goal state.

In summary, GPS embodied a set of quite sophisticated control decisions relating to difference reduction, sometimes called means-ends analysis. It also had the ability of feedback, or recursion, for repeatedly going back to find the most appropriate adjacent-state solution. However, its major reliance on control rather than intelligent procedures limited its problem-solving abilities.

Human and Computer Learning

10.7 ARTIFICIAL INTELLIGENCE AND LEARNING

Computer Learning—An Introduction

There are two questions that continuously fascinate any AI practitioner—"Can machines and computers learn?" and "Can machines and computers learn by themselves?" To answer one or both of these questions may require at least partial answers to several other questions, such as "What is learning?" and "What is the result of learning?" As a starting point, most humans possess the ability to *change* because of the information they input from observing or reading. In contrast, only very special computers or, specifically, computer programs can be said to change (themselves) because of the information they input.

The question can computers learn can also be approached from the perspective of cognitive psychology, using the concept of *cognitive understanding.* In general, normal human beings have the ability to process newly acquired information and conclude whether it does or does not make sense. This ability is quite amazing when one considers that even a very small child can absorb, process, and usually use, in a positive and fruitful way, all the sensations that he or she receives at home, on a busy street, at the supermarket, a busy county fair, etc. In contrast, most computers (programs) cannot successfully process (make sense of) this type of global information. Still, there certainly has been some progress in teaching computers to understand relationships and to learn enough from their inputs to somewhat alter their processes. This section will give some insights into the AI research concerning computer learning and examine examples of the techniques that were developed. It is also germane to now review the introductory section on human and machine thinking, especially the paragraph that discusses learning by "reflecting back" from both computer-program processing and human-brain processing.

Computer Learning—A Learning System Model

Figure 10.8 shows a comparison between a learning model (one of many) and a simple model for what is known as process control.

Using this type of model for the furnace example, the Reference Input is the thermostat's temperature setting. The "+," or adder circuit, combines the input from the thermostat with the Sensor signal and the Transfer Function processes these combined inputs for use with the Controller. The Controller will turn the plant (the furnace) on or off to keep the room temperature constant.

This same general model represents a paradigm for how humans and machines may learn. The outside world, here termed "ENV" for environment, will feed in the information to be used (learned) by the system. The Learning Element uses

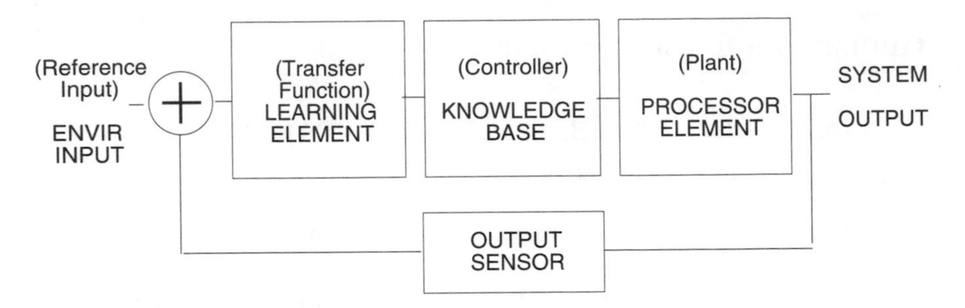

Figure 10.8 A simple feedback system used to compare the more familiar model of a plant controller with that of a learning system. In a plant, such as a common home furnace, the temperature is kept relatively constant by sensing the room temperature and comparing it with the setting of the thermostat. Likewise, the performance for the knowledge base is compared with the knowledge input from the environment. (See the text for more details.)

(transforms) this information to improve the system Knowledge Base that, in turn, is used to (control or) perform a task to reach the desired goal. The system's final Processor Element output (goal) is compared with the knowledge input to produce a system learning error. As with any classical feedback system, this error signal is used to improve the precision of the system's knowledge output in the same way that, in the furnace example, the feedback was used to improve the constancy of the room temperature.

The characteristics of the reference (input) knowledge are an important factor in the type of learning system required to perform certain tasks. This idea will be developed in the later examples of computer learning. If the input knowledge is high-level, it is quite abstract and thus may be relevant to a large class of problems. However, this high level of abstraction means that there are many missing details which, in turn, means that the Learning Element must transform these abstractions into detailed information that the Knowledge Base can use to perform the assigned task. If the reference information is low-level, the system may need to generalize and ignore some unimportant details. In the Learning system, this feedback error forces the system to continually guess or form a new hypotheses about the correct knowledge-system output required.

Examples of Computer Learning—Rote or Memorized Learning

One of the earliest successful AI programs that investigated the principles of learning was a checkers-playing program developed by Samuel in 1959 (see the Barr, et al. reference). It was based upon the premise that if this AI program can memorize the moves of many successful checker games, it, in turn, can play a successful game of checkers. Before each successive move, the program would use a tree search (see Figure 10.4) to look ahead several moves and attempt to evaluate the best possible course of action based on its memorized knowledge base. In every case, the program assumed that the opponent would counter with its wisest move. After each move, the board positions were stored for future reference. Thus, if a particular board configuration is repeated in a later game, the search and evaluation have been done before and need not be repeated.

After several iterations and improvements of the program, Samuel and other members of the AI community drew these conclusions including:

- To obtain maximum speed and efficiency, the knowledge base must be carefully organized to avoid redundant or ambiguous information. The search algorithm must produce rapid and efficient searches.
- At first, a rote learning program will continue to improve with additional information inputs but after a while, the additional memory requirements, coupled with the additional search time, tend to make further improvements impractical.

Examples of Computer Learning—Learning by Taking Advice

In some instances, it is almost as difficult to give advice to a computer program as it is to a human being. The major problem with the computer taking advice is its ability of understanding or interpretation. It is usually desirable to input the advice information in the form of the human's natural language—the normal language, such as English, that the human speaks and understands. However, for a computer, human natural language understanding is extremely difficult. In fact, natural language understanding is in itself a separate branch of AI. There are several other types of AI programs that involve the problem of understanding and interpreting the written symbols of human language we call words. The section on understanding concepts will also discuss this problem.

Early research on computer learning by taking advice followed two paths. The first type of program accepted high-level, abstract information and converted it to rigid rules that could be used as a guide by a system's final processor element. This was an open-loop system with no feedback or correction (sensor) path. The second type of advice-taking program was a closed-loop system since (at least at first) a human expert usually monitored the output conclusions and provided changes or updates—see Figure 10.9. Consequently, this second type of advice-taking program is usually called an expert system—see the later section on expert systems. Thus, although advice-taking systems, especially expert systems, have become one of the major success stories of AI, the creator must be aware of these major problems:

- The advice is usually presented in high-level, abstract form that necessitates very careful design of the techniques and rules used in the learning stage. In addition to producing rules from the input knowledge, there must also be *rules about rules* to ensure that, as new knowledge is received, it does not cancel or ambiguate items in the previous knowledge base.

- Since the advice is usually in natural-language form, there are added difficulties in producing an accurate and complete knowledge base. (See also the section on expert systems.)

Examples of Computer Learning—Learning from Examples

"For example" is a very common classroom phrase that usually introduces one of the most successful teaching and learning techniques. An example refers to specific knowledge or procedures. It is used in the hope that the student (or computer program) can generalize from this particular illustration and formulate rules or procedures that will be applicable to a number of similar situations. This process of progressing from the specific to the general is termed induction or inductive reasoning. Likewise, the process of going from many general facts or assumptions, such as your favorite literary detective would, to a specific instance (the villain), is called deduction or deductive reasoning.

The learning from examples processing includes these procedures:

- An example (sometimes termed an instance) is selected from the example space (the catalog of examples) using the induction in-terpreter. It is assumed that there may be more than one example and that more examples will be added later.

- If this example is considered appropriate after it is selected and processed by the interpreter, then it is ranked or graded for its applicability to the rules governing a particular problem. Using the reverse deduction planning and selection processor, another example is selected. After all the likely examples are selected and graded, the example with the highest score is selected.
- This two-space model assumes that the search processor can produce appropriate candidates.

- This also assumes that the program (teacher) can select examples that are not ambiguous. This is an excellent guide to the quality of training.

- To be successful, the interpreter must be able to intelligently process the examples so that they are classified into a proper and useful order.

Examples of Computer Learning—
Learning Concepts with Winston's Program

In the previous section, an example could be learned and successfully used without knowing what it really is or what it represents. To understand the "thingness" of an object or an idea requires a high level of abstraction for both a human and a

An Inductive Learning Model

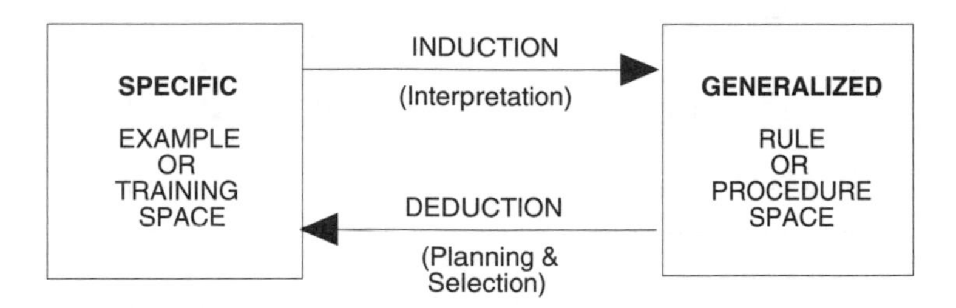

Figure 10.9 A two-space model of example learning. It is assumed that if the interpretation and inductive generalization processing is efficient, either the human or the computer can continue to expand its learning. Thus, its rule or procedure base will expand by being trained with additional rules.

computer, although humans are much more adept at this than computers. For instance, even a small child can soon learn to have some concept of time or what "tomorrow" means or the ideas of reward and punishment. Even the most sophisticated computer program has much difficulty with concepts like these. In 1970, Patrick H. Winston (later to become head of the MIT Artificial Intelligence Laboratory) wrote a landmark program that involved computer learning of concepts. In this program, using objects similar to children's blocks, an arch was taught to know that it is an arch (see Figure 10.10).

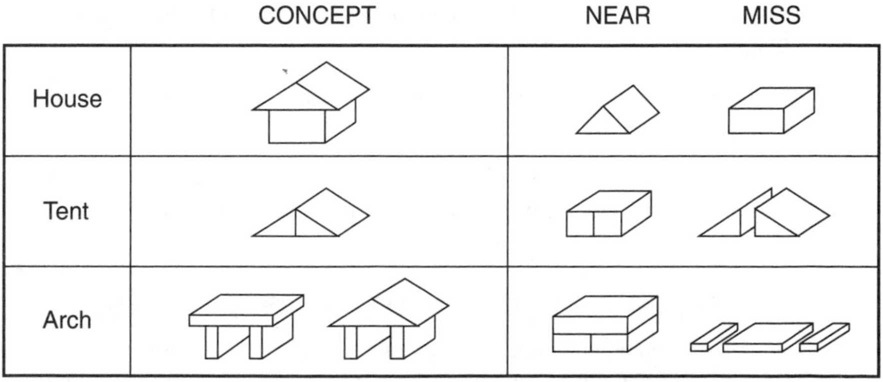

	CONCEPT	NEAR MISS
House		
Tent		
Arch		

Figure 10.10 Some hypothetical blocks like those that were used in Winston's program of learned concepts.

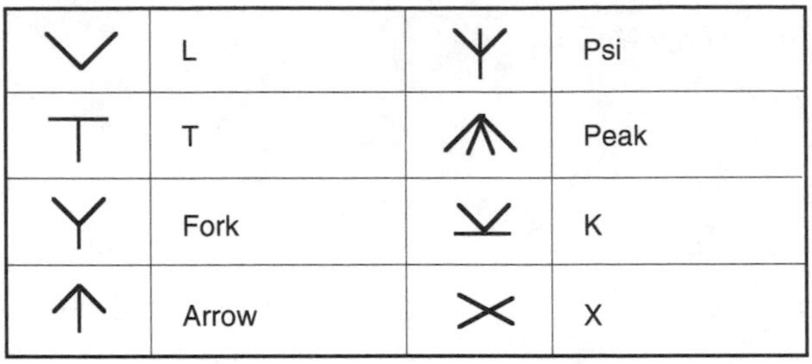

Shape	Name	Shape	Name
∨	L	Ψ	Psi
⊤	T	⩚	Peak
Y	Fork	K	K
↑	Arrow	X	X

Figure 10.11 Some elementary line shapes used by programs to describe three-dimensional objects.

Winston's program was built upon some of the following programming principles:

- Provide a good description of both the object and its boundaries. To do this required rules to describe such boundaries as an "L," a "T," a Fork, an Ar-row, a Psi, a Peak, a "K," and an "X."
- Devise rules that describe how these elementary line shapes are linked to create three-dimensional objects (see Figure 10.11). This defining and labeling of lines and boundaries was originated in works by Huffman and Clowes and extended by some very extensive work by D. Waltz. The Waltz expansion included shadows, cracks, and concave edges. (See "B5. Constraint Propagation in Interpreting Line Drawings" in vol. III of the Barr et al. reference.)
- Devise rules to show how object faces meet. This is especially difficult when the faces meet to form trick or optical illusion, nonrealizable objects—see Figure 10.12.
- Devise rules such as "to the right of it," "above it," "inside of it," etc., to establish object relationships.
- Devise rules that define concepts—in this case, the concepts (and the relationships to other items) of an arch, a brick, a wedge, etc.

Examples of Computer Learning—Learning Mathematical Concepts

The concepts of mathematics are some of the most, if not *the* most difficult objects of human thought. Quite soon after the happy marriage of digital computers and the research and insights that were later known as AI, investigators started probing the possibility of writing programs that would solve both elementary and advanced mathematical problems. Of the many attempts to write mathematical solution programs (some were successful and some certainly were not), the following two examples illustrate the research that provided the backgrounds for

Figure 10.12 A sketch of a physically nonrealizable object. Optical illusions that seem to trick the mind are the subject of many of the drawings of M. C. Escher. (See the Escher reference.)

such math-oriented programs as the contemporary spreadsheet programs (Lotus 1-2-3, Excel, etc.) as well as programs such as Mathematica, MathCad, and Theorist.

Examples of Computer Learning—Learning to Solve Calculus Problems

The original attempts, including SAINT and SIN (see vol. I of the Barr et al. reference) produced programs that would integrate by following the time-honored techniques of integral table lookup and integration by parts. Added to these rudimentary approaches were the AI-type methods of search and goal seeking. The flow of the goal-searching algorithms followed a typical tree-like structure. The first attempt was to find a standard integral form that would solve the problem. An example of a standard integral form would be $\int \cos x \rightarrow \sin x$. If this original search did not find a goal, a second search would look for algorithms such as factoring out a constant, decomposing into the sum of integrals, etc. A third search would rely upon AI-type heuristics such as an experienced human calculus expert might use including an attempt to find appropriate substitutions or integration by parts.

Examples of Computer Learning—AM and Discovery

Several of the previous programs learned abstract concepts to solve practical problems. These programs and similar later programs can be defined as simulating the valuable human function of conceptualizing. They seem to have advanced to a higher level and thus may be termed more intelligent than some of the earlier AI programs. Another example of a program that was certainly attempting to enter the more rarefied atmosphere of intelligence is a 1976 program by Douglas Lenat called AM. AM did not learn concepts to solve practical problems. Its purpose was to discover new and interesting concepts in mathematics. In fact, to the delight of its creator and his colleagues, AM once discovered a very sophisticated concept that, at the time, seemed new and unique. However, after a period of careful research, they found that a very famous Middle Eastern philosopher and mathematician had earlier discovered and documented it.

AM was one of the pioneering programs that was structured on a large knowledge base. (See more about knowledge-based programs in the following section on expert systems.) The original program started with a knowledge base of some 115 mathematical concepts. There were two overall goals in the program. First, using its task agenda, it would use the knowledge base to find examples of the concept. Likewise, it would, if possible, find more information about this given concept including specifications, generalizations, and analogies relating to other concepts. Thus, the program was creating more and more specific information about this given concept by searching its knowledge base. The second goal of the program was, with the information thus gathered and categorized, to postulate new concepts.

The creation of information about an existing concept and the creation of new concepts (and then the creation of information about that new concept) was very carefully ordered by the rules in the task agenda. This agenda included a very special feature termed "interestingness." When a concept was entered or discovered, it was continually graded on its interestingness. As more examples, specializations, generalizations, etc., were amassed, the interestingness or worth grade would go up or down. If the program found a particular concept to lack interest, it would go on to a more promising concept.

To summarize, this section on learning started with the elementary—but very important—idea of learning by rote. It concluded with the description of quite a sophisticated program for discovering concepts in mathematics. As time progressed, the emphasis in AI programs shifted from the development of clever control strategies (although some of these will always be important) to an emphasis on knowledge. Professor Randall Davis, of MIT, often states in his lectures and writings, "In the knowledge lies the power."

Finally, there is also a powerful technique in both human and artificial learning that is often overlooked—learning by mistakes. For instance, referring back to the search trees, many times the incorrect branch path will be selected which, if progress is to be made, necessitates falling back and starting again. Yale professor Rodger Schank (see the reference) notes that, whether it is natural or artificial learning, the "importance of failure" should not be disregarded.

Programs Based upon the Knowledge of Human Experts

10.8 KNOWLEDGE OR EXPERT SYSTEMS

Expert Systems—An Introduction

During the 1980s, a very powerful type of AI program technique developed, which became popularly known as an expert system. In contrast to the research that progressed from the broad control principles of GPS, expert systems migrated from the concept of a dominant control-rich structure to a dominant knowledge-rich structure. Expert system programs concentrated on a narrow, well-defined subject area rather than a general subject area.

Expert systems are built primarily using the knowledge and the problem-solving techniques of a human expert in a particular field. For most successful expert systems, the area of the human's expertise is purposely limited. Thus, although a general medical practitioner may have a very wide area of diagnostic ability, a typical expert system would only use his or her knowledge of blood diseases, stomach disorders, etc. Expert systems are interactive. A doctor or other medical professional can question the patient about his or her problems and enter this background medical information into the computer running the expert system. With this patient information, i.e., these observations, added to the existing knowledge base, the expert system, using the rules that were obtained by the original "expert" doctor, can (at least theoretically) provide a diagnosis and usually a recommended treatment for the patient's ailment. (It should be noted that although several medical diagnostic expert systems have been developed and they are often used as examples, their use in everyday medicine is extremely limited.)

The Elemental Structure of an Expert System—Rules

Expert systems are also sometimes termed rule-based systems. The fundamental rule structure is IF . . . THEN . . . . This is usually expanded to include IF . . . AND . . . OR . . . THEN. For instance, to diagnose a nonoperating flashlight, IF it has recently been in working condition, AND it has not been dropped, AND the batteries are a year old, THEN replace the batteries. If the flashlight had been dropped, it is assumed there would be a rule about replacing the bulb.

There are several classes of rules in an expert system including absolute rules, probabilistic rules, heuristic rules, and control rules (rules about rules).

1. Absolute rules indicate a strict relationship between the IF . . . and THEN . . . parts of the rule. For instance, IF the traffic light in your lane is red, THEN stop.

2. Probabilistic rules indicate either some stated degree of mathematical probability or some less precise indication of what chance, which is usually stated in terms of confidence. If the rules are derived from information that is believed to be statistically accurate, then the resulting rules can include some statistically appropriate factor. The Bayes' theorem, which is derived from probability theory, is often used to calculate the probability of particular events based upon a given set of observations. (See the Rich reference.)

   It is more common to use rules that reflect the general level of confidence of the original expert rather than some specific probability. For instance, the expert might have stated, "Under these given circumstances, such and such will happen in about six out of ten occurrences."

3. Heuristic rules are the fundamental rule structure for expert systems. Whether it be a skilled physician, a skilled auto mechanic, a skilled production-line worker, or any other expert, their well-distilled rules of thumb are the unique and most useful items in their knowledge base.
4. Although expert systems have substituted knowledge, in the form of rules and user inputs, for exotic control systems there is still a need for a control structure to propel the program toward its desired goals. Thus, some of the knowledge is in the form of rules about rules and how these rules relate and produce motion through the system.
5. There are also "why" rules that can be used to ascertain why and how a certain conclusion was reached. This ability to ask why is one of the outstanding characteristics of most expert systems. For instance, again using the example of a medical expert system, it is extremely important that the doctor not only know what the system's diagnosis is, but the procedure and the rationale that were used to derive it.

The Controls and Control Strategies Used in Expert Systems

As has been indicated, some control structure is required in expert systems. This structure is somewhat analogous to the operating system used in computers. Not only are there the aforementioned rules about rules, there are rules about inputs and rules about conclusions. There are also specific strategies designed to facilitate the progressive choice of the next proper rule. These operating systems are sometimes termed production systems, although the original meaning of a production system related to a system that attempted to duplicate one model of human mental operations. Another term that is often used to describe the prime mover of an expert system is the inference engine. This part of the overall knowledge base is used for knowledge acquisition from the user interface, for rule explanation, and, of course, for the choice of the proper rule for search and inference and for the choice strategies.

Two very important expert system control strategies are forward and backward chaining. These two strategies are somewhat akin to the earlier mentioned problem-solving methods of forward and backward reasoning and the examples of inductive and deductive reasoning.

A forward-chaining strategy starts by assuming that all beginning IF-rule clauses are true. Once it establishes these "true" facts, it then continues by finding out what other rules might be true. It then uses the facts from these new rules to find other possible rules. The process is repeated until the program reaches a goal or runs out of possibilities. Backward chaining starts with a goal rule to find if it is true. When a rule is thus found, the program uses this knowledge to again go backward to find another correct IF . . . THEN . . . rule. Eventually the back-chaining sequence ends when a question asked of the previously stored result is found. (Refer back to the who-done-it example in the section on reasoning.)

Gathering Knowledge—The Knowledge Engineer

An expert system usually reflects the training, experience, and intuition of one outstanding individual or, in rare cases, a small group. To produce an expert system, i.e., to duplicate these problem-solving skills in an interactive computer program, requires an individual who can glean the facts and procedures from this superior individual and transform them into a working knowledge base that includes the information that is contained in the IF . . . THEN . . . rules. It is, of course, much more complex than just the creation of these rules. It is usually very difficult for a knowledgeable problem-solver in any field to consciously know all the steps used in solving a problem. Even if most of the procedure can be pieced together, it may be a lengthy, sometimes agonizing process. Likewise, after most or all of the procedure is assembled, the expert's account must be transformed into the structure and flow of the computer expert system. The individual who possesses both the communication skills and patience to successfully interview the expert and the technical skill to translate the information into the complete knowledge base for the computer program is called the knowledge engineer. The task of the knowledge engineer may take weeks or months and thus he or she must have the ability to continue to ask more questions without overly disturbing the human expert.

Creating the Expert System—The Expert Shell

The knowledge engineer will enter the assimilated and translated knowledge into what is commonly known as an expert shell application program. This expert shell is roughly analogous to a computer spreadsheet. The shell is configured to accept the information and do some reformatting and reorganizing so that the rule base will allow for the proper execution of the program. Even though the shell does significantly aid in the information entry, building a complete, working expert system usually requires skill, almost infinite patience, and much reiteration.

Examples of Expert Systems

Expert shells vary from the very simple, used primarily for tutorial purposes and costing about $100, to very complex systems costing $10,000 or more. If one wishes to learn to create expert systems, adequate shells can be obtained from most bulletin boards. Some shells are quite open and detail the how, what, and why of all the steps required in creating and executing the expert system. Conversely, some shells are much more streamlined and take a general paint by the numbers approach. Likewise, the nomenclature used to denote the same general parts of the shell will vary from program to program.

To illustrate the general character of an open and complete expert shell, the following will outline the creation and use of a shell called EXPERT. This program was created by the Computer Science Department of Rutgers University and is patterned after a medical diagnostic program called MYCIN and, for that reason, tends to use medical-like terminology. EXPERT was designed to run on mainframe computers. The major structure of the program besides its rule base includes the hypotheses or conclusions the program makes, based upon the findings; the findings or observations gathered for the interactive inputs; and the suggested treatments. The findings are derived for a simple question-and-answer checklist that the operator is asked to use.

There are three types of rules that are created for the rule base. FH rules relate the findings to the hypotheses, HH rules relate one hypothesis (usually a preliminary conclusion) to another hypothesis (usually a later or final conclusion), and FF rules relate findings to findings. FF rules are usually housekeeping rules about rules and should be used sparingly. The syntax for the rules departs from normal English and resembles the symbols used in assembly language or symbolic computer languages like LISP.

There are two examples of degree of confidence. The hypotheses or conclusions can contain a "WT" (weighting) number from .001 to 1.0. The default value is 1.0. A WT factor of less than 1.0 does not connote a relative number in a strict statistical sense but rather a frequency of occurrence relationship to other related conclusions. EXPERT also permits putting a cost factor on questions ranging from the default of 1.0 to a maximum of 99. The cost factor on a question is used to flag the designer and the user that this question and its possible results dictate that it should only be asked down the list after other preliminary, less risky, questions are asked. For instance, "Have you had a blood test?" infers the possibility of risk and expense and requires pertinent background information before the question is raised.

The two possibilities of weighting can be used as part of an overall program strategy. For instance, the strategy used in the example to follow employs the following plan: [1] ask the least costly questions first, [2] examine the highest weighted hypotheses first, [3] examine hypotheses that are related to the user input findings, and [4] first try findings that can potentially increase the maximum absolute value of the hypothesis. Different shell models will offer different strategy-forming techniques. However, it is the responsibility of the expert system designer to use his or her experience and judgment to produce an expert program that will reflect the procedures and results of the human expert and still do it in a minimum amount of time. The designs of most practical expert systems require a number of iterations to produce a final, working system.

An Automobile Engine Diagnostic Program—An Example of a Simple Expert System

The following simple tutorial example is taken from the reference by Weiss et al. and is used by permission. Since the EXPERT shell was resident on a mainframe

Table 10.1 The Outline of the Categories Used by EXPERT

**HYPOTHESES	. . . The conclusion the program reaches based on the findings and rules. The asterisks are part of EXPERT's syntax.
*TREATMENTS	. . . The solution to the problems or problem descriptions selected by the operator.
**FINDINGS	. . . The general category for observations that the operator will enter.
*CHECKLIST	. . . The format used to allow the user a simple way to enter the problems with the car. There are several checklists relating to different problem categories.
**RULES	. . . The FH, HH, and FF (not used in this example) rules.
*END	. . . The search procedure will stop here and the program will then display its suggested treatment solutions.

computer, the procedures for creating and running the program entitled "Car" is different from a shell that is resident in a PC. Likewise, the specific syntax, although representative, is peculiar to EXPERT, as shown in Table 10.1. The construction and strategy must be carefully thought out before the program is written, using a word processor or text editor, and then compiled in the EXPERT shell. The outline of the program units follows.

Based on this general outline, the sample "Car" listing would be as follows:

Car—A Sample Model

**HYPOTHESES . . . The compiled program uses the following abbreviations. Note also the use of one or two periods for syntax and the use of items like (.5) to indicate weighting.

```
CWS    Car Won't Start
FUEL   .Fuel System Problems (.5)
FLOD   ..Car Flooded (.2)
CHOK   ..Choke Stuck (.2)
EMPT   ..No Fucl (.2)
FILT   ..Fuel Filter Clogged
ELEC   .Electrical System Problems (.5)
```

CAB ..Battery Cables Loose or Corroded (.2)
BATD ..Battery Discharged (.2)
STRT ..Starter Malfunction (.2)

*TREATMENTS
REP Car Repairs
WAIT .Wait 10 minutes or Depress Accelerator to Floor While Starting
OPEN .Remove Air Cleaner Ass'y and Manually Open Choke With Pencil
GAS .Put More Gasoline into Tank
RFLT .Replace Gas Filter
CLEN .Clean and Tighten Battery Cables
NSTR .Replace Starter

**FINDINGS

*CHECKLIST
Type of Problem:
FCWS Car Won't Start
FOTH Other Car Problems

*Multiple Choice
Odor of Gasoline in Carburetor:
NGAS None
MGAS Normal
LGAS Very Strong

*Checklist
Simple Checks:
DIM Headlights Are DIM
CFLT Fuel Filter Clogged
LCAB Battery Cables Loose/Corroded

*Checklist
Starter Data:
NCRN No Cranking
SCRN Slow Cranking
OCRN Normal Cranking
GRND Grinding Noise from Starter

*Numerical
Temp Outdoor Temperature (degrees F):

*Yes/No

EGAS Gas Gauge Reads EMPTY
**RULES
*FH Rules

F(LGAS,T) -> (FLOD,.8)
F(TEMP,0:50)&[1:F(SCRN,T),F(OCRN,T)] -> H(CHOK,.6)
. . . If the TEMP is between 0 & 50 degrees, and [1] the SCRN is clogged (True) or there is OCRN (slow cranking), then there is a 0.6 chance the choke is bad.
F(EGAS.T) -> H(EMPT,.9) . . . If the gauge reads empty, there is no fuel (9).
F(CFLT,T) -> H(FILT,1.0)
F(LCAB),T) -> H(CAB,.9)
F(SCRN,T) &F(DIM,T) -> H(BATD,.7)
F(NCRN,T) & F(DIM,T) -> H(BATD,.9)
F(NCRN,T) & F(DIM,F) -> H(STRT,.7) . . . The F, after DIM, stands for false
F(GRND,T) -> H(STRT,.9)
F(FCWS,F) -> H(CWS,-1)

Note that in many entries, the casual observer might question or outright disagree with the logic. For instance, in the item F(EGAS.T) -> H(EMPT,.9), the human expert evidently assumed a small chance of fuel gauge failure.

The foregoing is a listing of the program model before it is compiled in the EXPERT shell. Once it is compiled, the system is ready for a diagnostic consultation session. (The following session does not include all of the EXPERT options —see the Weiss et al. reference.)

@EXPERT
Enter the file name: Car
Type ? for a summary of valid responses to any question asked by the program.
CASE TYPE: (5) Demo
Enter Name or ID Number: Cadillac
Enter Initial Findings (Press return to begin questioning):

1. Type of Problem
   1) Car Won't Start
   2) Other Car Problems
   Checklist:
   1 . . . The asterisk "" is used as a prompt. The user entered a 1 or 2.

2. Simple Checks:
   1) Headlights Are Dim
   2) Fuel Filter Clogged
   3) Battery Cables Loose/Corroded
   Checklist:

*2

3. Starter Data:
1) No Cranking
2) Slow Cranking
3) Normal Cranking
4) Grinding Noise from Starter
Checklist:
*why ... The user wishes to know why this question was asked. The program replied:

The question was asked to help determine FH-Rule 9:
? Starter Data: Grinding Noise from Starter. GRND must be true; It is unset. ... See above for the use of the "?" When the "?" is answered, the systems returns to the last item.

3. Starter Data:
1) No Cranking
2) Slow Cranking
3) Normal Cranking
4) Grinding Noise from Starter
Checklist:
*Fix 2 ... The operator wishes to revise the answers in item 2.

Fix: Simple Checks:
1) Headlights Are Dim
2) Fuel Filter Clogged
3) Battery Cables Loose/Corroded
Checklist:
*1,2

3. Starter Data: ... Back in order again.
1) No Cranking
2) Slow Cranking
3) Normal Cranking
4) Grinding Noise from Starter
Checklist:
*1 ... Note the operator revised this answer again.

4. Gas Gauge Reads EMPTY
*no

5. Odor of Gasoline in Car:
   1) None
   2) Normal
   3) Very Strong
   Choose one:
   *2

Summary

NAME: Cadillac

Case 1: Visit 1 Date: 12/12/93

Type of Problem:
Car Won't Start

Odor of Gasoline in Car:
Normal

Simple Checks:
Headlights Are Dim
Fuel Filter Is Clogged

Starting Data:
No Cranking

Diagnostic Status:

1.00 Fuel Filter Clogged
0.91 Battery Discharged . . . The program has calculated some confidence factors.

Treatment Recommendations:

0.82 Replace Gas Filter
0.82 Charge or Replace Battery

This simple example illustrates the general workings of an expert system. The details of this large mainframe system certainly are not the same as one designed for a PC. However, one can see that several example sessions are required before a new design is attempted. The major points are that this is primarily a knowledge-based system and that any usable expert system shell (and thus the final program) will provide the user with an analysis of how and why the diagnosis and recommendations were determined. (The foregoing examples of the use of the expert system "CAR" were used with the permission of Prof. Casimir A. Kulikowski of Rutgers University, New Brunswick, New Jersey.)

10.9 NEURAL NETWORKS—ATTEMPTING TO SIMULATE THE PROBLEM-SOLVING TECHNIQUES OF THE HUMAN BRAIN IN SOFTWARE AND HARDWARE

Neural Networks—An Introduction

Just as expert systems, which seemed suddenly to be invented in the 1980s, were an outgrowth of the previous three decades of research (see Barr et al., vol. II), so do Neural Networks (NNs), the rising technology of the 1990s, have roots going back several decades. Even before the formulation of what is now called AI, researchers were investigating the possibility of structuring an artificial thinking machine (the term computer came later) based on proposed models of how the human brain operates. For instance, John Von Neumann, Marvin Minsky, and Frank Rosenblatt were all influenced by a 1943 paper by Warren McCulloch and Walter Pitts titled "A Logical Calculus of Ideas Imminent in Nervous Activity." Probably the most direct outcome of this early landmark paper was Rosenblatt's Perceptron (see later).

The problem-solving ability and information base of an expert system is gleaned from the education, skill, and experience of a human being. However, the problem-solving ability of a neural network is derived from many previous problem-solving examples. A neural network can be taught—it can learn. In general, the more (pertinent) examples that are used in the training process, the more accurately a neural network will solve a problem. A neural network, like an expert system, must be restricted to one problem-solving area. Neural networks are not general problem-solvers. Neural networks make excellent pattern recognition and classification systems. For instance, the disk that is included with the Nelson and Illingworth reference gives an example of a neural network system that will approve or reject applications for bank loans. This program, Loan Advisor, was created by using information from real loan applications at a local bank. Examples of neural networks that classify types of plants or plant diseases, types of leaves, etc., are legion.

Neural Networks—Neurons, Artificial Neurons, and Processing Elements

A neuron (Figure 10.13) is the model elemental building block for human and animal thinking—plants do not have neurons. When scientists attempt to produce artificial mechanical, electrical, chemical, or optical neurons, the resulting elementary artificial cells are often called processing elements (PEs)—see Figure 10.14. The four major parts of the natural neuron cells are mimicked in the artificial parts of the PE. Of these four elemental parts, the synapse and the nucleus are the most influential in the signal processing. The synapses are the weighted junctions between neurons that successively transfer the previous input signals to each neuron. The nucleus is the major processor. It sums the inputs from the preceding connected neurons and then modifies these combined signals in a manner similar to an electrical network transfer function. The basic equation of a transfer func-

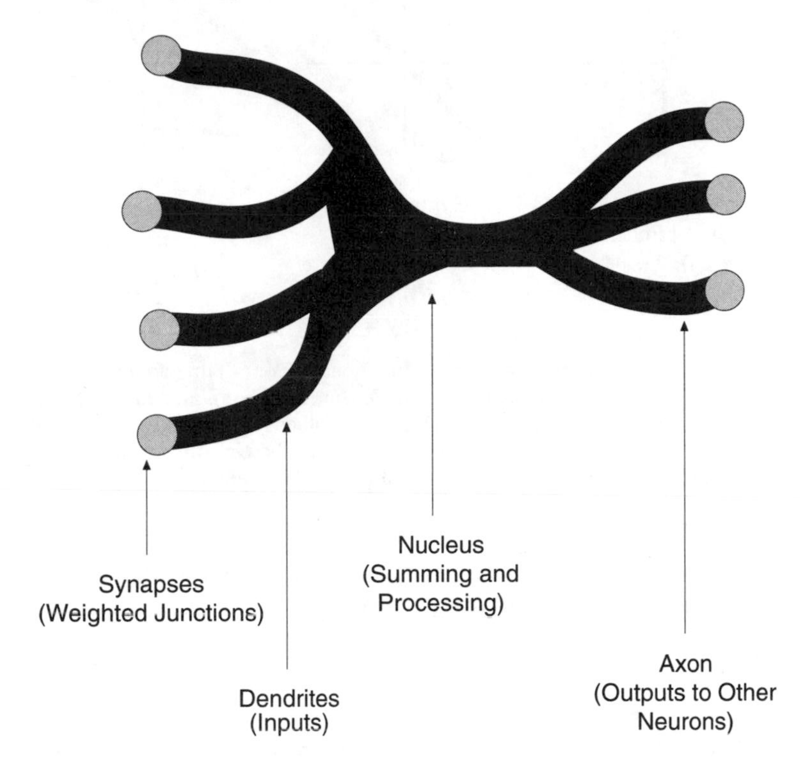

Figure 10.13 One common model of the natural neuron cell as found in the human brain.

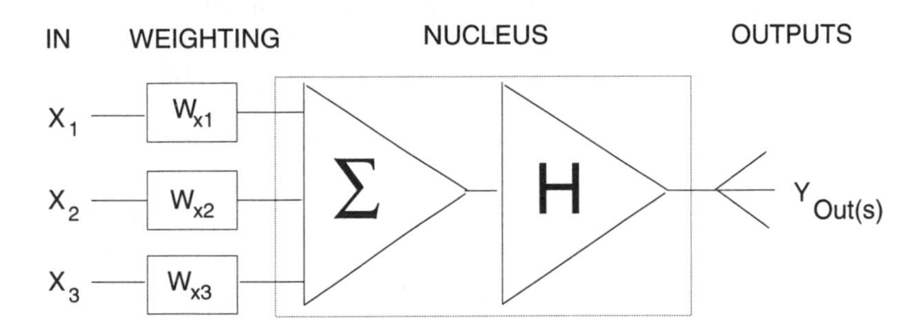

Figure 10.14 A model of an artificial neural cell, commonly termed a Processing Element. The items "X_n" simulate the inputting dendrites and the items "Wx_n" are the best-attempt duplications of the weighting synapses. The Σ (summing functions) and the H (transfer functions) simulate the processing done by the nucleus.

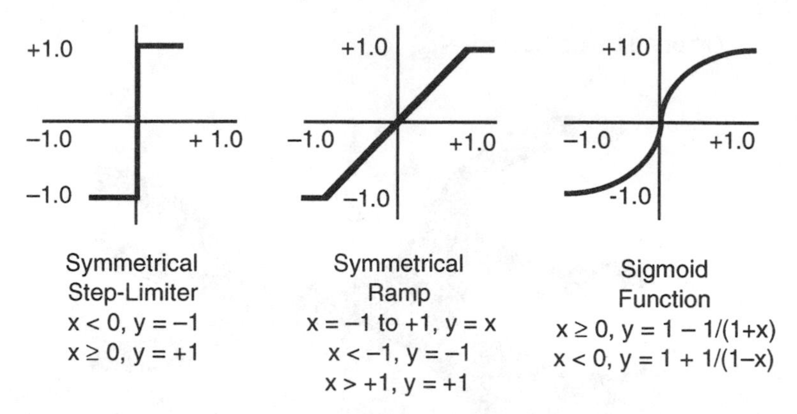

Figure 10.15 Some representative transfer functions (H) or (T) that are used for artificial neuron processing elements. Many of the transfer functions are nonlinear. Although all of the above functions are symmetrical about the x-axis, other symmetries, including functions operating only above the x-axis, are also common. (See the Nelson and Illingworth reference [Appendix C] for the transfer functions of other NN paradigms.)

tion relates the input(s) "x" to the processed output "y" with "H," or more commonly H(s) or H(jω), representing the transfer function.

$$Y_{out} = H X_{in} \text{ or } Y_{out} = H(j\omega)\, Xi_n \qquad \text{(E 10.1)}$$

Neural Networks—Combining Processing Elements into a Neural Network

An elementary tutorial neural network may contain from as few as four to perhaps twenty to forty PEs. Many real problem-solving neural networks may contain several hundred PEs and a network of several thousand PEs is possible. The number of inputs to the network will depend on the number of variables in the problem and the number of outputs will depend on the solution options required. As a starting rule of thumb, the number of processing elements should total about four times the number of inputs.

The processing elements are conventionally placed in groups or layers. (Unfortunately, it is common to represent the layers as vertical groups rather than layered horizontal groups—see Figure 10.16.) There is always an input layer and an output layer. The main working PEs are arranged in groups called hidden layers—since they are hidden from any input or output signals.

Neural Networks—An Introduction to Architecture and Weighting

The selection of an architecture for a neural network is somewhat analogous to choosing the external circuit for an operational amplifier—it depends on the problem. Research has produced a number of working NN architectures and more are being developed. Like amplifiers, NNs can be open loop or use (sometimes un-

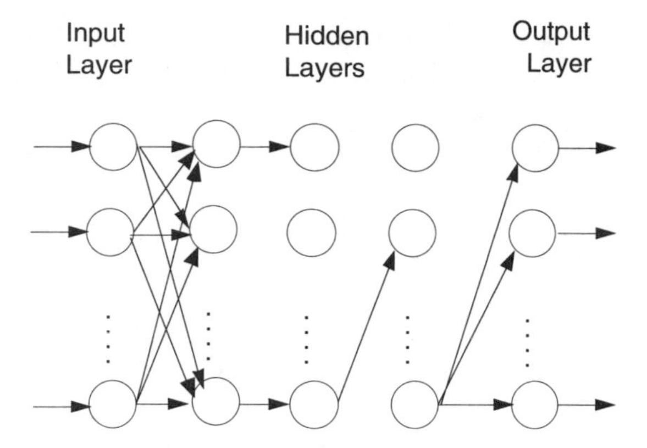

Figure 10.16 One representation of a simple neural network. The input and output layers often have weightings of 1.0 and thus are simply isolation buffers. In most cases, the output of each PE in a given layer connects to the inputs of all PEs in the next layer. (Many arrows were omitted for clarity.) Unfortunately, the literature has no uniform way to indicate the shape of the PEs or the position and action of the very important weighting elements. Usually it is tacitly assumed that the weighting factors are represented in the (usually arrowed) connection lines.

stable) feedback loops. Again like operational amplifiers, special resistor input networks can be used for summing and for gain selection or weighting. (The Nelson and Illingworth reference gives an excellent model of a processing element constructed with resistors and diodes, for nonlinearity, and operational amplifiers. Practically all neural networks can be simulated in hardware or in software.) Although there are twenty or so common paradigms for neural networks, the following four examples present a general summery of common NN models.

Neural Networks—The Perceptron

This is one of the first neural networks and is usually considered the godfather of electrical networks that attempt to simulate some of the actions of the human brain. The device, also termed a learning machine, was produced in 1958 by Frank Rosenblatt, a psychologist and neurophysicist. It was developed as an artificial (eye) retina that could learn to recognize patterns.

The Perceptron could be taught to recognize a limited number of patterns but, unfortunately, it did not live up to its initial predictions. There were some defects that doomed the Perceptron. First, compared to later but similar circuits, it lacked the facility to change the waiting in the extractor circuits. This critically limited its ability to learn. Likewise, there was one very important, commonly used logic function, the exclusive-or (XOR) function, that was theoretically impossible with the Perceptron configuration.

The derivation of the XOR for neural networks is a simple exercise that illustrates some of the principles of using weighting and layers and is very well pre-

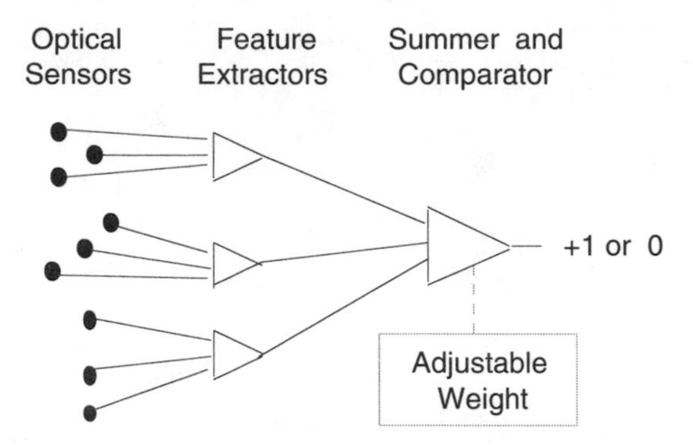

Figure 10.17 A simple Perceptron. The optical sensors are placed at random to pick up some characteristic of a given pattern. The extractor circuits combine these inputs, which are then modified with a nonadjustable weight to output a +1 or a 0. The summer/comparator will, by using an adjustable threshold (weight), likewise output a +1 or a 0.

sented in the Nelson and Illingworth reference. The first illustration will use a two-layer Perceptron-type network similar to Figure 10.17. The XOR logic requires two inputs, as shown in Figure 10.18.

If the inputs are 0,0, zero times anything is zero and thus the final cell will not fire and therefore its output will be 0, the desired output.

If the inputs are 1,0, the top input will be activated and, even though the lower cell is not activated, the sum of the top 1 and the bottom 0 will still equal 1 and the final output will be 1 as desired.

If the inputs are 0,1, the network action will be the same, but reversed, and again the desired output will be 1. If the inputs are 1,1, both input cells will be activated and therefore both cells will output a 1 and the final output will be 1, which is not what is desired.

A second XOR neural network attempt will incorporate a network that has a hidden layer and processing elements with selectable weighting. This neural network would require a training phase to establish the final, correct weightings.

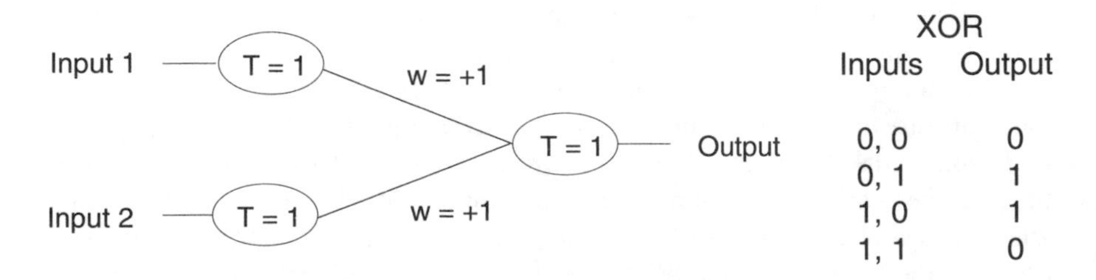

Figure 10.18 A first attempt at using a Perceptron-type NN to implement an XOR function. All weights are 1.0 and the ON/OFF threshold "T" is also set to 1. On the right is the desired XOR logic table.

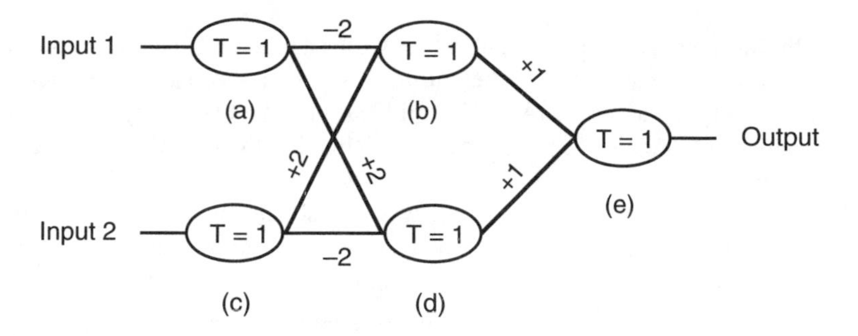

Figure 10.19 A second try at implementing the XOR function in a neural netwok. This attempt includes a hidden layer. The threshold "T" values are shown in the cells and the weighting factors are shown as simple "+" or "–" numbers. The details of the training that established the PE weights are not described. This is only one of several workable configurations.

If the inputs are 0,0, cells (a) and (c) will not fire and thus they will propagate zeros to cells (b) and (d). Again, their outputs will be zero. That will produce a zero output from (e)—the correct response.

If the inputs are 1,0, (a) will produce a summed value of –2 at (b) and a +2 at (d); (c) will produce a 0 at (b) and a 0 at (d). Now, combining at (b) the –2 and the 0 will not fire (b) and its output will be 0. The output of (b) is 0 times the +1 weighting—still 0. However, at (d), +2 and 0 will sum to a +2 and fire (d). At (e), the sum of the of 0 and +1 will override its threshold and the output of (e) will be the desired +1.

If the inputs are 0,1, the processes will reverse and again produce a +1 output.

If the inputs are 1,1, the inputs to both (b) and (d) will be the algebraic sum of +2 and –2 to produce a 0. Consequently, neither (b) or (d) will fire, which will produce a combined 0 input to (e) and a 0 output—the correct XOR value.

Neural Networks—ADALINE/MADALINE

The previous section about the Perceptron and its failure to execute the XOR function will serve as a foundation for the introduction of more advanced neural network paradigms. It is historically interesting to note that in the 1960s, two well-respected researchers, Marvin Minsky and Seymour Papert, published a work stating that because of the limitations, the Perceptron and neural computing in general, were not interesting subjects to study. Luckily, a few loyal, tenacious scientists kept working and later produced the extremely useful technology we now know as neural networks.

The ADALINE (ADAptive LINear Elements), developed by Bernard Wilson and Marcian Hoff, introduced several innovations that are used in many contemporary neural networks. Their network was three layers. Each PE could sum its inputs including a bias term. The output values were changed from the Perceptron. If the input sum was greater than 0, the output was a +1 and if it was less than or equal to 0, the output was –1. ADALINE could learn from a teacher by systematically comparing the outputs with the inputs and adjusting the neural weights until there was a match.

ADALINE, as shown in Figure 10.20, was a practical, working neural network. One of its early demonstrations was to show its ability to create adaptive feedback networks. Using a TV camera as an optical pickup, a system was developed to balance a vertical broomstick on a moving cart—a feat almost impossible for a human to do. (See the Nelson and Illingworth reference.) Likewise, early in its life, Professor Bernard Wilson used ADALINE to developed an adaptive neural network that would cancel echoes and noise on telephone lines. Later generations of this fundamental device are still used on every telephone call that is made. (See the Anderson and Rosenfeld reference.)

The MADALINE or Multiple Adaline neural network in Figure 10.21 runs several ADALINEs in parallel. By using different ADALINEs with different weighting algorithms, a much wider variety of patterns could be recognized.

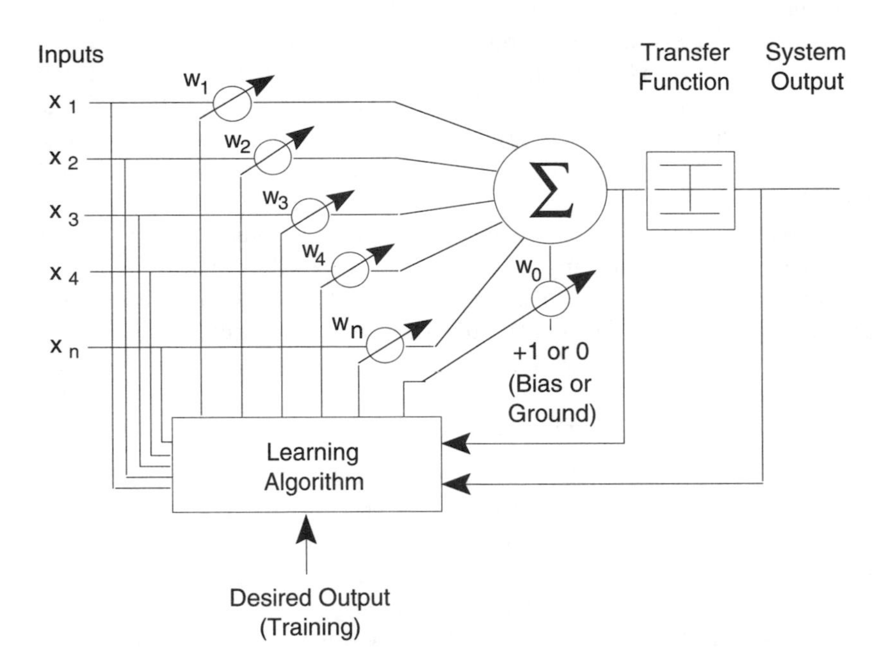

Figure 10.20 The ADALINE neural network processor.

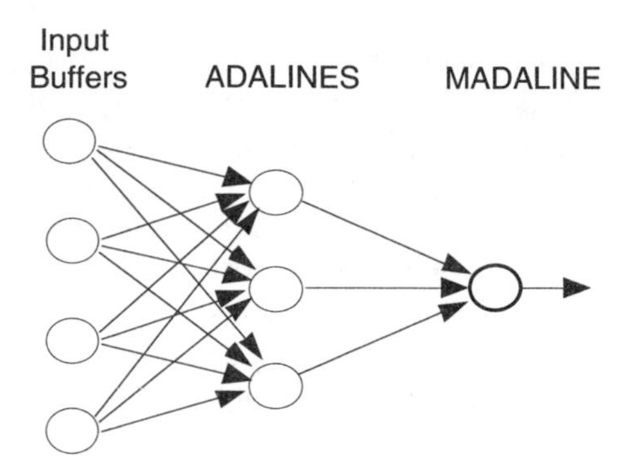

Figure 10.21 The MADALINE or Multiple Adaline neural processor.

How Does MADALINE/ADALINE Learn?

Referring to Figure 10.20, the learning or training phase requires a series of known pattern (data) inputs plus their corresponding correct training outputs. The input pattern is applied to the x1, x2, . . . x_n inputs and both the system outputs and the correct training outputs are applied to the Learning Algorithm box. When the system output does not match the training data output, the learning algorithm progressively changes the neuron weights until a closer match is obtained. The value of the final error, typically in the order of 1 percent, depends on the training technique (the algorithm), the number of training examples, and the time allowed for training.

Neural Networks—The Back-Propagation Network

Of the twenty or so contemporary neural network structures, back propagation is probably the most popular and certainly is the one that is generally used for an introduction to neural networks. It is versatile and solves most of the problems of the earlier approaches. The name back propagation is derived from the technique it uses to distribute the cause for errors. All back-propagation networks will have at least one hidden layer in addition to the input and output layers. During learning, information is propagated back through the network and used to update the connection weights.

Designing Neural Networks— Formatting the Input and Training Data

Both the input information, which is applied to the x1, x2, . . . x_n inputs, and the training-set data must usually be normalized to a range of zero to one. Likewise, the output(s) to the network are usually in the range of zero to one. These outputs may have to be rescaled if real-world values are again required. If the inputs come from a simple black and white pattern, the white values may be set to 0 and the black values may be set to 1—an easy and understandable arrangement. If some gray scale is assumed, there will be intermediate decimal fractions in the data representing the gradations from 0 to 1. If the NN is designed to choose your ideal or best compromise automobile, then such data items as cost, miles per gallon, size, and even style and color, must each be normalized to fit in the range of 0 to 1.

An exact understanding of the use of the input and the training data can best be understood by simulating a neural network with a common personal computer spreadsheet. The Nelson and Illingworth reference describes a technique of scaling data using Lotus 1-2-3. Likewise, the Jones and Hoskins reference presents a simulation that gives a real-world understanding of the input and output numbers of the hidden neurons.

Designing Neural Networks—Choosing the Neural Network Shell

The design of most neural networks may begin with a rough pencil and paper outline or some preliminary spreadsheet calculations but, eventually, the final design is usually accomplished by using a prepackaged neural network application or shell (this word is adopted from expert-system terminology) program. Neural network shell-type programs range in price from those derived from bulletin boards that are free or inexpensive shareware, to some very fine tutorials in the order of \$100 to \$200, to the full-featured models that can run in the thousands of dollars. Probably as important as the program itself is the documentation that accompanies it. Any program, including those for neural networks, that does not supply documentation giving some background, alternative approaches, and a considerable number of worked out application examples is not worth it, regardless of the price. Try to obtain an application program that will allow experimentation with at least three types of neural network paradigms—the Perceptron, ADALINE, back propagation, etc. Also, remember that there are also integrated circuit implementations of both experimental and application-specific neural networks. In fact, many, if not most, of the industrial applications of neural networks are in integrated hardware form—they are faster and cheaper.

Designing Neural Networks—Creating a Back-Propagation Network to Classify Species of Iris

As an example of building a simple but real-world neural network, a classification neural network will be constructed, step by step, using NeuralWare's NeuralWorks Explorer—see Figures 10.22 and 10.23

1. The Explorer program is opened and "BackProp Builder" (BPB) is chosen. This particular program also allows a choice of several other NN paradigms. The Backprop screen starts with no PEs or interconnections except node 1, which is labeled and has the function of the "Bias." There are several other network paradigms besides back propagation that are available in programs like NeuralWorks.

2. The BPB presents several choices. First, since there are four input sepals and petal size choices and three species categories, the Input Layer is set to four and the Output Layer is set to three. As a start, one hidden layer is chosen with four PEs.

3. Other start-up selections include:

   - Connect Bias—will connect the bias (usually considered the system ground) to all PEs.
   - Default Schedule—will set the learning and default schedule for all layers. With more advanced networks or with more experience, each layer would have itsown learning and recall schedule.
   - RMS Learning Graph—will present a graph that will show the error converging around zero as the network is trained.
   - Weight Histogram—will show the weights going to theoutput layer.
   - Delta Rule—sets the learning rule for this network. There are several learning algorithms that can be selected depending on the choice of network type.
   - Sigmoid—is the transfer function that will be used.
   - iris_tra—the name of the normalized database that is used for the learn (training) phase.
   - iris_tes—the database that is used to test the operation and accuracy of the final rework.

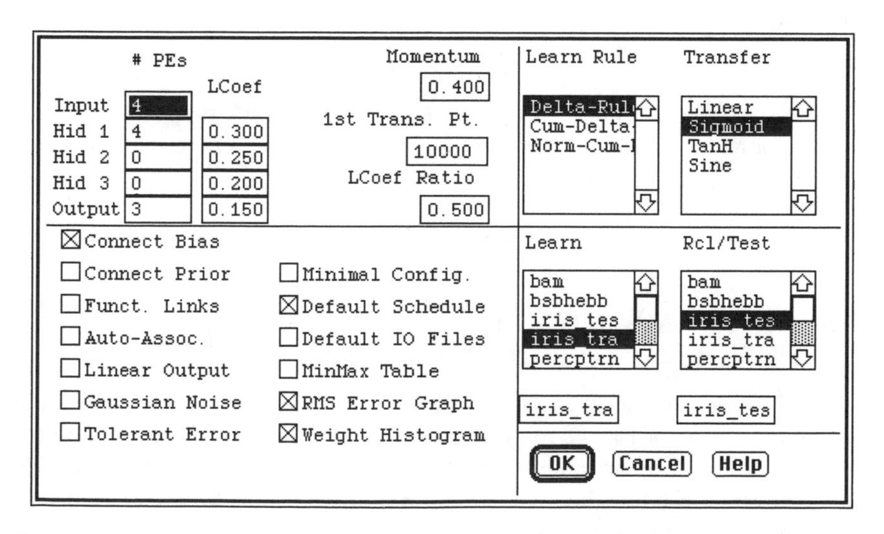

Figure 10.22 The NeuralWorks setup dialog box for a back-propagation network. Note that for clarity, there are a number of choices that are not detailed in this introductory tutorial outline.

A Neural Network Example:

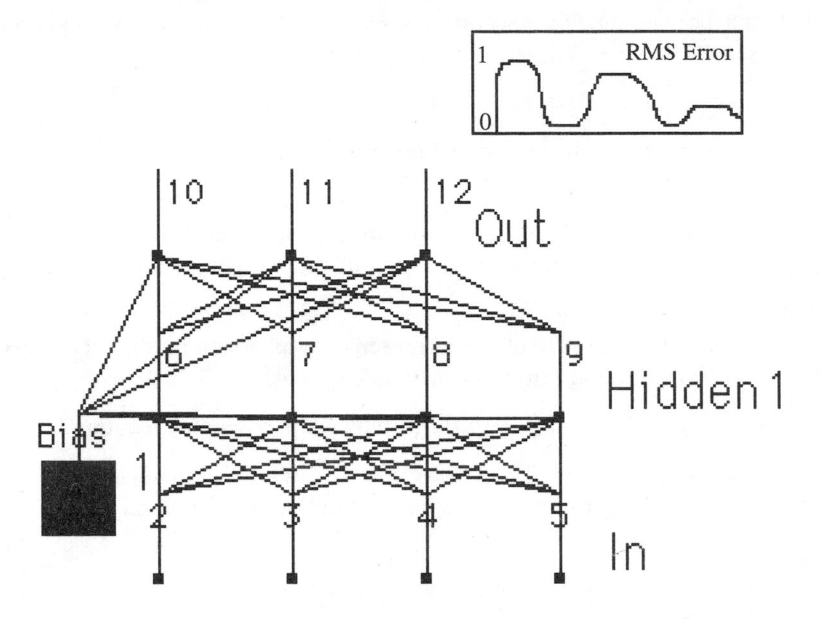

Figure 10.23 The network of the iris classifier. The box in the upper right, labeled RMS Error, shows how the vertical output error is decreasing as the horizontal training time is increased. Note that the error is approaching 0. There are a number of these graphs—NeuralWare calls them instruments—that can be added to aid the designer in following and understanding the training process.

(The preceding information and figures describing the back-propagation neural network for an iris classifier were used with the permission of NeuralWare, Inc., of Pittsburgh, Pennsylvania)

10.10 ARTIFICAIL INTELLIGENCE AND NEURAL NEWORKS—THE CHAPTER IN RETROSPECT

This final chapter tends to stand alone because it describes programming machines, such as calculators or robots, to perform processes that are based on presumed models of how the human brain functions. Since no one is certain about the precise processing—the thinking—of the brain, these attempts to model such processes are always open to scrutiny or criticism.

The Beginnings and the Development of Artificial Intelligence

Although AI and Neural Networks can claim some of the same roots, this chapter starts with the description and growth of AI and concludes with information about the now more popular NNs. Artificial intelligence, like neural networks, first developed as a way to make electromechanical devices operate or "think" in ways modeled after hypotheses of the functions of the human brain. As the new electronic computers began to proliferate, proponents of AI pushed to change computer architecture from that of the pioneering (and successful) efforts of John Von Neumann to the "symbolic processing" architecture proposed by Alan Turing. The main thrust of the AI advocates was to produce a better, more human-like problem solving machine.

As AI developed, two primary paradigms were followed as approaches to make the machines (computers) solve problems. One was to develop as series of procedures where the idea of control was dominant and the other was another series of procedures where knowledge was dominant. Both of these approaches are explained and developed and examples are given for the use of both. The program GPS is the dominant example for the use of control and the general category of programs known as expert systems represents the dominance of knowledge. These two classes of programs, or their antecedents, are shown to solve problems in describing objects, solving logic or game-type problems, solving searching or who-done-it problems, mathematical problems, etc. The work in AI on how humans and machines learn is also described and models representing computer learning are presented. In the concluding section on AI, expert systems are described in detail and examples are presented.

Neural Networks—Computer Programs That Truly Learn

The prime mover in NNs is the use of data about the activity that needs a solution to a problem. For instance, if one wished to produce an AI expert system that would help him or her pick the best common stocks, he or she would ask the best-known stock picker—the human expert—and use that knowledge to write a stock-selection expert system. However, if it was decided to design an NN that will aid in the selection of common stocks, one must first obtain all possible data about stocks in general, when and why they fluctuate in price, their price to earnings ratios, etc. These data can then be taught to a properly designed NN and, at least in theory, the NN, which has learned all about the stocks, can help you make a fortune on Wall Street.

From the above illustration, two important items can be gleaned. First, the NN must be properly planned, designed, and executed. Although there are several types of NNs, it is most common to first start with one called a back-propagation network. The general design rules and approaches for designing back-propagation networks as well as examples using this paradigm are given. Second, the training process is important. Again, training procedures and algorithms are given.

All of the design and programming of neural networks is based on models derived from investigations and measurements that physicians and physiologists have made in the human brain. Models are presented of both the natural, physical neurons as they seem to exist and artificial neurons that are used to configure artificial, computer neural networks. The discussions and the illustrations include the input synapses, with their weighting, the dendrites or input paths, the nucleus or processing elements, and the axons or output connections. The operation of NNs, which is explained and illustrated, is dependent upon the transfer functions in the artificial nuclei and the weighting factors, which are created by training, in the artificial dendrites.

Special Examples in the Development of Neural Networks

Both from the perspective of historical development and the understanding of the operation of NNs, additional special programs stand out and are described. First is the concept and programming of the Perceptron. This early Rosenblatt development was the foundation of contemporary NNs even though it could not perform the all-impotent XOR function. The importance of the XOR function is described, in the context of weighting with the threshold and transfer functions, and a second program is then developed that will perform this all-important NN function. The pioneering and versatile ADALINE network is also presented, since it was the first NN to be mass produced and used in telecommunications for echo cancellation. Likewise, it is shown that this program, which, like all NN models, is based upon the way we believe the brain functions, can perform some feats, using adaptive feedback, that are almost impossible for humans to do.

Archival and Cardinal References

Anderson, James A., and Rosenfeld, Edward (eds.) *Neurocomputing: Foundations of Research.* Cambridge, MA: MIT Press, 1988.

Cherry, Colin. *On Human Communication: A Review, a Survey, and a Criticism.* 2d ed. Cambridge, MA: MIT Press, 1966.

Escher, M. C. *M. C. Escher: 29 Master Prints.* New York: Harry N. Abrams, 1983.

Feigenbaum, Edward A., and Feldman, Julian (eds.) *Computers and Thought.* New York: McGraw-Hill, 1963.

Hofstadter, Douglas R. *Gödel, Escher, Bach: An Eternal Golden Band.* New York: Random House, 1980.

Minsky, Marvin, and Papert, Seymour. *Perceptrons.* Cambridge, MA: MIT Press, 1966.

Minsky, Marvin. *The Society of Mind.* New York: Simon & Schuster, 1986.

Schank, Rodger C., with Childers, Peter G. *The Cognitive Computer.* Reading, MA: Addison-Wesley, 1984.

Swade, Doron D. "Redeeming Charles Babbage's Mechanical Computer." *Scientific American,* February 1993, p. 86. (Historians have argued that Charles Babbage—circa 1830—was unable to build his vast mechanical computers because his conception exceeded the capacity of nineteenthth-century engineering. The construction in 1991 of a working, three-ton engine proves that his designs were well within the realm of possibility.)

Winograd, Terry. *Understanding Natural Language.* New York: Academic Press, 1972.

Contemporary References

Barr, Avron; Cohen, Paul R.; and Feigenbaum, Edward A. *The Handbook of Artificial Intelligence.* 3 vols. Los Alto, CA: William Kaufmann, 1981.

Bavarian, B. "Introduction to Neural Networks for Intelligent Control." *IEEE Control System Magazine,* vol. 8, no. 2, 1988.

Caudill, Maureen, and Butler, Charles. *Understanding Neural Networks: Computer Explorations.* 2 vols. Cambridge, MA: Bradford Books/MIT Press, 1992. (These fine tutorial workbooks come with two computer disks in editions for either the PC or the Macintosh.)

Cox, Earl. "How a Machine Reasons." (an 8 part series). *AI Expert,* August 1992–March 1993.

Harmon, Paul, and King, David. *Expert Systems: Artificial Intelligence in Business.* New York: John Wiley & Sons, 1985.

Jones , William P., and Hoskins, Josiah. "Back-Propagation—A Generalized Delta Learning Rule." *Byte*, October 1987, p. 155.

Kosko, B. *Neural Networks for Signal Processing.* Englewood Cliffs, NJ: Prentice Hall, 1992.

Liebowitz, Jay. *Expert Systems Applications to Telecommunications.* New York:John Wiley and Sons, 1988.

Nelson, Marilyn McCord, and Illingworth, W. T. *A Practical Guide to Neural Nets.* Reading, MA: Addison-Wesley, 1990.

NeuralWare, Inc. *Neural Computing—NeuralWorks Professional II/PLUS and NeuralWorks Explorer.* Pittsburgh, PA: NeuralWare, Inc., 1991a.

———. *Reference Guide—NeuralWorks Professional II/PLUS and NeuralWorks Explorer.* Pittsburgh, PA: NeuralWare, Inc., 1991b.

———. *Using Works—An Extended Tutorial for NeuralWorks Professional II/PLUS and NeuralWorks Explorer.* Pittsburgh, PA: NeuralWare, Inc., 1991c.

Reid, Krista, and Zeichick, Alan. "Neural Network Resource Guide." *AI Expert*, June 1992.

Rich, Elaine. *Artificial Intelligence.* New York: McGraw-Hill, 1983

Schafer, Dan. *Artificial Intelligence Programming for the Macintosh.* Indianapolis: Howard W. Sams & Co., 1986.

Shaw, Julie E. "Neural Network Resource Guide." *AI Expert*, February 1993.

Shaw, Julie E., and Zeichick, Alan. "Expert System Resource Guide." *AI Expert*, December 1992.

Stein, Roger. "Analysis of Financial Data Using Neural Nets." *AI Expert*, February 1993.

Weiss, Sholom M.; Kern, Kevin B.; Kulikowski, Casimir A.; and Uschold, Michael. *A Guide to the Use of the Expert Consultation System.* New Brunswick, NJ: Rutgers University, 1984.

Winston, Patrick H., Davis, Randall, and Horn, Berthold K. "Artificial Intelligence—An MIT Video Course." Cambridge, MA: Massachusetts Institute of Technology, 1984.

Winston, Patrick Henry. *Artificial Intelligence. 2d ed.* Reading, MA: Addison-Wesley, 1984.

General References

AI Expert (magazine) is one of the best sources for contemporary information and articles on Artificial Intelligence and Neural Networks. AI Expert, 600 Harrison St., San Francisco, CA 94107; (415) 905-2200.

IEEE Transactions on Neural Networks. The IEEE, 345 East 47 St., New York, NY 10017-4929; (212) 705-7900.

Neural Networks—The journal of the International Neural Network Society (INNS). New York: Pergamon Press.

Appendix–Noise

A.1 AN INTRODUCTION TO NOISE

What is Noise?

Noise is an annoyance to humankind. Static on the radio or the din of traffic disturbs the way we hear a conversation or a selection of music. "Snow" on television or blurred printing degrades our ability to see and comprehend. Noise is an aggravation. Unfortunately, noise cannot be completely relegated to the arena of pure science or mathematics. Since noise primarily affects humans, their reactions must be included in at least some of the definitions and measurements of noise. The spectrum and character of the noise determine how much it interferes with our aural concentration and reception. In TV, black noise, i.e., black specks in the picture, are much less distracting than white specks. Several principles of psychology, including the Weber-Fechner law, help to explain and define the way people react to different aural and visual disturbances. The following items only present an introduction to the subject of noise.

Thermal or Johnson Noise

In the 1920s, J. B. Johnson and H. Nyquist of Bell Telephone Laboratories studied the noise resulting from thermal agitation. (The molecules in any substance are always in temperature-driven motion.) This thermally generated noise produces a spectrum that has about the same energy for each cycle of bandwidth. Thus, whether it is the cycle from cycle number one to cycle number two or from one gigahertz to one gigahertz plus one cycle, the energy is the same. This equal-power-per-cycle noise is termed Gaussian or white noise. Johnson and Nyquist calculated that metallic (wire-wound) resistors produced an independent, open circuit voltage termed v_n as:

Average noise power versus temperature related to voltage:

$$v_n(RMS) = (4kTRB)^{1/2} \qquad \text{(E A.1)}$$

where:

RMS, or root-means-square, is the (square) root of all the mean (average) val ues squared T, again, is the temperature in °K
(room temperature = 68°F = 20°C = 293°K)
R is the resistance in ohms

Average noise power versus temperature:

$$P_n = kTB \qquad \text{(E A.2)}$$

where:

P_n is the average noise power
k is Boltzmann's constant. k is a number related to the energy in the molecules of a substance. In joules (related to watt-seconds). Boltzmann's constant = 1.38×10^{-23} joules/°Kelvin.
T is the temperature in °K (room temperature is 293°K)
B is the bandwidth in hertz

The noise power density spectrum:

The noise spectrum, S_n, for each cycle is:

$$S_n = kT \qquad \text{(E A.3)}$$

where:

S_n is in watts/hertz

In practical numbers, a 10K resistor will have an open circuit noise voltage of about 1.3 μv if the bandwidth is 10 kHz. This is a statistical number. Its exact value depends upon the laws of chance—see the Horowitz and Hill reference. It should be remembered that Johnson noise is a property of molecular (Brownian) motion and, therefore, is found in other electrical components such as semiconductors.

If wide-band noise passes through a filter with a transfer function given as $H(\omega)$, the output spectrum density will be:

$$S_n = |H(\omega)|^2 kT \qquad \text{(E A.4)}$$

The noise-power in a given bandwidth

$$P_n = S_n B \tag{E A.5}$$

where:

S_n is in watts/hertz
B is the bandwidth in Hz

Flicker, Current, and 1/f Noise

In addition to the fundamental Johnson noise, many devices exhibit a second noise phenomenon caused by the flow of electric current. Electron or charge flow (current) is not a continuous, well-behaved process. There is a randomness that produces an alternating current (random AC) on top of the main direct current flow. This noise is difficult to quantify and to measure. It becomes very random (more than normal statistics would predict) in loosely constructed carbon-composition resistors. This noise, in general, has a 1/f spectrum and is termed pink noise. Pink noise has equal power per decade of frequency. (Again, see the excellent Horowitz and Hill reference.)

Shot of Gaussian Noise

Shot noise is again based on the observation that electric current (change flow) does not flow in a uniform, well-behaved manner. This is especially true when electrons are expelled from a hot cathode in a vacuum tube or the emitting element of a semiconductor. In essence, a supposedly steady (DC) current will also include an alternating (AC) component. Shot noise is often termed "rain on the roof" noise. This noise is again usually quantified using the principles of probability and statistics. In general, the probable noise caused by fluctuations in a direct current are:

$$I_{noise}\,(RMS) = \left(2qI_{DC}\,B\right)^{1/2} \tag{E A.6}$$

where:

q is the charge on an electron (1.60×10^{-19} coulombs)
I_{DC} is the direct current in question
B is any given bandwidth

Shot noise is Gaussian or white noise with an equal-power-per-cycle spectrum.

Equivalent Noise Resistance

Some manufacturers specify a number (a fictitious component) termed an equivalent noise resistance. The use of this fictitious resistor usually takes one of two forms—both assume room temperature. One assumes a perfect, noiseless amplifier with an equivalent resistor noise generator of $4R_nKTB$ between the amplifier's input resistor and its input grid, base, etc. The other configuration combines a given generator resistance, such as 50 ohms, with the amplifier's input resistance in its Thevenin equivalent. If the specification "equivalent noise resistance" is used, be certain that it is defined.

Interference

In the contemporary world many of our electrical signals are in some way affected (usually degraded) by man made interference. Because of the abundance of these impediments, an entire industry has been built that will shield, filter, isolate, or reroute the original signal.

A.2 SIGNAL-TO-NOISE RATIO

One of the major indications of the quality of an amplifier or an entire communication system is its signal-to-noise ratio (SNR). The classical formula, derived from information theory, for the signal-to-noise ratio of a system is given as:

$$SNR = 10 \log_{10}\left(\frac{V_s^2}{V_n^2}\right) \qquad \text{(E A.7)}$$

This is actually stated in terms of a power ratio since power is proportional to voltage squared. Often, in specific scientific and technical disciplines, there may be some variations to this fundamental relationship. For instance, for some television calculations, the signal v_s is stated in peak-to-peak units while the noise V_n is stated in RMS units. Using these units of measure will give a result that is 9.03 dB higher than if both measurements were stated in RMS units.

A.3 ASSIGNING A FIGURE OF MERIT FOR THE NOISE IN AN AMPLIFIER

The noise performance of an amplifier, such as the tuner in a television set, can be compared and specified in several ways. In fact, references and manufacturers differ on the exact method or equation they use to define this figure of merit. One way is to call the factor F as the ratio of two signal-to-noise ratios as:

The noise factor of an amplifier:

$$F = \frac{\textit{available S/N ratio power at the input}}{\textit{available S/N ratio power at the output}} \qquad \text{(E A.8a)}$$

$$F = \frac{\left(\frac{P_{si}}{P_{ni}}\right)}{\left(\frac{P_{so}}{P_{no}}\right)} = \frac{\left(\frac{Input\,Signal\,Power}{Input\,Noise\,Power}\right)}{\left(\frac{Output\,Signal\,Power}{Output\,Noise\,Power}\right)} \qquad \text{(E A.8b)}$$

The bandwidth should be constant when using this formula.

Another frequently used figure of merit compares the noise performance of a normal, real-world amplifier with a similar but perfect or noiseless amplifier. It is assumed that both amplifiers have a normal, noise-generating resistor R_g connected across their inputs. If the resistor generates a Johnson noise v_n^2 (see E A.2), the noise figure (NF) will be:

The noise figure of an amplifier in dB:

$$NF = 10\log_{10}\left(\frac{4kR_gT + v_n^2}{4kR_gT}\right)\text{dB} \qquad \text{(E A.9a)}$$

$$= 10\log_{10}\left(1 + \frac{v_n^2}{4kR_gT}\right)\text{dB} \qquad \text{(E A.9b)}$$

Quite often, especially for amplifiers and antennas that are used for space communications, the equivalent noise temperature (T_e) is an important parameter. One way to derive T_e is as follows:

The derivation of equivalent noise temperature:

$$F = \frac{P_{no}}{G\,P_{ni}} \qquad \text{(E A.10a)}$$

where:

G is the available power gain and R_{no} and P_{ni} are defined in E A.8.

$$P_{no} = FGP_{ni} = FGkT_oB \qquad \text{(E A.10b)}$$

where:

T_o is the noise temperature in question, or,

$$P_{ni(total)} = FGkT_oB \qquad \text{(E A.10c)}$$

where:

$P_{ni(total)}$ is the total noise referred to the input and kT_oB is that portion contributed by the source.

Therefore,

$$\begin{aligned} P_{na} &= FkT_oB - k\,T_o\,B \\ &= (F-1)\,k\,T_o\,B \end{aligned} \qquad \text{(E A.10d)}$$

where:

This is the noise contribution of the amplifier.

Returning to the (fundamental) equation E A.2, it is now possible to perform one more mathematical manipulation to conclude that an equivalent noise temperature T_e can be derived as follows:

If, from E A.2, $P_n = kTB$ and from E A.10d, $P_{na} = (F-1)kT_0\,B$, then the two can be combined as:

T_e— The equivalent noise temperature of an amplifier input:

$$\begin{aligned} kT_oB &= (F-1)\,kT_oB \text{ or} \\ T_e &= (F-1)\,T_o \end{aligned} \qquad \text{(E A.11)}$$

Archival and Cardinal References

Schwartz, Mischa. *Information Transmission, Modulation, and Noise.* New York: McGraw-Hill, 1959. (The classic early reference on noise.)

Contemporary References

Horowitz, Paul, and Hill, Winfield. *The Art of Electronics.* Cambridge, MA: Cambridge University Press, 1983.

Roddy, Dennis, and Coolen, John. *Electronic Communications.* Reston, VA: Reston Publishing Company, 1984.

Bibliography

Adams, R. W. "Companded Predictive Delta Modulation: A Low-Cost Conversion Technique for Digital Recording." J. *Audio Eng. Soc.*, March 1984.

AI Expert (magazine) is one of the best sources for contemporary information and articles on Artificial Intelligence and Neural Networks. AI Expert, 600 Harrison St., San Francisco, CA 94107; (415) 905-2200.

Anderson, James A., and Rosenfeld, Edward (eds.) *Neurocomputing: Foundations of Research.* Cambridge, MA: MIT Press, 1988.

Anner, George E. *Elements of Television Systems.* New York: Prentice-Hall, 1951.

Bahar, H. *Analog and Digital Signal Processing.* New York: John Wiley & Sons, 1990.

Banks, Michael A. *The Modem Reference: The Complete Guide to Selection, Installation, and Applications.* New York: Brady/Simon & Schuster, 1991.

Barr, Avron; Cohen, Paul R.; and Feigenbaum, Edward A. *The Handbook of Artificial Intelligence,* 3 vols. Los Alto, CA: William Kaufmann, 1981.

Barton, D. K. and Ward, H. R., *Handbook of Radar Measurement.* Norwood, MA: Artech House Inc., 1984.

Basch, E. E. (ed.) *Optical Fiber Transmission.* Indianapolis, IN: Howard W. Sams & Co., 1987.

Bavarian, B. "Introduction to Neural Networks for Intelligent Control." *IEEE Control System Magazine,* vol. 8, no. 2, 1988.

Bell and Newell. *Computer Structures.* New York: McGraw-Hill, 1971.

Bellamy, John. *Digital Telephony.* New York: John Wiley & Sons, 1991.

Benson, K. Blair (ed.) *Audio Engineering Handbook.* New York: McGraw-Hill Book 1988.

Benson, K. Blair, and Fink, Donald G. *HDTV Advanced Television of the 1990s.* New York: McGraw-Hill, 1991.

Benson, K. Blair, and Whitaker, J. (eds.) *Television and Audio Handbook for Technicians and Engineers.* New York: McGraw-Hill, 1990.

———. *Television Engineering Handbook.* rev. ed., New York: McGraw-Hill, 1991.

Beranek, Leo L. *Acoustical Measurements.* New York: John Wiley & Sons, 1949.

Black, H. S. "Stabilized Feedback Amplifiers." Bell System Technical Journal, January, 1934.

Brewer, Bryan, and Key, Ed. *The Compact Disc Book.* New York: Harcort Brace Jovanovich, 1987.

Brooks, John. *Telephone: The First Hundred Years.* New York: Harper & Row, 1976.

Bruce III, Walter R. *Using PROCOMM PLUS.* Carmel, IN: Que Corporation, 1991.

Bushong, Stewart C. *Magnetic Resonance Imaging: Physical and Biological Principles.* St. Louis, MO: C. V. Mosby Co., 1988a.

———. *Radiologic Science for Technologists.* St. Louis, MO: C. V. Mosbyby Co., 1988b.

Campbell, Joe. *C Programmer's Guide to Serial Communications.* Carme, CAl: Howard W. Sams & Co., 1989. (This book contains an excellent bibliography.)

Caudill, Maureen, and Butler, Charles. *Understanding Neural Networks: Computer Explorations*, 2 vols. Cambridge, MA: Bradford Books/MIT Press, 1992. (These fine tutorial workbooks come with two computer disks in editions for either the PC or the Macintosh.)

Chassaing, Ralph, and Horning, Darrell W. *Digital Signal Processing with the TMS320C25,* New York: John Wiley & Sons, 1990.

Cherry, Colin. *On Human Communication: A Review, a Survey, and a Criticism*, 2d ed. Cambridge, MA: The MIT Press, 1966.

Chorafas, Dimitris N. *Local Area Network Reference.* New York: McGraw-Hill, 1989.

Cochran, L. A. "The XL-100 ColorTrack System." *The RCA Engineer,* July 1976.

Corrigan, Patrick H., and Guy. Aisling. *Building Local Area Networks with Novell's Netwate.* Redwood City, CA: M&T Books, 1989.

Corrington, M. S., and Kidd, M. C. "Amplitude and Phase Measurements in Loudspeaker Cones." *Proc. IRE*, September 1951.

Corrington, M. S., and Murakami, T. "Applications of the Fourier Integral in the Analysis of Color TV Systems." *IRE Trans. on Circuit Theory.* vol. CT-2, no. 3. September 1955.

Corrington, M. S. "Transient Testing of Loudspeakers." Audio Engineering, August 1950.

Cortright, Edgar M. *Exploring Space With a Camera.* Washington, DC: NASA, 1968.

Couch, L. W. *Digital and Analog Communication Systems.* New York: Macmillan, 1990.

Cox, Earl. "How a Machine Reasons." (an 8 part series). *AI Expert,* August 1992 –March 1993.

Curlander, John C., and McDonough, Robert N. *Synthetic Aperture Radar Systems and Signal Processing.* New York: John Wiley & Sons, 1991.

da Cruz, Frank, and Catchings, Bill. "Kermit: A File-Transfer Protocol for Universities." Parts 1 and 2. *Byte,* June and July 1984.

Davidson, Homer L. *Troubleshooting and Repairing Compact Disc Players.* Blue Ridge Summit, PA: Tab Books, 1989.

Dolby, Ray M. "An Audio Noise Reduction System." *Journal of the Audio Engineering Society*, October 1967.

Dorf, Richard C. (ed.) *The Electrical Engineering Handbook.* Boca Raton, FL: CRC Press, 1993.

Dortch, Michael. *The ABC's of Local Area Networks.* San Francisco: Sybex, 1990.

Ennes, Harold E. *Television Broadcasting: Equipment, Systems, and Operating Fundamentals.* Indianapolis, IN: Howard W. Sams & Co, 1971.

Escher, M. C. *M. C. Escher: 29 Master Prints.* New York: Harry N. Abrams, 1983.

FCC et al. The submissions of the various laboratories and organizations to be considered as an HDTV standard. 1990–1991.

Feigenbaum, Edward A., and Feldman, Julian (eds.) *Computers and Thought.* New York: McGraw-Hill, 1963.

Feldman, Leonard. *FM Multiplexing for Stereo.* Indianapolis: Howard W. Sams & Co., 1972.

Fink, Donald G., and Christiansen, Donald (eds.) *Electronics Engineers' Handbook.* New York: McGraw-Hill, 1989.

Fink, Donald G. *Television Engineering.* New York: McGraw-Hill, 1952.

Fortier, Paul J. *Handbook of LAN Technology.* New York: McGraw-Hill, 1989. (This book contains an excellent communications bibliography.)

Frayne, John G. and Wolf, Halley. *Elements of Sound Recording.* New York: John Wiley & Sons, 1949.

Freeman, Roger L. *Telecommunication System Engineering.* 2d ed. New York: John Wiley & Sons, 1989.

———. *Radio Systems Design for Telecommunication.* New York: John Wiley &Sons, 1987.

———. *Telecommunication System Engineering.* New York: John Wiley & Sons, 1981a.

———. *Telecommunication Transmission Handbook* 2d ed. New York: John Wiley & Sons, 1989b.

Gage, S. et al. *Optoelectronics/Fiber-Optics Applications Manual.* 2d ed. New York: Hewlett-Packard/McGraw-Hill, 1981

Gaonkar, R. S. *MicropProcessor Architecture, Programming and Applications with the 8085/8085A.* 2d ed. Columbus, OH: Merrill Publishing Co., 1984.

Gersho, A.and Gray, R. M. *Vector Quantization and Signal Compression.* Norwell, MA: Kluwer Academic Publishers, 1991.

Gilbert, E. N. "Information Theory after Eighteen Years." *Bell Telephone Monograph,* Bell Telephone Laboratories, 1965.

Gonzalez, Rafael C., and Wintz, Paul. *Digital Image Processing.* Reading MA: Addison-Wesley, 1978.

Gordon, Bernard M. "Linear Electronic Analog/Digital Conversion Architectures, Their Origins, Parameters, Limitations, and Applications." *IEEE Transactions on Circuits and Systems,* no. 7, July 1987.

Harmon, Paul, and King, David. *Expert Systems: Artificial Intelligence in Business.* New York: John Wiley & Sons, 1985.

Harwood, L. A. "A Chrominance Demodulator IC with Dynamic Fleshtone Correction." *IEEE Trans. on Consumer Electronics,* February 1976.

Herbert, Nick. *Quantum Reality Beyond the New Physics.* Garden City, NY: Anchor Press/ Doubleday, 1985.

Hofstadter, Douglas R. *Gödel, Escher, Bach: An Eternal Golden Band.* New York: Random House, 1980.

Horowitz, Paul, and Hill, Winfield. *The Art of Electronics.* Cambridge and New York: Cambridge University Press, 1983.

Hovanessian, S. A. *Radar Design and Analysis.* Norwood, MA: Artech House Inc., 1984.

Hunter, Colin and Banning, John. "DOS at RISC," *Byte*, November 1989.

IEEE. *IEEE Transactions of Acoustics, Speech, and Signal Processing.* The IEEE, 345 East 47 St., New York, NY 10017-4929; (212) 705-7900.

IEEE. *IEEE Transactions on Microwave Theory,* vol. 21, no. 4, 1973 (special issue on SAW filters).

IEEE Transactions on Neural Networks. The IEEE, 345 East 47 St., New York, NY 10017-4929; (212) 705-7900.

IEEE. *IEEE Trial-Use Standard on Video Signal Transmission Measurement of Linear Waveform Distortion.* New York: The Institute of Electrical and Electronics Engineers, 1974.

Immink, K. A. S. "Coding Methods for High-Density Optical Recording." *Phillips J. Res.,* vol. 41, 1986.

Inglis, Andrew F. *Satellite Technology: An Introduction.* Boston: Focal Press, 1991.

Intel. *Microprocessors.* Bellevue WA and Mt. Prospec ILt: Intel, 1990a.

———. *386SX Microprocessor Programmers Reference Manual.* Bellevue WA and Mt. Prospect IL: Intel, 1990b.

———. *486 SX Microprocessor, 487 Math Coprocessor* Bellevue WA and Mt. Prospect IL: Intel, 1991.

James, J. N. "The Voyage of Mariner II." *The Scientific American*, July 1963.

Jayant, N. "Signal Compression: Technology Targets and Research Directions." *IEEE Journal on Selected Areas in Communications*, vol. 10, no. 5, 1992.

Johnson , David E., and Hilburn, John L. *Rapid Practical Designs of Active Filters.* New York: John Wiley & Sons 1975.

Johnson, Thomas L. "The RISC/CISC Melting Pot." *Byte*, April 1987.

Jones, Don. "Delta Modulation for Voice Transmission." Application Note 607. Melbourne, FL: Harris Semiconductor Div., 1980.

Jones, William P., and Hoskins, Josiah. "Back-Propagation—A Generalized Delta Learning rule." *Byte*, October 1987, p. 155.

Jordan, Larry E. *System Integration for the IBM PS/2 and PC.* New York: Brady/Simon & Schuster, 1990. (One of the better refeence books. It includes a nice section of protocols.)

Journal of the Acoustical Society of America, 500 Sunnyside Blvd., Woodbury, NY 11797; (516) 349-7800.

Journal of the Audio Engineering Society. Audio Engineering Society, 60 East 42 St., Room 2520, NY 10065.

Journal of the Society of Motion Picture and Television Engineers, 595 West Hartsdale Ave., White Plains, NY 10607; (914) 761-1100.

Kak, A., and Slaney, M. *Principles of Computerized Tomographic Imaging*. New York: IEEE Press, 1988.

Kennedy, George. *Electronic Communication Systems.* New York: McGraw-Hill, 1985.

Kingsley, Simon, and Quegan, Shaun. *Understanding Radar Systems.* London and New York: McGraw-Hill, 1992.

Kloker, K. "The Architecture and Applications of the Motorola DSP56000 Digital Signal Processing Family." Proceedings of the IEEE International Conference ASSP, paper 13.13, April 1987, pp. 523–526.

Kosiur, David R. "Expanding the Conversation—An Overview of AppleTalk Network products." *MACWORLD,* May, 1988.

———. "File Service—How to Move and Share Files over an AppleTalk Network." *MACWORLD*, August 1990.

———. "Building a Better Network—How to Avoid Growing Pains As Your Network Needs Increase." *MACWORLD*, November 1991. (Read this before you design, let alone install, any network!)

———. "Going the Ethernet Route." *MACWORLD*, April 1991. (Excellent information on routers and router tests as well as Ethernet.)

———. "Network Connections." *MACWORLD*, November 1989. (AppleTalk and networking)

Kosko, B. *Neural Networks for Signal Processing*. Englewood Cliffs, NJ: Prentice Hall, 1992.

Krestel, Erich (ed.) *Imaging Systems for Medical Diagnostics.* Berlin and Munich: Siemens AG, 1990.

Landee, Robert W.; Davis, Donovan C.; and Albright, Albert P. *Electronic Designers Handbook.* New York: McGraw-Hill, 1957.

Langford-Smith, F. (ed.) *Radiotron Designer's Handbook.* Sydney, Australia: Wireless Press, 1952.

Larson, David G.; Rony, Peter R.; and Titus, Jonathan A. *The Bugbook III— Micro Computer Interfacing.* Derby, CT: E & L Instruments, Inc., 1975.

Le Galley, Donald P. (ed.) *Space Science.* New York: John Wiley & Sons, 1963.

Lee, William C. Y. *Mobile Cellular Telecommunications Systems.* New York: McGraw-Hill, 1989.

Liebowitz, Jay. Expert Systems Applications to Telecommunications. New York: John Wiley & Sons, 1988.

Leighton, Robert B. "The Photographs from Mariner IV." *The Scientific American*, April 1966.

———. "The Surface of Mars." *The Scientific American*, May 1970.

Libbey, R. L. *A Handbook of Circuit Mathematics for Technical Engineers.* Boca Raton, FL: CRC Press, 1991.

Lynn, Paul A., and Fuerst, Wolfgang. *Digital Signal Processing with Computer Applications.* Chichester and New York: John Wiley & Sons Ltd., 1989a.

———. *Introductory Signal Processing with Computer Applications.* Chichester and New York: John Wiley & Sons, 1989b.

Macovski, A. *Medical Imaging Systems.* Englewood Cliffs, NJ: Prentice Hall, 1983.

Marchant, A. B. *Optical Recording.* Reading: Addison-Wesley, 1990.

Markowitz, M. C. "EDN's 18th Annual mP/mC Chip Directory." *EDN,* November 21, 1991. and See also *EDN* November 26, 1992.

Marshall, Christopher (ed.) *The Physical Basis of Computed Tomography.* St. Louis, MO: Warren H. Green, 1982.

Marshall, Trevor. "A Calculating RISC." *Byte,* May 1990. (This article introduces the idea of a RISC coprocessor.)

Mead, William, "Multi-dimensional Audio for Stereo Television." *Broadcast News,* July 1987.

Meng, Brita. "Mac to VAX and Back." *MACWORLD,* May, 1990.

Minsky, Marvin. *The Society of Mind.* New York: Simon & Schuster, 1986.

Minsky, Marvin, and Papert, Seymour, *Perceptrons.* Cambridge, MA: MIT Press, 1966.

MIT Staff, *Applied Electronics.* New York: John Wiley & Sons, 1949.

MIT. *Radar System Engineering.* New York: McGraw-Hill, 1947. (Take time to see the wealth of material in these early, but still useful, landmark texts. For THE background material on the origins of radar, especially in the United States, see *Radar System Engineering,* vol. 1 of the 28-volume MIT Radiation Laboratory Series.)

Motorola. *DSP56000 Technical Summary—A 56-Bit General-Purpose Digital Signal Processor.* Phoenix, AZ: Motorola Semiconductor Products, 1986.

———. *DSP56000/DSP56001 Digital Signal Processor User's Manual.* Phoenix, AZ: Motorola Semiconductor Products, 1990a.

———. *MC68000 Family Reference Manual.* Phoenix, AZ: Motorola, Inc., 1990b.

———. *MC68000 Programmer's Reference Manual.* Phoenix, AZ: Motorola, Inc., 1989a.

———. *MC68020 32-Bit Microprocessor User's Manual.* Englewood Cliffs, NJ: Prentice Hall, Inc., 1990c.

———. *MC68020 32-Bit Virtual Memory Microprocessor,* Phoenix, AZ: Motorola, Inc., 1989b.

———. *MC68040 Third-Generation Microprocessor,* Phoenix, AZ: Motorola, Inc., 1989c.

Nance, Barry, *Network Programming in C.* Carmel, IN: Que Corporation, 1990.

Nathanson, F. E. et al. *Radar Design Principles* 2d ed. New York: McGraw-Hill, 1991.

Nelson, M. *The Data Compression Book.* San Mateo, CA: M & T Books, 1991.

Nelson, Marilyn McCord, and Illingworth, W. T. *A Practical Guide to Neural Nets.* Reading, MA: Addison-Wesley, 1990.

Netravali, A. N., and Haskell, B. G. *Digital Pictures: Representations and Compression.* New York: Plenum Press, 1988.

Neural Networks—The journal of the International Neural Network Society (INNS). New York: Pergamon Press.

NeuralWare, Inc. *Neural Computing—NeuralWorks Professional II/PLUS and NeuralWorks Explorer*. Pittsburgh, PA: NeuralWare, Inc., 1991a.

———. *Reference Guide—NeuralWorks Professional II/PLUS and NeuralWorks Explorer*. Pittsburgh, PA: NeuralWare, Inc., 1991b.

———. *Using Works—An Extended Tutorial for NeuralWorks Professional II/PLUS and NeuralWorks Explorer*. Pittsburgh, PA: NeuralWare, Inc., 1991c.

Nilson, Arthur R. and Hornung J. L. *Practical Radio Communication*. New York: McGraw-Hill, 1943.

Nilsson, James W. *Electric Circuits*. Reading, MA: Addison-Wesley Publishing Co., 1986.

Nyyquist, H. "Certain Topics in Telegraph Transmission Theory." *BSTJ*, April 1928.

Olson, Harry F. *Acoustical Engineering*. New York: D. Van Nostrand, 1947.

———. *Dynamical Analogies*. Princeton: D. Van Nostrand, 1958.

Osborne, Adam. *An Introduction to Microcomputers, Volume II*. Berkeley: Adam Osborne and Associates, 1976.

Pahlavan, K., and Holsinger, J. L. "Voice-Band Communication Modems: A Historical Review." *IEEE Communications Magazine*, January 1988.

Patterson, D., and Hennessy, J. *Computer Architecture, A Quantitative Approach*, San Mateo, CA: Morgan Kaufmann, 1989.

Pennebacker, William B., and Mitchell, Joan L. *JPEG Still Image Data Compression Standard*. New York: Van Nostrand Reinhold, 1993.

Pierce, John R., and Noll, A. Michael. *SIGNALS The Science of Telecommunications*. New York: Scientific American Library, 1990.

Pratt, W. K. *Digital Image Processing*. New York: John Wiley & Sons, 1991.

Prentiss, Stan. *HDTV High-Definition Television*. Blue Ridge Summit, PA: Tab Books, 1990.

Putman, Byron. *RS-232 Simplified*. Englewood Cliffs NJ: Prentice Hall, 1987.

Reid, Krista, and Zeichick, Alan. "Neural Network Resource Guide." *AI Expert*, June 1992.

Rich, Elaine. *Artificial Intelligence*. New York: McGraw-Hill, 1983

Rider, John F. *The Cathode-Ray Tube at Work*. New York: John F. Rider Publishing, 1945.

Robinson, Phillip. "How Much of a RISC." *Byte*, April 1987.

Rodgers, Mike, and Bare, Virginia. *Hands-On AppleTalk*. New York: Brady Books, 1989.

Roddy, Dennis, and Coolen, John. Electronic Communications. Reston, VA: Reston Publishing Co., 1984.

Sandbank, C. P. (ed.) *Digital Television*. Chichester: John Wiley & Sons Ltd., 1990.

Schade, Otto H. Sr. *Image Quality: A Comparison of Photographic and Television Systems*. Princeton, NJ: RCA Laboratories, 1975.

Schafer, Dan. *Artificial Intelligence Programming for the Macintosh*. Indianapolis, IN: Howard W. Sams & Co., 1986.

Schank, Rodger C., with Childers, Peter G. *The Cognitive Computer.* Reading, MA: Addison-Wesley, 1984.

Schwartz, Mischa. *Information Transmission, Modulation, and Noise.* New York: McGraw-Hill, 1959.

Segal, Mark L. "Toward Standardized Video Terminals." *Byte*, April 1984.

Shannon, C. E. "A Mathematical Theory of Communication." *Bell System Technical Journal,* July 1948.

Shaw, Julie E., and Zeichick, Alan. "Expert System Resource Guide." *AI Expert*, December 1992.

Shaw, Julie E. "Neural Network Resource Guide." *AI Expert*, February 1993.

———(ed.). *Radar Handbook* 2d ed. New York: McGraw-Hill, 1990.

Skolnik, Merrill I. *Introduction to Radar* 2d ed. New York: McGraw-Hill, 1980.

Staff of the Jet Propulsion Laboratory. *Mariner Mission to Venus.* New York: McGraw-Hill, 1963.

Stallings, William. *The Business Guide to Local Area Networks.* Carmel, IN: Howard W. Sams & Co., 1990.

Stark, David, and Bradley, William G. *Magnetic Resonance Imaging.* St. Louis, MO: C. V. Mosby Co., 1988.

Stein, Roger. "Analysis of Financial Data Using Neural Nets." *AI Expert*, February 1993.

Strum, Robert D., and Kirk, Donald E. *First Principles of Discrete Systems and Digital Signal Processing*. Reading MA: Addison-Wesley, 1988.

Swade, Doron D. "Redeeming Charles Babbage's Mechanical Computer." *Scientific American,* February 1993, p. 86.

Terman, Fredrick Emmonds. *Radio Engineering.* New York: McGraw-Hill, 1947.

Terrell, Trevor J. *Introduction to Digital Filters.* 2d ed. New York: John Wiley & Sons, 1988.

Texas Instruments. *TMS32010 User's Guide,* Dallas, TX: Texas Instruments, 1983.

Texas Instruments Digital Signal Processing Group. *Digital Signal Processing Applications with the TMS320 Family.* Dallas, TX: Texas Instruments, 1986.

———. *TMS320 Family Development Support Reference Guide.* Dallas, TX: Texas Instruments, 1992.

———. *TSS400 Family—Sensor Signal Processors.* Dallas and Germany: Texas Instruments, 1991.

Titus, John. "Build the Mark 8 Minicomputer." *Radio-Electronics*, July 1974. (This was THE article that started the personal computer.)

Tohlmann, Ken. *Principles of Digital Audio*. Indianapolis, IN: Howard W. Sams & Co., 1989. (Distributed by Prentice Hall)

Tremaine, Howard M. *Audio Cyclopedia.* Indianapolis, IN: Howard W. Sams & Co., 1969.

Veljkov, Mark D. *MacLANs Local Area Networking with the Macintosh.* Glenview, IL: Scott, Foresman and Co., 1988.

Watson, Scott. *White Knight.* Beaver Falls,PA: The FreeSoft Co., 1989. (THE instruction book for the Mac communication software—the successor to *Red Ryder.)*

Weiss, Daniel P. "Experiences with the AT&T DSP32 Digital Signal Processor in Digital Audio Applications." *The Proceedings of the Audio Engineering Society Seventh International Convention.* New York: The Audio Engineering Society, 1990.

Weiss, Sholom M.; Kern, Kevin B.; Kulikowski, Casimir A.; and Uschold, Michael. *A Guide to the Use of the Expert Consultation System.* New Brunswick, NJ: Rutgers University, 1984.

Winograd, Terry. *Understanding Natural Language*. New York: Academic Press, 1972.

Winston, Patrick H.; Davis, Randall; and Horn, Berthold K. "Artificial Intelligence—An MIT Video Course." Cambridge, MA: Massachusetts Institute of Technology., 1984.

Winston, Patrick Henry. *Artificial Intelligence.* 2d ed. Reading, MA: Addison-Wesley, 1984.

"An EitherPort in the Storm." *MACWORLD*, May 1988.

"Telecommunications Quality." *IEEE Communications Magazine*, October 1988 (special issue).

Qmodem. Cedar Falls: The Forbin Project, Inc., 1990. The instruction book for the PC communication software.

"Apple Sharing—Here's What You Need to Know to Work with the AppleTalk Filing Protocol." *Byte*, October 1991, p. 247.

CCITT, *Red Book(s).* (Recommendations) Malaga-Torremolinos.

Index